Senescence Processes in Plants

Senescence Processes in Plants

Edited by

SUSHENG GAN
Department of Horticulture
Cornell University
Ithaca
NY, USA

Blackwell
Publishing

Editorial Offices:
Blackwell Publishing Ltd, 9600 Garsington Road, Oxford OX4 2DQ, UK
Tel: +44 (0)1865 776868
Blackwell Publishing Professional, 2121 State Avenue, Ames, Iowa 50014-8300, USA
Tel: +1 515 292 0140
Blackwell Publishing Asia Pty Ltd, 550 Swanston Street, Carlton, Victoria 3053, Australia
Tel: +61 (0)3 8359 1011

First published 2007 by Blackwell Publishing Ltd

ISBN: 978-1-4051-3984-7

Library of Congress Cataloguing-in-Publication Data
Senescence processes in plants / edited by Susheng Gan.
p. cm.—(Annual plant reviews ; v. 26)
Includes bibliographical references and index.
ISBN: 978-1-4051-3984-7 (hardback : alk. paper)
1. Plants—Aging. I. Gan, Susheng.

QK762.5.S47 2007
571.8′782—dc22

2006025504

A catalogue record for this title is available from the British Library

Set in 10/12 pt Times
by TechBooks, New Delhi, India

For further information on Blackwell Publishing, visit our website:
www.blackwellpublishing.com

Contents

Contributors

Dr Ian Armstead Plant Genetics and Breeding Department, Institute of Grassland & Environmental Research, Plas Gogerddan, Aberystwyth, Ceredigion SY23 3EB, Wales, UK

Dr Marie-Jeanne Carp Department of Biology, Technion – Israel Institute of Technology, Haifa 32000, Israel

Dr Jen-Chih Chen Department of Plant Sciences, University of California, One Shields Drive, Davis, CA 95616, USA

Dr Paul P. Dijkwel Department of Molecular Biology of Plants, Groningen Biomolecular Sciences and Biotechnology Institute, University of Groningen, Kerklaan 30, 9751 NN, Haren, The Netherlands

Dr Iain Donnison Plant Genetics and Breeding Department, Institute of Grassland & Environmental Research, Plas Gogerddan, Aberystwyth, Ceredigion SY23 3EB, Wales, UK

Dr Andreas M. Fischer Department of Plant Sciences, 210 AgBioScience Facility, Montana State University, Bozeman, MT 59717, USA

Dr Isaac Galyuon Department of Molecular Biology and Biotechnology, School of Biological Sciences, University of Cape Coast, Cape Coast, Ghana

Dr Susheng Gan Department of Horticulture, G51 Emerson Hall, Cornell University, Ithaca, NY 14853, USA

Dr Shimon Gepstein Department of Biology, Technion – Israel Institute of Technology, Haifa 32000, Israel

Dr James J. Giovannoni USDA-ARS Plant, Soil and Nutrition Lab and Boyce Thompson Institute for Plant Research, Cornell University Campus, Tower Road, Ithaca, NY 14853, USA

Dr Yongfeng Guo Department of Horticulture, G51 Emerson Hall, Cornell University, Ithaca, NY 14853, USA

Dr Jacques Hille Department of Molecular Biology of Plants, Groningen Biomolecular Sciences and Biotechnology Institute, University of Groningen, Kerklaan 30, 9751 NN, Haren, The Netherlands

Marianne Hopkins Department of Biology, University of Waterloo, Waterloo, Ontario, Canada N2L 3G1

Dr Stefan Hörtensteiner Institute of Plant Sciences, University of Bern, Altenbergrain 21, CH-3013 Bern, Switzerland

Dr Catherine Howarth Plant Genetics and Breeding Department, Institute of Grassland & Environmental Research, Plas Gogerddan, Aberystwyth, Ceredigion SY23 3EB, Wales, UK

Dr Hai-Chun Jing Wheat Pathogenesis Programme, Plant-Pathogen Interactions Division, Rothamsted Research, Harpenden AL5 2JQ, UK

Dr Hyo Jung Kim Division of Molecular Life Sciences, Pohang University of Science and Technology (POSTECH), Pohang, Kyungbuk 790-784, South Korea

Dr David W. Lee Department of Biological Sciences, Florida International University, Miami, FL 33199, USA

Dr Amnon Lers Department of Postharvest Science of Fresh Produce, Agricultural Research Organization, The Volcani Center, PO Box 6, Bet Dagan 50250, Israel

Dr Pyung Ok Lim Department of Science Education, Cheju National University, Cheju 690-756, Korea

Dr Hong Gil Nam Division of Molecular Life Sciences, Pohang University of Science and Technology (POSTECH), Pohang, Kyungbuk 790-784, South Korea

Linda McNamara Department of Biology, University of Waterloo, Waterloo, Ontario, Canada N2L 3G1

Dr Helen Ougham Plant Genetics and Breeding Department, Institute of Grassland & Environmental Research, Plas Gogerddan, Aberystwyth, Ceredigion SY23 3EB, Wales, UK

Professor Michael S. Reid Department of Plant Sciences, University of California, One Shields Drive, Davis, CA 95616, USA

Jos H.M. Schippers Department of Molecular Biology of Plants, Groningen Biomolecular Sciences and Biotechnology Institute, University of Groningen, Kerklaan 30, 9751 NN, Haren, The Netherlands

Catherine Taylor Department of Biology, University of Waterloo, Waterloo, Ontario, Canada N2L 3G1

Professor Dr Howard Thomas Plant Genetics and Breeding Department, Institute of Grassland & Environmental Research, Plas Gogerddan, Aberystwyth, Ceredigion SY23 3EB, Wales, UK

Dr John Thompson Department of Biology, University of Waterloo, Waterloo, Ontario, Canada N2L 3G1

Tzann-Wei Wang Department of Biology, University of Waterloo, Waterloo, Ontario, Canada N2L 3G1

Dr Ulrike Zentgraf ZMBP, General Genetics, University of Tübingen, Auf der Morgenstelle 28, 72076 Tübingen, Germany

Preface

The importance of research into plant senescence cannot be overemphasized. Senescence processes are unique developmental programs that involve unique mechanisms. For example, unlike many other developmental processes in plants that involve cell division, cell differentiation, and/or cell growth (enlargement), leaf senescence is achieved by a massive operation of programmed cell death and nutrient recycling. It is known that new gene expression is required in order for leaf cells to destroy themselves and to recycle nutrients. The cell has to maintain its machinery necessary for new gene expression and nutrient transport while its subcellular structure and macromolecules are being dismantled by some of the new gene products. How gene expression is regulated and how this complex process operates are currently among the most significant biological questions.

Senescence has a tremendous impact on agriculture. Leaves are the primary organs that absorb light energy from the sun and convert it to chemical energy in the form of sugars via photosynthesis. With the onset of senescence, the photosynthetic capability of a leaf declines sharply. Therefore, leaf senescence limits crop yield and biomass production, and contributes substantially to postharvest loss in vegetable and ornamental crops during transportation, storage and on shelves. In addition, proteins, antioxidants and other nutritional compounds are degraded during senescence. Senescing tissues also become more susceptible to pathogen infection, and some of the pathogens may produce toxins, rendering food unsafe. Mitotic senescence may also determine sizes of leaves, fruits and whole plants.

This scientific and economic significance means that much effort has been made to understand the senescence processes in plants and to devise means of manipulating them agriculturally. During the past few years there has been significant progress in this regard, especially in the molecular, genetic and genomic aspects.

This volume summarizes recent progress in the physiology, biochemistry, cell biology, molecular biology, genomics, proteomics, and biotechnology of plant senescence. The term *senescence* has been used by both plant and animal biologists, but it may describe completely different processes. Beginning with senescence-related terminology and our current knowledge of mitotic senescence in plants (a less well-studied area, Chapter 1), the book focuses on post-mitotic senescence, including senescence of leaves (Chapters 2 through 10), flowers (Chapter 11), and fruits (Chapter 12). This research has led to the development of various new biotechnologies for manipulating the senescence processes of fruit (Chapter 12) and leaves (Chapter 13), some of which are approaching commercialization.

Senescence Processes in Plants will be a very useful reference book for senescence researchers based in academia and industry. It may also serve as a textbook for advanced undergraduate students and graduate students.

I would like to thank all the authors for their excellently written chapters and the publishers for their enthusiasm.

Susheng Gan

Annual Plant Reviews

A series for researchers and postgraduates in the plant sciences. Each volume in this series focuses on a theme of topical importance and emphasis is placed on rapid publication.

Editorial Board:

Titles in the series:

1. **Arabidopsis**
   Edited by M. Anderson and J.A. Roberts
2. **Biochemistry of Plant Secondary Metabolism**
   Edited by M. Wink
3. **Functions of Plant Secondary Metabolites and their Exploitation in Biotechnology**
   Edited by M. Wink
4. **Molecular Plant Pathology**
   Edited by M. Dickinson and J. Beynon
5. **Vacuolar Compartments**
   Edited by D.G. Robinson and J.C. Rogers
6. **Plant Reproduction**
   Edited by S.D. O'Neill and J.A. Roberts
7. **Protein–Protein Interactions in Plant Biology**
   Edited by M.T. McManus, W.A. Laing and A.C. Allan
8. **The Plant Cell Wall**
   Edited by J.K.C. Rose
9. **The Golgi Apparatus and the Plant Secretory Pathway**
   Edited by D.G. Robinson
10. **The Plant Cytoskeleton in Cell Differentiation and Development**
    Edited by P.J. Hussey
11. **Plant–Pathogen Interactions**
    Edited by N.J. Talbot
12. **Polarity in Plants**
    Edited by K. Lindsey
13. **Plastids**
    Edited by S.G. Moller
14. **Plant Pigments and their Manipulation**
    Edited by K.M. Davies
15. **Membrane Transport in Plants**
    Edited by M.R. Blatt
16. **Intercellular Communication in Plants**
    Edited by A.J. Fleming
17. **Plant Architecture and Its Manipulation**
    Edited by C. Turnbull
18. **Plasmodeomata**
    Edited by K.J. Oparka
19. **Plant Epigenetics**
    Edited by P. Meyer

20. **Flowering and Its Manipulation**
    Edited by C. Ainsworth
21. **Endogenous Plant Rhythms**
    Edited by A.J.W. Hall and H.G. McWatters
22. **Control of Primary Metabolism in Plants**
    Edited by W.C. Plaxton and M.T. McManus
23. **Biology of the Plant Cuticle**
    Edited by M. Riederer and C. Müller
24. **Plant Hormone Signaling**
    Edited by P. Hedden and S.G. Thomas
25. **Plant Cell Separation and Adhesion**
    Edited by J.A. Roberts and Z. Gonzalez-Carranza
26. **Senescence Processes in Plants**
    Edited by S. Gan
27. **Seed Development, Dormancy and Germination**
    Edited by K. Bradford and H. Nonogaki
28. **Plant Proteomics**
    Edited by C. Finnie
29. **Regulation of Transcription in Plants**
    Edited by K.D. Grasser
30. **Light and Plant Development**
    Edited by G.C. Whitelam and K.J. Halliday
31. **Plant Mitochondria**
    Edited by D.C. Logan

1 Mitotic senescence in plants

Susheng Gan

1.1 Introduction

The word *senescence* derives from two Latin words: *senex* and *senescere*. *Senex* means 'old'; this Latin root is shared by 'senile', 'senior', and even 'senate'. In ancient Rome the 'Senatus' was a 'council of elders' that was composed of the heads of patrician families. *Senescere* means 'to grow old'. The Merriam-Webster online dictionary defines *senescence* as 'the state of being old or the process of becoming old'. Aging is also the process of getting older. Therefore, aging has been regarded as a synonym of senescence, and the two words have often been used interchangeably, which, in some cases, is fine but in some other cases causes confusion. This chapter will first briefly discuss the terminology of senescence, and then will review the literature related to mitotic senescence, a topic that has not been well discussed in the plant senescence research area.

1.2 Terminology and types of senescence

Senescence is a universal phenomenon in living organisms, and the word *senescence* has been used by scientists working on a variety of systems, such as yeast, fruit fly, worm, human being and plants. However, the meaning of the word *senescence* to scientists working on different organisms can be different, and the difference can be subtle in some cases and very obvious in some other cases. Here I try to clarify the term at cellular and organismal levels to avoid possible confusion.

At the cellular level, as shown in the cartoon in Figure 1.1, a cell's life history consists of mitotic and postmitotic processes (Gan, 2003). A cell may undergo a certain number of mitotic divisions to produce daughter cells. After a limited number of divisions (e.g. about 40 divisions in human fibroblasts), the cell can no longer divide mitotically. Once a cell ceases mitotic division permanently, it is called mitotic senescence. In the literature concerning yeast, germline cells and mammalian cells in culture, this type of senescence is often referred to as cellular senescence, replicative senescence, proliferative senescence or, sometimes, replicative aging (Sedivy, 1998; Takahashi *et al.*, 2000; Ben-Porath and Weinberg, 2005; Patil *et al.*, 2005). If a cell keeps dividing and fails to undergo mitotic senescence (e.g. cancer cells), it is said to be 'immortalized'. Therefore, mitotic senescence is a mechanism to suppress cancer development. If a cell stops mitosis temporarily due to unfavorable conditions but retains its mitotic capacity and can re-enter mitotic cycles to produce

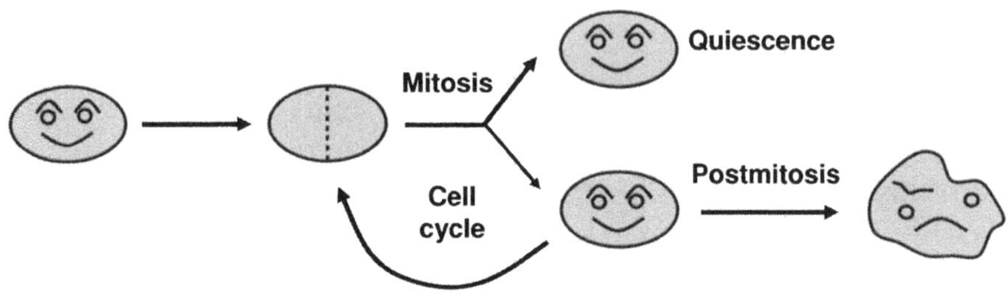

Figure 1.1 Illustration of a cell's life history consisting of both mitotic and postmitotic processes. When the cell stops dividing, it is called mitotic senescence or replicative senescence or proliferative senescence. The active degenerative and attrition process of the cell that can no longer undergo cell division is postmitotic senescence. If a cell stops dividing due to, for example, adverse conditions, but will resume division, the status of the cell is called cell quiescence.

more daughter cells, the temporarily undividing or resting status or process is called cell quiescence (Stuart and Brown, 2006). Although a mitotically senescent cell is not dead it may undergo degenerative process leading to death. If the degeneration is solely a function of age, 'aging' is the right word to describe it. In animal literature, the term 'cell(ular) aging' or 'postmitotic aging', or 'postmitotic senescence' is used for this process. If the degeneration is an active yet quick process, it is a form of 'apoptosis' or 'programmed cell death'. It however should be noted that mitotically senescent mammalian cells in culture are resistant to 'apoptosis'. Most of the postmitotic cells are somatic in nature (e.g. brain, neuron, and muscle cells); the term *somatic senescence* is therefore also used in literature concerning animals. As will be discussed below, postmitotic senescence also occurs in plant somatic tissues such as leaves, flowers and fruits. Compared with postmitotic senescence in animals, leaf/flower/fruit senescence (that involves an active but slow degenerative process) and hypersensitive response (involving an active yet very quick degenerative process) are typical postmitotic senescence processes in plants.

At the organismal level, when an organism's ability to respond to stress declines, its homeostasis becomes increasingly imbalanced, and its risk of disease increases with age, which leads to the ultimate death of the whole organism. This is the aging of the whole organism, and is often referred to as organismal senescence. Although cellular senescence may contribute to organismal senescence (Ben-Porath and Weinberg, 2005), the latter is much more inclusive, for example many age-related diseases, such as Alzheimer's disease, are parts of organismal aging. In literature concerning plants, organismal senescence is senescence of the whole plant. Among the most studied whole plant senescence processes is monocarpic senescence. Annuals (e.g. *Arabidopsis*), biennials (e.g. wheat) and some perennials (e.g. bamboo) possess a monocarpic life pattern, which is characterized by only a single reproductive event in the life cycle. After flowering (and setting seeds or fruits), the whole plant will senesce and die. Monocarpic senescence includes three coordinated processes: (a) senescence of somatic organs and tissues such as leaves (a form of postmitotic senescence, see below), (b) arrest of shoot apical meristems (SAM), a form of mitotic senescence or proliferative senescence (see below), and (c) permanent suppression of axillary buds to prevent formation of new shoots/branches. This third aspect of whole plant senescence has not received enough attention in the senescence research community.

1.3 Plants exhibit mitotic senescence, postmitotic senescence and cell quiescence

Plants exhibit both types of senescence (Figure 1.2). An example of mitotic senescence in plants is the arrest of apical meristem; the meristem consists of nondifferentiated, germline-like cells that can divide finite times to produce cells that will be then differentiated to form new organs such as leaves and flowers. The arrest of apical meristem is also called proliferative senescence in plant literature (Hensel *et al.*, 1994). This is similar to replicative senescence in yeast and animal cells in culture, as discussed above. Another example of mitotic senescence is the arrest of mitotic cell division at early stages of fruit development. Fruit size is a function of cell number, cell size and intercellular space, and cell number is the major factor. Cell number is determined at the very early stage of fruit development and remains unchanged thereafter (Tanksley, 2004). Postmitotic senescence occurs in some plant organs, such as leaves and floral petals. Once formed, cells in these organs rarely undergo cell division; their growth is mainly contributed by cell expansion; thus, their senescence, unlike mitotic senescence, is not due to an inability to divide. This type of senescence involving predominantly somatic tissues is very similar to that

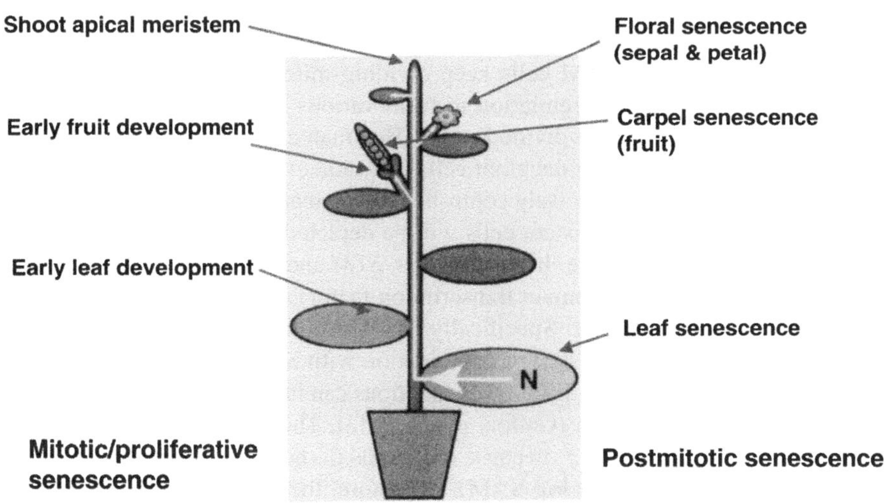

Figure 1.2 Mitotic and postmitotic senescence in plants. Mitotic senescence occurs in SAM, in fruits and leaves that are at very early stages of development. In contrast, postmitotic senescence occurs in leaves, flowers and fruits that are at late stages of development (thus leaf senescence, flower senescence and fruit senescence, respectively).

of such animal model systems as *Drosophila* and *Caenorhabditis elegans* whose adult bodies, with exception of germline, are postmitotic (Gan, 2003).

Cell quiescence also occurs in plants. Cells of apical meristems will stop dividing under unfavorable conditions. For example, the apical meristems of many trees will stop proliferative process when they perceive the short-day photoperiod signal; short day often means that the winter season is coming. These meristem cells retain their division capability during winters and will resume division activity when spring is coming. Therefore, the short-day-induced cell quiescence is an evolutionary fitness strategy. A recent study shows that ethylene and abscisic acid may play a role in regulating the temporary 'arrest' of tree meristem (Ruonala *et al.*, 2006).

1.4 Mitotic senescence: arrest of SAM

1.4.1 Initiation of SAM

SAM is a dome-shaped structure at the tip of a stem that consists of small isodiametric cells with thin-wall and dense protoplasm. It is formed at the globular stage during embryogenesis, and at least three genes, *SHOOT MERISTEM LESS* (*STM*), *CUP-SHAPED COTYLEDONS* (*CUC*)*1* and *CUC2*, are required for SAM initiation, because mutation in *STM* or in both *CUC1* and *CUC2* results in no formation of SAM (Bowman and Eshed, 2000). *STM* encodes a homeodomain transcription factor and *CUC1* and *CUC2* encode duplicated NAC family transcription factors.

1.4.2 Maintenance of SAM

SAM is responsible for generating above-ground postembryonic organs such as leaves and flowers. The SAM cells keep dividing mitotically, and some of their daughter cells undergo differentiation to form various aerial organs while others remain as stem cells that can divide further (Bowman and Eshed, 2000). The balance between the numbers of daughter cells that remain as meristem cells and that undergo differentiation is precisely controlled; if too many daughter cells enter differentiation, the pool of meristem cells will be depleted. Several genes have been shown to regulate this balance. In *Arabidopsis, STM* and *WUSCHEL (WUS,* a gene that also encodes a homeodomain transcription factor) are necessary to keep cells undifferentiated and dividing. Specifically, WUS produces a noncell autonomous signal that activates cell division in combination with STM (Gallois *et al.*, 2002). On the other hand, combined *WUS/STM* functions can initiate the progression from stem cells to organ initiation (Gallois *et al.*, 2002). The balance is also regulated by *CLAVATA* (*CLV*)*1*, *2* and *3*, because mutations in these genes lead to too many cells in the SAM (thus a too big SAM). Therefore, these three genes may inhibit cell division or promote cell differentiation in the SAM. *CLV1* encodes a receptor kinase and *CLV2* a receptor-like protein. *CLV3* encodes a small protein that may act as a ligand for the CLV1/2 receptor heterodimer complex. *STM* and *CLV* may function independently in regulating *SAM*, and *WUS* may act downstream of the

CLV pathway. Recent studies show that a transcription factor complex consisting of C-, D-, and E-type MADS-box proteins controls the stem cell population in the floral meristem (Ferrario *et al.*, 2006). In addition, the homeodomain/leucine zipper transcription factor REVOLUTA (Otsuga *et al.*, 2001) may also control the relative growth of apical (and nonapical) meristems in *Arabidopsis* (Talbert *et al.*, 1995).

1.4.3 Arrest of SAM: a mitotic senescence in nature

After producing certain number of organs (leaves and flowers), the SAM cells cease dividing. The loss of cell division capability of SAM is called the arrest of SAM. The arrest is a proliferative senescence process (Hensel *et al.*, 1994). Figure 1.3 shows an arrested primary inflorescence apex compared with a proliferating one in *Arabidopsis*.

1.4.3.1 Physiological regulation

Reproductive development appears to play an important role in regulating proliferative senescence in plants, which is especially true in many monocarpic plants. Hensel *et al.* (1994) found that meristems of all inflorescence branches in the wild-type *Arabidopsis* ecotype Landsberg *erecta* (L*er*) ceased to produce flowers coordinately, but such a coordinated proliferative arrest did not occur in the wild-type L*er* plants with their fruits surgically removed. Similarly, meristem arrest was not observed in a male-sterile line that never sets seeds. This result suggests that the arrest of inflorescence meristems is regulated by developing fruits/seeds (Hensel *et al.*, 1994). Hensel *et al.* further proposed two models to explain the effect of developing fruits on the mitotic activity of meristems. One model is that a factor necessary for sustaining mitotic activity at the SAM is gradually taken and eventually depleted by developing fruits, resulting in arrest. The other model is that developing fruits produce a negative regulator of mitotic activities, and that the negative regulator is transferred to and accumulated in the SAM to a threshold level so that the SAM is arrested. The factor, either positive or negative, is unknown.

Like postmitotic senescence that is hormonally regulated (Chapter 7), SAM arrest is controlled by plant hormones. It is known that both cytokinins and auxin can promote cell division (Trehin *et al.*, 1998). A mitotic cycle consists of G1 → S → G2 → M (and then back to G1). DNA is synthesized during S while mitosis occurs during M. Tissue culture studies have shown that auxin appears to promote advancement from G1 to S by up-regulating the expression of a cyclin-dependent kinase (CDK) gene. Cytokinins can advance the cycle through M, likely by maintaining cyclin homeostasis (Lee *et al.*, 2006). These data were largely obtained by tissue culture experiments. Whether these mechanisms are involved in the regulation of the cell division in SAM is unknown. In contrast to cytokinins and auxin, mitotic drugs also cause meristem arrest. For example, oryzalin, a chemical that can depolymerize microtubules, can very rapidly lead to meristem cell division arrest in *Arabidopsis* (Grandjean *et al.*, 2004).

In addition, many environmental factors, especially extreme conditions, regulate meristem arrest. For example, broccoli normally develops a ramified inflorescence

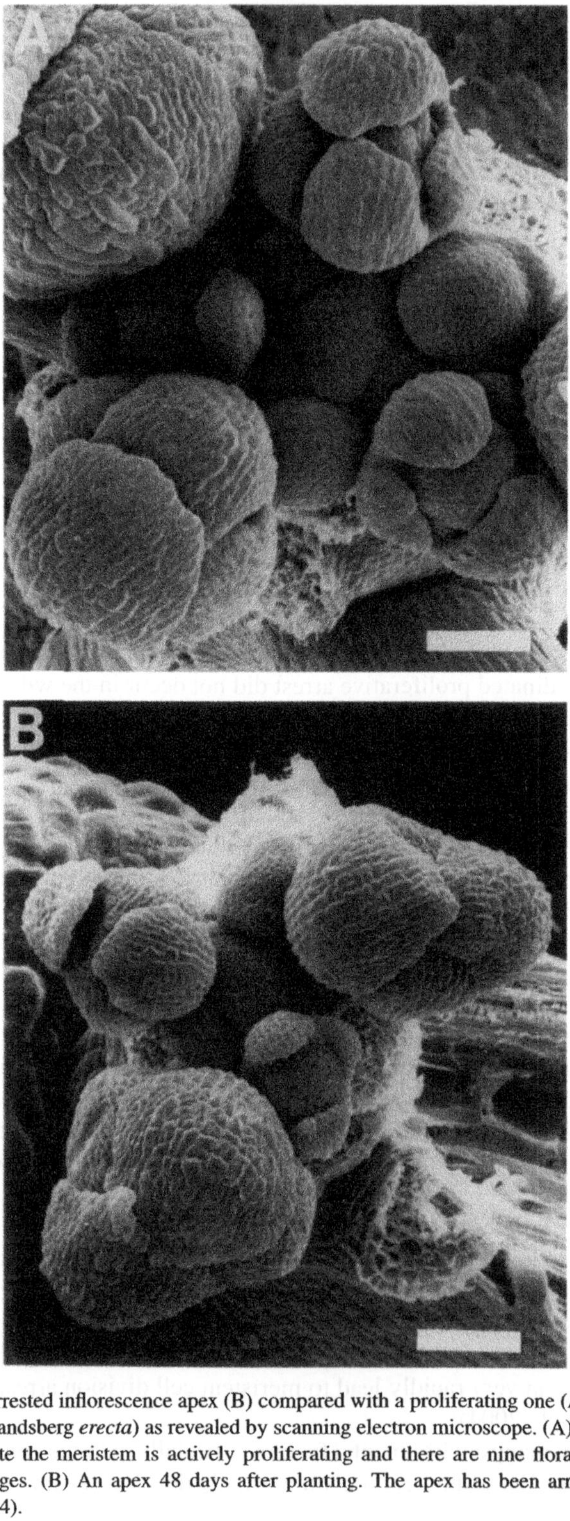

Figure 1.3 An arrested inflorescence apex (B) compared with a proliferating one (A) in *Arabidopsis thaliana* (strain: Landsberg *erecta*) as revealed by scanning electron microscope. (A) An apex 25 days after planting. Note the meristem is actively proliferating and there are nine floral buds at various developmental stages. (B) An apex 48 days after planting. The apex has been arrested for 1 week (Hensel *et al.*, 1994).

without flower bud development. After a certain period, meristems begin to make flower buds instead of more inflorescences. The meristem will be arrested at this transition if the temperature is too high (Bjorkman and Pearson, 1998). The temperature-sensitive arrest of meristem has also been observed in *Arabidopsis* (Pickett *et al.*, 1996).

1.4.3.2 Genetic regulation

In contrast to the initiation and growth, the arrest of apical meristems may be regulated in part by *FIREWORKS* (*FIW*). During a course of screening for mutants that exhibit premature cessation of inflorescence growth in *Arabidopsis*, Nakamura *et al.* (2000) isolated a novel mutant line named *fireworks* (*fiw*) that displayed earlier cessation of flower formation and inflorescence stem elongation. The recessive mutant *fiw*/*fiw* displayed an inflorescence meristem arrest 7 days earlier than wild-type *Arabidopsis* plants. Otherwise the vegetative growth and development in the mutant line were normal, and the mutant plants produced normal flowers and set fully matured siliques, although the flowers and siliques were clustered at the top of the inflorescence, looking like fireworks (thus so named). The early arrest in the *fiw*/*fiw* plants occurred globally, not only in the primary inflorescence but also in the lateral inflorescences (Nakamura *et al.*, 2000). In addition to the early mitotic senescence phenotype, the mutant plants also exhibited accelerated rosette leaf senescence (Nakamura *et al.*, 2000), suggesting that FIW may also have a role in regulating postmitotic senescence. The *fiw* mutation was mapped on the lower arm of chromosome 4 but the corresponding gene has not been cloned yet. The cloning and characterization of *FIW* will help us understand how a single gene may control both mitotic and postmitotic senescence.

1.5 Role of telomere and telomerase in mitotic senescence

1.5.1 Telomere

Telomeres are specialized structures consisting of proteins and highly repeated DNA at the ends of the linear eukaryotic chromosomes. The repeated sequences are relatively conserved, for example, the repeated sequence in vertebrates is TTAGGG, but the length of the telomere varies among different species, different individuals, different tissues and even among different chromosomes (Bekaert *et al.*, 2004). In humans, the telomere may be 3–20 kb in length. In yeasts, the repeated sequence is $T_{1-4}G_{1-4}$, not as highly conserved as that of humans. In many higher plants, the repeated sequence is TTTAGGG. Telomeres can serve as caps to prevent chromosomes from fusion with each other.

1.5.2 Telomerase

Chromosomal DNAs are replicated during S phase by DNA polymerases. DNA polymerases move from the 3′ to 5′ direction (polymerizing in the 5′ to 3′ direction), so at a replication fork there are two new DNA strands: one is the leading strand that will have no problem to replicate the DNA to the end of the template, and the other is the lagging strand. The lagging strand will have problem to replicate the very end of the linear template DNA sequence. Therefore, the DNA sequence at the very end of a chromosome will be lost each time the chromosome is replicated. This is called telomere shortening.

Telomerases are special reverse transcriptases that add telomere DNA to chromosome ends. A telomerase contains both RNA and protein components. The RNA component is approximately 150 nucleotides long and contains about 1.5 copies of a specific telomeric repeat. The RNA component serves as a template to synthesize the corresponding telomeric repeat DNA sequence. In general, germ cells contain high telomerase activity and telomere length in the germ cells is maintained relatively stable because of the telomerase activity. In contrast, somatic cells in animals lack telomerase activity, which prevents somatic cells, such as skin cells, from developing into cancer cells, because the telomeres will be shortened after each division (Bekaert *et al.*, 2004).

1.5.3 Telomere shortening and replicative senescence in animals

In mammalian cells in culture, there is a molecular clock of senescence or aging that counts cell division numbers (Sedivy, 1998; Sherr and DePinho, 2000; Bekaert *et al.*, 2004). The nature of the molecular clock appears to be the telomere shortening. The length and amount of telomeric DNA in human fibroblasts decrease as a function of serial passage (division) during aging *in vitro* and possibly *in vivo* (Harley *et al.*, 1990). When the telomeres become very short, the DNA ends will be open, and the cell will perceive it as damaged DNA, and consequently the senescence process will be triggered. One strong line of evidence that supports this replicative senescence model involves the overexpression of a telomerase (Bodnar *et al.*, 1998). Normal human cells in culture undergo a certain number of mitotic divisions and then start replicative senescence. When the cells overexpressed *hTRT* that encodes the human telomerase catalytic subunit via transfection, the telomeres in these cells were elongated, and the cells kept dividing vigorously even after the control cells had entered nondividing status. The *hTRT*-overexpressing cells had a significantly prolonged replicative life span (Bodnar *et al.*, 1998).

1.5.4 Telomere biology in plants

The telomere length remains constant throughout the life cycle of, for example, *Arabidopsis* and *Silene latifolia* (Riha *et al.*, 1998; Fitzgerald *et al.*, 1999), although the exception has been reported in barley: there is a significant reduction (50 kb) in telomere length during embryogenesis (Kilian *et al.*, 1995). Telomere shortening in the SAM is likely trivial because the meristem cells, like stem cells in animals, possess telomerase activity. On the basis of homology to the human telomerase reverse transcriptase (hTERT), an *Arabidopsis thaliana* cDNA named *AtTERT* was cloned (Fitzgerald *et al.*, 1999; Oguchi *et al.*, 1999). The cDNA contains an open reading frame of 3372 bp, encoding a protein with a predicted size of 131 kDa and isoelectric point of 9.9. The AtTERT protein contains the conserved reverse transcriptase motifs 1, 2 and A-E as well as the TERT-specific T motif. Reverse transcription polymerase chain reaction analysis and an assay of telomerase activity revealed that both *AtTERT* mRNA and telomerase activity are abundant in the SAM but are not detectable in rosette leaves. However, it should be noted that no detailed analysis of changes in telomere length in young versus senescent SAM has been reported, perhaps due to technical difficulties in collecting enough meristem tissues for analysis.

The cell culture system, like in animals, has been employed for the studies of telomere in plants, although the mitotic senescence process in the cultured plant cells has not been well characterized. Opposite to the situation in cultured animal cells, the telomere length in cultured plant cells does not shorten but increases upon a prolonged culture (Kilian *et al.*, 1995; Riha *et al.*, 1998).

The effect of telomere shortening on plant growth and development has been analyzed in *Arabidopsis* mutant plants in which the telomerase gene was knocked out due to T-DNA insertion. The telomerase-null plants displayed a slow loss of telomeric DNA, ~500 bp per generation (the *Arabidopsis* telomeres are about 2–5 kb), which is 10 times slower than that observed in telomerase-deficient mice (Fitzgerald *et al.*, 1999). The first several generations of the telomerase-null plants developed normally. The later generations, beginning in the sixth generation, exhibited an extended life span compared with wild-type plants. However, the later generations also displayed some developmental abnormalities including altered phyllotaxy, abnormal leaf shape and reduced fertility (Riha *et al.*, 2001). Therefore, the extended life span might have resulted from reduced fertility; as discussed above, the SAM of male-sterile plants had much longer proliferative longevity (Hensel *et al.*, 1994). The meristems of telomerase-null plants of very late generations were enlarged (however disorganized) and, in some cases, dedifferentiated into a callusoid mass, and failed to produce leaves and/or flowers (Riha *et al.*, 2001). Only a few mutants were able to survive into the ninth generation and none survived later than the tenth generation (Riha *et al.*, 2001) because of genome instability (Siroky *et al.*, 2003; McKnight and Shippen, 2004).

It is therefore unlikely that telomere shortening plays an important role in controlling proliferative senescence in plants (Gan, 2003).

1.6 Closing remarks

The term *senescence* has been used by both plant and animal scientists, but the exact meanings of the term could be different. This chapter tried to clarify the difference. At the cellular level, there are two types of senescence: mitotic and postmitotic senescence. Although plants exhibit both these types of senescence, mitotic or replicative or proliferative senescence in plants has been much less studied than the comparable processes in yeasts, animals, and humans, and postmitotic senescence in plants. Nonetheless, reasonable progresses have been made toward the understanding of physiological, molecular and genetic mechanisms of mitotic senescence in plants. It is known that many environmental stresses and fruit development can promote mitotic senescence in SAM, and that, unlike in animals, telomere and telomerase play little role in modulating plant mitotic senescence.

Acknowledgment

I thank Dr Richard Amasino for introducing me to the senescence research field. This research on plant senescence has been supported by grants from the National Science Foundation, US Department of Energy Basic Bioscience Program, USDA NRI, US-Israel BARD and Cornell Genomics Initiative. I thank Dr Yongfeng Guo for carefully reading the manuscript, and the past and current members of the Gan Laboratory for useful discussions.

References

Bekaert, S., Derradji, H. and Baatout, S. (2004) Telomere biology in mammalian germ cells and during development. *Dev Biol* **274**, 15–30.

Ben-Porath, I. and Weinberg, R.A. (2005) The signals and pathways activating cellular senescence. *Int J Biochem Cell Biol* **37**, 961–976.

Bjorkman, T. and Pearson, K.J. (1998) High temperature arrest of inflorescence development in broccoli (*Brassica oleracea* var. italica L.). *J Exp Bot* **49**, 101–106.

Bodnar, A.G., Ouellette, M., Frolkis, M., *et al.* (1998) Extension of life-span by introduction of telomerase into normal human cells. *Science* **279**, 349–352.

Bowman, J.L. and Eshed, Y. (2000) Formation and maintenance of the shoot apical meristem. *Trends Plant Sci* **5**, 110–115.

Ferrario, S., Shchennikova, A.V., Franken, J., Immink, R.G. and Angenent, G.C. (2006) Control of floral meristem determinacy in petunia by MADS-box transcription factors. *Plant Physiol* **140**, 890–898.

Fitzgerald, M.S., Riha, K., Gao, F., Ren, S., McKnight, T.D. and Shippen, D.E. (1999) Disruption of the telomerase catalytic subunit gene from *Arabidopsis* inactivates telomerase and leads to a slow loss of telomeric DNA. *Proc Natl Acad Sci U S A* **96**, 14813–14818.

Gallois, J.-L., Woodward, C., Reddy, G.V. and Sablowski, R. (2002) Combined SHOOT MERISTEMLESS and WUSCHEL trigger ectopic organogenesis in *Arabidopsis*. *Development* **129**, 3207–3217.

Gan, S. (2003) Mitotic and postmitotic senescence in plants. *Sci Aging Knowledge Environ* **2003**, RE7.

Grandjean, O., Vernoux, T., Laufs, P., Belcram, K., Mizukami, Y. and Traas, J. (2004) In vivo analysis of cell division, cell growth, and differentiation at the shoot apical meristem in *Arabidopsis*. *Plant Cell* **16**, 74–87.

Harley, C.B., Futcher, A.B. and Greider, C.W. (1990) Telomeres shorten during ageing of human fibroblasts. *Nature* **345**, 458–460.

Hensel, L.L., Nelson, M.A., Richmond, T.A. and Bleecker, A.B. (1994) The fate of inflorescence meristems is controlled by developing fruits in *Arabidopsis*. *Plant Physiol* **106**, 863–876.

Kilian, A., Stiff, C. and Kleinhofs, A. (1995) Barley telomeres shorten during differentiation but grow in callus culture. *Proc Natl Acad Sci U S A* **92**, 9555–9559.

Lee, H., Auh, C.K., Kim, D., Lee, T.K. and Lee, S. (2006) Exogenous cytokinin treatment maintains cyclin homeostasis in rice seedlings that show changes of cyclin expression when the photoperiod is rapidly changed. *Plant Physiol Biochem* **44**, 248–252.

McKnight, T.D. and Shippen, D.E. (2004) Plant telomere biology. *Plant Cell* **16**, 794–803.

Nakamura, M., Mochizuki, N. and Nagatani, A. (2000) Isolation and characterization of an *Arabidopsis* mutant, fireworks (fiw), which exhibits premature cessation of inflorescence growth and early leaf senescence. *Plant Cell Physiol* **41**, 94–103.

Oguchi, K., Liu, H., Tamura, K. and Takahashi, H. (1999) Molecular cloning and characterization of AtTERT, a telomerase reverse transcriptase homolog in *Arabidopsis thaliana*. *FEBS Lett* **457**, 465–469.

Otsuga, D., DeGuzman, B., Prigge, M.J., Drews, G.N. and Clark, S.E. (2001) REVOLUTA regulates meristem initiation at lateral positions. *Plant J* **25**, 223–236.

Patil, C.K., Mian, I.S. and Campisi, J. (2005) The thorny path linking cellular senescence to organismal aging. *Mech Ageing Dev* **126**, 1040–1045.

Pickett, F.B., Champagne, M.M. and Meeks-Wagner, D.R. (1996) Temperature-sensitive mutations that arrest *Arabidopsis* shoot development. *Development* **122**, 3799–3807.

Riha, K., Fajkus, J., Siroky, J. and Vyskot, B. (1998) Developmental control of telomere length and telomerase activity in plants. *Plant Cell* **10**, 1691–1698.

Riha, K., McKnight, T.D., Griffing, L.R. and Shippen, D.E. (2001) Living with genome instability: plant responses to telomere dysfunction. *Science* **291**, 1797–1800.

Ruonala, R., Rinne, P.L., Baghour, M., Moritz, T., Tuominen, H. and Kangasjarvi, J. (2006) Transitions in the functioning of the shoot apical meristem in birch (*Betula pendula*) involve ethylene. *Plant J* **46**, 628–640.

Sedivy, J.M. (1998) Can ends justify the means? Telomeres and the mechanisms of replicative senescence and immortalization in mammalian cells. *Proc Natl Acad Sci U S A* **95**, 9078–9081.

Sherr, C.J. and DePinho, R.A. (2000) Cellular senescence: mitotic clock or culture shock? *Cell* **102**, 407–410.

Siroky, J., Zluvova, J., Riha, K., Shippen, D.E. and Vyskot, B. (2003) Rearrangements of ribosomal DNA clusters in late generation telomerase-deficient *Arabidopsis*. *Chromosoma* **112**, 116–123.

Stuart, J.A. and Brown, M.F. (2006) Energy, quiescence and the cellular basis of animal life spans. *Comp Biochem Physiol A Mol Integr Physiol* **143**, 12–23.

Takahashi, Y., Kuro-o, M. and Ishikawa, F. (2000) Aging mechanisms. *Proc Natl Acad Sci U S A* **97**, 12407–12408.

Talbert, P.B., Adler, H.T., Parks, D.W. and Comai, L. (1995) The REVOLUTA gene is necessary for apical meristem development and for limiting cell divisions in the leaves and stems of *Arabidopsis thaliana*. *Development* **121**, 2723–2735.

Tanksley, S.D. (2004) The genetic, developmental, and molecular bases of fruit size and shape variation in tomato. *Plant Cell* **16**(Suppl.), S181–S189.

Trehin, C., Planchais, S., Glab, N., Perennes, C., Tregear, J. and Bergounioux, C. (1998) Cell cycle regulation by plant growth regulators: involvement of auxin and cytokinin in the re-entry of Petunia protoplasts into the cell cycle. *Planta* **206**, 215–224.

2 Chlorophyll catabolism and leaf coloration

Stefan Hörtensteiner and David W. Lee

2.1 Introduction

Chlorophyll (chl) breakdown is an integral part of the senescence syndrome, characterized by physiological and biochemical changes that aim at the recycling of nutrients from senescing tissues, like leaves and fruits. Thus, worldwide, an estimated 10^9 tons of chl is degraded every year, but the fate of chl was enigmatic for a long time (Hendry *et al.*, 1987). Only 15 years ago, the first final degradation product could be identified as a linear tetrapyrrolic, nonfluorescent chlorophyll catabolite (NCC) (Kräutler *et al.*, 1991), and a pathway involved in the formation of NCCs has been elucidated gradually since then. Most helpful for the elucidation of breakdown intermediates and reactions (Table 2.1) was the availability of stay-green mutants that are affected in chl catabolic steps. Most of the reactions of chl breakdown are now known, and genes for some of the catabolic enzymes have been cloned recently. The current knowledge will be outlined in this review.

Autumnal leaf coloration in deciduous trees is a most spectacular phenomenon that attracts millions of people every year (Hendry *et al.*, 1987). The loss of chl and unmasking of retaining carotenoids together with the new synthesis of anthocyanins represent the biochemical basis of the polychromatic beauty of autumnal leaves. Whereas the chemical structures and the biosynthetic pathways of the involved pigments are rather well established, the biological function of leaf coloration is poorly understood. Several hypotheses have been presented in the literature and will be discussed here.

2.2 Chlorophyll catabolites

2.2.1 *Green catabolites*

2.2.1.1 *Chlorins*

Green-colored pigments that are derived from chl have been identified in a number of different species and include chlides, pheides, 13^2-hydroxy chl, pyropheide and pyropheophytin (Schoch *et al.*, 1981; Ziegler *et al.*, 1988). Their importance for a chl degradation pathway that ultimately leads to the disappearance of green color has not been unequivocally established for all of them. Whereas the occurrence of pigments like pheide, chlide and pheophytins well fits the concept of chl breakdown ending in the formation of NCCs (Figure 2.1), colorless derivatives of pyro forms or of 13^2-hydroxylated forms of chl have so far escaped detection. Arguably, the latter

Table 2.1 Overview over chl catabolic enzymes

	Enzyme	Abbreviation	Identification/gene locus in *Arabidopsis*	Localization	Reference
I	Chlorophyll *b* reductase	CBR	Enzyme activity	Plastids, thylakoid	Scheumann *et al.* (1998, 1999)
II	Hydroxychlorophyll *a* reductase	CAR	Enzyme activity	Plastids	Scheumann *et al.* (1998)
III	Chlorophyllase	CLH	*AtCLH1*: At1g19670 *AtCLH2*: At5g43860	Plastids, vacuole?	Tsuchiya *et al.* (1999)
IV	Mg-dechelatase (metal-chelating substance)	MCS	Enzyme activity	Plastids	Suzuki and Shioi (2002)
V	Pheophorbide *a* oxygenase	PAO	At3g44880	Plastids, inner envelope	Pružinská *et al.* (2003)
VI	Red chlorophyll catabolite reductase	RCCR	At4g37000	Plastids, stroma	Wüthrich *et al.* (2000)
VII	Catabolite exporter (ATP-hydrolyzing)		Activity	Plastids, envelope	Matile *et al.* (1992)
VIII	13^2-Demethylase (pheophorbidase)	—	Enzyme activity	Cytosol?	Suzuki *et al.* (2002)
IX	$C8^2$ Hydroxylase	—	—	—	
X	C3 Hydroxylase	—	—	—	
XI	Glucosyltransferase	—	—	—	
XII	Malonyltransferase	—	Enzyme activity	Cytosol	Hörtensteiner (1998)
XIII	ABC transporter	—	*AtMRP2*: At2g34660 *AtMRP3*: At3g13080	Tonoplast	Lu *et al.* (1998); Tommasini *et al.* (1998)

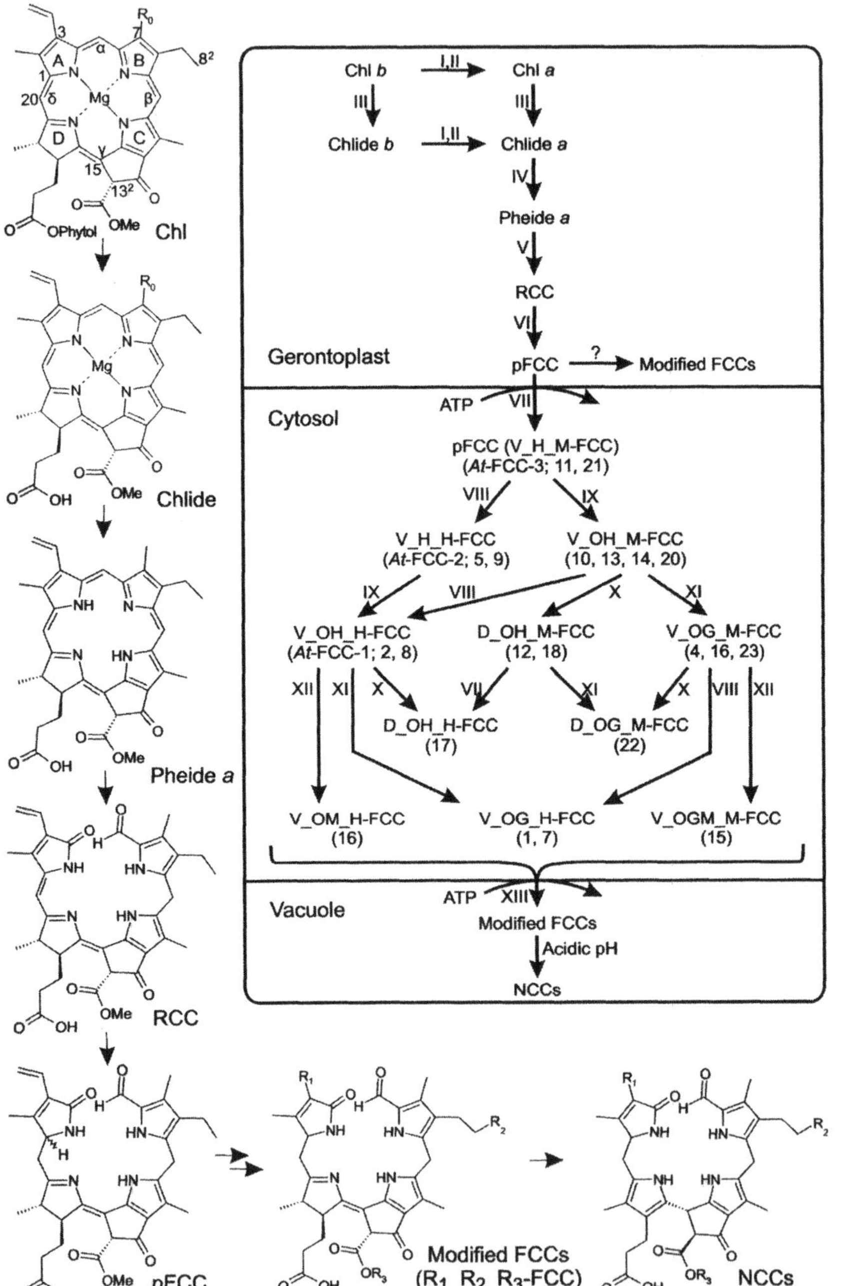

Figure 2.1 Chemical structures of chl and chl catabolites, and topographical model of chl breakdown. Reactions and enzymes are indicated with roman numbers according to Table 2.1. Putative reactions are indicated with a question mark. Pyrrole rings (A–D), methine bridges (α–δ) and relevant carbon atoms are labeled in chl. $R_0 = CH_3$, chl *a*; $R_0 = CHO$, chl *b*. R_1–R_3 in modified FCCs and NCCs indicate side-chain groups as follows: D, dihydroxyethyl; H, hydrogen; M, methyl; OG, *O*-glucosyl; OGM, *O*-glucosyl-malonyl; OH, hydroxyl; OM, *O*-malonyl; V, vinyl. Note that only in two cases have modified FCCs been structurally analyzed. The occurrence of other modified FCCs is implied from the identification of respective NCCs as indicated in arabic numbers according to Table 2.2.

ones may represent products of oxidative activities or artifacts of tissue extraction, not related to 'natural' chl breakdown during leaf senescence. Thus, pyro forms of chl-derived red bilins, isolated from nitrogen-deprived *Chlorella protothecoides* cultures (Engel *et al.*, 1991; Iturraspe *et al.*, 1993), were shown to be formed during extraction (Engel *et al.*, 1996). Analysis of mutants that are defective in chl catabolic steps was helpful in the identification of intermediates. Some of these mutants, such as Bf 993 from *Festuca pratensis* (Thomas *et al.*, 1989) or a *Lolium* introgression line containing the mutated gene from *F. pratensis* (Thomas *et al.*, 1999), exhibit a stay-green phenotype. The retention of chl in these mutants is accompanied by the accumulation of chlides and pheide *a*, but not of one of the other above-mentioned green pigments (Thomas *et al.*, 1989; Roca *et al.*, 2004). Likewise, mutants that are defective in pheophorbide *a* oxygenase (PAO) (see below) retain chl and accumulate pheide *a* (Pružinská *et al.*, 2003, 2005; Tanaka *et al.*, 2003). Collectively, these findings suggest that chlide and pheide *a* represent true intermediates of senescence-related chl breakdown, whereas the significance of 13^2-hydroxyl or pyro forms of chl remains to be established.

2.2.1.2 *Phytol*

After removal from chl, phytol is nearly quantitatively retained within senescing chloroplasts (gerontoplasts) (Peisker *et al.*, 1989). A large proportion of phytol accumulates in plastoglobules in either a free or an esterified form. Reutilization for the synthesis of α-tocopherol has been suggested, which requires the activation of phytol to phytyl pyrophosphate, the cosubstrate for condensation with homogentisate (Soll, 1987). Increasing evidence supports an important role for plastoglobules in tocopherol biosynthesis (F. Kessler and P. Dörmann, personal communication). In addition, several tocopherol biosynthetic genes are upregulated during senescence (Buchanan-Wollaston *et al.*, 2005), coinciding with an increase in plastoglobule number and size during chloroplast to gerontoplast transition (Matile, 1992). Thus, reuse of chl-derived phytol for tocopherol synthesis would be reasonable, but so far the evidence supporting this idea is limited; in particular, the postulated kinase necessary for phytol activation has not been identified (Collakova and DellaPenna, 2003).

2.2.2 *Catabolites with a tetrapyrrolic structure*

2.2.2.1 *Red chlorophyll catabolites*

When *C. protothecoides* cells were grown under heterotrophic and nitrogen-limiting conditions, red, chl-derived pigments were excreted into the surrounding medium (Oshio and Hase, 1969). A common feature of the structures of these compounds is a linear tetrapyrrole oxygenolytically opened between pyrrole rings A and B (Engel *et al.*, 1991, 1996; Iturraspe *et al.*, 1993). Although, red derivatives of chl do not accumulate during plant senescence, *in vitro* studies have demonstrated that a red chl catabolite (RCC, Figure 2.1) with a structure similar to that of the *C. protothecoides* bilines is the first nongreen product of chl degradation in higher plants (Rodoni *et al.*, 1997). This finding is corroborated by the analysis of *acd2-2*, which is deficient in red chlorophyll catabolite reductase (RCCR) and in which RCC accumulates upon dark-induced senescence induction (A. Pružinská and S. Hörtensteiner, unpublished; see below).

2.2.2.2 *Fluorescent chlorophyll catabolites*

Reduction of the C20/C1 double bond of RCC leads to the formation of primary fluorescent chlorophyll catabolite (pFCC), which occurs in two isomeric forms, pFCC-1 and pFCC-2, due to the introduction of a new stereocenter at C1 (Figure 2.1). Interestingly, within a given plant species, only one of the two possible isomers is formed. Thus, e.g. in *Arabidopsis* or canola pFCC-1 occurs, whereas pFCC-2 is formed in pepper or tomato (Hörtensteiner *et al.*, 2000). The specificity is determined by stereospecificity of the respective RCCRs from these species (see below). The chemical structures of pFCC-1 and -2 were elucidated after *in vitro* synthesis from pheide *a* (Mühlecker *et al.*, 1997, 2000). Although pFCCs have long been claimed as true intermediates of chl breakdown (Ginsburg and Matile, 1993), only recently the *in vivo* occurrence of pFCC-1 in senescent *Arabidopsis* leaves could be confirmed by mass spectroscopic means (Pružinská *et al.*, 2005). This investigation also corroborates other earlier findings: In extracts of senescent canola cotyledons or *F. pratensis* leaves, trace amounts of several FCCs were detected, which were more polar than pFCC, as judged by their retention on reversed-phase high-performance liquid chromatography (HPLC) (Ginsburg *et al.*, 1994). Indeed, in *Arabidopsis* at least two FCCs have been proven by mass spectroscopy to accumulate during chl breakdown (Pružinská *et al.*, 2005). In both *Arabidopsis* FCCs, the C13-methylester is hydrolyzed. One of them, *At*-FCC-1, carries an additional hydroxyl group, most probably at $C8^2$. The demonstration of the accumulation of modified FCCs *in vivo* strongly supports the model that side-chain modifications, as found in NCCs, are introduced into FCCs before nonenzymatic conversion to the respective NCC occurs after their import into the vacuole (Oberhuber *et al.*, 2003) (see below).

2.2.2.3 *Nonfluorescent chlorophyll catabolites*

NCCs have been found in all higher plant species analyzed for the presence of colorless chl breakdown products and a convenient nomenclature has been proposed for distinction (Ginsburg and Matile, 1993). A total of 23 NCCs from different species have been structurally characterized (Table 2.2). Except for one recently elucidated NCC from *Arabidopsis*, *At*-NCC-3, (Müller *et al.*, 2006), all NCCs structurally analyzed so far are derived from chl *a* and share a common basic tetrapyrrolic structure (Figure 2.1). They differ from each other by modifications of three peripheral side chains, i.e. dihydroxylation of the vinyl group of pyrrole A (R_1 in Figure 2.1), modification with various residues of the $C8^2$ position (R_2), and hydrolysis of the C13 carboxymethylester (R_3). These modifications occur separately or simultaneously in a given NCC (Table 2.2). Due to the occurrence of stereoisomeric pFCCs, the derived NCCs fall into two stereochemical groups; thus, the structurally identical NCCs, *Hv*-NCC-1 and *So*-NCC-2, are isomeric at C1 (Oberhuber *et al.*, 2001).

Table 2.2 Structures of NCCs from higher plants

	Name[a]	R_1[b]	R_2[b]	R_3[b]	C1-chemistry[c]	Reference
1	*At*-NCC-1	Vinyl	*O*-glucosyl	H	1	Pružinská *et al.* (2005)
2	*At*-NCC-2	Vinyl	OH	H	1	Pružinská *et al.* (2005)
3	*At*-NCC-3	Vinyl	OH[d]	H	1	Pružinská *et al.* (2005)
4	*At*-NCC-4	Vinyl	*O*-glucosyl[d]	CH_3	1	Pružinská *et al.* (2005)
5	*At*-NCC-5	Vinyl	H	H	1	Pružinská *et al.* (2005)
6	*Bn*-NCC-1	Vinyl	*O*-malonyl	H	1	Mühlecker and Kräutler (1996)
7	*Bn*-NCC-2	Vinyl	*O*-glucosyl	H	1	Mühlecker and Kräutler (1996)
8	*Bn*-NCC-3	Vinyl	OH	H	1	Mühlecker and Kräutler (1996)
9	*Bn*-NCC-4	Vinyl	H	H	1	Berghold (2005)
10	*Cj*-NCC-1	Vinyl	H	CH_3	2	Curty and Engel (1996)
11	*Cj*-NCC-2	Vinyl	H	CH_3	2	Oberhuber *et al.* (2003)
12	*Hv*-NCC-1	Dihydroxyethyl	OH	CH_3	1	Kräutler *et al.* (1991)
13	*Lo*-NCC-1	Vinyl	OH	CH_3	Nd	Iturraspe *et al.* (1995)
14	*Ls*-NCC-1	Vinyl	OH	CH_3	Nd	Iturraspe *et al.* (1995)
15	*Nr*-NCC-1	Vinyl	*O*-glucosyl-malonyl	CH_3	2	Berghold *et al.* (2004)
16	*Nr*-NCC-2	Vinyl	*O*-glucosyl	CH_3	2	Berghold *et al.* (2004)
17	*So*-NCC-1	Dihydroxyethyl	OH	H	2	Berghold *et al.* (2002)
18	*So*-NCC-2	Dihydroxyethyl	OH	CH_3	2	Oberhuber *et al.* (2001)
19	*So*-NCC-3	Vinyl	OH	H	2	Berghold *et al.* (2002)
20	*So*-NCC-4	Vinyl	OH	CH_3	2	Berghold *et al.* (2002)
21	*So*-NCC-5	Vinyl	H	CH_3	2	Berghold *et al.* (2002)
22	*Zm*-NCC-1	Dihydroxyethyl	*O*-glucosyl	CH_3	2	Berghold (2005)
23	*Zm*-NCC-2	Vinyl	*O*-glucosyl	CH_3	2	Berghold (2005)

[a] A nomenclature for NCCs (and FCCs) has been defined (Ginsburg and Matile,1993) in which a prefix indicates the plant species and a suffix number indicates decreasing polarity on reversed-phase HPLC. *At*, *Arabidopsis thaliana*; *Bn*, *Brassica napus*; *Cj*, *Cercidiphyllum japonicum*; *Hv*, *Hordeum vulgare*; *Lo*, *Liquidambar orientalis*; *Ls*, *Liquidambar styraciflua*; *Nr*, *Nicotiana rustica*; *So*, *Spinacia oleracea*; *Zm*, *Zea mays*.

[b] R_1–R_3 indicate residues at C3, $C8^2$ and $C13^2$ sidepositions, respectively, of NCCs as shown in Figure 2.1.

[c] C1 stereochemistry refers to the type of pFCC, i.e. pFCC-1 or pFCC-2, formed in the respective species; nd, not determined.

[d] In *At*-NCC-3, the site of hydroxylation is indicated to be $C7^1$ (rather than $C8^2$); in *At*-NCC-4, the site of attachment of the glucose moiety is not yet defined (S. Moser, T. Müller, S. Hörtensteiner and B. Kräutler, unpublished).

2.2.2.4 *Are NCCs degraded further?*

For some species, the amount of accumulating NCCs were shown to account for the entire amount of degraded chl (Mühlecker and Kräutler, 1996; Pružinská *et al.*, 2005). This was interpreted as NCCs representing the final catabolites of chl breakdown, but a few recent reports indicate degradation of chl beyond the stage of NCCs. Thus, a derivative of *Hv*-NCC-1 was identified in which the formyl group attached to pyrrole B is absent (Losey and Engel, 2001). Degradation of chl *in planta* to monopyrrolic oxygenation products, such as ethylmethyl maleimide and hematinic acid, has been proposed earlier (Brown *et al.*, 1991). Recently this concept has received new support by the group of Shioi who identified chl-derived monopyrroles during leaf senescence (Suzuki and Shioi, 1999).

2.3 The chlorophyll degradation pathway

2.3.1 Chlorophyll cycle

Chl *b* is a component of antenna complexes of the photosystems and occurs at variable ratios to chl *a*. Notably, all but one of the NCCs identified so far from higher plants are derivatives of chl *a*, which is explained by the pheide *a* specificity of PAO. Thus, conversion to chl *a* is a prerequisite of chl *b* breakdown. This is corroborated by the identification of a chl *b*-derived NCC (with an *a* configuration) in barley (Folley and Engel, 1999). A chl cycle has been defined that is able to interconvert chl(ide) *a* and *b* (see Rüdiger, 2002, for a recent review). The oxidative half of the cycle is catalyzed by chlide *a* oxygenase (CAO), a Rieske-type iron–sulfur oxygenase (Tanaka *et al.*, 1998; Oster *et al.*, 2000). Thereby, chlide *a* is oxidized to chlide *b* by two successive hydroxylations, followed by the spontaneous loss of water (Oster *et al.*, 2000). *b* to *a* conversion occurs on both chlide or chl with C7-hydroxy chl(ide) *a* as a stable intermediate. Two enzymes are involved, NADPH-dependent chl(ide) *b* reductase (CBR) and ferredoxin (Fd) dependent hydroxyl-chl(ide) *a* reductase (CAR) (Rüdiger, 2002). Neither the enzymes nor the genes of these reductases have been molecularly identified so far. The requirement of the reductive reactions of the chl(ide) cycle for chl *b* degradation is substantiated by a marked increase of CBR activity during dark-induced senescence of barley leaves (Scheumann *et al.*, 1999). CBR is the only enzyme of chl breakdown that localizes to the thylakoid membrane (Figure 2.1), qualifying it as the first enzyme of chl *b* degradation.

2.3.2 Reactions on green pigments

2.3.2.1 Chlorophyllase

Chlorophyllase (CLH) catalyzes the conversion of chl to chlide and phytol. It acts preferentially on chl *a* (Benedetti and Arruda, 2002), but also accepts chl *b* and pheophytins as substrates, but not porphyrins with an oxidized pyrrole ring D (see Hörtensteiner (1999) and references therein). CLH was considered to be a membrane protein of chloroplasts, and in barley and *Citrus* it has been localized on the inner envelope membrane (Matile *et al.*, 1997). Related to this is a remarkable functional latency; i.e. CLH is active only after solubilization in the presence of detergents or acetone (Trebitsh *et al.*, 1993). Surprisingly, the cloning of CLH genes from different species (Jakob-Wilk *et al.*, 1999; Tsuchiya *et al.*, 1999; Tang *et al.*, 2004; Arkus *et al.*, 2005) did not locate putative transmembrane domains from the deduced proteins and recombinant CLHs were active in the absence of detergents. This raises questions about the true localization of CLH (Takamiya *et al.*, 2000). A localization of CLH outside thylakoid membranes is rationalized by a spatial separation of CLH from its substrates, but implies the requirement of a specific carrier protein for chl (Matile *et al.*, 1999). Members of the family of water-soluble chl proteins, which are able to remove chl from pigment–protein complexes (Satoh *et al.*, 1998, 2001), were proposed candidates for such a chl shuttle (Matile *et al.*, 1999), but recently a role as pigment carrier in chl biosynthesis was demonstrated (Reinbothe *et al.*, 2004a).

A typical N-terminal transit peptide is missing in some of the cloned CLHs, such as *Arabidopsis* CLH1 or CaCLH1 from *Chenopodium album*. Instead, CaCLH1 seems to be glycosylated and has motifs suggesting localization in the vacuole (Tsuchiya *et al.*, 1999). This absence was explained by a possible second pathway localized inside the vacuole, where CLH together with unknown oxidases would catabolize chl (Takamiya *et al.*, 2000). The finding of a mass exodus of chl-containing globules from senescent chloroplasts (Guiamét *et al.*, 1999) supported this idea. But so far neither the oxidases nor catabolites of such a pathway have been found. Experiments on subcellular localization and the analysis of mutants will be required to elucidate the *in vivo* role of CLHs. Notably, downregulation of AtCLH1 did not cause an obvious senescence-related phenotype (Benedetti and Arruda, 2002; Kariola *et al.*, 2005), and a role of AtCLH1 in regulating defense pathways in plants through the detoxification of free chl occurring upon tissue damage was suggested (Kariola *et al.*, 2005). Furthermore, it can be argued that not all predicted CLHs might hydrolyze chl *in vivo*, since recombinant CLH from wheat efficiently cleaved the ester bonds of hydrophobic esters different from chl (Arkus *et al.*, 2005).

AtCLH1 was originally identified as a gene induced by coronatine, a methyl jasmonate analog (Benedetti *et al.*, 1998), and consequently, *AtCLH1* mRNA levels increased after methyl jasmonate treatment (Tsuchiya *et al.*, 1999). Furthermore, expression of *Citrus CLH1* is highly upregulated by ethylene (Jakob-Wilk *et al.*, 1999). These findings are consistent with CLH regulation by senescence-accelerating hormones (Drazkiewicz, 1994), but other CLH genes, as *AtCLH2*, are not hormonally controlled (Tsuchiya *et al.*, 1999; Benedetti and Arruda, 2002; Tang *et al.*, 2004).

2.3.2.2 Mg dechelation

In the past, two different catalytic activities were identified that released Mg^{2+} from chlide, i.e. a heat-stable low-molecular-weight compound tentatively named metal-chelating substance (MCS) (Shioi *et al.*, 1996a; Suzuki and Shioi, 2002) and a heat-labile protein termed *Mg dechelatase* that was associated with chloroplast membranes (Vicentini *et al.*, 1995b). Recent investigations from the Shioi laboratory have readdressed this issue. Thus, Mg dechelatase acts only on the frequently used artificial substrate chlorophyllin (Vicentini *et al.*, 1995b), but MCS was required to remove Mg^{2+} from chlide (Suzuki and Shioi, 2002; Kunieda *et al.*, 2005; Suzuki *et al.*, 2005). It can be assumed, therefore, that MCS acts as a Mg-dechelating factor *in vivo*, but elucidation of the molecular nature is required to consolidate the function of MCS and to answer questions on localization and regulation. Interestingly, the size of MCS is different in *Chenopodium album* (<400 Da) and strawberry (2180 + 20 Da) (Costa *et al.*, 2002; Suzuki *et al.*, 2005).

2.3.3 *Loss of green color*

Identification of pFCC as an intermediate of chl breakdown allowed the elucidation of an *in vitro* assay with pheide *a* as substrate (Hörtensteiner *et al.*, 1995). The assay required the simultaneous presence of gerontoplast membranes and stromal proteins. Thus, two enzymes, PAO and RCCR, were shown to jointly catalyze conversion of pheide *a* to pFCC with the formation of an unstable intermediate, RCC (Rodoni *et al.*, 1997). Biochemical properties of the two enzymes have been reviewed repeatedly (Matile *et al.*, 1996, 1999; Hörtensteiner, 1999; Kräutler and Matile, 1999; Hörtensteiner and Kräutler, 2000; Kräutler, 2003). Therefore, the following section will mainly focus on the recent molecular identification of the genes of these two chl catabolic proteins.

2.3.3.1 *Pheophorbide a oxygenase*

The biochemical properties of PAO as an envelope-localized nonheme iron monooxygenase allowed the identification of PAO candidate genes from *Arabidopsis*. After expression in *Escherichia coli*, one of them, *Accelerated cell death 1* (*Acd1*) (At3g44880), exhibited PAO activity with properties similar to native PAO (Pružinská *et al.*, 2003). ACD1/AtPAO is orthologous to lethal leaf spot 1 (LLS1) from maize (Gray *et al.*, 1997; Yang *et al.*, 2004). The maize *lls1* mutant and a knockout line in At3g44880, *pao1*, were subsequently analyzed to confirm their *in vivo* role of ACD1 and LLS1 as PAO (Pružinská *et al.*, 2003, 2005). Thus, both mutants retained chl during dark-induced senescence and were devoid of PAO activity. They showed a light-dependent cell death phenotype in leaves, which was caused by the accumulation of photoreactive pheide *a*.

In *Arabidopsis*, PAO belongs to a small family of five Rieske-type iron–sulfur oxygenases (Gray *et al.*, 2002), of which two other members are involved in chl metabolism as well, i.e. CAO (Oster *et al.*, 2000) and a proposed protochlorophyllide *a* oxygenase (Reinbothe *et al.*, 2004b). Like other Rieske-type oxygenases, the plant proteins require Fd as the obligate source of electrons (Tanaka *et al.*, 1998; Pružinská *et al.*, 2003). PAO-like proteins are highly conserved within oxygenic photosynthetic organisms, but are absent from anoxygenic photosynthetic prokaryotes (Gray *et al.*, 2004) (S. Hörtensteiner, unpublished). This indicates that the capacity to degrade chl to colorless catabolites coevolved with the invention of oxygenic photosynthesis (see below).

Regulation of PAO was inferred from its senescence-specific activity (Hörtensteiner *et al.*, 1995; Vicentini *et al.*, 1995a), and a regulation at the post-transcriptional level had been suggested (Pružinská *et al.*, 2003). A recent detailed comparison of PAO activity with PAO mRNA and protein abundance during senescence indicate that in *Arabidopsis*, PAO is exclusively regulated at the expression level (Pružinská *et al.*, 2005). *Pao* is also upregulated rapidly upon wounding (Yang *et al.*, 2004) and is highly expressed in flowers and siliques (Pružinská *et al.*, 2005). Microarray analysis using the Genevestigator tool (Zimmermann *et al.*, 2004) indicates upregulation of PAO in response to various stress conditions, such as osmotic stress and pathogen infection (S. Hörtensteiner, unpublished). It is likely, therefore, that the PAO pathway is activated not only during senescence, but also under other conditions that cause chl degradation.

2.3.3.2 Red chlorophyll catabolite reductase

RCCR is a stromal protein, which has an intriguing specificity towards reduction of the C20/C1 double bond of RCC. Hence, the source of RCCR defines which of the two possible C1 isomers of pFCC is formed. Screening the RCCR activities of more than 60 plant species indicated that within a family all genera and species produce the same isomer (Hörtensteiner *et al.*, 2000). *Rccr* genes were recently cloned and functional identity was confirmed after expression of *Arabidopsis Rccr* in *E. coli* (Wüthrich *et al.*, 2000). In *Arabidopsis*, the enzyme is encoded by a single gene (At4g37000) and is identical to ACD2 (Mach *et al.*, 2001). RCCR is a novel protein, but is distantly related to a family of bilin reductases. These include phytochromobilin synthase and different bilin reductases required for phycobilin biosynthesis (Frankenberg *et al.*, 2001). Like the bilin reductases, RCCR is Fd dependent, but appears to lack a metal or flavin cofactor, indicating that electrons are directly transferred from Fd to the respective substrates (Frankenberg and Lagarias, 2003; Kräutler, 2003). RCCR may have the role of a 'chaperone' rather than a catalytic reductase, i.e. it mediates an efficient interaction between Fd and RCC (still bound to PAO), thereby enabling a fast, regio- and stereoselective reduction at C1 (Kräutler, 2003; Hörtensteiner, 2004).

The *acd2* mutants were shown to be deficient in RCCR, and RCC was suggested to be responsible for the observed light-dependent cell death phenotype (Mach *et al.*, 2001). Indeed, RCC accumulates during senescence in *acd2-2* (A. Pružinská and S. Hörtensteiner, unpublished).

In contrast to PAO, RCCR is not regulated during senescence (Wüthrich *et al.*, 2000; Mach *et al.*, 2001; Pružinská *et al.*, 2005) and *Rccr* expression is rather constant throughout plant development and under stress conditions (S. Hörtensteiner, unpublished). Surprisingly, both PAO and RCCR are present in roots and other non-green tissues (Wüthrich *et al.*, 2000; Yang *et al.*, 2004), but their roles in these tissues have not been elucidated.

2.3.4 Reactions on pFCC

The diversity of NCCs found in different species (Table 2.2) allows drawing a network of reactions that occur after the (common) formation of pFCC (Figure 2.1). Altogether, six reactions can be distinguished: dihydroxylation of the vinyl group of pyrrole A, hydroxylation at $C8^2$, followed by glucosylation and/or malonylation, $C13^2$ demethylation and, finally, tautomerization of FCCs to NCCs. Except for the $C8^2$ hydroxylation and tautomerization, which are reactions common to all investigated species, other side-chain modifications occur in a species-specific manner.

2.3.4.1 Hydroxylation

Chl catabolite hydroxylation and subsequent modification with glucose resembles features of detoxification of herbicides and xenobiotics (Kreuz *et al.*, 1996). Thus, hydroxylation is probably catalyzed by cytochrome P_{450}-type enzymes, which are known to be involved in detoxification processes. Yet, hydroxylating activity on chl catabolites has not been demonstrated to date. Furthermore, the presence in *Arabidopsis* of more than 250 P_{450}-type oxygenases (Schuler and Werck-Reichhart, 2003) renders the elucidation of a respective protein in the near future rather unlikely.

2.3.4.2 Glucosylation

Most likely, glucosylation is catalyzed by a glucosyltransferase, but activity has not yet been detected (Hörtensteiner, 1998). More than 100 glucosyltransferases are encoded in the *Arabidopsis* genome. They have been cloned and expressed in *E. coli* in order to elucidate their role by functional screening. This strategy was successfully employed, e.g. for elucidation of transferases that act on cytokinins (Hou *et al.*, 2004), and may also be helpful to elucidate the glucosyltransferase(s) that is responsible for the formation of glycosylated NCCs in *Arabidopsis*.

2.3.4.3 Malonylation

The predominant NCCs from canola and *Nicotiana rustica*, *Bn*-NCC-1 and *Nr*-NCC-1, respectively, are malonylated. Respective transferases that require malonyl-CoA as cosubstrate are present in both systems (Hörtensteiner, 1998; Berghold *et al.*, 2004) and *in vitro* have been shown to act on NCCs. As judged from the cytosolic localization of the canola transferase (Hörtensteiner, 1998), and because FCC to NCC conversion occurs after import into the vacuole (see below), it can be assumed that FCCs are the natural substrates for transfer of the malonyl moiety.

2.3.4.4 Demethylation

Pheophorbidase has been shown to catalyze the demethylation of the C13 carboxymethyl group of pheide *in vitro*. Because of the presence in respective assays of high acetone concentrations, the free carboxyl group was spontaneously lost and, consequently, pyro pheide was identified as the final product of the reaction (Shioi *et al.*, 1996b). The distribution of pheophorbidase within different species (Suzuki *et al.*, 2002) matches the occurrence of $C13^2$-demethylated NCCs, indicating that pheophorbidase could be responsible for their formation. Pheophorbidase most likely localizes to the cytosol (Shioi *et al.*, 1996b); thus, it can be argued that its native substrates are FCCs rather then pheide. Indeed, an enzyme has been described from canola that is able to demethylate pFCC (Hörtensteiner, 1999) (S. Hörtensteiner, unpublished). Pheophorbidase has been purified to near homogeneity (Suzuki *et al.*, 2002) and has been cloned recently (Y. Shioi, personal communication).

2.3.4.5 Tautomerization

FCCs are intermediates of chl breakdown, which ultimately are converted to NCCs. This tautomerization results in the deconjugation of pyrroles C and D. Localization of NCCs inside the vacuole (Matile *et al.*, 1988; Hinder *et al.*, 1996) suggested a nonenzymatic reaction under acidic conditions (Matile *et al.*, 1999; Kräutler, 2002). Indeed, pure pFCC-2 was shown to rapidly convert into the respective NCC at acidic pH (Oberhuber *et al.*, 2003). The involved mechanism was proposed to be a two-step protonation/deprotonation reaction that leads to the formation of an instable NCC in which the C13^2 side group and pyrrole D are *cis* to each other. This intermediate isomerizes to the final NCC, thereby establishing a *trans* configuration of C15 and C13^2 (Figure 2.1) (Oberhuber *et al.*, 2003).

2.4 Chlorophyll catabolic mutants

Stay-green mutants in which senescence is apparently delayed by a visible retention of chl have been classified into four groups (Thomas and Howarth, 2000). Type C stay-greens most likely represent mutants that have a defect in a catalytic or regulatory gene of chl breakdown, since chl retention in these mutants is uncoupled from other, normally proceeding, senescence processes. Among these, Bf 993 of *F. pratensis* has been best characterized (Thomas and Stoddart, 1975). Bf 993 exhibits a significant reduction of PAO activity, and accumulates chlides and pheide *a* during senescence. In order to identify the mutated gene, *sid*, the stay-green character of Bf 993 has been transferred from *Festuca* into *Lolium* species by intergeneric hybridization. Repeated backcrossing has reduced the size of the introgressed segment and enabled the localization of the mutated gene by genomic *in situ* hybridization (Thomas *et al.*, 1997, 1999). Subsequently, *Festuca*-specific polymorphisms in the *Lolium* background allowed molecular tagging of the *sid* locus (Moore *et al.*, 2005). A close synteny of the *Festuca* genome with the sequenced rice genome provides an additional tool for cloning of *sid* in the future.

A biochemical lesion in PAO activity has also been shown for Gregor Mendel's green peas (Thomas *et al.*, 1996), but the affected gene has not been identified so far. Molecular mapping of two other stay-green mutations, i.e. pepper *CL* (Efrati *et al.*, 2005) and tomato *GF* (Akhtar *et al.*, 1999), on homologous chromosomes indicated the mutation of orthologous genes in these two mutants. Furthermore, CL mapped on different chromosomes than *Pao* and three *CLH* tomato genes, implying that *CL/GF* corresponds to another catabolic or regulatory gene of chl breakdown (Efrati *et al.*, 2005). Recently, the mutated gene, *sgr*(t), of a type C stay-green mutant in rice was mapped (Cha *et al.*, 2002) and *Sgr* was cloned subsequently (GenBank accession no. AY850134; J.-W. Yu, S.-Y. Park, J. Li, J.-S. Park, H.-J. Koh and N.-C. Paek, unpublished). Highly homologous sequences are widely distributed in higher plant species, and the *Arabidopsis* SGR1 protein (At4g22920) is specifically expressed during senescence (S. Hörtensteiner, unpublished). A function for SGR has not been elucidated so far.

2.5 Significance of chlorophyll breakdown

2.5.1 Topology of chlorophyll breakdown

The localization of NCCs inside the vacuole (Matile *et al.*, 1988; Hinder *et al.*, 1996) indicates the requirement of respective transport systems at the gerontoplast envelope and the tonoplast. A primary active transport system has been detected in barley vacuoles, which has a particularly high affinity for FCCs (Hinder *et al.*, 1996). This, together with a nonenzymatic FCC–NCC tautomerization catalyzed by the acidic vacuolar sap (Oberhuber *et al.*, 2003), implies that FCCs rather than NCCs are the natural substrates for transport across the tonoplast. Two members of the multidrug-resistance-associated protein (MRP) subfamily of ATP-binding cassette (ABC) transporters, AtMRP2 and AtMRP3, were shown to be capable of transporting NCCs after expression in yeast (Lu *et al.*, 1998; Tommasini *et al.*, 1998), but the molecular identity of the *in vivo* transport system remains to be elucidated.

Export of FCCs from isolated gerontoplasts depended on the hydrolysis of ATP (Matile *et al.*, 1992), but the nature of the transport system at the plastid envelope is still unknown. Chloroplast proteomics (Ferro *et al.*, 2002) as well as *in silico* prediction of envelope transporters (Koo and Ohlrogge, 2002) may enable the identification of candidate transporters in the future.

Modifications and subsequent disposal of FCCs in the vacuole are reminiscent of the stepwise detoxification in plants of herbicides and xenobiotics. After hydroxylation and conjugation with different moieties to increase their water solubility, these compounds are detoxified by export from the metabolically active cytoplasm. From this point of view, chl degradation in higher plants can be regarded as chl detoxification required to safely dispose this potential phototoxin during senescence (see below). In contrast, *C. protothecoides* does not need such a mechanism and the first water-soluble product of porphyrin macrocycle cleavage, RCC, is excreted into the surrounding medium. Consequently, 'invention' of RCCR for the intracellular metabolism of RCC can be regarded a prerequisite for land colonization of plants during evolution.

2.5.2 Chl breakdown and cell death

The reactions of PAO and RCCR cause the loss of green pigment color and, therefore, are most important for the detoxification of chl during senescence. The importance of chl breakdown is corroborated by the analysis of respective mutants. Thus, mutations in, or antisense expression of, *Pao* and *Rccr* induce light-dependent cell death phenotypes (Greenberg and Ausubel, 1993; Greenberg *et al.*, 1994; Gray *et al.*, 1997, 2002; Mach *et al.*, 2001; Spassieva and Hille, 2002; Pružinská *et al.*, 2003, 2005; Tanaka *et al.*, 2003), and accumulation of pheide *a* was demonstrated to be responsible for cell death induction in the case of PAO mutants (Pružinská *et al.*, 2003, 2005; Tanaka *et al.*, 2003). In *acd2-2*, RCC accumulates during senescence, and contents of RCC have been shown to correlate with the progression of cell death (A. Pružinská and S. Hörtensteiner, unpublished). In addition, treatment of leaves or of isolated protoplasts with proto IX induced cell death in the mutant (Yao *et al.*, 2004). Thus it remains to be shown whether *in vivo* RCC and/or proto IX is responsible for the lesion mimic phenotype of *acd2*.

Nevertheless, removal of potentially phototoxic chl intermediates appears to be crucial for plant survival. This phenomenon is corroborated by the analysis of mutants and transgenic plants with defects in chl biosynthetic steps (Grimm, 1998; Hu *et al.*, 1998; Ishikawa *et al.*, 2001; Meskauskiene *et al.*, 2001) where cell death is triggered by the accumulation of photoreactive porphyrins. The mechanism by which cell death is executed in chl catabolic mutants has not been elucidated so far. In *Arabidopsis*, changes in the mitochondrial membrane potential have been suggested to play an important role in programmed cell death (Yao *et al.*, 2004). In the *Arabidopsis flu* mutant, light absorption by protochlorophyllide caused the production of singlet oxygen (op den Camp *et al.*, 2003), which in turn induced a cell-death-signaling pathway that was different from cell death programs induced by other reactive oxygen species. EXECUTER1 has been identified as an (early) component of cell death signaling, but the role of the EXECUTER1 protein has not been elucidated (Wagner *et al.*, 2004). It can be argued that light-dependent singlet oxygen production by other tetrapyrrolic compounds, such as pheide *a* and RCC/proto IX, respectively, in PAO and RCCR mutants, might induce the EXECUTER1-dependent pathway as well, and analysis of respective double mutants will answer this question in the future.

Altogether, it can be concluded that functional chl metabolism, i.e. biosynthesis and breakdown, is vitally important for plant development and plant survival in order to prevent the accumulation of phototoxic intermediates. The pathways have to be tightly regulated and PAO plays a major role in chl degradation. Such regulation has also been proposed for chl biosynthesis (Eckhardt *et al.*, 2004), and it is likely that chl catabolic enzymes are arranged in high-molecular-weight complexes thereby minimizing the risk of release and photoactivation of intermediates. Notably, there is biochemical evidence for a close interaction between PAO and RCCR, and metabolic channeling of RCC (Rodoni *et al.*, 1997).

2.5.3 Chl breakdown and nitrogen economy

Retention of chl in stay-green mutants and the maize PAO mutant, *lls1*, is accompanied by a retention of chl-binding proteins, such as LHCIIb polypeptides, but not of soluble plastid proteins (Thomas and Smart, 1993; Pružinská *et al.*, 2003). This is explained by the requirement of chl for stabilization of chl-binding proteins (Horn and Paulsen, 2004), and it can be argued that removal of chl is a prerequisite for the degradation of chl-binding proteins. In Bf 993, LHCIIb undergoes some proteolytic cleavage, which was assumed to remove the N-terminal region facing the stroma (Thomas *et al.*, 2001). This finding points to an interplay between chl catabolic and proteolytic activities during senescence. Thereby, Chl(ide) *b* reductase (CBR) could play a critical role. *In vitro* assembly of LHCIIb requires the presence of chl *b* in stoichiometric rates (Horn and Paulsen, 2004), and lack of chl *b* in chl *b*-less mutants causes a fast turnover of LHCIIb (Harrison *et al.*, 1993). Arguably, the activation/synthesis of the thylakoid-located CBR during senescence reduces the amount of chl *b* of individual chl-protein complexes, which may destabilize the complexes and make them accessible for proteases.

The proteases responsible for degradation of chl-binding proteins during senescence are largely unknown. One exception is the D1 protein of the PS II reaction center, which has a rapid turnover rate. D1 degradation occurs in two steps, an initial cleavage into two fragments by DegP-type proteases, followed by complete degradation through members of chloroplast-localized FtsH proteases (Adam and Clarke, 2002). In contrast, knowledge about the mechanisms of LHCIIb degradation is rather scarce and contradicting. During high light acclimation, serine/cysteine-type proteases were shown to degrade LHCIIb (Tziveleka and Argyroudi-Akoyunoglou, 1998), and a thylakoid-located activity was identified, which cleaved a N-terminal peptide of defined length from Lhcb1 (Forsberg *et al.*, 2005). In contrast, a (zinc-binding) metalloprotease acting toward Lhcb3 was identified (Zelisko and Jackowski, 2004) which, judged from its enzymatic properties, could be a member of the FtsH family. The activity was integrally associated with thylakoid membranes, but susceptibility of Lhcb3 for proteolytic attack required the removal of unknown protective factors in a senescence-dependent manner.

2.6 The pigments of senescing leaves

Depending on the rate and timing of chl breakdown, and the dynamics of accessory pigments, leaves take on dramatically different coloration during senescence. In a few cases, such as leaves of nitrogen-fixing papilionoid legume trees, chl is not completely degraded, and leaves retain at least some green color (P. del Tredici, personal communication). Plants also vary in color changes in different biomes or even in different locations within a biome. Fewer leaves of tropical forest species senesce bright colors (Lee and Collins, 2001) compared to leaves of temperate deciduous forests (Lee *et al.*, 2003). Color production may be more pronounced in deciduous broadleaf forests of northeastern North America compared to similar forests in temperate Asia and Europe (Hoch *et al.*, 2001). However, in each of these biomes one can find examples of leaves that senesce pale green, yellow, orange and red.

These patterns of color change are due to the production and/or retention of different pigments and the scattering and path-lengthening effects within leaf tissue (Vogelmann, 1993). Light greens are due to partial breakdown of chl, sometimes present as small areas of darker color. Yellows are due to the retention of carotenoid pigments. These pigments are also broken down during senescence (Lee *et al.*, 2003), but more of them are retained than the chls, resulting in yellow coloration. Senescing leaves of at least some species have high concentrations of β-carotene, which suggests a considerable amount of change in carotenoid composition during senescence. We need to learn more about which carotenoid pigments are present during senescence. In a few cases red carotenoids, as anhydroeschscholtzxanthin, may accumulate to produce red leaves in boxwoods (*Buxus* sp.) (Ida *et al.*, 1995) or orange carotenoids in fiddlewood (*Citharexylum fruticosum*) (Lee and Collins, 2001). In ginkgo the bright-yellow color during leaf senescence is due to carotenoids and also a yellow-fluorescent molecule, 6-hydroxykynurenic acid (Matile, 1994).

In temperate mixed deciduous forests of northeastern North America, most species turn orange or red during leaf senescence; at Harvard Forest in Central Massachusetts, about 70% of the woody species (62/89) turned these colors during senescence (Lee *et al.*, 2003). Virtually all this color change is caused by the *de novo* synthesis of anthocyanins in the leaves; these are not present in the mature leaves, as is the case for carotenoids. There are a few cases where other flavonoid and carotenoid pigments may cause this color change, and even examples where the nitrogenous betalain pigments produce red colors in senescing herbs of species in the Caryophyllales, as in *Amaranthus* spp. (D.W. Lee, unpublished observations). However, the vast majority of red leaf senescence is due to the synthesis of anthocyanins in the cytoplasm and their sequestration in cell vacuoles when the breakdown of chl molecules has begun. In a sample of tree species at the Harvard Forest, anthocyanin synthesis occurred when about half of the chl concentration present at the mid-summer peak had been removed (Lee *et al.*, 2003). In this sample, almost all of the anthocyanin accumulation was in the chlorenchymatous mesophyll cells, particularly in the palisade layer. In herbs, anthocyanins accumulated in the adaxial epidermal cells in many species (D.W. Lee, unpublished observations).

In the large majority of red-senescing species, the principal (and often the only) anthocyanin produced is cyanidin-3-glucoside. It may be that more careful surveys with modern instrumentation may reveal a bit more variation, in the Aceraceae (Ji *et al.*, 1992), but it is clear that the anthocyanins accumulated during leaf senescence are a small portion of those present in flowers and fruits. This anthocyanin has a peak absorbance of 525–530 nm, broadened considerably *in vivo* because of the scattering of light within the mesophyll of the leaf.

After a century or so of speculation about the potential physiological and ecological functions of anthocyanins in vegetative tissues, considerable progress has been made in the past decade. A photoprotective function of anthocyanins had been advocated among the physiological anatomists during the late nineteenth century, and Büning had later argued for a specific photoprotective defense against UV radiation (Lee and Gould, 2002). However, anthocyanins do not absorb strongly in the UV-B regions that could damage plant tissue and are poorly placed to provide such protection compared to other flavonoids and simpler phenolic acids (Lee *et al.*, 1987). In the past 10 years, subsequent to the development of techniques of the measurement of fluorescent kinetics and ideas about photoinhibition and photodamage in photosynthetic tissues, evidence of photoinhibitory protection by anthocyanins has accumulated (Gould *et al.*, 1995; Krol *et al.*, 1995; Smillie and Hetherington, 1999; Merzlyak and Chivkunova, 2000; Steyn *et al.*, 2002; Close and Beadle, 2003). Simultaneously, with the increased interest in antioxidative activity in components of foods, particularly fruits, anthocyanins have been demonstrated to have potent antioxidative capability (Yamasaki, 1997; Lila, 2004). Such activity has been detected *in vivo* in leaves and in sweet potatoes (Gould *et al.*, 2002a,b; Philpott *et al.*, 2004). However, anthocyanins are produced in the cytoplasm and actively pumped into the vacuole. Other than the breakdown of H_2O_2, as produced during tissue damage, anthocyanins seem unlikely to protect sites of singlet oxygen production, as the chloroplasts. Since anthocyanins may be colorless in the more pH-neutral cytoplasm, they may still provide such protection there, but there is presently no supportive evidence. At any rate, anthocyanins clearly have physiological roles within vegetative tissues, in addition to the important biological functions (i.e. pollination and dispersal) they perform in flowers and fruits.

2.7 The function of anthocyanins in leaf senescence

The challenge in understanding the functions of accessory pigments, particularly anthocyanins that are synthesized *de novo*, is to justify the metabolic expense of producing pigments at the end of the leaf life span, particularly in seasonally cold climates (Huner *et al.*, 1998). Previous research and speculation about such pigment function in vegetative organs was focused on pigment production at the beginning of the life span, in early development. Thus, anthocyanins could physiologically protect the photosynthetic tissue during development in expanding ('flushing') leaves, particularly common in the tropics (Lee and Collins, 2001), or protect against photo-oxidative stress in mature leaves (Gould *et al.*, 2002a). Biologically, coloration could protect developing leaves through camouflage (Juniper, 1993) or the greater apparency of insect herbivores on the colored foliage (Lev-Yadun *et al.*, 2004). Recently, two separate (and competing) hypotheses have been published, one physiological and one biological, that provide mechanisms for protecting senescing leaves to aid plants the following year.

2.7.1 *Physiological explanations*

A physiological answer to this conundrum came in a review by Hoch *et al.* (2001). They argued that the protective function of anthocyanins during leaf senescence, particularly under conditions of low temperature and high solar irradiance during the autumn, would be to screen chloroplasts during the chl breakdown process, help prevent the formation of reactive oxygen species and allow leaves to more efficiently resorb (or more efficiently translocate) nitrogen back into woody tissues for use during the *next* growing season. Differences in efficiency of retention could be detected by the ratio of nitrogen present in green leaves (resorption efficiency) and more directly as differences in the amount of residual nitrogen in falling leaves (resorption proficiency) (Killingbeck, 1996). Such protection could be due to the screening effect in blue-green wavelengths or to an antioxidative mechanism. Four studies have appeared consistent with this hypothesis. Feild *et al.* (2001) demonstrated a strong photoprotective function of anthocyanins in the mesophyll of red-osier dogwood (*Cornus stolonifera*), but they were not able to demonstrate a difference in residual nitrogen in red versus yellow senescent leaves. Among leaves of tree species during autumn senescence at the Harvard Forest, the amount of anthocyanin detectable in falling leaves was negatively correlated with the amount of residual nitrogen, consistent with this hypothesis (Lee *et al.*, 2003). In sugar maple (*Acer saccharum*), an important tree in more northerly deciduous forests, trees and even individual branches vary considerably in the amounts of anthocyanins and the color brilliance of leaves. Schaberg *et al.* (2003) showed that leaves with low foliar nitrogen concentrations turned red more quickly and intensely than leaves with higher nitrogen.

Redder leaves also had higher starch and sugar concentrations. Although these comparative studies are consistent with the hypothesis of nitrogen resorption, it must be mentioned that nutrient stress often induces anthocyanin production in leaves of crop plants (Chalker-Scott, 1999), and variations in color among trees during the autumn may be due to subsurface geology, which could, in turn, influence nutrient uptake (Canney *et al.*, 1979).

The best support for the physiological hypothesis is the experimental work of Hoch and colleagues (2003) who tested the degrees of photoinhibition, nitrogen levels and anthocyanin concentrations in wild types and anthocyanin-lacking mutants of three shrubs, *Cornus sericea*, *Vaccinium elliottii* and *Viburnum sargentii*. The mutants were more photoinhibited after exposure to bright light, and recovered more slowly; and falling leaves had significantly lower nitrogen content than did the wild types with anthocyanin. These independent papers report results consistent with this hypothesis, but more research among a greater diversity of taxa is certainly needed.

2.7.2 *Ecological explanations*

Excitement about the possible selective advantages of color production during leaf senescence was aroused from recent reviews (Hamilton and Brown, 2001; Archetti and Brown, 2004). Hamilton's reputation in evolutionary theory gave this hypothesis (now called a theory) all the more impact, and it was published nearly coincident with his untimely death in Africa. Hamilton and Brown postulated that brightly colored leaves during autumn could be a handicap signal, warning egg-laying insects against increased defensive compounds in those trees and reducing herbivory by newly hatched individuals the following year. They tested the hypothesis by a meta-analysis of the literature on aphid herbivory on trees. They found a significant repellent effect of yellow colors against aphids in the autumn, but not a significant association with red anthocyanic leaves. Evidence consistent with predictions of the hypothesis has been reported in two studies. In the mountain birch (*Betula pubescens*), Hagen and colleagues found a significant negative association between intensity of yellow color in tree crowns and insect damage, assessed by the degree of leaf damage and an analysis of fluctuating asymmetry (indicative of stress during leaf development) during the following year (Hagen *et al.*, 2003). They also showed a decrease in reproductive investment (i.e. fitness) associated with the degree of leaf damage. However, they did not reveal any chemical basis for a reduction in palatability associated with color. Moreover, the difference in color production was really the difference between the timing of leaf senescence and fall; earlier senescing trees were avoided by these insects. In addition, herbivorous insects were not identified, nor was their egg-laying activity determined. Archetti and Leather (2005) observed visits to a stand of trees of the bird cherry (*Prunus padus*) by the aphid *Rhopalosiphun padi* in southern England. Visits to leaves and the degree of coloration, a combination of yellow and red appearing on individual leaves and on individual tree crowns, were negatively correlated. This pattern of color made it impossible to determine whether insects avoided the red or yellow coloration; perhaps any color different from the green preferred by insects was effective. Aphids, however, grew little the following spring and trees were therefore not significantly different in visits and herbivory. Again, no evidence of differences in defensive compounds or palatability among the individual trees was shown. Thus, many of the predictions of this hypothesis have not been tested, but there appear to be visitation preferences during autumn leaf senescence, at least for some insects and some trees.

An additional hypothesis addressing leaf coloration and visitation by dispersal agents should be mentioned: the fruit flag hypothesis (Stiles, 1982). When fruits are retained on trees later in the autumn, to be picked and dispersed by migrating songbirds, brightly colored autumn foliage may 'advertise' the presence of these fruits. Although there is supporting evidence for this hypothesis in a few cases, as among some sumac species, most trees lose their fruits well before autumn senescence.

2.7.3 Reconciling these explanations

The publication of these two conflicting hypotheses on accessory pigment function during leaf senescence has generated some controversy (Holopainen and Peltonen, 2002; Schaefer and Wilkinson, 2004). The questions concern the sufficiency of evidence for either hypothesis, and whether they are mutually exclusive. Apparency is easily achieved in a green canopy by degrading chl early and leaving residual carotenoids to produce a bright yellow color. So the apparency hypothesis does not really explain the most interesting accessory pigment change during leaf senescence: the *de novo* synthesis of anthocyanin. It may well be that the apparency hypothesis may better explain the timing of senescence among certain species, and the production of yellow colors. The physiological hypothesis may better explain the selective advantage of the synthesis of anthocyanins during senescence, particularly in the autumn for temperate trees.

2.8 Conclusions and perspectives

The elucidation of structures of chl catabolites and the identification of chl catabolic enzymes at the molecular level during the last 15 years has considerably extended our knowledge on the biochemical reactions of chl breakdown. Despite these efforts, many questions remain to be solved such as (i) by which mechanism is chl turned over at the steady state? (ii) what are the regulatory mechanisms of chl breakdown during senescence? (iii) do different catabolic enzymes work in protein complexes to prevent phototoxic effects from colored intermediates? (iv) do chl catabolites have physiological functions, e.g. in intracellular signaling, as has been shown for chl biosynthetic intermediate? Answering these questions requires further direct investigations on chl breakdown, in particular elucidation of missing steps in the pathway and cloning of the affected genes of stay-green mutants. In addition, other aspects of senescence and leaf coloration will have to be uncovered, such as the mechanisms of protein degradation or regulation of autumnal leaf pigmentation.

References

Adam, Z. and Clarke, A.K. (2002) Cutting edge of chloroplast proteolysis. *Trends Plant Sci* **7**, 451–456.

Akhtar, M.S., Goldschmidt, E.E., John, I., Rodoni, S., Matile, P. and Grierson, D. (1999) Altered patterns of senescence and ripening in *gf*, a stay-green mutant of tomato (*Lycopersicon esculentum* Mill.). *J Exp Bot* **50**, 1115–1122.

Archetti, M. and Brown, S.P. (2004) The coevolution theory of autumn colours. *Proc Royal Soc Lond B* **271**, 1219–1223.

Archetti, M. and Leather, S.R. (2005) A test of the coevolution theory of autumn colours: colour preference of *Rhopalosiphum padi* on *Prunus padus*. *Oikos* **110**, 339–343.

Arkus, K.A.J., Cahoon, E.B. and Jez, J.M. (2005) Mechanistic analysis of wheat chlorophyllase. *Arch Biochem Biophys* **438**, 146–155.

Benedetti, C.E. and Arruda, P. (2002) Altering the expression of the chlorophyllase gene *ATHCOR1* in transgenic *Arabidopsis* caused changes in the chlorophyll-to-chlorophyllide ratio. *Plant Physiol* **128**, 1255–1263.

Benedetti, C.E., Costa, C.L., Turcinelli, S.R. and Arruda, P. (1998) Differential expression of a novel gene in response to coronatine, methyl jasmonate, and wounding in the *Coi1* mutant of *Arabidopsis*. *Plant Physiol* **116**, 1037–1042.

Berghold, J. (2005) *Chemische und biochemische Untersuchungen zum Chlorophyllabbau*. University of Innsbruck, Innsbruck, 194 pp.

Berghold, J., Breuker, K., Oberhuber, M., Hörtensteiner, S. and Kräutler, B. (2002) Chlorophyll breakdown in spinach: on the structure of five nonfluorescent chlorophyll catabolites. *Photosynth Res* **74**, 109–119.

Berghold, J., Eichmüller, C., Hörtensteiner, S. and Kräutler, B. (2004) Chlorophyll breakdown in tobacco: on the structure of two nonfluorescent chlorophyll catabolites. *Chem Biodivers* **1**, 657–668.

Brown, S.B., Houghton, J.D. and Hendry, G.A.F. (1991) Chlorophyll breakdown. In: *Chlorophylls* (ed. Scheer, H.). CRC Press, Boca Raton, FL, pp. 465–489.

Buchanan-Wollaston, V., Page, T., Harrison, E., *et al.* (2005) Comparative transcriptome analysis reveals significant differences in gene expression and signalling pathways between developmental and dark/starvation-induced senescence in *Arabidopsis*. *Plant J* **42**, 567–585.

Canney, F.C., Cannon, H.L., Cathrall, J.B. and Robinson, K. (1979) Autumn colors, insects, plant disease, and prospecting. *Econ Geol* **74**, 1673–1676.

Cha, K.W., Lee, Y.J., Koh, H.J., Lee, B.M., Nam, Y.W. and Paek, N.C. (2002) Isolation, characterization, and mapping of the stay green mutant in rice. *Theor Appl Genet* **104**, 526–532.

Chalker-Scott, L. (1999) Environmental significance of anthocyanins in plant stress responses. *Photochem Photobiol* **70**, 1–9.

Close, D.C. and Beadle, C.L. (2003) The ecophysiology of foliar anthocyanin. *Bot Rev* **69**, 149–161.

Collakova, E. and DellaPenna, D. (2003) The role of homogentisate phytyltransferase and other tocopherol pathway enzymes in the regulation of tocopherol synthesis during abiotic stress. *Plant Physiol* **133**, 930–940.

Costa, M.L., Civello, P.M., Chaves, A.R. and Martinez, G.A. (2002) Characterization of Mg-dechelatase activity obtained from *Fragaria* × *ananassa* fruit. *Plant Physiol Biochem* **40**, 111–118.

Curty, C. and Engel, N. (1996) Detection, isolation and structure elucidation of a chlorophyll *a* catabolite from autumnal senescent leaves of *Cercidiphyllum japonicum*. *Phytochemistry* **42**, 1531–1536.

Drazkiewicz, M. (1994) Chlorophyllase: occurrence, functions, mechanism of action, effects of external and internal factors. *Photosynthetica* **30**, 321–331.

Eckhardt, U., Grimm, B. and Hörtensteiner, S. (2004) Recent advances in chlorophyll biosynthesis and breakdown in higher plants. *Plant Mol Biol* **56**, 1–14.

Efrati, A., Eyal, Y. and Paran, I. (2005) Molecular mapping of the *chlorophyll retainer* (*cl*) mutation in pepper (*Capsicum* spp.) and screening for candidate genes using tomato ESTs homologous to structural genes of the chlorophyll catabolism pathway. *Genome* **48**, 347–351.

Engel, N., Curty, C. and Gossauer, A. (1996) Chlorophyll catabolism in *Chlorella protothecoides*. 8. Facts and artefacts. *Plant Physiol Biochem* **34**, 77–83.

Engel, N., Jenny, T.A., Mooser, V. and Gossauer, A. (1991) Chlorophyll catabolism in *Chlorella protothecoides*. Isolation and structure elucidation of a red bilin derivative. *FEBS Lett* **293**, 131–133.

Feild, T.S., Lee, D.W. and Holbrook, N.M. (2001) Why leaves turn red in autumn. The role of anthocyanins in senescing leaves of red-osier dogwood. *Plant Physiol* **127**, 566–574.

Ferro, M., Salvi, D., Riviere-Rolland, H., *et al.* (2002) Integral membrane proteins of the chloroplast envelope: identification and subcellular localization of new transporters. *Proc Natl Acad Sci U S A* **99**, 11487–11492.

Folley, P. and Engel, N. (1999) Chlorophyll *b* to chlorophyll *a* conversion precedes chlorophyll degradation in *Hordeum vulgare* L. *J Biol Chem* **274**, 21811–21816.

Forsberg, J., Strom, J., Kieselbach, T., *et al.* (2005) Protease activities in the chloroplast capable of cleaving an LHCII N-terminal peptide. *Physiol Plant* **123**, 21–29.

Frankenberg, N. and Lagarias, J.C. (2003) Phycocyanobilin:ferredoxin oxidoreductase of *Anabaena* sp. PCC 7120. *J Biol Chem* **278**, 9219–9226.

Frankenberg, N., Mukougawa, K., Kohchi, T. and Lagarias, J.C. (2001) Functional genomic analysis of the HY2 family of ferredoxin-dependent bilin reductases from oxygenic photosynthetic organisms. *Plant Cell* **13**, 965–978.

Ginsburg, S. and Matile, P. (1993) Identification of catabolites of chlorophyll porphyrin in senescent rape cotyledons. *Plant Physiol* **102**, 521–527.

Ginsburg, S., Schellenberg, M. and Matile, P. (1994) Cleavage of chlorophyll-porphyrin. Requirement for reduced ferredoxin and oxygen. *Plant Physiol* **105**, 545–554.

Gould, K.S., Kuhn, D.N., Lee, D.W. and Oberbauer, S.F. (1995) Why leaves are sometimes red. *Nature* **378**, 241–242.

Gould, K.S., McKelvie, J. and Markham, K.R. (2002a) Do anthocyanins function as antioxidants in leaves? Imaging of H_2O_2 in red and green leaves after mechanical injury. *Plant Cell Environ* **25**, 1261–1269.

Gould, K.S., Neill, S.O. and Vogelmann, T.C. (2002b) A unified explanation for anthocyanins in leaves? In: *Advances in Botanical Research*, Vol. 37 (eds Gould, K.S. and Lee, D.W.). Academic Press, London, pp. 167–192.

Gray, J., Close, P.S., Briggs, S.P. and Johal, G.S. (1997) A novel suppressor of cell death in plants encoded by the *Lls1* gene of maize. *Cell* **89**, 25–31.

Gray, J., Janick-Bruckner, D., Bruckner, B., Close, P.S. and Johal, G.S. (2002) Light-dependent death of maize *lls1* cells is mediated by mature chloroplasts. *Plant Physiol* **130**, 1894–1907.

Gray, J., Wardzala, E., Yang, M., Reinbothe, S., Haller, S. and Pauli, F. (2004) A small family of LLS1-related non-heme oxygenases in plants with an origin amongst oxygenic photosynthesizers. *Plant Mol Biol* **54**, 39–54.

Greenberg, J.T. and Ausubel, F.M. (1993) *Arabidopsis* mutants compromised for the control of cellular damage during pathogenesis and aging. *Plant J* **4**, 327–341.

Greenberg, J.T., Guo, A., Klessig, D.F. and Ausubel, F.M. (1994) Programmed cell death in plants: a pathogen-triggered response activated coordinately with multiple defense functions. *Cell* **77**, 551–563.

Grimm, B. (1998) Novel insights in the control of tetrapyrrole metabolism of higher plants. *Curr Opin Plant Biol* **1**, 245–250.

Guiamét, J.J., Pichersky, E. and Noodén, L.D. (1999) Mass exodus from senescing soybean chloroplasts. *Plant Cell Physiol* **40**, 986–992.

Hagen, S.B., Folstad, I. and Jakobsen, S.W. (2003) Autumn colouration and herbivore resistance in mountain birch (*Betula pubescens*). *Ecol Lett* **6**, 807–811.

Hamilton, W.D. and Brown, S.P. (2001) Autumn tree colours as a handicap signal. *Proc Royal Soc Lond B* **268**, 1489–1493.

Harrison, M.A., Nemson, J.A. and Melis, A. (1993) Assembly and composition of the chlorophyll *a-b* light-harvesting complex of barley (*Hordeum vulgare* L.): immunochemical analysis of chlorophyll *b*-less and chlorophyll *b*-deficient mutants. *Photosynth Res* **38**, 141–151.

Hendry, G.A.F., Houghton, J.D. and Brown, S.B. (1987) Chlorophyll degradation. A biological enigma. *New Phytol* **107**, 255–302.

Hinder, B., Schellenberg, M., Rodoni, S., *et al.* (1996) How plants dispose of chlorophyll catabolites. Directly energized uptake of tetrapyrrolic breakdown products into isolated vacuoles. *J Biol Chem* **271**, 27233–27236.

Hoch, W.A., Singsaas, E.L. and McCown, B.H. (2003) Resorption protection. Anthocyanins facilitate nutrient recovery in autumn by shielding leaves from potentially damaging light levels. *Plant Physiol* **133**, 1296–1305.

Hoch, W.A., Zeldin, E.L. and McCown, B.H. (2001) Physiological significance of anthocyanins during autumnal leaf senescence. *Tree Physiol* **21**, 1–8.

Holopainen, J.K. and Peltonen, P. (2002) Bright autumn colours of deciduous trees attract aphids: nutrient retranslocation hypothesis. *Oikos* **99**, 184–188.

Horn, R. and Paulsen, H. (2004) Early steps in the assembly of light-harvesting chlorophyll *a/b* complex – time-resolved fluorescence measurements. *J Biol Chem* **279**, 44400–44406.

Hörtensteiner, S. (1998) NCC malonyltransferase catalyses the final step of chlorophyll breakdown in rape (*Brassica napus*). *Phytochemistry* **49**, 953–956.

Hörtensteiner, S. (1999) Chlorophyll breakdown in higher plants and algae. *Cell Mol Life Sci* **56**, 330–347.

Hörtensteiner, S. (2004) The loss of green color during chlorophyll degradation – a prerequisite to prevent cell death? *Planta* **219**, 191–194.

Hörtensteiner, S. and Kräutler, B. (2000) Chlorophyll breakdown in oilseed rape. *Photosynth Res* **64**, 137–146.

Hörtensteiner, S., Rodoni, S., Schellenberg, M., *et al.* (2000) Evolution of chlorophyll degradation: the significance of RCC reductase. *Plant Biol* **2**, 63–67.

Hörtensteiner, S., Vicentini, F. and Matile, P. (1995) Chlorophyll breakdown in senescent cotyledons of rape, *Brassica napus* L.: enzymatic cleavage of phaeophorbide *a in vitro*. *New Phytol* **129**, 237–246.

Hou, B.K., Lim, E.K., Higgins, G.S. and Bowles, D.J. (2004) N-glucosylation of cytokinins by glycosyltransferases of *Arabidopsis thaliana*. *J Biol Chem* **279**, 47822–47832.

Hu, G., Yalpani, N., Briggs, S.P. and Johal, G.S. (1998) A porphyrin pathway impairment is responsible for the phenotype of a dominant disease lesion mimic mutant of maize. *Plant Cell* **10**, 1095–1105.

Huner, N.P.A., Oquist, G. and Sarhan, F. (1998) Energy balance and acclimation to light and cold. *Trends Plant Sci* **3**, 224–230.

Ida, K., Masamoto, K., Maoka, T., Fujiwara, Y., Takeda, S. and Hasegawa, E. (1995) The leaves of the common box, *Buxus sempervirens* (Buxaceae), become red as the level of a red carotenoid, anhydroeschscholtzxanthin, increases. *J Plant Res* **108**, 369–376.

Ishikawa, A., Okamoto, H., Iwasaki, Y. and Asahi, T. (2001) A deficiency of coproporphyrinogen III oxidase causes lesion formation in *Arabidopsis*. *Plant J* **27**, 89–99.

Iturraspe, J., Engel, N., Matzinger, P., Mooser, V. and Gossauer, A. (1993) Chlorophyll catabolism. Part 3: Structure elucidation and partial synthesis of a new red bilin derivative from *Chlorella protothecoides*. *Photochem Photobiol* **58**, 116–119.

Iturraspe, J., Moyano, N. and Frydman, B. (1995) A new 5-formylbilinone as the major chlorophyll *a* catabolite in tree senescent leaves. *J Org Chem* **60**, 6664–6665.

Jakob-Wilk, D., Holland, D., Goldschmidt, E.E., Riov, J. and Eyal, Y. (1999) Chlorophyll breakdown by chlorophyllase: isolation and functional expression of the *Chlase1* gene from ethylene-treated *Citrus* fruit and its regulation during development. *Plant J* **20**, 653–661.

Ji, S.B., Yokoi, M., Saito, N. and Mao, L.S. (1992) Distribution of anthocyanins in Aceraceae leaves. *Biochem Syst Ecol* **20**, 771–781.

Juniper, B. (1993) Flamboyant flushes: a reinterpretation of non-green flush colours in leaves. *Int Dendrol Soc Yearbook* 49–57.

Kariola, T., Brader, G., Li, J. and Palva, E.T. (2005) Chlorophyllase 1, a damage control enzyme, affects the balance between defense pathways in plants. *Plant Cell* **17**, 282–294.

Killingbeck, K.T. (1996) Nutrients in senesced leaves: keys to the search for potential resorption and resorption proficiency. *Ecology* **77**, 1716–1727.

Koo, A.J.K. and Ohlrogge, J.B. (2002) The predicted candidates of *Arabidopsis* plastid inner envelope membrane proteins and their expression profiles. *Plant Physiol* **130**, 823–836.

Kräutler, B. (2002) Unravelling chlorophyll catabolism in higher plants. *Biochem Soc Trans* **30**, 625–630.

Kräutler, B. (2003) Chlorophyll breakdown and chlorophyll catabolites. In: *The Porphyrin Handbook*, Vol. 13 (eds Kadish, K.M., Smith, K.M. and Guilard, R.). Elsevier Science, New York, pp. 183–209.

Kräutler, B., Jaun, B., Bortlik, K.-H., Schellenberg, M. and Matile, P. (1991) On the enigma of chlorophyll degradation: the constitution of a secoporphinoid catabolite. *Angew Chem Int Ed Engl* **30**, 1315–1318.

Kräutler, B. and Matile, P. (1999) Solving the riddle of chlorophyll breakdown. *Acc Chem Res* **32**, 35–43.

Kreuz, K., Tommasini, R. and Martinoia, E. (1996) Old enzymes for a new job. Herbicide detoxification in plants. *Plant Physiol* **111**, 349–353.

Krol, M., Gray, G.R., Hurry, V.M., Oquist, G., Malek, L. and Huner, N.P.A. (1995) Low-temperature stress and photoperiod affect an increased tolerance to photoinhibition in *Pinus banksiana* seedlings. *Can J Bot* **73**, 1119–1127.

Kunieda, T., Amano, T. and Shioi, Y. (2005) Search for chlorophyll degradation enzyme, Mg-dechelatase, from extracts of *Chenopodium album* with native and artificial substrates. *Plant Sci* **169**, 177–183.

Lee, D.W., Brammeier, S. and Smith, A.P. (1987) The selective advantages of anthocyanins in developing leaves of mango and cacao. *Biotropica* **19**, 40–49.

Lee, D.W. and Collins, T.M. (2001) Phylogenetic and ontogenetic influences on the distribution of anthocyanins and betacyanins in leaves of tropical plants. *Int J Plant Sci* **162**, 1141–1153.

Lee, D.W. and Gould, K.S. (2002) Anthocyanins in leaves and other vegetative organs: an introduction. In: *Advances in Botanical Research*, Vol. 37 (eds, Gould, K.S. and Lee, D.W.). Academic Press, London, pp. 1–16.

Lee, D.W., O'Keefe, J., Holbrook, N.M. and Feild, T.S. (2003) Pigment dynamics and autumn leaf senescence in a New England deciduous forest, eastern USA. *Ecol Res* **18**, 677–694.

Lev-Yadun, S., Dafni, A., Flaishman, M.A., *et al.* (2004) Plant coloration undermines herbivorous insect camouflage. *BioEssays* **26**, 1126–1130.

Lila, M.A. (2004) Plant pigments and human health. In: *Plant Pigments and Their Manipulation. Annual Plant Reviews* (ed. Davies, K.). Blackwell Publishing, Oxford, pp. 248–274.

Losey, F.G. and Engel, N. (2001) Isolation and characterization of a urobilinogenoidic chlorophyll catabolite from *Hordeum vulgare* L. *J Biol Chem* **276**, 27233–27236.

Lu, Y.-P., Li, Z.-S., Drozdowicz, Y.-M., Hörtensteiner, S., Martinoia, E. and Rea, P.A. (1998) AtMRP2, an *Arabidopsis* ATP binding cassette transporter able to transport glutathione *S*-conjugates and chlorophyll catabolites: functional comparisons with AtMRP1. *Plant Cell* **10**, 267–282.

Mach, J.M., Castillo, A.R., Hoogstraten, R. and Greenberg, J.T. (2001) The *Arabidopsis*-accelerated cell death gene ACD2 encodes red chlorophyll catabolite reductase and suppresses the spread of disease symptoms. *Proc Natl Acad Sci U S A* **98**, 771–776.

Matile, P. (1992) Chloroplast senescence. In: *Crop Photosynthesis: Spatial and Temporal Determinants* (eds. Baker, N.R. and Thomas, H.). Elsevier Science, Amsterdam, pp. 413–440.

Matile, P. (1994) Fluorescent idioblasts in autumn leaves of *Ginkgo biloba*. *Bot Helv* **104**, 87–92.

Matile, P., Ginsburg, S., Schellenberg, M. and Thomas, H. (1988) Catabolites of chlorophyll in senescing barley leaves are localized in the vacuoles of mesophyll cells. *Proc Natl Acad Sci U S A* **85**, 9529–9532.

Matile, P., Hörtensteiner, S. and Thomas, H. (1999) Chlorophyll degradation. *Annu Rev Plant Physiol Plant Mol Biol* **50**, 67–95.

Matile, P., Hörtensteiner, S., Thomas, H. and Kräutler, B. (1996) Chlorophyll breakdown in senescent leaves. *Plant Physiol* **112**, 1403–1409.

Matile, P., Schellenberg, M. and Peisker, C. (1992) Production and release of a chlorophyll catabolite in isolated senescent chloroplasts. *Planta* **187**, 230–235.

Matile, P., Schellenberg, M. and Vicentini, F. (1997) Localization of chlorophyllase in the chloroplast envelope. *Planta* **201**, 96–99.

Merzlyak, M.N. and Chivkunova, O.B. (2000) Light-stress-induced pigment changes and evidence for anthocyanin photoprotection in apples. *J Photochem Photobiol B* **55**, 155–163.

Meskauskiene, R., Nater, M., Goslings, D., Kessler, F., op den Camp, R. and Apel, K. (2001) FLU: a negative regulator of chlorophyll biosynthesis in *Arabidopsis thaliana. Proc Natl Acad Sci U S A* **98**, 12826–12831.

Moore, B.J., Donnison, I.S., Harper, J.A., *et al.* (2005) Molecular tagging of a senescence gene by introgression mapping of a stay-green mutation from *Festuca pratensis. New Phytol* **165**, 801–806.

Mühlecker, W. and Kräutler, B. (1996) Breakdown of chlorophyll: constitution of nonfluorescing chlorophyll-catabolites from senescent cotyledons of the dicot rape. *Plant Physiol Biochem* **34**, 61–75.

Mühlecker, W., Kräutler, B., Moser, D., Matile, P. and Hörtensteiner, S. (2000) Breakdown of chlorophyll: a fluorescent chlorophyll catabolite from sweet pepper (*Capsicum annuum*). *Helv Chim Acta* **83**, 278–286.

Mühlecker, W., Ongania, K.-H., Kräutler, B., Matile, P. and Hörtensteiner, S. (1997) Tracking down chlorophyll breakdown in plants: elucidation of the constitution of a 'fluorescent' chlorophyll catabolite. *Angew Chem Int Ed Engl* **36**, 401–404.

Müller, T., Moser, S., Ongania, K.-H., Pružinská, A., Hörtensteiner, S. and Kräutler, B. (2006) A divergent path of chlorophyll breakdown in the model plant *Arabidopsis thaliana. Chem BioChem* **7**, 40–42.

Oberhuber, M., Berghold, J., Breuker, K., Hörtensteiner, S. and Kräutler, B. (2003) Breakdown of chlorophyll: a nonenzymatic reaction accounts for the formation of the cololess 'nonfluorescent' chlorophyll catabolites. *Proc Natl Acad Sci U S A* **100**, 6910–6915.

Oberhuber, M., Berghold, J., Mühlecker, W., Hörtensteiner, S. and Kräutler, B. (2001) Chlorophyll breakdown – on a nonfluorescent chlorophyll catabolite from spinach. *Helv Chim Acta* **84**, 2615–2627.

op den Camp, R.G., Przybyla, D., Ochsenbein, C., *et al.* (2003) Rapid induction of distinct stress responses after release of singlet oxygen in *Arabidopsis. Plant Cell* **15**, 2320–2332.

Oshio, Y. and Hase, E. (1969) Studies on red pigments excreted by cells of *Chlorella protothecoides* during the process of bleaching induced by glucose or acetate. I: Chemical properties of the red pigments. *Plant Cell Physiol* **10**, 41–49.

Oster, U., Tanaka, R., Tanaka, A. and Rudiger, W. (2000) Cloning and functional expression of the gene encoding the key enzyme for chlorophyll *b* biosynthesis (CAO) from *Arabidopsis thaliana. Plant J* **21**, 305–310.

Peisker, C., Düggelin, T., Rentsch, D. and Matile, P. (1989) Phytol and the breakdown of chlorophyll in senescent leaves. *J Plant Physiol* **135**, 428–432.

Philpott, M., Gould, K.S., Lim, C. and Ferguson, L.R. (2004) In situ and in vitro antioxidant activity of sweetpotato anthocyanins. *J Agric Food Chem* **52**, 1511–1513.

Pružinská, A., Anders, I., Tanner, G., Roca, M. and Hörtensteiner, S. (2003) Chlorophyll breakdown: pheophorbide *a* oxygenase is a Rieske-type iron-sulfur protein, encoded by the *accelerated cell death 1* gene. *Proc Natl Acad Sci U S A* **100**, 15259–15264.

Pružinská, A., Tanner, G., Aubry, S., *et al.* (2005) Chlorophyll breakdown in senescent *Arabidopsis* leaves: characterization of chlorophyll catabolites and of chlorophyll catabolic enzymes involved in the de-greening reaction. *Plant Physiol* **139**, 52–63.

Reinbothe, C., Satoh, H., Alcaraz, J.P. and Reinbothe, S. (2004a) A novel role of water-soluble chlorophyll proteins in the transitory storage of chorophyllide. *Plant Physiol* **134**, 1355–1365.

Reinbothe, S., Quigley, F., Gray, J., Schemenewitz, A. and Reinbothe, C. (2004b) Identification of plastid envelope proteins required for import of protochlorophyllide oxidoreductase A into the chloroplast of barley. *Proc Natl Acad Sci U S A* **101**, 2197–2202.

Roca, M., James, J., Pružinská, A., Hörtensteiner, S., Thomas, H. and Ougham, H. (2004) Analysis of the chlorophyll catabolism pathway in leaves of an introgression senescence mutant of *Lolium temulentum. Phytochemistry* **65**, 1231–1238.

Rodoni, S., Mühlecker, W., Anderl, M., *et al.* (1997) Chlorophyll breakdown in senescent chloroplasts. Cleavage of pheophorbide *a* in two enzymic steps. *Plant Physiol* **115**, 669–676.

Rüdiger, W. (2002) Biosynthesis of chlorophyll *b* and the chlorophyll cycle. *Photosynth Res* **74**, 187–193.

Satoh, H., Nakayama, K. and Okada, M. (1998) Molecular cloning and functional expression of a water-soluble chlorophyll protein, a putative carrier of chlorophyll molecules in cauliflower. *J Biol Chem* **273**, 30568–30575.

Satoh, H., Uchida, A., Nakayama, K. and Okada, M. (2001) Water-soluble chlorophyll protein in Brassicaceae plants is a stress-induced chlorophyll-binding protein. *Plant Cell Physiol* **42**, 906–911.

Schaberg, P.G., Van den Berg, A.K., Murakami, P.F., Shane, J.B. and Donnelly, J.R. (2003) Factors influencing red expression in autumn foliage of sugar maple trees. *Tree Physiol* **23**, 325–333.

Schaefer, H.M. and Wilkinson, D.M. (2004) Red leaves, insects and coevolution: a red herring? *Trends Ecol Evol* **19**, 616–618.

Scheumann, V., Schoch, S. and Rüdiger, W. (1998) Chlorophyll *a* formation in the chlorophyll *b* reductase reaction requires reduced ferredoxin. *J Biol Chem* **273**, 35102–35108.

Scheumann, V., Schoch, S. and Rüdiger, W. (1999) Chlorophyll *b* reduction during senescence of barley seedlings. *Planta* **209**, 364–370.

Schoch, S., Scheer, H., Schiff, J.A., Rüdiger, W. and Siegelman, H.W. (1981) Pyropheophytin *a* accompanies pheophytin *a* in darkened light grown cells of *Euglena*. *Z Naturforsch* **36C**, 827–833.

Schuler, M.A. and Werck-Reichhart, D. (2003) Functional genomics of P450s. *Annu Rev Plant Biol* **54**, 629–667.

Shioi, Y., Tomita, N., Tsuchiya, T. and Takamiya, K. (1996a) Conversion of chlorophyllide to pheophorbide by Mg-dechelating substance in extracts of *Chenopodium album*. *Plant Physiol Biochem* **34**, 41–47.

Shioi, Y., Watanabe, K. and Takamiya, K. (1996b) Enzymatic conversion of pheophorbide *a* to a precursor of pyropheophorbide *a* in leaves of *Chenopodium album*. *Plant Cell Physiol* **37**, 1143–1149.

Smillie, R.M. and Hetherington, S.E. (1999) Photoabatement by anthocyanin shields photosynthetic systems from light stress. *Photosynthetica* **36**, 451–463.

Soll, J. (1987) α-Tocopherol and plastochinone synthesis in chloroplast membranes. *Methods Enzymol* **148**, 383–392.

Spassieva, S. and Hille, J. (2002) A lesion mimic phenotype in tomato obtained by isolating and silencing an *Lls1* homologue. *Plant Sci* **162**, 543–549.

Steyn, W.J., Wand, S.J.E., Holcroft, D.M. and Jacobs, G. (2002) Anthocyanins in vegetative tissues: a proposed unified function in photoprotection. *New Phytol* **155**, 349–361.

Stiles, E.W. (1982) Fruit flag: two hypotheses. *Am Nat* **120**, 500–509.

Suzuki, T., Kunieda, T., Murai, F., Morioka, S. and Shioi, Y. (2005) Mg-dechelation activity in radish cotyledons with artificial and native substrates, Mg-chlorophyllin *a* and chlorophyllide *a*. *Plant Physiol Biochem* **43**, 459–464.

Suzuki, T. and Shioi, Y. (2002) Re-examination of Mg-dechelation reaction in the degradation of chlorophylls using chlorophyllin *a* as substrate. *Photosynth Res* **74**, 217–223.

Suzuki, Y., Doi, M. and Shioi, Y. (2002) Two enzymatic reaction pathways in the formation of pyropheophorbide *a*. *Photosynth Res* **74**, 225–233.

Suzuki, Y. and Shioi, Y. (1999) Detection of chlorophyll breakdown products in the senescent leaves of higher plants. *Plant Cell Physiol.* **40**, 909–915.

Takamiya, K., Tsuchiya, T. and Ohta, H. (2000) Degradation pathway(s) of chlorophyll: what has gene cloning revealed? *Trends Plant Sci* **5**, 426–431.

Tanaka, A., Ito, H., Tanaka, R., Tanaka, N.K., Yoshida, K. and Okada, K. (1998) Chlorophyll *a* oxygenase (*CAO*) is involved in chlorophyll *b* formation from chlorophyll *a*. *Proc Natl Acad Sci U S A* **95**, 12719–12723.

Tanaka, R., Hirashima, M., Satoh, S. and Tanaka, A. (2003) The *Arabidopsis-accelerated cell death* gene *ACD1* is involved in oxygenation of pheophorbide *a*: inhibition of pheophorbide *a* oxygenase activity does not lead to the 'stay-green' phenotype in *Arabidopsis*. *Plant Cell Physiol* **44**, 1266–1274.

Tang, L., Okazawa, A., Itoh, Y., Fukusaki, E. and Kobayashi, A. (2004) Expression of chlorophyllase is not induced during autumnal yellowing in *Ginkgo biloba*. *Z Naturforsch C* **59**, 415–420.

Thomas, H., Bortlik, K.-H., Rentsch, D., Schellenberg, M. and Matile, P. (1989) Catabolism of chlorophyll *in vivo*: significance of polar chlorophyll catabolites in a non-yellowing senescence mutant of *Festuca pratensis* Huds. *New Phytol* **111**, 3–8.

Thomas, H., Evans, C., Thomas, H.M., *et al.* (1997) Introgression, tagging and expression of a leaf senescence gene in *Festulolium*. *New Phytol* **137**, 29–34.

Thomas, H. and Howarth, C.J. (2000) Five ways to stay green. *J Exp Bot* **51**, 329–337.

Thomas, H., Morgan, W.G., Thomas, A.M. and Ougham, H.J. (1999) Expression of the stay-green character introgressed into *Lolium temulentum* Ceres from a senescence mutant of *Festuca pratensis*. *Theor Appl Genet* **99**, 92–99.

Thomas, H., Ougham, H. and Hörtensteiner, S. (2001) Recent advances in the cell biology of chlorophyll catabolism. In: *Advances in Botanical Research*, Vol. 35 (ed. Callow, J.A.) Academic Press, San Diego, CA, pp. 1–52.

Thomas, H., Schellenberg, M., Vicentini, F. and Matile, P. (1996) Gregor Mendel's green and yellow pea seeds. *Bot Acta* **109**, 3–4.

Thomas, H. and Smart, C.M. (1993) Crops that stay green. *Ann Appl Biol* **123**, 193–219.

Thomas, H. and Stoddart, L. (1975) Separation of chlorophyll degradation form other senescence processes in leaves of a mutant genotype of medow fescue (*Festuca pratensis* L.). *Plant Physiol* **56**, 438–441.

Tommasini, R., Vogt, E., Fromenteau, M., *et al.* (1998) An ABC transporter of *Arabidopsis thaliana* has both glutathione-conjugate and chlorophyll catabolite transport activity. *Plant J* **13**, 773–780.

Trebitsh, T., Goldschmidt, E.E. and Riov, J. (1993) Ethylene induces *de novo* synthesis of chlorophyllase, a chlorophyll degrading enzyme, in *Citrus* fruit peel. *Proc Natl Acad Sci U S A* **90**, 9441–9445.

Tsuchiya, T., Ohta, H., Okawa, K., *et al.* (1999) Cloning of chlorophyllase, the key enzyme in chlorophyll degradation: finding of a lipase motif and the induction by methyl jasmonate. *Proc Natl Acad Sci U S A* **96**, 15362–15367.

Tziveleka, L.A. and Argyroudi-Akoyunoglou, J.H. (1998) Implications of a developmental-stage-dependent thylakoid-bound protease in the stabilization of the light-harvesting pigment–protein complex serving photosystem II during thylakoid biogenesis in red kidney bean. *Plant Physiol* **117**, 961–970.

Vicentini, F., Hörtensteiner, S., Schellenberg, M., Thomas, H. and Matile, P. (1995a) Chlorophyll breakdown in senescent leaves: identification of the biochemical lesion in a *stay-green* genotype of *Festuca pratensis* Huds. *New Phytol* **129**, 247–252.

Vicentini, F., Iten, F. and Matile, P. (1995b) Development of an assay for Mg-dechelatase of oilseed rape cotyledons, using chlorophyllin as the substrate. *Physiol Plant* **94**, 57–63.

Vogelmann, T.C. (1993) Plant-tissue optics. *Annu Rev Plant Physiol Plant Mol Biol* **44**, 231–251.

Wagner, D., Przybyla, D., op den Camp, R., *et al.* (2004) The genetic basis of singlet oxygen-induced stress responses of *Arabidopsis thaliana*. *Science* **306**, 1183–1185.

Wüthrich, K.L., Bovet, L., Hunziker, P.E., Donnison, I.S. and Hörtensteiner, S. (2000) Molecular cloning, functional expression and characterisation of RCC reductase involved in chlorophyll catabolism, *Plant J* **21**, 189–198.

Yamasaki, H. (1997) A function of colour. *Trends Plant Sci* **2**, 7–8.

Yang, M., Wardzala, E., Johal, G.S. and Gray, J. (2004) The wound-inducible *Lls1* gene from maize is an orthologue of the *Arabidopsis Acd1* gene, and the LLS1 protein is present in non-photosynthetic tissues. *Plant Mol Biol* **54**, 175–191.

Yao, N., Eisfelder, B.J., Marvin, J. and Greenberg, J.T. (2004) The mitochondrion – an organelle commonly involved in programmed cell death in *Arabidopsis thaliana*. *Plant J* **40**, 596–610.

Zelisko, A. and Jackowski, G. (2004) Senescence-dependent degradation of Lhcb3 is mediated by a thylakoid membrane-bound protease. *J Plant Physiol* **161**, 1157–1170.

Ziegler, R., Blaheta, A., Guha, N. and Schönegge, B. (1988) Enzymatic formation of pheophorbide and pyropheophorbide during chlorophyll degradation in a mutant of *Chlorella fusca* SHIRIA et KRAUS. *J Plant Physiol* **132**, 327–332.

Zimmermann, P., Hirsch-Hoffmann, M., Hennig, L. and Gruissem, W. (2004) GENEVESTIGATOR. *Arabidopsis* microarray database and analysis toolbox. *Plant Physiol* **136**, 2621–2632.

3 Membrane dynamics and regulation of subcellular changes during senescence

Marianne Hopkins, Linda McNamara, Catherine Taylor, Tzann-Wei Wang and John Thompson

3.1 Introduction

Cellular membranes are important constituents of all plant cells, not only in the context of enabling compartmentalization of cell function, but also through provision of specialized transport and signaling capabilities that are essential for normal cell function. The ability to serve as a permeability barrier is conferred upon membranes by their lipids, whereas their more specialized functions are normally mediated by proteins embedded in the lipid bilayer. One of several exceptions to this general principle is that certain membrane lipids, for example derivatives of phosphatidylinositol, are involved in signaling.

Lipids are a major chemical component of all membranes. The predominant lipids in most membranes are phospholipids, which are amphipathic and spontaneously form bilayers in an aqueous environment. The exception is the thylakoid membranes of chloroplasts, which contain primarily galactolipids. However, galactolipids are also amphipathic and, like phospholipids, are most stable in a bilayer configuration.

The amphipathic character of membrane lipids and their consequent propensity to form bilayers are underlying principles of cellular life in that they enable the formation of sealed membranous compartments, including the cell itself. More specifically, the amphipathic nature of membranes permits the formation of membranous sheets, and the self-sealing nature of bilayers ensures annealing of the edges of a membranous sheet into a sealed compartment. This is evident from theoretical considerations of the properties of amphipathic lipids in an aqueous environment, and has also been amply demonstrated empirically over the years. For example, suspensions of phospholipids in aqueous buffer result in the formation of liposomes (Lee, 2000). Similarly, during tissue homogenization continuous sheets of membrane, such as the plasmalemma, are disrupted into small fragments that immediately self-seal, forming small vesicles termed *microsomes* (Yao *et al.*, 1991).

The most important functional attribute of membranes stemming from the propensity of their lipids to form bilayers is that they constitute permeability barriers. Indeed, liposomes, which are lipid bilayers free of protein, are essentially impermeable to polar solutes. Thus, the lipid bilayer constitutes the permeability barrier of membranes. However, cellular membranes are not typically impermeable; rather, they are selectively permeable, a feature that is conferred by specialized proteins that traverse the bilayer. Proteins are the second major component of membranes, and they can be either amphipathic integral proteins that extend across the bilayer, or peripheral proteins attached to the surfaces of membranes by hydrophilic interaction with the polar ends of transmembrane proteins. Transmembrane integral proteins thus interrupt the continuous nature of the lipid bilayer. Specialized transmembrane proteins known as transporters enable translocation of ions and metabolites into and out of membranous compartments and thus confer upon membranes their selectively permeable nature.

3.2 Loss of membrane structural integrity during senescence

Loss of membrane structural integrity is a seminal feature of senescence that is initiated early in the cell death cascade. This is perhaps most clearly evident from permeability studies indicating increased leakage of solutes at the onset of senescence, whether it occurs naturally or is induced prematurely by environmental stress. For example, in senescing carnation flowers, changes in permeability reflecting membrane deterioration are initiated before the climacteric rise in ethylene production (Eze *et al.*, 1986). It is also apparent from ultrastructural studies that the different types of cellular membranes do not undergo senescence synchronously. For example, recent analysis of senescing barley and maize leaves revealed that thylakoids are the first membranes to undergo degradative changes. This is followed by structural alterations in the internal mitochondrial membranes and, finally, dismantling of the chloroplast envelope (Kolodziejek *et al.*, 2003). That cellular membranes are not all degraded simultaneously during senescence is in keeping with the fact that catabolism of macromolecules and the attendant dismantling of membranes and organelles are metabolically coupled to energy production and reallocation of carbon, nitrogen and minerals to growing parts of the plant (Matile, 1992). For this metabolic coupling to be successful, certain cellular membranes have to retain their structural integrity well into the late stages of senescence.

3.2.1 Senescence-associated changes in the molecular organization of membrane lipid bilayers

It is now well accepted that the molecular components of membranes are degraded during senescence. This includes extensive catabolism of membrane phospholipids. Indeed, declining lipid phosphate levels reflecting phospholipid catabolism are often evident well before morphological manifestations of tissue aging (Paliyath and Thompson, 1990) and have been demonstrated for a variety of senescing tissues including leaves, cotyledons, flower petals and ripening fruit (Thompson *et al.*, 1998). Additional senescence-associated changes in the chemical composition of lipid bilayers include a decrease in fatty acid content leading to enhanced relative concentrations of sterol within membranes and, in some cases, a selective depletion of unsaturated fatty acids (Fobel *et al.*, 1987; Thompson *et al.*, 1998). Of particular interest is the finding that these chemical changes in lipid bilayers can account for the onset of membrane leakiness during the early stages of senescence. Indeed, there is now evidence from studies of a number of different senescing tissues that de-esterified fatty acids phase separate from bulk membrane lipid, which is liquid-crystalline in nature, forming discrete domains of gel phase lipid. This in turn causes the bilayer to become leaky because of packing imperfections at the phase boundaries (Barber and Thompson, 1980; Barber and Thompson, 1983; Thompson, 1988; Yamane *et al.*, 1993). Lipid phase separation in membrane bilayers prompted by de-esterification of membrane fatty acids also appears to be a characteristic of premature senescence induced by environmental stress. For example, chilling injury has been shown to induce lipid phase separations in the membranes of tomato fruit in a manner that is temporally correlated with increased membrane leakage (Sharom *et al.*, 1994). That de-esterification of membrane lipids is both an early and seminal feature of senescence is further supported by the finding that overexpression of SAG101, a lipolytic acyl hydrolase isolated from *Arabidopsis*, using an inducible promoter, induces precocious leaf senescence (He and Gan, 2002).

These observations raise an interesting question, namely how is the structural integrity of at least some cellular membranes, or even discrete regions within a cellular membrane, retained into the late stages of senescence in the face of net degradation of membrane lipids? This is essential for the production of energy to support senescence and for nutrient recruitment and translocation. In part, it is achieved through asynchronous dismantling of the different types of cellular membranes, orchestrated to ensure that membrane functions essential for the execution of senescence are retained into its late stages. One clear illustration of this is the finding for senescing carnation flowers that lipid phase separations engendering membrane leakiness, which can be visualized by freeze-fracture electron microscopy, are evident in the membranes of the endoplasmic reticulum prior to the climacteric rise in ethylene production and petal inrolling. Yet, there is no evidence of lipid phase separations in the plasmalemma at this early stage of senescence (Paliyath and Thompson, 1990). Indeed, lipid phase separations in the plasmalemma are not discernible until a much later stage of petal senescence (Paliyath and Thompson, 1990). Thus, the selectively permeable nature of the plasmalemma is retained during the period when there is a need for nutrient translocation out of senescing cells. Preservation of this integrity is essential because many sugar and amino acid transporters are functionally dependent upon maintenance of ion or metabolite gradients across the membrane, which would not be possible if the lipid bilayer were leaky. It is also apparent from freeze-fracture electron microscopic studies that even at the molecular level within the plane of the membrane, senescence-related degradation is highly regulated such that regions of the bilayer retain their structural integrity, while other regions of the same membrane become leaky. This arises because gel phase domains reflecting lipid phase separations are not uniformly distributed within the bilayer (Paliyath and Thompson, 1990). Thus, during the early stages of carnation petal senescence, for example, some regions of the endoplasmic reticulum are leaky, whereas other regions in which there is no gel phase lipid are still structurally intact. Indeed, the development of lipid phase separations appears to occur slowly, allowing for gradual loss of compartmentalization. Also, when gel phase lipid domains form in membrane bilayers, integral membrane proteins are displaced laterally through the plane of the membrane into adjacent liquid-crystalline regions rather than being immediately degraded (Thompson *et al.*, 1998). Thus, even though the bilayer may be leaky, it continues to serve as a scaffold for membrane proteins that remain functional.

Degradation of the energy-producing membranes of the cell, thylakoids of the chloroplast and the inner membrane of mitochondria, constitutes another clear example of the asynchronous nature of membrane deterioration in senescing tissues. One of the first manifestations of leaf senescence is loss of chlorophyll, which reflects dismantling of thylakoids. By contrast, deterioration of the inner membranes of mitochondria does not occur until late in the leaf senescence cascade (Kolodziejek *et al.*, 2003). Thus the capacity for chloroplastic energy production is eliminated during the early stages of senescence, but mitochondrial energy production is preserved until much later. Why is chloroplast energy production sacrificed so early in the process? The fatty acids of thylakoid membranes are the most abundant source of carbon in leaves, and presumably capture of this carbon during the early stages of senescence, while much of the cell metabolic machinery is still intact, helps to ensure its successful conversion to phloem-mobile sucrose. Moreover, there is provision for conversion of fatty acids originating from thylakoids to substrate for mitochondrial ATP formation through β-oxidation (Buchanan-Wollaston, 1997; Charlton *et al.*, 2005), thereby ensuring a supply of energy to support the metabolic reactions of senescence.

3.2.2 Role of lipases

The alterations in lipid composition that engender lipid phase separations in senescing membranes appear to be enzymatically mediated. Indeed, comprehensive analyses of changes in gene expression during senescence have indicated that phospholipases and lipases are abundant components of the genes upregulated in senescing tissues (Gepstein *et al.*, 2003; Guo *et al.*, 2004). A recently developed transcriptome for *Arabidopisis* leaf senescence yielded approximately 6200 expressed sequence tags that correspond to 2491 unique genes (Guo *et al.*, 2004). Among the proteins encoded by these unique genes are 11 lipases/acyl hydrolases, 6 phospholipases, 2 lipoxygenases, and 9 β-oxidation enzymes, all of which could be expected to be involved in aspects of membrane lipid catabolism. Thus, dismantling of membrane bilayers and the conversion of their fatty acid equivalents to energy or phloem-mobile sucrose represents a significant aspect of senescence.

Among the lipase genes upregulated at the onset of senescence are those encoding lipolytic acyl hydrolases, which de-esterify fatty acids at both the *sn*-1 and *sn*-2 positions of phospholipids. Indeed, rapid de-esterification of membrane fatty acids is a characteristic feature of natural senescence (Thompson *et al.*, 1998). It is also a feature of postharvest senescence of, for example, cut flowers (Hong *et al.*, 2000) and broccoli florets (Page *et al.*, 2001). Two senescence-associated genes (SAGs) encoding proteins that exhibit lipolytic acyl hydrolase activity have been isolated and characterized, one from senescing carnation petals (Hong *et al.*, 2000) and the other from senescing *Arabidopsis* rosettes (He and Gan, 2002). The carnation lipolytic acyl hydrolase gene is not expressed in young flowers, but is strongly upregulated coincident with the onset of petal inrolling, the first morphological manifestation of carnation flower senescence. In addition, expression of the gene in flower petals is induced by treatment with ethylene at concentrations that invoke early flower senescence. The same gene is also expressed in leaves of carnation plants that were treated with ethylene (Hong *et al.*, 2000). The protein contains the lipase consensus sequence, ITFAGHSLGA, and overexpression of the corresponding cDNA in *Escherichia coli* yielded a recombinant protein of the expected molecular weight that proved capable of de-esterifying fatty acids from *p*-nitrophenylpalmitate, trilinolein, soybean phospholipids and Tween in both *in situ* and *in vitro* assays of enzyme activity. This is consistent with the protein being a lipolytic acyl hydrolase (Hong *et al.*, 2000). The lipolytic acyl hydrolase gene isolated from senescing *Arabidopsis* leaves, termed *SAG101* (He and Gan, 2002), is also not expressed in young tissue, but is strongly upregulated coincident with the onset of leaf senescence. Moreover, antisense suppression of its expression in transgenic plants delayed the onset of leaf senescence, and overexpression induced premature leaf senescence (He and Gan, 2002). This latter finding supports the contention that de-esterification of membrane lipids is both an early and seminal feature of senescence. There is also evidence for upregulation of phospholipase D activity in senescing tissues (Ryu and Wang, 1995). The action of phospholipase D on phospholipids results in the formation of diacylglycerol (DAG), and thus it may be relevant that *SAG101* and its counterpart in carnation petals appear to be lipolytic acyl hydrolases rather than phospholipase A-type enzymes inasmuch as the former are capable of de-esterifying both phospholipids and DAG (Galliard, 1980).

3.2.2.1 *Initial fate of de-esterified fatty acids in senescing membranes*

It is important to appreciate that fatty acids de-esterified from phospholipids of senescing membranes are not immediately removed from the membrane bilayer. Indeed, free fatty acids are charged at physiological pH rendering them highly amphipathic and, when liberated from complex membrane lipids, have a strong propensity to remain within the bilayer rather than partition into surrounding aqueous compartments. This is evident from the finding that senescence is often accompanied by an increase in the free–esterified fatty acid ratio within membranes (Thompson *et al.*, 1997).

The tendency for de-esterified fatty acids to reside within membrane bilayers presents a logistical problem. On one hand, free fatty acids act like detergents (Barclay and McKersie, 1994) and have a severely disruptive effect on bilayer structure, and on the other, there is a need to ensure that loss of membrane structural integrity in a senescing tissue is not precipitous but, rather, carefully controlled. Comparative chemical analyses of the lipid composition of microsomes, which are small vesicles formed from various types of cellular membranes during tissue homogenization, have indicated that one response to this dilemma is the conversion of free fatty acids to metabolites that are more easily accommodated within the bilayer structure. Specifically, although there is some accumulation of free fatty acids in

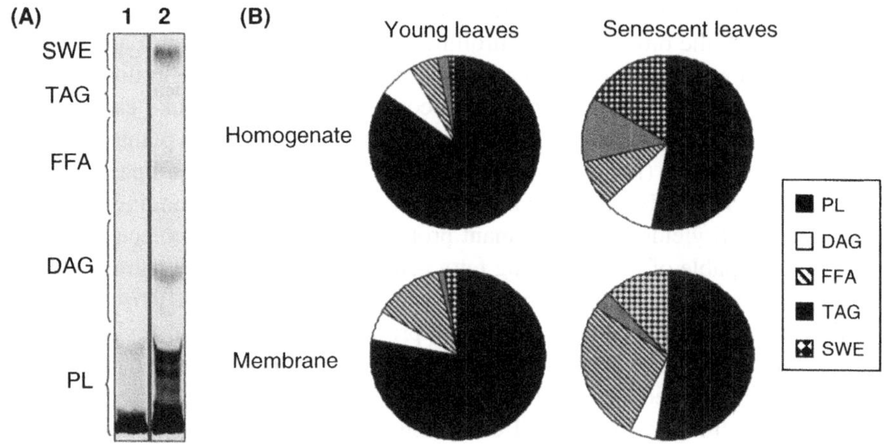

Figure 3.1 Formation of membrane lipid catabolites during senescence of *Arabidopsis* rosette leaves. During senescence, membrane lipids are catabolized releasing free fatty acids, which are in turn sequestered in triacylglycerols and steryl/wax esters. (A) Thin-layer chromatogram of total membrane lipid extracted from homogenates of young *Arabidopsis thaliana* rosette leaves (lane 1) and senescing *Arabidopsis thaliana* rosette leaves (lane 2). The lipid was visualized with iodine vapor. There is a clear increase in diacylglycerol (DAG), free fatty acids (FFA), triacylglycerols (TAG) and steryl/wax esters (SWE) as senescence progresses. Although the polar lipid (PL) fraction actually decreases during senescence, it appears darker in the chromatogram for senescing leaves (lane 2) than in the chromatogram for young leaves, because the enhanced levels of pigment in the senescing leaves comigrate with the PL. (B) Lipid classes expressed as a percentage of the total fatty acid complement. The lipids were measured by gas chromatography/mass spectrometry (GCMS) of total lipid extracts from homogenate and microsomal membranes for young ($n = 3$) and senescent ($n = 3$) rosette leaves. The amount of PL decreases in the senescing leaves, and there is a corresponding increase in nonpolar lipid catabolites.

senescing membranes, there is also a large increase in the levels of steryl and wax esters as well as triacylglycerol (TAG) (Figure 3.1; Thompson *et al.*, 1997). Thus, it would appear that senescing cells cope with the progressive increase in membrane free fatty acids in part by converting them to steryl and wax esters or assimilating them into TAG, molecular species that can be accommodated within the bilayer with minimal structural perturbation. Additionally, free fatty acids phase separate within the bilayer (Thompson *et al.*, 1998), and this also minimizes their disruptive, detergent-like action on bilayer structure.

3.2.2.2 Autocatalytic nature of membrane fatty acid de-esterification

Since free fatty acids formed in senescing membranes induce lipid phase separations leading to bilayer leakiness, it seems clear that de-esterification of membrane lipids must be temporally regulated in order to ensure retention of sufficient membrane functionality throughout the course of senescence for energy production and nutrient translocation. This appears to be achieved in part by the fact that lipase-mediated de-esterification of fatty acids in lipid bilayers is autocatalytic. The underlying basis for this autocatalysis is that perturbed phospholipid bilayers are more effective substrates for de-esterifying lipases than are their unperturbed counterparts, and cellular membranes become increasingly perturbed as free fatty acids accumulate (Goormaghtigh *et al.*, 1981). In senescing membranes, perturbation of the bilayer is not only attributable to the free fatty acids themselves, but also to indirect consequences of phospholipid de-esterification including lipid phase changes and the formation of nonbilayer lipid configurations, specifically inverted micelles (Thompson, 1988). The outcome of this progressive development of perturbing influences within senescing membranes would appear to be temporal regulation of fatty acid de-esterification and ensuing orchestration of the onset of membrane leakiness leading to loss of cellular compartmentalization.

3.2.3 Role of galactolipases

Thylakoids are by far the most abundant membranes in leaves, and during leaf senescence the fatty acids of galactolipids, the major lipid of thylakoids, are also de-esterified. There is some question, though, as to whether this is achieved by galactosidase-mediated conversion of galactolipids to DAG and ensuing de-esterification of DAG fatty acids, or by galactolipase-mediated direct de-esterification of galactolipids (Figure 3.2). Recently, for example, a gene encoding α-galactosidase, designated *SAG-Osh69,* was isolated from senescing rice leaves (Lee *et al.*, 2004). The cognate protein was shown to be localized in chloroplasts and to be upregulated during natural leaf senescence, by darkness or wounding, and by methyl jasmonate and salicylic acid. The authors propose that during senescence the α-galactosidase encoded by *SAG-Osh69* converts digalactosyl diacylglycerol (DGDG) to monogalactosyl diacylglycerol (MGDG), which is in turn metabolized to DAG and free fatty acids by β-galactosidase and lipolytic acyl hydrolase, respectively.

There is also evidence that galactolipids in senescing leaves are directly de-esterified by galactolipase (Engelmann-Sylvestre *et al.*, 1989; Kim *et al.*, 2001). As well, Matos *et al.* (2001) recently characterized a cDNA clone isolated from drought-stressed cowpea that encodes a patatin-like protein with strong galactolipid acyl hydrolase activity. Moreover, the same protein hydrolyzes phospholipids only very slowly and appears to be incapable of de-esterifying TAG (Matos *et al.*, 2001). This gene does not have a consensus chloroplast-targeting sequence, although the authors contend that this does not necessarily preclude localization in the chloroplast. In addition, El-Hafid *et al.* (1989) have biochemically characterized an MGDG hydrolytic activity in cotton (*Gossypium hirsutum)* leaves that appears to be attributable to an acyl hydrolase. This activity is strongly upregulated under drought conditions and hence may be involved in premature senescence induced by stress (El-Hafid *et al.*, 1989). Proteins with galactolipid-hydrolyzing ability have also been purified from the leaves of *Phaseolus multiflorus* (Sastry and Kates, 1964), pole bean (Anderson *et al.*, 1974), and wheat (O'Sullivan *et al.*, 1987). These enzymes proved capable of catalyzing the removal of fatty acids from both *sn* positions of MGDG and DGDG and were also active against phospholipids but not TAGs. They were thus classified as galactosyl diglyceride acyl hydrolases or galactolipases.

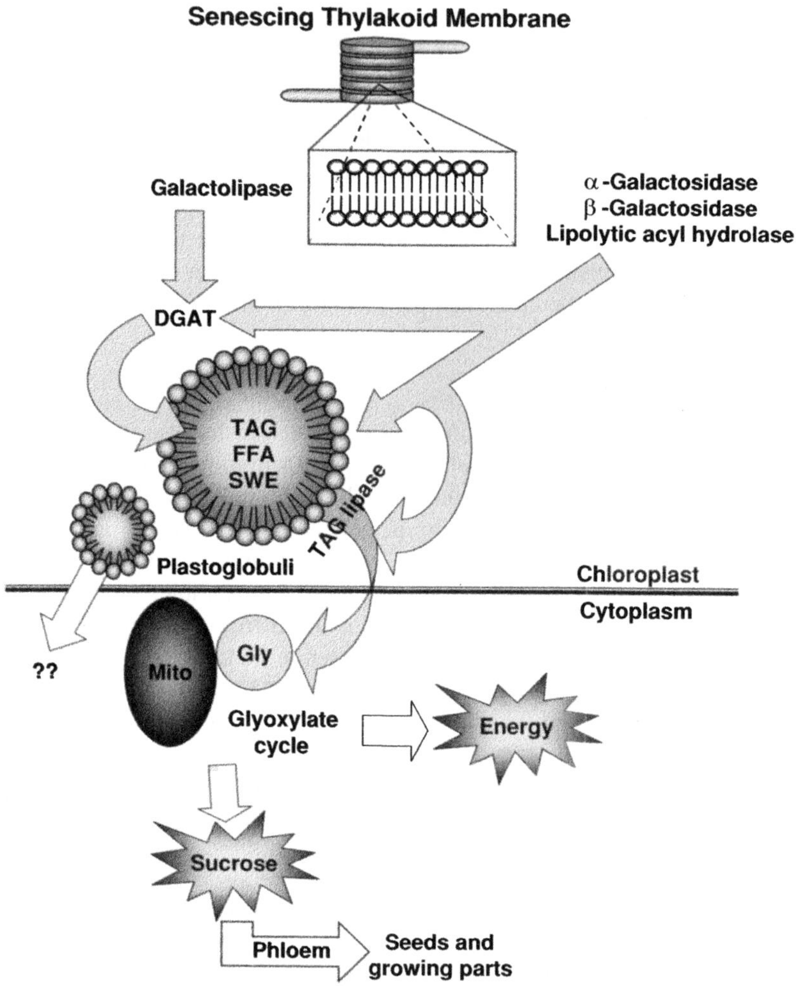

Figure 3.2 Schematic illustration of the fate of thylakoid fatty acids during leaf senescence. The galactolipid fatty acids of senescing thylakoid membranes are a source of energy and phloem-mobile sucrose. Galactolipid fatty acids are thought to be de-esterified by either galactolipases or galactosidases and lipolytic acyl hydrolases (indicated as grey arrows). These free fatty acids accumulate within the membrane or are sequestered within the membrane as triacylglycerol (TAG) and steryl/wax esters (SWE). The free fatty acids together with TAG and SWE are subsequently voided from the thylakoid surface, forming plastoglobuli. The plastoglobuli act as temporary storage reservoirs, and their TAG and SWE components are further catabolized by lipases (TAG lipase and lipolytic acyl hydrolase) within the chloroplast. It is not unequivocally clear how the free fatty acids of plastoglobuli reach glyoxysomes (gly), but plastoglobuli may be extruded into the cytoplasm and targeted to glyoxysomes.

The putative lipase gene, Accession number At2g30550, which was identified in a recent transcriptome analysis for *Arabidopsis* leaf senescence (Guo *et al.*, 2004), also appears to be a chloroplastic galactolipase. This gene has a chloroplast-targeting sequence, and Northern analyses using the corresponding full-length cDNA isolated

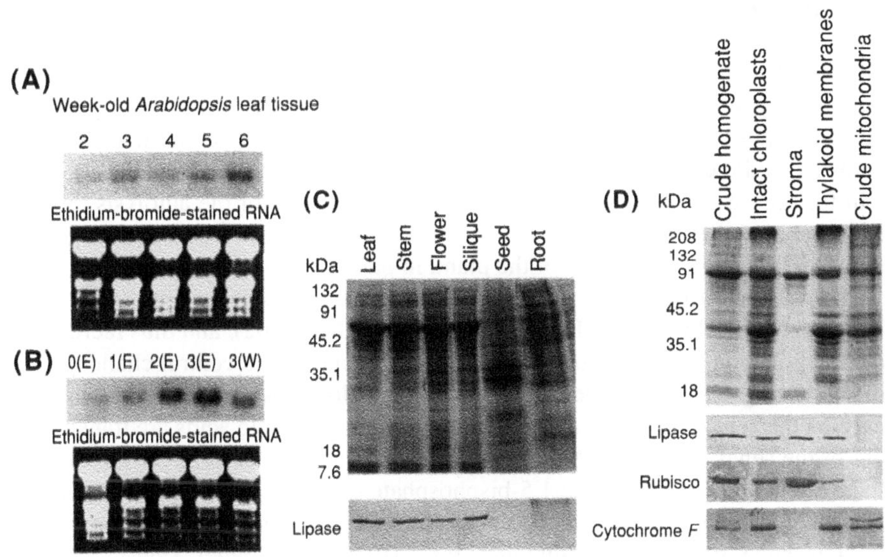

Figure 3.3 Expression of the *Arabidopsis* putative galactolipase gene, At2g30550, during development and ethephon treatment, and localization of its cognate protein. (A) Northern blot of total RNA extracted from 2- to 6-week-old *Arabidopsis* rosette leaves. (B) Northern blot of total RNA extracted from 4-week-old *Arabidopsis* rosette leaves that were detached and floated on 100-mM ethephon (E) or water (W) for 0, 1, 2, or 3 days. For (A) and (B), 10 mg of RNA for each sample was fractionated by electrophoresis through a formaldehyde denaturing gel and transferred to a nylon membrane. The membrane was hybridized overnight with α-^{32}P-labeled cDNA for At2g30550, washed and exposed to film. The corresponding ethidium-bromide-stained formaldehyde gels are illustrated. (C) Western blot of total protein extracted from *Arabidopsis* rosette leaves, stem, flower, silique, seeds or roots. Each lane contained 20 mg of protein, and the blot was probed with antibody raised against a synthetic peptide corresponding to a unique region (aa 494 to aa 510) of the putative galactolipase protein. (D) Western blot of *Arabidopsis* leaf homogenate and isolated subcellular fractions. Each lane contained 20 mg of protein, and the blot was probed with antibody raised against a synthetic peptide corresponding to a unique region (aa 494 to aa 510) of the putative galactolipase protein, stripped and probed with antibody against Rubisco (a stromal marker), stripped and probed with antibody against cytochrome *f* (a thylakoid membrane marker). For (C) and (D), anti-rabbit IgG coupled to alkaline phosphatase was used as the secondary antibody. The corresponding Coomassie-stained SDS-PAGE gels are illustrated.

from an *Arabidopsis* leaf senescence cDNA library as a probe have indicated that it is upregulated in naturally senescing *Arabidopsis* rosettes and in response to ethephon (Figure 3.3). In addition, Western analyses indicate that the protein is detectable only in green tissues and is present in purified stroma and in association with thylakoid membranes (Figure 3.3). The protein contains the recognized 10 amino acid lipase/esterase consensus sequence, ITVTGHSLGG (Derewenda and Derewenda, 1991), indicating that it is capable of de-esterifying fatty acids from complex lipids, and the corresponding recombinant protein has been shown capable of de-esterifying fatty acids from MGDG, although not reproducibly (C. Taylor and J.E. Thompson, unpublished data).

3.3 Role of proteolysis in membrane senescence

Of the 2491 unique genes identified as components of the transcriptome for *Arabidopsis* leaf senescence, 116 are predicted to be involved in some aspect of protein degradation. Of these, 75 appear to encode elements of the ubiquitin–proteasome pathway, and the remainder to be proteases (Guo *et al.*, 2004). Proteolysis is seminal for senescence in that it not only is a central feature of the autolysis inherent in programmed cell death, but also enables recruitment of cellular nitrogen for translocation to growing parts of the plant, in particular developing seeds.

It has been estimated that as much as 75% of the total cellular nitrogen in leaves is localized in chloroplasts (Hoertensteiner and Feller, 2002), and thus recruitment and reallocation of these nitrogen equivalents is one of the central challenges for plants during leaf senescence. It is clear that chloroplast proteolysis is initiated early in the senescence cascade, and that the resultant amino acids are translocated out of the senescing tissue (Matile *et al.*, 1996; Masclaux *et al.*, 2000). Moreover, there is evidence that ribulose-1,5-bisphosphate carboxylase/oxygenase (Rubisco) and other stromal proteins can be catabolized within isolated chloroplasts, which implies the existence of plastidial proteases (Thoenen and Feller, 1998).

A key question is how the nitrogen equivalents of thylakoids are harvested during the nutrient recruitment phase of senescence. Thylakoids are highly enriched in proteins, the most abundant being the apoproteins of chlorophyll. Indeed, the apoprotein, LHCP II, together with stromal Rubisco are the major sources of cellular nitrogen in senescing leaves (Matile, 1992). Recent evidence indicates that color changes during foliar senescence, which are reflective of chlorophyll catabolism, are directly related to the regulation of nitrogen mobilization and resorption from thylakoid proteins (Ougham *et al.*, 2005). Chlorophyll is degraded through a metabolic pathway that is specifically activated during senescence. Many of the chlorophyll-catabolizing enzymes, together with the genes that encode them, have been identified and characterized, and it has been shown that genetic interventions in this pathway lead to disruptions in LHCP II mobilization during foliar senescence (Ougham *et al.*, 2005). Thus LHCP II and chlorophyll degradation in senescing leaves appear to be codependent.

One of the most elegant demonstrations of this codependence has come from studies of a stay-green mutant of *Phaseolus vulgaris* (Bachmann *et al.*, 1994). Typically, Rubisco and LHCP II are degraded at comparable rates during foliar senescence. The stay-green mutant of *P. vulgaris* has a lesion in the chlorophyll degradation pathway and not only retains most of its chlorophyll as senescence progresses, but also retains LHCP II long past the point when it is degraded in corresponding wild-type plants. By contrast, the metabolism of Rubisco in the mutant temporally mirrors its degradation in wild-type plants (Bachmann *et al.*, 1994).

There is strong upregulation of a number of proteases in senescing leaves (Guo *et al.*, 2004). In fact, the most abundantly expressed transcript in the transciptome for senescing *Arabidopsis* leaves is *SAG12* (Guo *et al.*, 2004), which encodes a cysteine protease (Lohman *et al.*, 1994). Indeed, transcripts for the entire family of cysteine proteases proved to be the most abundant components of the senescing leaf transcriptome (Guo *et al.*, 2004), which is reflective of the complexity of proteolysis and recruitment of nitrogen during foliar senescence. Yet, it is not clear which, if any, of these proteases are involved in degrading LHCP II and other thylakoid proteins.

There is increasing evidence that membrane proteins, like their cytosolic counterparts, have to be targeted or undergo some kind of conformational change to become effective protease substrates. For example, protein removal is typically accomplished through the ubiquitin/26S proteasome pathway in which ubiquitins covalently attached to a protein serve as a signal that targets it for degradation by the 26S proteasome (Hershko and Ciechanover, 1998; Callis and Vierstra, 2000). There is some evidence that the ubiquitin-proteasome pathway is active in senescing tissue. For example, the ORE9 F-box protein, a component of the SCF (Skp1/Cul1/F-box protein) complex in the ubiquitin pathway, appears to positively regulate leaf senescence in *Arabidopsis,* possibly by facilitating the ubiquitin-mediated degradation of a transcriptional repressor of SAGs (Woo *et al.*, 2001). As well, pSEN3, an *Arabidopsis* cDNA clone encoding ubiquitin, is upregulated in senescing leaves (Park *et al.*, 1998), as are At1g14400 (ubiquitin carrier protein), At1g53750 (26S proteasome ATPase subunit) and At2g21950 (SKP1 interacting partner 6) (Gepstein *et al.*, 2003).

Although there is no evidence to date that proteins of senescing membranes are ubiquitinated prior to proteolysis, there are indications that changes in the conformation of membrane proteins may initiate their proteolysis. One of the best documented incidences of this is in respect of the D-1 protein of thylakoids, a photosynthetic protein that is rapidly turned over in a light-dependent manner. This rapid turnover appears to be dependent upon the presence of a 14-amino-acid α-helix, termed a *destabilization sequence*, that is adjacent to the putative cleavage domain. Data obtained by using radical scavengers suggest that light-dependent D-1 degradation is initiated by activated oxygen, likely the hydroxyl radical, which causes it to undergo a change in conformation (Sopory *et al.*, 1990).

That nitrogen mobilization and chlorophyll degradation in senescing thylakoids are codependent (Thomas and Donnison, 2000) represents a further illustration of the likelihood that membrane proteolysis is prompted by protein conformational changes. This codependence has been interpreted as indicating that chlorophyll-binding proteins like LHCP II are stabilized by their intimate association with chlorophyll and, further, that proteolysis does not occur as long as the pigment–protein complexes are intact (White and Green, 1987). In light of the fact that LHCP II is so tightly associated with chlorophyll, it is reasonable to assume that degradation and removal of chlorophyll would induce conformational changes in LHCP II that render it prone to proteolytic attack.

It is clear that membrane proteins are metabolized during senescence. In the event of foliar senescence, for example, there is complete degradation of thylakoids and their proteins (Kolodziejek *et al.*, 2003). There are also reports of decreased levels of plasmalemma proteins in senescing carnation petals (Borochov *et al.*, 1994), daylily tepals (Lay-Yee *et al.*, 1992) and iris tepals (Celikel and Doorn, 1995). However, transmembrane proteins, unlike cytosolic proteins, are largely embedded in lipid and disinclined to move from this environment because of their predominantly hydrophobic nature. Thus, whether a membrane protein is prone to proteolysis during senescence may be regulated as much by the accessibility of the particular protein as by the upregulation of proteases. There is evidence, for example, that senescence-associated changes in the lipid fluidity of membranes impact membrane proteins and render them prone to proteolysis. More specifically, the formation of gel phase lipid in the bilayers of senescing membranes not only engenders lateral displacement of proteins out of the developing gel phase domains (Paliyath and Thompson, 1990; Ismail *et al.*, 1999), but also displaces free sterols, campesterol, stigmasterol, sitosterol, isofucosterol and cholesterol, into adjacent liquid-crystalline phase lipid. The impact of this is a dramatic increase in the sterol–phospholipid ratio within the remaining liquid-crystalline phase lipid, the same lipid domains that also continue to solvate all the membrane proteins, including those displaced from the forming gel phase domains. Sterols are known to restrict the inherent rotational mobility of phospholipid fatty acids (Shinitzky and Inbar, 1976), and, accordingly, this increase in sterol–phospholipid ratio results in a dramatic decrease in the fluidity of the residual liquid-crystalline domains in senescing membranes. This has been demonstrated using primarily two physical techniques, fluorescence depolarization and electron spin resonance, for membranes from a number of different senescing tissues including cotyledons (Thompson *et al.*, 1987), flowers (Fobel *et al.*, 1987), leaves (Lesham *et al.*, 1984) and ripening fruit (Paliyath *et al.*, 1984).

Earlier work with mammalian cells has demonstrated that a decrease in bilayer fluidity results in displacement of membrane proteins toward membrane surfaces (Shinitzky, 1984). This displacement is thought to expose portions of membrane proteins that are normally in a hydrophobic lipid environment to a more hydrophilic environment, resulting in conformational changes. For example, an increase in sterol–phospholipid ratio has been correlated with altered protein conformation and receptor function (Kirby and Green, 1980). It follows, therefore, that the decrease in bulk lipid fluidity of membranes during senescence is likely to alter the properties of membrane proteins. Indeed, spin-labeling studies of microsomal membrane isolated from senescing cotyledons have shown that the decrease in bilayer fluidity accompanying senescence is of sufficient magnitude to alter the conformation of membrane proteins (Duxbury *et al.*, 1991b).

The enhanced relative concentration of free sterols in membranes may also render proteins more susceptible to proteolysis in a manner quite independent of their effect on bilayer fluidity. The reason for this is that phospholipid bilayers have a limited capability to solvate sterols. Physical studies with phospholipid liposomes have shown that free sterols associate with phospholipids at a molar stoichiometry of 1 sterol/2 phospholipid (McKersie and Thompson, 1979). Thus at concentrations that exceed this 2:1 ratio, sterols in a bilayer are not fully solvated. For senescing plant membranes, sterol concentrations as high as 45 mol% have been reported (Duxbury *et al.*, 1991a). At this concentration, all phospholipid molecules will be associated with sterol unless their affinity for proteins exceeds that for sterols. This is likely to induce conformational changes in proteins, rendering them prone to proteolysis, because the hydrophobic regions of membrane proteins normally require a close association with the hydrophobic fatty acid side chains of phospholipids, and sterols are usually excluded from this phospholipid annulus (Warren *et al.*, 1975). This contention is supported by freeze-fracture electron microscopy studies showing that sterol concentrations in excess of 27 mol% induce protein aggregation (Schneider *et al.*, 1982; Legge and Brown, 1988).

3.4 Dismantling of membranes in senescing tissue

Ultrastructural studies have provided compelling evidence for complete dismantling of cellular membranes as senescence progresses (Kolodziejek *et al.*, 2003). The thylakoid membranes of the chloroplast appear to be among the first membranes to be degraded during foliar senescence (Matile, 1992). However, the dismantling of membranes, whether it be the energy-producing thylakoids or the less complex phospholipid-containing membranes, presents a logistical problem because the lipids and proteins of membranes are mainly amphipathic, insoluble in aqueous environments and exhibit a strong propensity to remain within the bilayer. From a theoretical perspective, one possibility for overcoming this thermodynamic obstacle to dismantlement of membranes in senescing tissues would be to invoke metabolism of membrane lipids and proteins to their respective water-soluble metabolites, acetate and amino acids, within the milieu of the membrane. These metabolites would then partition into the aqueous compartments of the cell in accordance with their hydrophobic–hydrophilic partition coefficients. However, there is no evidence that this occurs, despite the presence of proteases in membranes such as that mediating the rapid turnover of the D-1 protein in thylakoids (Shipton *et al.*, 1990). Rather, there is growing evidence that the dismantling of senescing membranes entails formation of bilayer-destabilizing macromolecules, both lipid and protein, which move laterally through the plane of the membrane to form destabilizing domains within the bilayer that subsequently void from the membrane surface (Figure 3.4).

3.4.1 Plastoglobuli

By far the best documented incidence of this is in respect of thylakoid membranes where it seems clear that the dismantling of thylakoids during foliar senescence is accompanied by a parallel and commensurate formation of plastoglobuli (Matile, 1992; Kaup *et al.*, 2002). This has led to the assumption that plastoglobuli serve as temporary storage reservoirs for molecular components liberated during the dissolution of thylakoids (Lichtenthaler and Weinert, 1970; Figure 3.2). Indeed, plastoglobuli isolated from senescing leaves do contain thylakoid lipids and their catabolites including plastoquinone, α-tocopherol, TAG, carotenoids and free fatty acids (Steinmuller and Tevini, 1985; Kaup *et al.*, 2002). Moreover, there are indications that plastoglobuli are not of uniform lipid composition (Kessler *et al.*, 1999), which may be reflective of the fact that they bleb from different regions along the plane of thylakoid membranes.

Another class of stromal particles, termed *lipid-protein particles*, has also been characterized, and the distinguishing feature of these particles is that they contain catabolites of thylakoid proteins as well as galactolipids and galactolipid catabolites,

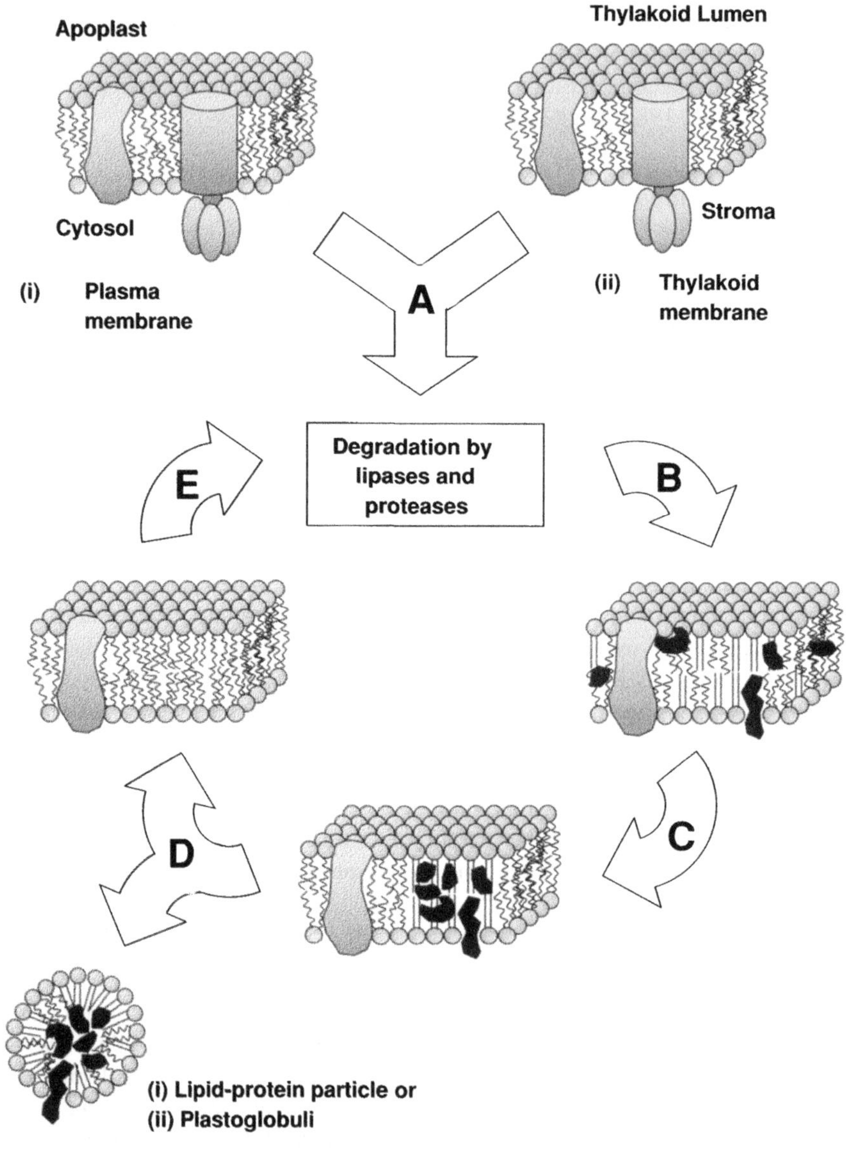

Figure 3.4 Conceptual model for dismantling of (i) plasma membrane and (ii) thylakoid membranes during senescence. (A) As degradative enzymes are upregulated at the onset of senescence, the lipid and protein constituents of cellular membranes begin to be catabolized. (B) Lipid catabolites (hydrocarbon chains indicated in bold) and protein catabolites (darkly shaded shapes) form at various sites throughout the bilayer. (C) These catabolites phase separate within the plane of the membrane, to form discrete domains. (D) These domains are voided from the membranes, giving rise to cytosolic lipid-protein particles in the case of the plasma membrane and plastoglobuli in the case of thylakoids. (E) The cycle is ongoing as more lipids and proteins are degraded in the membranes and released as particles.

in particular free fatty acids (Ghosh *et al.*, 1994). Transmission electron microscopy of these lipid-protein particles indicated that they bear morphological resemblance to plastoglobuli, specifically in that they are globular rather than microvesicular in nature. In a more recent study, the compositions of plastoglobuli, which are commonly isolated by flotation centrifugation of the stroma, and of a higher density class of stromal lipid-protein particles isolated by ultrafiltration were compared (Smith *et al.*, 2000). Both were found to contain the plastoglobuli-specific protein, PAP, indicating that the higher density lipid-protein particles are plastoglobuli like. PAP is thought to maintain the structural integrity of plastoglobuli in a manner analogous to that for oleosin associated with oil bodies (Rey *et al.*, 2000). Consistent with the presumed thylakoid origin of plastoglobuli, small amounts of PAP are also associated with thylakoids (Smith *et al.*, 2000). Moreover, constitutive overexpression of PAP has been shown to enhance the abundance of plastoglobuli in chloroplasts (Rey *et al.*, 2000), an observation that is consistent with a role for PAP in the formation of plastoglobuli.

Both types of particles, plastoglobuli isolated by flotation centrifugation and plastoglobuli-like particles, also contained thylakoid-specific fatty acids, further confirming that they originated from thylakoids, but they were clearly distinguishable on the basis of not only differences in buoyant density, but also differences in protein and lipid composition (Smith *et al.*, 2000). Of particular interest, though, is the finding that the plastoglobuli and the higher density lipid-protein particles both also contained catabolites of the thylakoid-specific protein, cytochrome *f* (Smith *et al.*, 2000). Cytochrome f is a major protein of the thylakoid membrane and a central component of the photosynthetic electron transport chain. It is anchored in the thylakoid through a single transmembrane α-helix with a 15 amino acid C terminus extending into the stroma and a large globular N terminus protruding into the thylakoid lumen (Gray, 1992). Of particular interest is the finding that the catabolites of cytochrome *f* in the plastoglobuli and plastoglobuli-like particles were not recognized by antibodies raised against synthetic peptides corresponding to the C terminus or the N terminus of the full-length protein (Smith *et al.*, 2000). This indicates that the catabolites are formed by proteolytic cleavage at both ends of the protein, events that presumably render the protein nonfunctional and possibly changed in conformation, and cause it to be voided from the thylakoid membrane.

These observations, together with the fact that plastoglobuli are formed in temporal lock-step with thylakoid degradation during foliar senescence, are consistent with the view that dismantling of thylakoids entails progressive voiding of plastoglobuli enriched in both lipid and protein catabolites (Figure 3.4). They are also consistent with an earlier proposal (Thomas and Hilditch, 1987) that thylakoid proteins are released from the membrane bilayer in association with lipid during normal thylakoid turnover. The question remains – what happens to the plastoglobuli, and how are the nitrogen and carbon equivalents of their macromolecular components recruited for subsequent reallocation to developing seeds (Figure 3.2)? There is one report of chlorophyll-containing plastoglobuli being exuded through the chloroplast envelope into the cytoplasm in senescing leaves (Guiamet *et al.*, 1999). Further to this, Matile *et al.* (1996) have reported that chlorophyllase, the enzyme that catalyzes the removal of the phytol chain from chlorophyll producing chlorophyllide, is localized in the inner chloroplast envelope membrane. This apparent spatial separation of chlorophyll and chlorophyllase implies that chlorophyll degradation is dependent upon its translocation from the thylakoids to the chloroplast envelope, which would be achieved if plastoglobuli containing chlorophyll were targeted to the envelope (Matile *et al.*, 1996). Further to the possibility of extraplastidial catabolism of chlorophyll, chlorophyllase genes encoding proteins that are deduced to be soluble and in the cytosol have been identified in *Arabidopsis* (Tsuchiya *et al.*, 1999) and citrus (Jakob-Wilk *et al.*, 1999).

3.4.2 Cytosolic lipid-protein particles

There is also evidence for the existence of cytosolic lipid-protein particles derived from lipoprotein membranes such as the plasmalemma that appear to be counterparts of plastoglobuli (Figure 3.4). These cytosolic lipid-protein particles are nonsedimentable and isolated by ultrafiltration or flotation centrifugation of a post-microsomal supernatant (Yao *et al.*, 1991; Hudak and Thompson, 1996). Generally, they are spherical, 30–300 nm in diameter and uniformly osmiophilic in thin section (Hudak and Thompson, 1996). That they originate from membranes is supported by the observation that they contain phospholipids, which may be in a monolayer conformation analogous to the phospholipid monolayer of seed oil bodies (Murphy, 1993), and can be generated *in vitro* from microsomal membranes under conditions in which phospholipid catabolism has been activated (Yao *et al.*, 1991; Hudak and Thompson, 1996). Moreover, these cytosolic lipid-protein particles are enriched in free fatty acids and steryl/wax esters, the same lipid metabolites that accumulate and phase separate in senescing lipoprotein membranes, as well as TAG. Indeed, quantitative measurements have indicated that cytosolic lipid-protein particles are enriched by six- to ten-fold in free fatty acids relative to phospholipids by comparison with corresponding microsomal membranes, and by >100-fold in steryl and wax esters on the same basis (McKegney *et al.*, 1995; Hudak and Thompson, 1996). There is also evidence that these cytosolic lipid-protein particles are enriched in peroxidized lipids of membrane origin, which are known to accumulate in senescing membranes (Yao and Thompson, 1993). Moreover, they also contain catabolites of the plasmalemma H^+-ATPase (Hudak and Thompson, 1996). Thus, it seems reasonable to propose that, as for thylakoids, dismantling of cellular lipoprotein membranes such as the plasmalemma and endoplasmic reticulum is also achieved by progressive blebbing of lipid-protein particles from the membrane surfaces (Figure 3.4).

3.4.2.1 Sites of cytosolic lipid-protein particle ontogeny

That cytosolic lipid-protein particles have a unique lipid composition clearly distinguishable from that of their membranes of origin suggests that they are formed at specific sites along the plane of the membrane where there is a disproportionately high concentration of phospholipid metabolites. Indeed, at least some of these metabolites, specifically free fatty acids and steryl/wax esters, have been shown to have a propensity to phase separate within membrane bilayers (Jain and Wu, 1977; McKersie and Thompson, 1979; Barber and Thompson, 1980; Yao *et al.*, 1991;

Welti and Glaser, 1994). It is therefore possible that the lipid metabolite-rich domains engendered by this phase separation become sites of lipid-protein particle formation along the plane of the membrane. In fact, voiding of lipid-protein particles from these sites is likely to be facilitated by the packing imperfections at the phase boundaries. This contention is supported by the finding that deformations in bilayer structure engendered by phase separations of lipids, particularly those with high spontaneous curvature, promote both protrusion of a bleb and its expulsion from the membrane (Dobereiner *et al.*, 1993; Lipowsky, 1993)

The recent isolation from tissue homogenates of a subpopulation of microsomes enriched in membrane lipid and protein catabolites constitutes further evidence for the concept of domain separation in senescing membrane bilayers (Madey *et al.*, 2001). Microsomal fractions were isolated from homogenates of both canola leaves and carnation petals, and the microsomal subpopulation was immunoprecipitated from these fractions using region-specific antibodies raised against a recombinant polypeptide of the plasma membrane H^+-ATPase. The immunopurified vesicles proved to be highly enriched in lipid catabolites relative to plasma membrane purified in an aqueous dextran–polyethylene glycol two-phase system, a standard technique used for purifying plant plasma membrane (Larsson *et al.*, 1994), indicating that they represented a unique population of plasma membrane vesicles formed during tissue homogenization. In particular, they were enriched in free fatty acids and steryl/wax esters, lipid metabolites that have been shown to phase separate in membrane bilayers (Yao *et al.*, 1991), and they also contained catabolites of the plasma membrane H^+-ATPase (Madey *et al.*, 2001). These findings are consistent with the view that lipid and protein metabolites formed within the plasma membrane come together to form discrete domains by moving laterally through the plane of the membrane. During tissue homogenization, these domains would be prone to breaking away as fragments that would then spontaneously reseal to form microsomal vesicles. In addition, however, *in situ* these domains may well have a propensity to void from the bilayer as lipid-protein particles because of packing imperfections at their boundaries. This would in turn enable dismantling of the plasmalemma in senescing tissues (Figure 3.4).

The other major lipid formed in response to fatty acid de-esterification in senescing membranes is TAG (Figure 3.1), and this too is found in cytosolic lipid-protein particles (Yao *et al.*, 1991). It is well established that oil bodies of seeds are formed by voiding TAG from endoplasmic reticulum membrane at sites along the plane of the membrane where there has been an accumulation of this neutral lipid into discrete domains (Hills *et al.*, 1993). It seems reasonable, therefore, to assume that this same principle underlies voiding of TAG into cytosolic lipid-protein particles in senescing tissues.

3.5 Role of autophagy

There is increasing evidence that autophagy may play a role in nutrient recycling during senescence. Autophagy employs the vacuole as a lytic organelle. In the case of proteolysis through autophagy, the proteins to be degraded are not individually targeted as in the case for degradation through the ubiquitin/26S proteasome pathway. Rather, autophagy entails encapsulation of portions of the cytoplasm within double-membrane vesicles termed *autophagosomes* that are thought to originate from the endoplasmic reticulum. The autophagosomes are subsequently targeted to the tonoplast, and the outer membrane of the autophagosome fuses with the tonoplast, releasing the internal vesicle into the vacuole where its cargo is degraded (Marty, 1999; Kim and Klionsky, 2000; Klionsky and Emr, 2000). Thus, degradation of proteins by autophagy is indiscriminate and, for this reason, thought to be invoked when there is a need for rapid mobilization and resorption of nutrients as in senescence and stress responses (Doelling *et al.*, 2002).

Through studies with yeast, two conjugating pathways culminating in autophagy have been identified, one mediated by APG8 and another by APG12. These two proteins become attached to other cell factors in a manner that resembles ubiquitination (Mizushima *et al.*, 1998; Ichimura *et al.*, 2000). Of interest is the fact that potential orthologs for all the components of the APG8 and APG12 yeast conjugation pathways are present in *Arabidopsis* (Doelling *et al.*, 2002). The contention that autophagy plays a key role in senescence is supported by the finding that one of these orthologs, APG7, has been shown to be required for normal leaf senescence (Doelling *et al.*, 2002). APG7 is encoded by a single gene and is an ATP-dependent activating enzyme that initiates both conjugation pathways. Disruption of *APG7* by insertional mutation has been shown to have no effect on growth and development of *Arabidopsis*, but it does engender hypersensitivity to nutrient-limiting conditions as well as premature leaf senescence (Doelling *et al.*, 2002). In addition, APG7 transcript preferentially accumulates in senescing *Arabidopsis* leaves (Doelling *et al.*, 2002). Further evidence supporting the involvement of autophagy in senescence comes from the finding that *AtATG18a,* one of eight members of the *AtATG18* family in *Arabidopsis* exhibiting sequence homology with the yeast autophagy gene, *ATG18,* is also upregulated in senescing leaves (Xiong *et al.*, 2005). Moreover, suppression of *AtTG18a* using RNA interference disrupts the normal development of autophagosomes and again results in premature leaf senescence (Xiong *et al.*, 2005).

The role of autophagy in senescence is thought to be of particular importance as a means of degrading cytosolic proteins (Dunn, 1994; Marty, 1999; Kim and Klionsky, 2000). The question of whether autophagy plays a role in the degradation of membrane proteins, such as LHCP II of thylakoids or even Rubisco which resides in the chloroplast stroma, is not resolved. There have been reports that organelles, including chloroplasts, can be engulfed by autophagosomes and degraded in the vacuole (Wittenbach *et al.*, 1982; Dunn, 1994). However, the finding that loss of chlorophyll is accelerated, rather than delayed, in *Arabidopsis* plants with impaired autophagic function has been interpreted as indicating that, at least in senescing tissue, chloroplasts are not normally degraded by autophagy (Doelling *et al.*, 2002).

At face value, these observations render questionable the prospect that autophagy plays any role whatsoever in the breakdown of thylakoid proteins or Rubisco, which collectively comprise the major source of recyclable nitrogen captured from senescing leaves. However, there are other observations that temper this point of view. In particular, evidence indicating that plastoglobuli contain catabolites of photosynthetic proteins (Ghosh *et al.*, 1994; Smith *et al.*, 2000), that chlorophyll-bearing plastoglobuli may be exuded through the chloroplast envelope into the cytosol in senescing leaves (Guiamet *et al.*, 1999), and that there may be an extraplastidial form of chlorophyllase (Jakob-Wilk *et al.*, 1999; Tsuchiya *et al.*, 1999) support the possibility posed by Hoertensteiner and Feller (2002) that catabolites generated inside chloroplasts may be further metabolized after release across the chloroplast envelope. Should this be the case, it is likely that this further metabolism within the cytosol is achieved through autophagy. Moreover, recent immunocytochemical electron microscopy has indicated that in senescing leaves of wheat, Rubisco and/or its degradation products are localized in small spherical particles within the cytosol and in the vacuole (Chiba *et al.*, 2003). These particles also contain glutamine synthase, another stromal protein, and were most abundant in senescing leaves. The authors interpret their findings as suggesting that degradation of Rubisco in senescing leaves may occur outside of the chloroplast, and should this be the case it could well be mediated by autophagy. It is conceivable as well that the cytosolic lipid-protein particles formed during senescence-related dismantling of lipoprotein membranes (Thompson *et al.*, 1998) are also degraded by autophagy.

3.6 Metabolism of membrane fatty acids in senescing tissues

The fatty acids of membranes are a rich source of carbon, and there is now compelling evidence that this carbon fuels the formation of ATP during senescence and, at least in some plants, is converted to phloem-mobile sucrose for transport to developing seeds (Wanner *et al.*, 1982, 1991; Froman *et al.*, 2000; Page *et al.*, 2001; Cornah and Smith, 2002). It is also clear that the first step in this mobilization is de-esterification of fatty acids from complex membrane lipids including phospholipids and galactolipids (Thompson *et al.*, 1998). These fatty acids in turn undergo β-oxidation in glyoxysomes forming acetyl-CoA for energy production and in many plants, but apparently not all (Charlton *et al.*, 2005), conversion through the glyoxylate cycle and the TCA cycle to oxaloacetate, leading to gluconeogenesis and the formation of sucrose (DeBellis *et al.*, 1990).

There is now good evidence for the conversion of leaf peroxisomes to glyoxysomes as senescence is engaged (DeBellis *et al.*, 1990). What is less clear is how de-esterified membrane fatty acids are translocated from their membrane of origin to glyoxysomes to initiate the metabolism that converts their carbon equivalents to energy or sucrose. Indeed, from a theoretical perspective this is a daunting task, in part because free fatty acids are not soluble and do not readily partition out of membrane bilayers and also because they act like detergents (Thomas, 1982) and tend to destabilize the structure of membrane bilayers. There is some accumulation of free fatty acids in senescing membranes (Fobel *et al.*, 1987), but by far the most prominent change in lipid composition is a dramatic increase in steryl and wax esters as well as TAG (Figure 3.1). Moreover, the formation of both steryl/wax esters and TAG requires fatty acids. Thus it seems reasonable to assume that the initial fate of many fatty acids upon their de-esterification is metabolic assimilation into either steryl/wax esters or TAG, which are more inert in terms of their impact on bilayer structure than are de-esterified fatty acids. This is illustrated, for example, by the finding that levels of steryl/wax ester fatty acids increase by upwards of 250%, relative to phospholipid fatty acids, as membranes senesce (McKegney *et al.*, 1995).

The contention that free fatty acids of senescing membranes are assimilated into steryl/wax esters and TAG is consistent with the fact that the enzymes required for this assimilation appear to be membrane associated. There is evidence, for example, that enzymes mediating the formation of both steryl esters and wax esters are associated with microsomal membranes (Garcia and Mudd, 1978). Diacylglycerol acyltransferase (DGAT; EC 2.3.1.20), which mediates the final acylation step in the synthesis of TAG, is also membrane associated (Lu *et al.*, 2003). DGAT forms TAG by acylating the *sn*-3 position of DAG, and it is noteworthy that DAG and de-esterified fatty acids, the substrates for this reaction, are both products of the action of lipolytic acyl hydrolase on either phospholipids or galactolipids. Moreover, TAG, together with steryl/wax esters, is present in both plastoglobuli and cytosolic lipid-protein particles, putative vehicles for voiding lipid catabolites from senescing membranes during their dissolution (Martin and Wilson, 1984; Ghosh *et al.*, 1994; McKegney *et al.*, 1995; Figure 3.4).

3.6.1 Galactolipid fatty acids

Thylakoids are the most abundant membranes in nature (Lee, 2000) and hence a rich source of membrane fatty acid carbon for energy production and carbon recycling during senescence. Moreover, there is a progressive accumulation of TAG coincident with the dismantling of thylakoids during foliar senescence, implicating TAG in the metabolism of thylakoid fatty acids (Kaup *et al.*, 2002). Indeed, several lines of evidence indicate that the TAG accumulating in senescing leaves is synthesized within chloroplasts and sequesters fatty acids de-esterified from thylakoids. First, DGAT1 transcript and protein are upregulated in senescing *Arabidopsis* leaves, and immunoblots have indicated that the upregulated DGAT1 protein is associated with chloroplast membranes, both thylakoid membranes and envelope membranes (Kaup *et al.*, 2002). Second, the TAG of senescing *Arabidopsis* leaves is enriched in hexadecatrienoic acid, which is uniquely associated with galactolipids, and linolenic acid, which, although present to some degree in phospholipids, is the most abundant fatty acid of galactolipids (Kaup *et al.*, 2002; Miquel *et al.*, 1998). Third, the enhanced synthesis of TAG in senescing leaves correlates temporally with an increase in both size and abundance of plastoglobuli (Kaup *et al.*, 2002), suggesting that the incremental TAG in the senescing leaves is primarily localized in plastoglobuli. It is, in fact, well established that plastoglobuli contain TAG (Steinmuller and Tevini, 1985).

These observations are all consistent with the notion that de-esterified galactolipid fatty acids are initially converted to TAG by DGAT within the thylakoid membrane, and that TAG, along with other lipid and possibly protein catabolites, then moves laterally through the plane of the membrane to form discreet domains that are subsequently voided into the stroma, giving rise to plastoglobuli (Figure 3.2). This would mean that sequestering of free fatty acids derived from galactolipids into TAG constitutes an intermediate step in the conversion of thylakoid fatty acids to energy and phloem-mobile sucrose in senescing leaves. Recent evidence suggesting that plastoglobuli of senescing chloroplasts are exuded through the envelope into the cytoplasm (Guiamet *et al.*, 1999) lends further credence to this possibility in that it provides a plausible explanation for how the fatty acid equivalents of plastoglobuli could gain access to glyoxysomes for β-oxidation (Figure 3.2).

There is growing evidence that plastoglobuli are analogous to oil bodies. In particular, they both store TAG, and plastoglobuli appear to be coated with a structural protein termed *fibrillin* or plastid lipid-associated protein that is analogous to oil body oleosin (Pozueta-Romero *et al.*, 1997; Kessler *et al.*, 1999; Rey *et al.*, 2000). As well, fibrillin is thought to prevent coalescence of plastoglobuli in the same manner that oleosin precludes coalescence of oil bodies (Huang, 1996; Rey *et al.*, 2000). In addition, modulation of fibrillin expression in transgenic plants has indicated that it regulates the formation of plastoglobuli in much the same way that oleosin is thought to regulate the formation of oil bodies (Huang, 1992). Of particular interest as well is the fact that just as oil bodies have never been detected in the lumen of the endoplasmic reticulum, plastoglobuli are not present in the thylakoid lumen. Thus in both cases blebbing is vectorial. Furthermore, it has been demonstrated that there is close contact between oil bodies and glyoxysomes in the cytoplasm of oil seed cells, which may facilitate the transfer of TAG, or fatty acids cleaved from TAG, to the glyoxysomes (Frederick *et al.*, 1968; Vigil, 1969, 1970; Mollenhauer and Toten, 1970; Wanner *et al.*, 1982). Given the similarities between plastoglobuli and oil bodies, this raises the possibility that plastoglobuli may also be targeted to glyoxysomes (Figure 3.2).

3.6.2 Fate of thylakoid fatty acids during stress-induced senescence

There is also evidence for the metabolic conversion of galactolipid fatty acids to TAG in response to a variety of plant stresses. For example, a decrease in MGDG, accompanied by a parallel increase in TAG, has been observed in leaves of plants subjected to rust infection (Loesel and Lewis, 1974; Loesel, 1978), cold hardening (Nordby and Yelenosky, 1984) as well as freezing and thawing (Nordby and Yelenosky, 1985). This phenomenon has been examined in detail using ozone-fumigated spinach leaves (Sakaki *et al.*, 1990a,b). Ozone fumigation of leaves engenders a large reduction in galactolipids, which, remarkably, is accompanied by a marked increase in TAG in the absence of a change in leaf fatty acid composition (Sakaki *et al.*, 1990b). Radiolabeling studies with [1-^{14}C]acetate indicated that the formation of TAG induced by ozonation reflected conversion of MGDG to DAG and subsequent acylation of DAG at the *sn*-3 position (Sakaki *et al.*, 1990a). This was further confirmed by demonstrating that [1-^{14}C]linolenic acid applied to disks from ozone-fumigated leaves was incorporated into TAG, but not DAG (Sakaki *et al.*,

1990a). These findings prompted the authors to conclude that leaf cells are able to sequester free fatty acids liberated from thylakoids into TAG, which is structurally and metabolically inert in comparison with free fatty acids. Indeed, free fatty acids have been shown to be strong inhibitors of the Hill reaction, to uncouple photophosphorylation and to disrupt the structure of thylakoid membranes (McCarty and Jagendorf, 1965; Siegenthaler, 1973; Okamoto *et al.*, 1977). The acylation of DAG resulting in the formation of TAG is mediated by DGAT, and there is significant upregulation of DGAT in chloroplast membranes coincident with the onset of natural leaf senescence (Kaup *et al.*, 2002). Accordingly, there may also be upregulation of chloroplastic DGAT in the event of premature senescence induced by ozonation and other types of stresses. Environmental stress has also been shown to induce upregulated synthesis of fibrillin, which is thought to regulate the formation of plastoglobuli from thylakoids (Rey *et al.*, 2000). Thus it is apparent that de-esterified thylakoid fatty acids are temporarily stored as TAG during both natural senescence and stress-induced senescence. This likely serves two purposes. First, it converts the free fatty acids into a nonamphipathic form, enabling their release from the thylakoid bilayer for subsequent conversion to energy and sucrose. Second, it would prevent any significant accumulation of free fatty acids in the thylakoids, which might otherwise engender uncontrolled dismantling of these membranes and a consequent ineffective capture of their fatty acid carbon equivalents for energy production and sucrose formation.

The finding that radiolabeled linolenic acid is actively incorporated into TAG in leaf disks excised from young actively growing leaves (Sakaki *et al.*, 1990a) indicates that the capacity to convert free fatty acids to TAG is operative in leaves under normal conditions and presumably upregulated in the event of stress leading to premature senescence. Indeed, chloroplastic DGAT, the enzyme that mediates this conversion, is known to be upregulated at the onset of natural leaf senescence (Kaup *et al.*, 2002). However, sublethal stress, which is much more common than lethal stress, does not lead directly to senescence. Rather, it is thought to invoke inhibition of growth and concomitant preparedness for senescence. That de-esterification of thylakoid fatty acids may be part of this inhibition of growth and preparedness for senescence is suggested by the finding that suppression of the *Arabidopsis* gene (Accession number At2g30550) encoding a putative *Arabidopsis* galactolipase (Figure 3.3) results in enhanced growth and seed yield relative to wild-type plants under conditions of sublethal stress (C. Taylor and J.E. Thompson, unpublished data). This is illustrated in Figure 3.5, and raises the possibility that conversion of de-esterified thylakoid fatty acids to TAG may be part of the 'preparedness for senescence' response to sublethal stress. Indeed, studies with algae have demonstrated that TAG serves as a reservoir of polyunsaturated fatty acids that can be used for the rapid formation of chloroplastic lipids (Cohen *et al.*, 2000). These observations collectively raise the interesting possibility that coincident with the onset of stress, de-esterified galactolipid fatty acids are stored temporarily as TAG and, in the event the stress proves to be nonlethal, are subsequently reutilized to form chloroplastic lipids. In the event the stress proves to be lethal, the TAG fatty acids would be converted to energy and sucrose as in normal senescence.

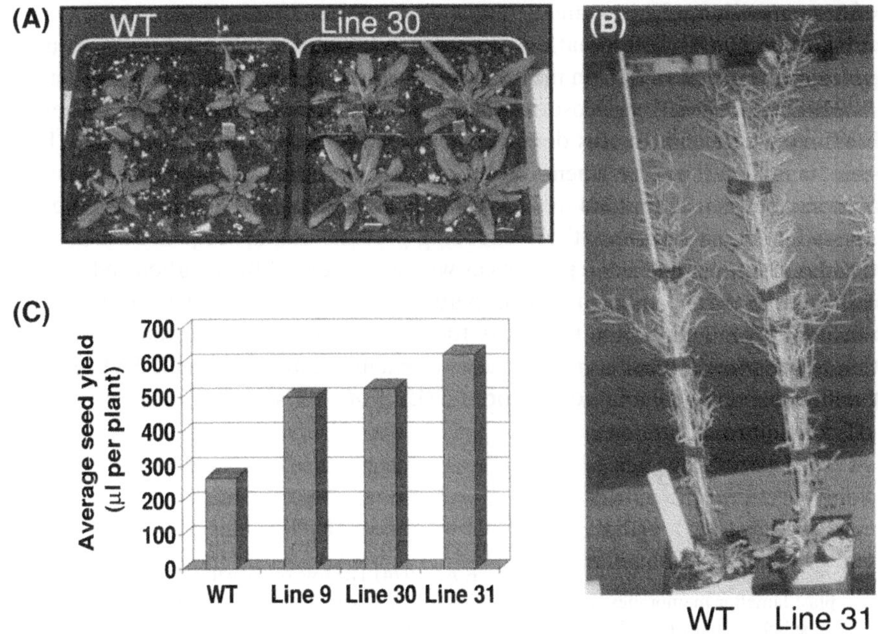

Figure 3.5 Phenotype of *Arabidopsis* plants with suppressed putative galactolipase, At2g30550. The plants were grown in low-nutrient soil to impose chronic sublethal stress. (A) Enhanced rosette size of T2 transgenic plants of Line 30 (right panel) compared with wild-type control plants (left panel) at 4.7 weeks after planting. (B) Taller bolts with more branches for transgenic Line 31 (T1 generation; right panel) compared with wild-type control plant (left panel) at 7 weeks after planting. (C) Seed yield for transgenic plants, Lines 9, 30 and 31, compared with the yield for wild-type (WT) control plants.

3.7 Translational regulation of senescence

It is clear that senescence entails changes in gene expression leading to the synthesis of new proteins. That this occurs on a massive scale is implicit in the finding that the transcriptome for senescing *Arabidopsis* leaves contains 2491 unique genes (Guo *et al.*, 2004). Recent evidence suggests that selective recruitment of these transcripts by a senescence-associated isoform of eukaryotic translation initiation factor 5A (eIF5A) may be an important step in controlling the onset of senescence. The first indication that eIF5A might be involved in senescence came from experiments showing that transcripts for eIF5A and deoxyhypusine synthase (EC 2.5.1.46), which is required for the posttranslational activation of eIF5A, are upregulated in parallel in senescing fruit, flowers and leaves, and also coincident with the onset of premature leaf senescence induced by temperature and osmotic stress (Wang *et al.*, 2001; Wang *et al.*, 2003; Thompson *et al.*, 2004). A clearer indication of the involvement of eIF5A in senescence has been obtained by suppressing its activation in transgenic plants. This results in inhibition of leaf senescence (Wang *et al.*, 2003, 2005) as well as delayed postharvest senescence and spoilage of fruit (Wang *et al.*, 2005). Studies with mammalian cells have indicated that activated eIF5A facilitates translation by acting as a shuttle protein that, in conjunction with CRM1 (the general nuclear export receptor), participates in the selective translocation of mRNA species from the nucleus to the cytoplasm (Rosorius *et al.*, 1999). Further support for this contention has come from experiments demonstrating that activated eIF5A selectively binds specific mRNAs (Xu and Chen, 2001; Xu *et al.*, 2004). In light of this, it is conceivable that eIF5A participates in senescence by recruiting specific mRNAs required for its execution, including those encoding proteins that are required for degradation of membrane lipids and proteins as well as capture of their carbon and nitrogen equivalents for translocation to other parts of the plant. Alternatively, or perhaps in addition, eIF5A may recruit mRNAs for senescence-associated transcription factors. It has recently been demonstrated, for example, that human eIF5A1 regulates translation of the transcription factor p53 (Li *et al.*, 2004).

References

Anderson, M.M., McCarty, R.E. and Zimmer, E.A. (1974) The role of galactolipids in spinach chloroplast lamellar membranes. I: Partial purification of a bean leaf galactolipid lipase and its action on subchloroplast particles. *Plant Physiol* **53**, 699–704.

Bachmann, A., Fernandes-Lopex, J., Ginsburg, S., *et al.* (1994) *Stay-green* genotypes of *Phaseolus vulgaris* L.: chloroplast proteins and chlorophyll catabolites during foliar senescence. *New Phytol* **126**, 593–600.

Barber, R.F. and Thompson, J.E. (1980) Senescence-dependent increase in the permeability of liposomes prepared from bean cotyledon membranes. *J Exp Bot* **31**, 1305–1313.

Barber, R.F. and Thompson, J.E. (1983) Neutral lipids rigidify unsaturated acyl chains in senescing membranes. *J Exp Bot* **34**(140), 268–276.

Barclay, K.D. and McKersie, B.D. (1994) Peroxidation reactions in plant membranes: effects of free fatty acids. *Lipids* **29**(12), 877–882.

Borochov, A., Cho, M. and Boss, W.F. (1994) Plasma membrane lipid metabolism of petunia petals during senescence. *Physiol Plant* **90**, 279–284.

Buchanan-Wollaston, V. (1997) The molecular biology of leaf senescence. *J Exp Biol* **48**, 181–199.

Callis, J. and Vierstra, R.D. (2000) Protein degradation in signaling. *Curr Opin Plant Biol* **3**, 381–386.

Celikel, F.G. and Doorn, W.G.v. (1995) Solute leakage, lipid peroxidation, and protein degradation during the senescence of Iris tepals. *Physiol Plant* **94**, 515–521.

Charlton, W.L., Johnson, B., Graham, I.A. and Baker, A. (2005) Non-coordinate expression of peroxisome biogenesis, β-oxidation and glyoxylate cycle genes in mature *Arabidopsis* plants. *Plant Cell Rep* **23**, 647–653.

Chiba, A., Ishida, H., Nishizawa, N.K., Makino, A. and Mae, T. (2003) Exclusion of ribulose-1,5-bisphosphate carboxylase/oxygenase from chloroplasts by specific bodies in naturally senescing leaves of wheat. *Plant Cell Physiol* **44**(9), 914–921.

Cohen, Z., Khozin-Goldberg, I., Adlerstein, D. and Bigogno, C. (2000) The role of triacylglycerol as a reservoir of polyunsaturated fatty acids for the rapid production of chloroplastic lipids in certain microalgae. *Biochem Soc Trans* **28**(6), 740–743.

Cornah, J.E. and Smith, S.M. (2002) Synthesis and function of glyoxylate cycled enzymes. In: *Plant Peroxisomes* (eds Baker, A. and Graham, I.A.). Kluwer Academic, London, pp. 57–101.

DeBellis, L., Picciarelli, P., Pistelli, L. and Alpi, A. (1990) Localization of glyoxylate-cycle marker enzymes in peroxisomes of senescent leaves and green cotyledons. *Planta* **180**, 435–439.

Derewenda, Z.S. and Derewenda, U. (1991) Relationships among serine hydrolase: evidence for a common structural motif in triacylglyceride lipases and esterases. *Biochem Cell Biol* **69**(12), 842–851.

Dobereiner, H.G., Kas, J., Noppl, D., Sprenger, I. and Sackmann, E. (1993) Budding and fission of vesicles. *Biophys J* **65**(4), 1396–1403.

Doelling, J.H., Walker, J.M., Friedman, E.M., Thompson, A.R. and Vierstra, R.D. (2002) The APG8/12-activating enzyme APG7 is required for proper nutrient recycling and senescence in *Arabidopsis thaliana*. *J Biol Chem* **277**(36), 33105–33114.

Dunn, W.A.J. (1994) Autophagy and related mechanisms of lysosome-mediated protein degradation. *Trends Cell Biol* **4**(4), 139–143.

Duxbury, C.L., Legge, R.L., Paliyath, G., Barber, R.F. and Thompson, J.E. (1991a) Alterations in membrane protein conformation in response to senescence-related changes. *Phytochemistry* **30**(1), 63–68.

Duxbury, C.L., Legge, R.L., Paliyath, G., and Thompson, J.E. (1991b) Lipid breakdown in smooth microsomal membranes from bean cotyledons alters membrane proteins and induces proteolysis. *J Exp Bot* **42**, 103–112.

El-Hafid, L., Pham, A.T., Zuily-Fodil, Y. and da Silva, J.V. (1989) Enzymatic breakdown of polar lipids in cotton leaves under water stress. I: Degradation of monogalactosyl-diacylglycerol. *Plant Physiol Biochem* **27**(4), 495–502.

Engelmann-Sylvestre, I., Bureau, J.-M., Tremolieres, A. and Paulin, A. (1989) Changes in membrane phospholipids and galactolipids during the senescence of cut carnations. Connection with ethylenic rise. *Plant Physiol Biochem* **27**(6), 931–937.

Eze, J.M.O., Mayak, S., Thompson, J.E. and Dumbroff, E.B. (1986) Senescence in cut carnation flowers: temporal and physiological relationship among water status, ethylene, abscisic acid and membrane permeability *Physiol Plant* **68**, 323–328.

Fobel, M., Lynch, D.V. and Thompson, J.E. (1987) Membrane deterioration in senescing carnation flowers. *Plant Physiol* **85**, 204–211.

Frederick, S.E., Newcomb, E.H., Vigil, E.L. and Wergin, W.P. (1968) Fine-structural characterization of plant microbodies. *Planta* **81**, 229–252.

Froman, B.E., Edwards, P.C., Bursch, A.G. and Dehesh, K. (2000) ACX3, a novel medium-chain Acyl-Coenzyme A oxidase from *Arabidopsis*. *Plant Physiol* **123**, 733–742.

Galliard, T. (1980) Degradation of acyl lipids: hydrolytic and oxidative enzymes. In: *The Biochemistry of Plants*, Vol. 4 (ed. Stumpf, P.K.). Academic Press, New York, pp. 85–116.

Garcia, R.E. and Mudd, J.B. (1978) Fatty acid and sterol specificity in the biosynthesis of steryl esters by enzyme preparations from spinach leaves. *Arch Biochem Biophys* **190**(1), 315–321.

Gepstein, S., Sabehi, G., Carp, M.J., *et al.* (2003) Large-scale identification of leaf senescence-associated genes. *Plant J* **36**(5), 629–642.

Ghosh, S., Hudak, K.A., Dumbroff, E.B. and Thompson, J.E. (1994) Release of photosynthetic protein catabolites by blebbing from thylakoids. *Plant Physiol* **106**, 1547–1553.

Goormaghtigh, E., van Campenhoud, M. and Ruysschaert, J.-M. (1981) Lipid phase separation mediates binding of porcine pancreatic phospholipase A2 to its substrate. *Biochem Biophys Res Commun* **101**, 1410–1418.

Gray, J.C. (1992) Cytochrome *f*: structure, function and biosynthesis. *Photosynth Res* **34**, 359–374.

Guiamet, J.J., Pichersky, E. and Nooden, L.D. (1999) Mass exodus from senescing soybean chloroplasts. *Plant Cell Physiol* **40**, 986–992.

Guo, Y., Cai, Z. and Gan, S. (2004) Transcriptome of *Arabidopsis* leaf senescence. *Plant Cell Environ* **27**, 521–549.

He, Y. and Gan, S. (2002) A gene encoding an acyl hydrolase is involved in leaf senescence in *Arabidopsis*. *Plant Cell* **14**, 805–815.

Hershko, A. and Ciechanover, A. (1998) The ubiquitin system. *Annu Rev Biochem* **67**, 425–479.

Hills, M.H., Watson, M.D. and Murphy, D.J. (1993) Targeting of oleosin to the oil bodies of oilseed rape (*Brassica napus* L.). *Planta* **189**, 24–29.

Hoertensteiner, S. and Feller, U. (2002) Nitrogen metabolism and remobilization during senescence. *J Exp Bot* **53**(370), 927–937.

Hong, Y., Wang, T.-W., Hudak, K.A., Schade, F., Froese, C.D. and Thompson, J.E. (2000) An ethylene-induced cDNA encoding a lipase expressed at the onset of senescence. *Proc Natl Acad Sci U S A* **97**(15), 8717–8722.

Huang, A.H.C. (1992) Oil bodies and oleosins in seeds. *Annu Rev Plant Physiol Plant Mol Biol* **43**, 177–200.

Huang, A.H.C. (1996) Oleosins and oil bodies in seeds and other organs. *Plant Physiol* **110**, 1055–1061.

Hudak, K.A. and Thompson, J.E. (1996) Floatation of lipid–protein particles containing triacylglycerol and phospholipid from the cytosol of carnation petals. *Physiol Plant* **98**, 810–818.

Ichimura, Y., Kirisako, T., Takao, T., *et al.* (2000) A ubiquitin-like system mediates protein lipidation. *Nature* **408**(6811), 488–492.

Ismail, A.M., Hall, A.E. and Close, T.J. (1999) Purification and partial characterization of a dehydrin involved in chilling tolerance during seedling emergence of cowpea. *Plant Physiol* **120**, 237–244.

Jain, M.K. and Wu, N.M. (1977) Effect of small molecules on the dipalmitoyl lecithin liposomal bilayer. *J Membr Biol* **34**, 157–201.

Jakob-Wilk, D., Holland, D., Goldschmidt, E.E., Riov, J. and Eyal, Y. (1999) Chlorophyll breakdown by chlorophyllase: isolation and functional expression of the Chlase1 gene from ethylene-treated citrus fruit and its regulation during development. *Plant J* **20**, 653–661.

Kaup, M.T., Froese, C.D. and Thompson, J.E. (2002) A role for diacylglycerol acyltransferase during leaf senescence. *Plant Physiol* **129**, 1616–1626.

Kessler, F., Schnell, D. and Blobel, G. (1999) Identification of proteins associated with plastoglobules isolated from pea (*Pisum sativum* L.) chloroplasts. *Planta* **108**, 107–113.

Kim, J. and Klionsky, D.J. (2000) Autophagy, cytoplasm-to-vacuole targeting pathway, and pexophagy in yeast and mammalian cells. *Annu Rev Biochem* **69**, 303–342.

Kim, M.-J., Oh, J.-M., Cheon, S.-H., *et al.* (2001) Thermal inactivation kinetics and application of phospho- and galactolipid-degrading enzymes for evaluation of quality changes in frozen vegetables. *J Agric Food Chem* **49**, 2241–2248.

Kirby, C.J. and Green, C. (1980) Erythrocyte membrane cholesterol levels and their effects on membrane proteins. *Biochim Biophys Acta* **598**(2), 422–425.

Klionsky, D.J. and Emr, S.D. (2000) Autophagy as a regulated pathway of cellular degradation. *Science* **290**(5497), 1717–1721.

Kolodziejek, I., Koziol, J., Waleza, M. and Mostowska, A. (2003) Ultrastructure of mesophyll cells and pigment content in senescing leaves of maize and barley. *J Plant Growth Regul* **22**, 217–227.

Larsson, C., Sommarin, M. and Widell, S. (1994) Isolation of highly purified plant plasma membranes and separation of inside-out and right-side-out vesicles. *Methods Enzymol* **228**, 451–469.

Lay-Yee, M., Stead, A.D. and Reid, M.S. (1992) Flower senescence in daylily. (Hemrocallis). *Physiol Plant* **86**, 308–314.

Lee, A.G. (2000). Membrane lipids: it's only a phase. *Curr Biol* **10**, R377–R380.

Lee, R.H., Lin, M.C. and Chen, S.C. (2004) A novel alkaline alpha-galactosidase gene is involved in rice leaf senescence. *Plant Mol Biol* **55**(2), 281–295.

Legge, R.L. and Brown, R.M. (1988) Modification of protoplast cell wall regeneration by membrane perturbation. *Protoplasma* **143**, 38–42.

Lesham, Y.Y., Sridhara, S. and Thompson, J.E. (1984) Involvement of calcium and calmodulin in membrane deterioration during senescence of pea foliage. *Plant Physiol* **75**, 329–335.

Li, A.-L., Li, H.-Y., Jin, B.-F., *et al.* (2004) A novel eIF5A complex functions as a regulator of p53 and p53-dependent apoptosis. *J Biol Chem* **279**(47), 49251–49258.

Lichtenthaler, H.K. and Weinert, H. (1970) The correlation between lipoquinone accumulation and plastoglobuli formation in the chloroplasts of *Ficus elastica* Roxb. *Z Naturforsch* **25b**, 619–623.

Lipowsky, R. (1993) Domain-induced budding of fluid membranes. *Biophys J* **64**, 1133–1138.

Loesel, D.M. (1978) Lipid metabolism of leaves of *Poa pratensis* during infection by *Puccina poarum*. *New Phytol* **80**, 167–174.

Loesel, D.M. and Lewis, D.H. (1974) Lipid metabolism in leaves of *Tussilago farfara* during infection by *Puccinia poarum*. *New Phytol* **73**, 1157–1169.

Lohman, K.N., Gan, S., John, M.C. and Amasino, R.M. (1994) Molecular analysis of natural leaf senescence in *Arabidopsis thaliana*. *Physiol Plant* **92**, 322–328.

Lu, C.L., de Noyer, S.B., Hobbs, D.H., *et al.* (2003) Expression pattern of diacylglycerol acyltransferase-1, and enzyme involved in triacylglycerol biosynthesis, in *Arabidopsis thaliana*. *Plant Mol Biol* **52**, 31–41.

Madey, E., Nowack, L.M., Su, L., Hong, Y., Hudak, K.A. and Thompson, J.E. (2001) Characterization of plasma membrane domains enriched in lipid metabolites. *J Exp Bot* **52**(357), 669–679.

Martin, B.A. and Wilson, R.F. (1984) Subcellular localization of triacylglycerol synthesis in spinach leaves. *Lipids* **19**(2), 117–121.

Marty, F. (1999) Plant vacuoles. *Plant Cell* **11**(4), 587–600.

Masclaux, C., Valadier, M.-H., Brugiere, N., Morot-Gaudry, J.-F. and Hirel, B. (2000) Characterization of the sink/source transition in tobacco (*Nicotiana tabacum* L.) shoots in relation to nitrogen management and leaf senescence. *Planta* **211**, 510–518.

Matile, P. (1992) Chloroplast senescence. In: *Crop Photosynthesis: Spatial and Temporal Determinants* (eds. Baker, N.R. and Thomas, H.). Elsevier Science B.V., Amsterdam, pp. 413–440.

Matile, P., Hoertensteiner, S., Thomas, H. and Kraeutler, B. (1996) Chlorophyll breakdown in senescent leaves. *Plant Physiology* **112**, 1403–1409.

Matos, A.R., d'Arcy-Lameta, A., Franca, M., *et al.* (2001) A novel patatin-like gene stimulated by drought stress encodes a galactolipid acyl hydrolase. *FEBS Lett* **491**, 188–192.

McCarty, R.E. and Jagendorf, A.T. (1965) Chloroplast damage due to enzymatic hydrolysis of endogenouse lipids. *Plant Physiol* **40**, 725–735.

McKegney, G., Yao, K., Ghosh, S., Huff, A., Mayak, S. and Thompson, J.E. (1995) The lipid composition of cytosolic particles isolated from senescing bean cotyledons. *Phytochemistry* **39**, 1335–1345.

McKersie, B.D. and Thompson, J.E. (1979) Phase properties of senescing plant membranes: role of neutral lipids. *Biochim Biophys Acta* **550**, 48–58.

Miquel, M., Cassagne, C. and Browse, J. (1998) A new class of *Arabidopsis* mutants with reduced hexadecatrienoic acid fatty acid levels. *Plant Physiol* **117**, 923–930.

Mizushima, N., Noda, T., Yoshimori, T., *et al.* (1998) A protein conjugation system essential for autophagy. *Nature* **395**, 395–398.

Mollenhauer, H.H. and Toten, C. (1970) Studies on seeds V. Microbodies, glyoxysomes, and ricinosomes of castor bean endosperm. *Plant Physiol* **46**, 794–799.

Murphy, D.J. (1993) Structure, function and biogenesis of storage lipid bodies and oleosins in plants. *Prog Lipid Res* **32**, 247–280.

Nordby, H.E. and Yelenosky, G. (1984) Effects of cold hardening on acyl lipids of citrus tissue. *Phytochemistry* **23**(1), 41–45.

Nordby, H.E. and Yelenosky, G. (1985) Change in citrus leaf lipids during freeze-thaw stress. *Phytochemistry* **24**(8), 1675–1679.

Okamoto, T., Katoh, S. and Murakami, S. (1977) Effects of linolenic acid on spinach chloroplast structure. *Plant Cell Physiol* **18**, 551–560.

O'Sullivan, J.N., Warwick, N.W.M. and Dalling, M.J. (1987) A galactolipase activity associated with the thylakoids of wheat leaves (*Triticum aestivum* L.). *J Plant Physiol* **131**, 393–404.

Ougham, H.J., Morris, P. and Thomas, H. (2005) The colors of autumn leaves as symptoms of cellular recycling and defenses against environmental stresses. *Curr Top Dev Biol* **66**, 135–160.

Page, T., Griffiths, G. and Buchanan-Wollaston, V. (2001) Molecular and biochemical characterization of postharvest senescence in broccoli. *Plant Physiol* **125**(2), 718–727.

Paliyath, G., Poovaiah, B.W., Munske, G.R. and Magnuson, J.A. (1984) Membrane fluidity in senescing apples. Effects of temperature and calcium. *Plant Cell Physiol* **25**, 1083–1087.

Paliyath, G. and Thompson, J.E. (1990) Evidence for early changes in membrane structure during postharvest development of cut carnation flowers. *New Phytol* **114**, 555–562.

Park, J.-H., Oh, S.A., Kim, Y.H., Woo, H.R. and Nam, H.G. (1998) Differential expression of senescence-associated mRNAs during leaf senescence induced by different senescence-inducing factors in *Arabidopsis*. *Plant Mol Biol* **37**, 445–454.

Pozueta-Romero, J., Rafia, F., Houlné, G., *et al.* (1997) A ubiquitous plant housekeeping gene, PAP, encodes a major protein component of bell pepper chromoplasts. *Plant Physiol* **115**, 1185–1194.

Rey, P., Gillet, B., Roemer, S., *et al.* (2000) Over-expression of a pepper plastid lipid-associated protein in tobacco leads to changes in plastid ultrastructure and plant development upon stress. *Plant J* **21**(5), 483–494.

Rosorius, O., Reichart, B., Kratzer, F., Heger, P., Dabauvalle, M.C. and Hauber, J. (1999) Nuclear pore localization and nucleocytoplasmic transport of eIF-5A: evidence for direct interaction with the export receptor CRM1. *J Cell Sci* **112**(Pt 14), 2369–2380.

Ryu, S.B. and Wang, X. (1995) Expression of phospholipase D during castor bean leaf senescence. *Plant Physiol* **108**(2), 713–719.

Sakaki, T., Kondo, N. and Yamada, M. (1990a) Pathway for the synthesis of triacylglycerols from monogalacotsyldiacylglycerols in ozone-fumigated spinach leaves. *Plant Physiol* **94**, 773–780.

Sakaki, T., Saito, K., Kawaguchi, A., Kondo, N. and Yamada, M. (1990b) Conversion of monogalactosyldiacylglycerols to triacylglycerols in ozone-fumigated spinach leaves. *Plant Physiol* **94**, 766–772.

Sastry, P.S. and Kates, M. (1964) Lipid components of leaves. V: Galactolipids, cerebrocides and lecithin of runner-bean leaves. *Biochemistry* **3**, 1280–1287.

Schneider, H., Hochli, M. and Hackenbrock, C.R. (1982) Relationship between the density distribution of intramembrane particles and electron transfer in the mitochondrial inner membrane as revealed by cholesterol incorporation. *J Cell Biol* **94**, 387–393.

Sharom, M., Willemot, C. and Thompson, J.E. (1994) Chilling injury induces lipid phase changes in membranes of tomato fruit. *Plant Physiol* **105**, 305–308.

Shinitzky, M. (1984) Membrane fluidity and cellular functions. In: *Physiology of Membrane Fluidity* (ed. Shinitzky, M.). CRC Press, Boca Raton, FL, pp. 1–51.

Shinitzky, M. and Inbar, M. (1976) Microviscosity parameters and protein mobility in biological membranes. *Biochim Biophys Acta* **433**(1), 133–149.

Shipton, C.A., Marder, J.B. and Barber, J. (1990) Determination of catabolism of the photosystem II D1 subunit by structural motifs in the polypeptide sequence. *Z Naturforsch* **45**(5), 388–394.

Siegenthaler, P.A. (1973) Change in pH dependence and sequestial inhibition of photosynthetic activity in chloroplasts by unsaturated fatty acids. *Biochim Biophys Acta* **305**, 153–162.

Smith, M.D., Licatalosi, D.D. and Thompson, J.E. (2000) Co-association of cytochrome *f* catabolites and plastid-lipid-associated protein with chloroplast lipid particles. *Plant Physiol* **124**, 211–221.

Sopory, S.K., Greenberg, B.M., Mehta, R.A., Edelman, M. and Mattoo, A.K. (1990) Free radical scavengers inhibit light-dependent degradation of the 32kDa photosystem II reaction center protein. *Z Naturforsch* **45c**, 412–417.

Steinmuller, D. and Tevini, M. (1985) Composition and function of plastoglobuli. II: Lipid composition of leaves and plastoglobuli during beech leaf senescence. *Planta* **163**, 91–96.

Thoenen, M. and Feller, U. (1998) Degradation of glutamine synthetase in intact chloroplasts isolated from pea (*Pisum sativum*) leaves. *Aust J Plant Physiol* **25**, 279–286.

Thomas, H. (1982) Control of chloroplast demolition during leaf senescence. In: *Plant Growth Substances*(ed. Waring, P.F.). Academic Press, London, pp. 559–567.

Thomas, H. and Donnison, I. (2000) Back from the brink: plant senescence and its reversibility. In: *Programmed Cell Death in Animals and Plants* (eds Bryant, J.A., Hughes, S.G. and Garland, J.M.). Oxford, pp. 149–162.

Thomas, H. and Hilditch, P. (1987) Metabolism of thylakoid membrane proteins during foliar senescence. In: *Plant Senescence: Its Biochemistry and Physiology* (eds Thompson, W.W., Nothnagel, E.A. and Huffaker, R.C.). The American Society of Plant Physiologists, Rockville, MD, pp. 114–122.

Thompson, J.E. (1988) The molecular basis for membrane deterioration during senescence. In: *Senescence and Aging in Plants* (eds Nooden, L.D. and Leopold, A.C.). Academic Press, San Diego, CA, pp. 51–83.

Thompson, J.E., Froese, C.D., Hong, Y., Hudak, K.A. and Smith, M.D. (1997) Membrane deterioration during senescence. *Can J Bot* **75**(6), 867–879.

Thompson, J.E., Froese, C.D., Madey, E., Smith, M.D. and Hong, Y. (1998) Lipid metabolism during plant senescence. *Prog Lipid Res* **37**(2/3), 119–141.

Thompson, J.E., Hopkins, M.T., Taylor, C. and Wang, T.-W. (2004) Regulation of senescence by eukaryotic translation initiation factor 5A: implications for plant growth and development. *Trends Plant Sci* **9**(4), 174–179.

Thompson, J.E., Legge, R.L. and Barber, R.F. (1987) The role of free radicals in senescence and wounding. *New Phytol* **105**, 317–344.

Tsuchiya, T., Ohta, H., Okawa, K., *et al.* (1999) Cloning of chlorophyllase, the key enzyme in chlorophyll degradation: finding of a lipase motif and the induction by methyl jasmonate. *Proc Natl Acad Sci USA* **96**, 15362–15367.

Vigil, E.L. (1969) Intracellular localization of catalase (peroxidatic) activity in plant microsomes. *J Histochem Cytochem* **17**, 425–427.

Vigil, E.L. (1970) Cytochemical and developmental changes in microbodies (glyoxysomes) and related organelles of castor bean endosperm. *J Cell Biol* **46**, 435–454.

Wang, T.W., Lu, L., Wang, D. and Thompson, J.E. (2001) Isolation and characterization of senescence-induced cDNAs encoding deoxyhypusine synthase and eucaryotic translation initiation factor 5A from tomato. *J Biol Chem* **276**(20), 17541–17549.

Wang, T.W., Lu, L., Zhang, C.G., Taylor, C.A. and Thompson, J.E. (2003) Pleiotropic effects of suppressing deoxyhypusine synthase expression in *Arabidopsis thaliana*. *Plant Mol Biol* **52**, 1223–1235.

Wang, T.W., Zhang, C.G., Wu, W., Nowack, L.M., Madey, E. and Thompson, J.E. (2005) Antisense suppression of deoxyhypusine synthase in tomato delays fruit softening and alters growth and development. *Plant Physiol* **138**(3), 1372–1382.

Wanner, G., Vigil, E.L. and Theimer, R.R. (1982) Ontogeny of microbodies (glyoxysomes) in cotyledons of dark-grown watermelons (*Citrullus vulgaris* Shrad.) seedlings. *Planta* **156,** 314–325.

Wanner, L., Keller, F. and Matile, P. (1991) Metabolism of radiolabeled galactolipids in senescent barley leaves. *Plant Sci* **78**, 199–206.

Warren, G.B., Houslay, M.D., Metcalfe, J.C. and Birdsall, N.J. (1975) Cholesterol is excluded from the phospholipid annulus surrounding an active calcium transport protein. *Nature* **255**(5511), 684–687.

Welti, R. and Glaser, M. (1994) Lipid domains in model and biological membranes. *Chem Phys Lipid* **73**, 121–137.

White, M.J. and Green, B. R. (1987) Polypeptides belonging to each of the three major chlorophyll $a + b$ protein complexes are present in a chlorophyll-*b*-less barley mutant. *Eur J Biochem* **165**, 531–535.

Wittenbach, V.A., Lin, W. and Habert, R.R. (1982) Vacuolar localization of proteases and degradation of chloroplast mesophyll protoplasts from senescing primary wheat leaves. *Plant Physiol* **69**, 98–102.

Woo, H.R., Chung, K.M., Park, J.-H., *et al.* (2001) ORE9, an F-box protein that regulates leaf senescence in *Arabidopsis*. *Plant Cell* **13**, 1779–1790.

Xiong, Y., Contento, A.L. and Bassham, D.C. (2005) AtATG18a is required for the formation of autophagosomes during nutrient stress and senescence in *Arabidopsis thaliana*. *Plant J* **42**(4), 535–546.

Xu, A. and Chen, K.Y. (2001) Hypusine is required for a sequence-specific interaction of eukaryotic initiation factor 5A with postsystematic evolution of ligands by exponential enrichment RNA. *J Biol Chem* **276**(4), 2555–2561.

Xu, P., Rogers, S.J. and Roossinck, M.J. (2004) Expression of antiapoptotic genes bcl-xL and ced-9 in tomato enhances tolerance to viral-induced necrosis and abiotic stress. *Proc Natl Acad Sci U S A* **101**(44), 15805–15810.

Yamane, K., Abiru, S., Fujishige, N., Sakiyama, R. and Ogata, R. (1993) Export of soluble sugars and increase in membrane permeability of gladiolus florets during senescence. *J Jpn Soc Hort Sci* **62**(3), 575–580.

Yao, K., Paliyath, G., Humphrey, R.W., Hallett, F.R. and Thompson, J.E. (1991) Identification and characterization of nonsedimentable lipid–protein microvesicles. *Proc Natl Acad Sci U S A* **88**, 2269–2273.

Yao, K. and Thompson, J.E. (1993) Enrichment of proteinase activity in deteriosomes, a new class of microvesicles. *FEBS Lett* **323**(1–2), 99–103.

4 Oxidative stress and leaf senescence

Ulrike Zentgraf

4.1 Introduction

Agricultural crop losses in the field as well as during transportation from the producer to the consumer are mainly caused by natural or harvest-induced senescence. Many different agriculturally important traits like number and quality of seeds, timing of seed set, fruit ripening, etc., are affected by senescence. Enhanced growth of fungi and other microorganisms on senescing tissue can also have an important impact on food quality and might be avoided by delayed senescence and increased longevity. In addition, senescence can be triggered by climatic extremes; abiotic stress is estimated to be the primary cause of crop loss worldwide, with the potential to cause a reduction of more than 50% in the average yield of the main crops. Vegetables harvested before full adolescence are also exposed to enormous stress by the sudden interruption of the energy and nutrient supply. Products like asparagus and broccoli show very fast postharvest senescence during storage and have a very short shelf life. Many changes observed during the storage of green vegetables, like the loss of chlorophyll, damage to cellular structures and finally cell death, exhibit similarities with the changes during developmental-dependent senescence. It could be shown that genes induced during leaf senescence are also expressed in stored broccoli (Buchanan-Wollaston *et al.*, 2003). Therefore, while considering global climate changes also, development of crop plants that cope better with changing environmental conditions will be one of our future challenges. Plants with altered senescence programs might be helpful tools to reach this goal.

Despite the importance of the senescence processes, our knowledge on the regulatory mechanisms of senescence is still poor. However, senescence is not a chaotic breakdown but an orderly loss of normal cell functions, which is under the control of the nucleus. A massive change in gene expression can be observed during leaf senescence and one can estimate from different expression analyses that approximately 12–16% of the *Arabidopsis* genes change their expression during leaf senescence (Buchanan-Wollaston *et al.*, 2003; Guo *et al.*, 2004; Zentgraf *et al.*, 2004). As senescence primarily serves the mobilization of nutrients and minerals out of the senescing tissue into the developing parts of the plant, genes involved in degradation and mobilization of macromolecules are switched on, whereas genes related to photosynthesis are turned off. There is an ongoing debate whether senescence is a form of programmed cell death or a developmental process with different features, and some evidence has been found for both views (Thomas *et al.*, 2003; van Dorn and Woltering, 2004). Temperature-induced leaf senescence was dramatically delayed in bcl-xL and ced-9 transgenic plants, and high levels of anthocyanins accumulated, possibly limiting oxidative stress. Hence, expression of these animal antiapoptotic genes was able to improve plant survival and delay senescence indicating common regulatory mechanisms between animal apoptosis and plant senescence (Xu *et al.*, 2004). However, the capacity of *Nicotiana* leaves to re-green indicates that cell death is not an inevitable consequence of senescence (Zavaleta-Mancera *et al.*, 1999a,b). Although leaf senescence normally results in cell death, senescence and cell death can be uncoupled: in nutrient-deficient, but not in fertilized, tobacco plants with delayed senescence (P_{SAG12}-IPT), necrotic lesions were detected in old, but otherwise green, leaves. By contrast, wild-type leaves of the same age were yellow, but not necrotic (Wingler *et al.*, 2005).

Senescence can be initiated by exogenous and endogenous triggers. Pathogen attack as well as many abiotic stress conditions like heat, cold, drought, high light, etc., are well known to activate senescence processes and to be coupled with an oxidative burst. The most important endogenous factors inducing senescence are the ages of the leaves and the age and developmental stage of the plant. The leaves of annual plants show a continuous decrease in their photosynthesis rate after full expansion (Batt and Woolhause, 1975; Hensel *et al.*, 1993). The model plant *Arabidopsis thaliana* is an example of a plant with extremely fast aging leaves. Under continuous light conditions its photosynthetic capacity decreases by 50% within 4–6 days of full leaf expansion (Hensel *et al.*, 1993). It is assumed that a decline in photosynthetic activity under a certain threshold may act as a senescence-inducing signal (Smart, 1994; Matile *et al.*, 1996). Although this is still an open question, there is some evidence to support this theory. It is known that elevated sugar content represses photosynthesis-related genes (Rolland *et al.*, 2002). In wheat, removal of reproductive 'sink' after anthesis delays the rate of flag leaf senescence. Wheat plants lacking a reproductive sink showed decreased oxidative stress and lower lipid peroxidation and maintained higher protein, oxidatively damaged proteins and nitrogen levels as compared to plants with reproductive sink during monocarpic senescence. Thus, the influence of the reproductive sink was due to its ability to drive forward the nitrogen mobilization process through high reactive oxygen species (ROS) levels, which mediated damage to the proteins and influenced proteolytic activities (Srivalli and Khanna-Chopra, 2004). There are several lines of evidence that ROS trigger leaf senescence (Ye *et al.*, 2000; Navabpour *et al.*, 2003; Barth *et al.*, 2004; Miao *et al.*, 2004; Zimmermann *et al.*, 2006).

Obviously, oxidative stress resistance and potential life span seem to be correlated in many organisms, ranging from *Caenorhabditis* to mammals (Harman 1956; 1998; Orr and Sohal, 1994; Martin *et al.*, 1996; Sohal and Weindruch, 1996). The loss of the antioxidative capacity during progression of senescence has also been reported for different plants (Chia *et al.*, 1981; Dhindsa *et al.*, 1981; McRae and Thompson, 1983; Pauls and Thompson, 1984; Pastori and del Rio, 1994; Jiménez *et al.*, 1998; Panavas and Rubinstein; 1998; Orendi *et al.*, 2001), implying that this may be a more general phenomenon for many aerobic organisms. Investigation of different late-flowering mutants of *A. thaliana* revealed that flowering time and longevity in *Arabidopsis* are also tightly correlated with the resistance to oxidative stress (Kurepa *et al.*, 1998). However, the relationship between life span and oxidative stress tolerance in plants is poorly understood.

4.2 Antioxidative capacity, oxidative stress and life span

Aerobic organisms use molecular oxygen as a terminal oxidant during respiration because it is relatively harmless and not very reactive. However, it has the potential to be reduced incompletely to toxic intermediates, like singlet oxygen ($^1O_2$), the superoxide radical ($O_2^{\bullet -}$), the hydroperoxyl radical ($HO_2^{\bullet -}$), hydrogen peroxide (H_2O_2) and the hydroxyl radical ($^{\bullet}OH$). These molecules are called ROS. All ROS are extremely reactive and are able to oxidize biological molecules, such as DNA, proteins or lipids (reviewed in Dat *et al.*, 2000). The most reactive ROS is the hydroxyl radical, which affects all kinds of macromolecules leading to massive impairments of all cellular components, often resulting in irreparable metabolic misfunction and cell death (Knox and Dodge, 1985). Specific amino acids, like histidine, methionine and tryptophane, can be oxidized by superoxide radicals. Hydrogen peroxide oxidizes SH groups and differs from the other ROS through its relatively long half-life and its diffusibility. In high concentrations, it may trigger programmed cell death.

In contrast to animal systems, chloroplasts are the main source of ROS in plants. During photosynthesis, light energy is absorbed by a series of redox reactions and transferred to the reaction centers of the photosystems. Thereby, the electrons are transmitted to CO_2. However, in most plant species, the rate of CO_2 fixation is not high enough to convert more than 50% of the light energy (Baker, 1991); therefore, alternative electron acceptors like molecular oxygen are used, leading to the formation of superoxide radicals ($O_2^{\bullet -}$). In addition, the chloroplasts can form significant amounts of singlet oxygen ($^1O_2$). Normally, the excited singlet status of the chlorophyll serves the transfer of energy or electrons. To emit energy, chlorophyll uses either fluorescence or conversion to the triplet status, which can, in combination with oxygen, lead to the formation of singlet oxygen (Arora *et al.*, 2002). Another source for ROS formation, especially for H_2O_2, is the photorespiration in the peroxisomes. During CO_2 fixation, ribulose-1,5-bisphosphate carboxylase (Rubisco) uses CO_2 to carboxylate ribulose-1,5-bisphosphate. This enzyme can also use molecular oxygen to oxygenate ribulose-1,5-bisphosphate (Foyer, 1996). During this reaction, glycolate is formed and transported from the chloroplasts into the peroxisomes. The glycolate is then oxidized, and H_2O_2 is formed as a by-product. Mitochondria are also an important source of reactive oxygen. The mitochondrial electron transport chain consists of several dehydrogenase complexes that reduce a common pool of ubiquinone (Millenaar and Lambers, 2003). Cytochrome *c* oxidase or an alternative oxidase (AOX) serves as terminal electron acceptor. Here, the superoxide radical is mainly produced by ubiquinone and the NADH dehydrogenases, namely by autooxidation of the reduced components of the respiration chain (Richter and Schweizer, 1997).

Oxidative stress arises from an imbalance between generation and elimination of ROS, often leading to cell death. Oxidative stress occurs when this critical balance is disrupted because of depletion of antioxidants or excess accumulation of ROS. Regardless of how or where they are generated, an increase in intracellular oxidants results in two very important effects: damage to various cell components and activation of specific signaling pathways, both of which influence numerous cellular processes. However, oxidative damage in plant tissues is especially important during senescence and is characterized by a notable increase in the metabolism of activated oxygen species (Kar and Feierabend, 1984; Thompson *et al.*, 1987; Halliwell and Gutteridge, 1989; see also Chapter 6 in this volume). In addition, the loss of the antioxidative capacity during senescence has been reported for many different organisms (Harman, 1956, 1998; Chia *et al.*, 1981; Dhindsa *et al.*, 1981; McRae and Thompson, 1983; Pauls and Thompson, 1984; Orr and Sohal, 1994; Pastori and del Rio, 1994; Martin *et al.*, 1996; Sohal and Weindruch, 1996; Jimenez *et al.*, 1998; Panavas and Rubinstein, 1998; Orendi *et al.*, 2001). Analyses of different late-flowering/extended-longevity mutants suggested that control of longevity and oxidative stress tolerance are tightly linked in *Arabidopsis* (Kurepa *et al.*, 1998). Furthermore, the detached leaves of the delayed leaf senescence mutants of *Arabidopsis*, *ore1, ore3* and *ore9*, exhibit increased tolerance to various types of oxidative stress. The *ore1, ore3* and *ore9* mutants were also more tolerant to oxidative stress at the level of the whole plant, as determined by measuring physiological and molecular changes associated with oxidative stress. However, the activities of antioxidant enzymes were similar or lower in the mutants, as compared to wild type. These results suggest that the increased resistance to oxidative stress in the *ore1, ore3* and *ore9* mutants is not due to enhanced activities of these antioxidant enzymes, and provide genetic evidence that oxidative stress tolerance is linked to control of leaf longevity in plants (Woo *et al.*, 2004).

4.3 Antioxidants

Hydrogen peroxide is most likely the most important ROS. In contrast to other ROS it has a relatively long half-life. It can be produced in all cell compartments but is also a diffusible molecule, which can pass membranes also. However, the endogenous H_2O_2 contents of plant cells can be much higher than those found in animals and bacteria; plant cells happily survive H_2O_2 levels that would kill animal cells. This tolerance is linked to the presence of an extensive antioxidant system, in which, besides the enzymatic H_2O_2 scavenging systems catalase and ascorbate peroxidase, ascorbic acid and glutathione fulfill crucial roles (Noctor and Foyer, 1998). Ascorbate peroxidase (APX; EC 1.11.1.11) is the most important enzyme scavenging H_2O_2 produced in the chloroplast, and uses ascorbate to reduce H_2O_2 to water, whereby dehydroascorbate (DHA) is produced. To maintain a certain level of ascorbate, it has to be regenerated by dehydroascorbate reductase (DHAR; EC 1.8.5.1), which reduces DHA to ascorbate by oxidizing reduced glutathione (GSH). Glutathione reductase (GR; EC 1.6.4.2) regenerates GSH via the reduction of oxidized glutathione (GSSG), using $NADPH^+$ (Bowler *et al.*, 1992). The ascorbate–glutathione cycle and the respective enzyme driving these reactions are also present in peroxisomes and mitochondria, and the participation of this cycle in the control of H_2O_2 concentration in both the cell organelles has been proposed (Jiménez *et al.*, 1997, 1998). During senescence, strong oxidative damage takes place, and in this situation the mitochondrial and peroxisomal ascorbate–glutathione cycle could help to inhibit any enhancement of activated oxygen species production.

However, the ascorbate (ASC) pool could be reduced by oxidative stress when the capacity of regenerative systems is exceeded (Foyer *et al.*, 1994; Smirnoff and Pallanca, 1996). In addition, the capacity of leaves to produce ascorbate declines with leaf age (Foyer, 2004). The content of GSH also significantly decreases in pea leaf mitochondria during senescence, which is probably due to the decrease in GR activity that took place under the same conditions (Jiménez *et al.*, 1998). Decreases in total glutathione have also been linked to nodule senescence (Dalton *et al.*, 1993; Evans *et al.*, 1999; Puppo *et al.*, 2005).

The redox coupling between the ascorbate and glutathione pools linked to ascorbate peroxidation is a necessary part of an H_2O_2 detoxification pathway. However, whereas glutathione will always reduce DHA, the degree of coupling between the ascorbate and glutathione redox couples varies greatly between different cellular compartments. The flexibility of coupling between these antioxidant pools is crucial to differential signaling by ascorbate and glutathione (Noctor *et al.*, 2002; Foyer *et al.*, 2005). Recently, evidence has been provided for the presence of a glutathione-independent pathway of ascorbate regeneration from DHA in tobacco BY-2 cell cultures (Potters *et al.*, 2004), and similar pathways may occur in other cell types. Like ascorbate, glutathione is a multifunctional compound with functions that extend beyond the antioxidative system (May *et al.*, 1998; Noctor and Foyer, 1998). Ascorbate and glutathione are also required for the operation of the cell cycle (Potters *et al.*, 2004).

Exposure to stress can result in changes in antioxidants levels, particularly in glutathione and ascorbate. In many plant organs, altered levels of these compounds and the ratio of their reduced to oxidized forms act as a signal to trigger specific cellular responses (Noctor and Foyer, 1998; Pastori *et al.*, 2003). Metabolic modeling was used as a new tool to analyze the network of redox reactions composing the superoxide dismutase–ascorbate–glutathione cycle. These kinds of modeling approaches cannot yet make exact prediction but can contribute to the theoretical understanding of the functioning of antioxidant systems by pointing out questions that need to be validated (Polle, 2001).

In the peroxisomes, catalases are responsible for detoxification of higher concentrations of H_2O_2. Whereas APX has a high affinity for H_2O_2 and is able to detoxify low concentrations of H_2O_2, catalase (CAT) has a high reaction rate, but a low affinity for H_2O_2. However, besides its role in the elimination of peroxisomal H_2O_2, catalase action appears to be critical for maintaining the redox balance during oxidative stress. It is also indispensable for stress defense in some C3 plants (Willekens *et al.*, 1997).

In addition, plant peroxidases have more functions than a Swiss army knife has and are involved in many physiological processes during plant life cycle. Because of two possible catalytic cycles, peroxidative and hydroxylic, peroxidases can generate ROS, polymerize cell wall compounds and regulate H_2O_2 levels. Their activity and expression is modulated by internal and external stimuli and is probably regulated by a fine-tuning that has yet to be elucidated and that meets the demands of plants during stress conditions and senescence (Passardi *et al.*, 2005). Higher plant metallothioneins are suspected of reducing metal-induced oxidative stress and binding copper and zinc cofactor metals for delivery to important apometalloproteins (Thomas *et al.*, 2005). Furthermore, additional nonenzymatic low-molecular-weight antioxidants are involved in redox balancing. Nitric oxide (NO) is also a bioactive molecule which is able to scavenge ROS, and can strongly counteract many ROS-mediated cytotoxic processes in plants (Beligni and Lamattina, 1999). Selenium (Se) is regarded as an antioxidant in animals and plants, even though considered as a nonessential element in plants, and might play a role as an antioxidative protectant in soybean during senescence (Djanaguiraman *et al.*, 2005). Tocopherols are also well-known low-molecular-weight antioxidants. The content of α-tocopherol as well as γ-tocopherol increased significantly in leaves of aging *A. thaliana* plants (Hollander-Czytko *et al.*, 2005). Taken together, the ROS levels in different cell compartments appear to be tightly regulated by various antioxidative systems during senescence. Tobacco leaves of plants with enhanced glutathione reductase activity or with autoregulated senescence-induced production of cytokinins show that the capacity of the antioxidative system to scavenge radicals is sufficiently balanced with the plant metabolism, and its decline with increasing age is not the cause, but a consequence, of senescence and aging in plants (Dertinger *et al.*, 2003). However, there is increasing evidence that ROS are also involved in the regulation of senescence as signaling molecules.

4.4 ROS signaling

ROS were originally considered to be exclusively detrimental to cells, but it is now recognized that redox regulation involving ROS plays a key role in the modulation of critical cellular functions. In addition to induction or repression of antioxidant defense genes, ROS are known to similarly affect expression of a variety of other genes involved in different signaling pathways in microbes, yeasts, plants and animals. It has become clear that there are far more genes and gene clusters responding to ROS than was previously thought, and that ROS likely participate far more in cellular activities than anticipated. In yeast, transcriptome analyses revealed that the response to oxidative stress involves about one-third of the genome (Gasch *et al.*, 2000). The expression programs following H_2O_2 or $O_2^{\bullet-}$ treatment were essentially identical, despite the fact that different ROS are involved. There was a strong induction of genes known to be involved in detoxification of H_2O_2 and $O_2^{\bullet-}$, such as catalase, superoxide dismutase and glutathione peroxidase, as well as genes involved in oxidative and reductive reactions (e.g. thioredoxin, glutathione reductase and glutaredoxin). Recently, it has also been demonstrated that H_2O_2 activates the Sty1 (stress-activated mitogen-activated protein (MAP) kinase) pathway in *Schizosaccharomyces pombe* in a dose-dependent manner via two sensing mechanisms (Quinn *et al.*, 2002). However, cellular ROS-sensing mechanisms are not well understood, but a number of transcription factors that regulate the expression of antioxidant genes are well characterized in prokaryotes and yeast, e.g. the H_2O_2-response regulator OxyR and the superoxide response transcription factor SoxRS of *E. coli* or ACE1, MAC1, YAP1, YAP2, HAP1 and HAP2/3/4 of yeast (Ruis and Schuller, 1995; Scandalios, 2002). In higher eukaryotes, oxidative stress responses are more complex and modulated by many regulators. Little is known about how the H_2O_2 signal is perceived and transduced in plant cells. In alfalfa, an oxidative-stress-activated MAP triple kinase 1 (OMTK1) has been identified, which is exclusively activated by hydrogen peroxide and directly interacts with a MAPK (MMK3) and induces cell death (Nakagami *et al.*, 2004). H_2O_2 can also activate a specific *Arabidopsis* mitogen-activated protein kinase kinase kinase, ANP1, which initiates a phosphorylation cascade involving two stress MAPKs, AtMPK3 and AtMPK6 (Kovtun *et al.*, 2000). In addition, the kinase activity of AtMAPK3 and AtMAPK6 can also be induced by chitin, and several WRKY transcription factors, namely *WRKY22*, *WRKY29*, *WRKY33* and *WRKY53*, were also up-regulated by these treatments. Expression of the tobacco MAPKK *NtMEK2* active mutant in transgenic *Arabidopsis* plants also induced *WRKY33* and *WRKY53* expression, suggesting a potential role for these WRKY transcription factors in transducing the signal from the MAPK cascade to downstream genes (Wan *et al.*, 2004). Besides pathogen response, wounding and trichome development, the family of WRKY transcription factors is involved in senescence regulation (Eulgem *et al.*, 2000; Hinderhofer and Zentgraf, 2001; Robatzek and Somssich, 2002; Miao *et al.*, 2004). Microarray analyses revealed that NAC and WRKY factors constitute the two largest groups of transcription factors of the senescence transcriptome (Guo, *et al.*, 2004) and *WRKY53* might be a key regulator in leaf senescence of *Arabidopsis* (Miao *et al.*, 2004). However, the expression of *WRKY53* can be induced by hydrogen peroxide, and hydrogen peroxide measurements in *Arabidopsis* revealed that the hydrogen peroxide content in leaves increased during development of plants exactly at the time point when plants start to bolt and *WRKY53* expression is highly induced and switched from being leaf age dependent to plant age dependent (Hinderhofer and Zentgraf, 2001; Miao *et al.*, 2004). This hydrogen peroxide peak appears to be caused by a fine-tuned regulation of the hydrogen-peroxide-scavenging enzymes, catalase and ascorbate peroxidase. CAT2 activity decreased at a very early stage during the time of bolting, whereas CAT3 activity increased with plant age. Since CAT3 is the isoform that reacts on the oxidative stress on activity level (Orendi *et al.*, 2001), it is most likely that CAT3 activity was induced by the increased hydrogen peroxide level, which could be determined after the decrease of CAT2 activity. The increase of the hydrogen peroxide level could be enforced by a decrease of APX1 activity at the same time point (Ye *et al.*, 2000; Zimmermann *et al.*, 2006). However, APX1 was not down-regulated on the transcriptional level during the time of bolting (Panchuk *et al.*, 2005) but hydrogen peroxide can lead to the inactivation of APX (Miyake and Asada, 1996; Orendi *et al.*, 2001). This inactivation of APX is dependent on the plant developmental stage, and plants are most sensitive during the time of bolting (Zimmermann *et al.*, 2006). This suggests a feedback amplification loop: CAT2 down-regulation appears to be the initial step to produce an elevated level of hydrogen peroxide, which then might lead to the inactivation of APX activity, which in turn increases the hydrogen peroxide level. This increased hydrogen peroxide level then leads to the induction of *CAT3* expression and activity, which then lowers the hydrogen peroxide level again and leads to a restoration of APX1 activity. This coordinated regulation of the hydrogen-peroxide-scavenging enzymes on the transcriptional and posttranscriptional level creates a distinct increase of the hydrogen peroxide right at that time point when the plants start to bolt and a coordinated senescence process of all rosette leaves should be induced (Zimmermann *et al.*, 2006). In addition, timing of senescence is also altered in the ascorbate-deficient *Arabidopsis* mutant *vtc1* (Barth *et al.*, 2004). This strongly indicates that ROS might be involved not only in degradation processes during leaf senescence but also in regulatory processes of senescence. Furthermore, the expression of many other senescence-enhanced genes was also found to be induced by the increase in ROS (Miller *et al.*, 1999; Navabpour *et al.*, 2003). Treatment of *Arabidopsis* cotyledon and leaf tissues with the catalase inhibitor, 3-amino-1,2,4-triazole, or with silver nitrate results in the enhanced expression of senescence-induced genes, e.g. a metallothionein (LSC54). Combined treatments with quenchers of ROS, such as ascorbate, tiron and benzoic acid, revealed that increased levels of ROS are responsible for the enhanced expression. However, many genes that showed induced expression during natural senescence were not expressed by these treatments, e.g. expression of the senescence-specific cysteine protease *SAG12* was not induced. Overall, about half of the senescence-enhanced genes that were tested showed induced expression in tissue treated with silver nitrate, indicating that ROS signaling is not sufficient to induce full senescence-associated gene expression (Navabpour *et al.*, 2003). Whether hydrogen-peroxide-induced expression of senescence-associated gene is transduced by MAPK signaling or directly by redox-sensitive transcription factor still has to be elucidated. However, the redox-sensitive zinc-finger DNA-binding domain of the WRKY proteins in which two cysteines together with two histidines interact electrostatically with a zinc atom to form a 'zinc finger' makes it an excellent candidate for direct redox regulation (Arrigo, 1999). There is some evidence that pathogen-induced senescence is also regulated by ROS signaling. Fungal infection of tomato leaves triggers significant changes in the peroxisomal antioxidant system leading to a collapse of the protective mechanism at advanced stages of infection. These changes appear to be partly the effect of pathogen-promoted leaf senescence (Kuzniak and Sklodowska, 2005). ROS, ascorbate and glutathione decline in a regulated manner during nodule development and senescence. This does not necessarily cause oxidative stress but all components might be involved in signaling processes or a development-related shift in redox-linked metabolite cross talk during nodule senescence (Groten *et al.*, 2005). In barley, expression of the *hpd* gene during senescence is most likely related to oxidative stress (Falk *et al.*, 2002).

Expression profiling by microarray analyses revealed that large changes in the transcriptome are induced by senescence and oxidative stress and that there is a remarkable overlap. A large-scale cDNA microarray analysis of the *Arabidopsis* transcriptome during oxidative stress identified 175 nonredundant expressed sequence tags from a sample of 11 000 that are regulated by H_2O_2. Of these, 62 are repressed and 113 are induced; and RNA blots showed that some of the H_2O_2-regulated genes are also modulated by other signals known to involve oxidative stress. Of the 175 genes identified as H_2O_2 responsive, most have no obvious direct role in oxidative stress but may be linked to stress or developmental signaling functions explaining their sensitivity to H_2O_2 (Desikan *et al.*, 2001). Recently, an analysis of changes in global gene-expression patterns during developmental leaf senescence using full genome chips in *Arabidopsis* has identified more than 800 genes that show a reproducible increase in transcript abundance. Gene expression analysis in senescing leaves of plant lines defective in signaling pathways involving salicylic acid, jasmonic acid and ethylene has shown that these three pathways are all required for expression of many genes during developmental senescence (Buchanan-Wollaston *et al.*, 2005). Despite some limitations, it has become clear through the use of microarrays that there are far more genes responding to ROS and/or senescence than was previously thought (Desikan *et al.*, 2001; Scandalios, 2002; Buchanan-Wollaston, 2005). Leaf senescence can also contribute to plant survival under drought stress conditions (Munne-Bosch and Alegre, 2004). For the future, the fundamental challenge will be to integrate the information now being obtained on gene-expression patterns with structural and functional parameters and interactions of the various proteins encoded by ROS- and senescence-responsive regulons.

4.5 Role of different cell compartments

4.5.1 Peroxisomes

Peroxisomes and ROS generated in these organelles were shown to play a central role in natural and dark-induced senescence in pea. Different experimental evidences have suggested that peroxisomes have a ROS-mediated cellular function in leaf senescence and in stress situations induced by xenobiotics and heavy metals. Peroxisomes could also have a role in plant cells as a source of signal molecules like NO, $O_2^{\bullet-}$, H_2O_2 and possibly S-nitrosoglutathione (GSNO) (Pastori and del Rio, 1997; del Rio *et al.*, 1998, 2003a). Whereas the superoxide- and hydrogen-peroxide-generating enzymes like xanthine oxidase, urate oxidase or Mn-SOD and the NADPH-dependent generation of superoxide on the membranes of these organelles increased during leaf senescence (Pastori and del Rio 1997; del Rio *et al.*, 1998), catalase activity almost completely decreased (Pastori and del Rio, 1994, 1997). The enzymes of the ascorbate–glutathione cycle in the peroxisomes were also notably affected by progression of senescence, and the reduced and oxidized glutathione pools were considerably increased in peroxisomes (Jiménez *et al.*, 1998). Therefore, hydrogen peroxide levels and lipid peroxidation rate significantly increased in these organelles during senescence. Moreover, it is very likely that the peroxisomal NADH-dependent production of $O_2^{\bullet-}$ radicals is intensified by the reverse transition of leaf peroxisomes to glyoxysomes which occurs during senescence (Landolt and Matile, 1990; Pistelli *et al.*, 1996; Pastori and del Río, 1997), since more NADH would be available as a result of the induction of fatty acid beta-oxidation and the glyoxylate cycle (Jiménez *et al.*, 1998). Recently, the generation of NO by pea peroxisomes was reported, but, in contrast to ROS, NO production is clearly down-regulated during leaf senescence. Corpas and coworkers could show by confocal laser scanning microscopy analyses with 4,5,-diaminofluorescein diacetate in pea leaf sections that endogenous NO was predominantly detected in the vascular tissues and suggested that it could be involved in long-distance communication (Corpas *et al.*, 2004). Taken together, these data suggest that peroxisomes could act as subcellular sensors of plant stress and senescence by releasing nitric oxide, superoxide and hydrogen peroxide as signaling molecules to the cytosol and thereby triggering a specific gene expression (Corpas *et al.*, 2001; del Rio *et al.*, 2002; Corpas *et al.*, 2004). The differential response to senescence of the mitochondrial and peroxisomal ascorbate–glutathione cycle suggests that mitochondria could be affected by oxidative damage earlier than peroxisomes, which may participate in the cellular oxidative mechanism of leaf senescence longer than mitochondria (Jiménez *et al.*, 1998). Moreover, mitochondrial and peroxisomal Mn-SOD expression is regulated differently. The expression of mitochondrial Mn-SOD is induced during the senescence of pea leaves, whereas peroxisomal Mn-SOD could be posttranslationally activated. Previously described results showing decreased mitochondrial Mn-SOD activity and increased peroxisomal Mn-SOD activity may be reflective of posttranslational events regulating enzymatic activity during leaf senescence (del Rio *et al.*, 2003b). In addition, diverse stresses that generate H_2O_2 as a signaling molecule result in peroxisome proliferation via the up-regulation of components (*PEX* genes) required for biogenesis of the organelle and import of proteins (Lopez-Huertas *et al.*, 2000). Whether the number of peroxisomes also increases during leaf senescence still has to be elucidated.

4.5.2 Chloroplasts

The first visible sign of senescence is the onset of chloroplast degradation, which coincides with a decrease in the amount of chlorophyll, the degradation products of which are transported into the vacuole (Thomas and Stoddart, 1980; Matile *et al.*, 1996; Gan and Amasino, 1997). The loss of chloroplast integrity can be observed in the very early stages of senescence. Electron microscopy revealed that the chloroplasts of senescing leaves show an increased number of enlarged plastoglobuli, a disorientation of the grana stacks and a swelling of the thylakoids. It is assumed that the formation of plastoglobuli is associated with the degradation of the thylakoids (Smart, 1994). The term 'gerontoplast' was established to describe the organelle of a senescing, formerly green tissue (Parthier, 1988). The conversion of chloroplasts to gerontoplasts in leaves is reversible in some, possibly all, higher plants (Zavaleta-Mancera *et al.*, 1999a,b; Thomas *et al.*, 2003). From a physiological point of view, the activity of the membrane-associated electron transport of photosystems I and II decreases continuously during senescence (Thomas and Stoddart, 1980; Smart, 1994), while composition and fluidity of the thylakoid membrane is not changed. However, the chloroplasts may play a regulatory role during leaf senescence, similar to that of the mitochondria during animal programmed cell death (PCD). In animal PCD, mitochondria integrate signals of proapoptotic and antiapoptotic proteins regulating the release of cytochrome *c* and the production of ROS that direct subsequent apoptotic processes (Green and Reed, 1998; Jones, 2000; Ferri and Kroemer, 2001; Dufur and Larsson, 2004). Here, complex I is the main site for the production of $O_2^{\bullet -}$; complexes II and III are involved to a lesser extent (Dufur and Larsson, 2004). In chloroplasts, the Ndh complex regulates the redox level of cyclic electron transporters by providing electrons that are removed by the Mehler reaction and the coordinated action of SOD and peroxidase when transporters become over-reduced. Transgenic tobacco with a knockout of the plastid *ndhF* gene (ΔndhF) shows low levels of the plastid Ndh complex and more than a 30-day delay in leaf senescence with respect to wild-type tobacco plants (Zapata *et al.*, 2005). The level and activity of the Ndh complex increase during leaf senescence (Zapata *et al.*, 2005), which strikingly parallels the increased transcription of the mitochondrial complex I during human aging (Pich *et al.*, 2004). Cytochrome *c* release from mitochondria and the decrease of Calvin cycle activity in chloroplasts both lead to an increased generation of ROS in the respective organelle. In addition, the decrease of SOD activity (Casano *et al.*, 1994; Orr and Sohal, 1994; Jimenez *et al.*, 1998) in both organelles would amplify the levels of ROS most likely triggering further PCD or senescence processes. Chloroplastic control of leaf senescence provides an unexpected role of the plastid *ndh* genes that are present in most higher plants. The regulation of leaf senescence by chloroplasts opens up the question whether plastids are targets for factors similar to proapoptotic and antiapoptotic proteins affecting mitochondria in animal PCD.

4.5.3 Mitochondria

During plant senescence, the function of the mitochondria is maintained during the gradual breakdown of the cell up to a late time point. This is essential for the cell to gain energy via ATP synthesis during respiration (Thomas and Stoddart, 1980). It is therefore unlikely that plant mitochondria trigger plant senescence in the same way they do in animal PCD. Electron microscopy studies show that senescence in pea leaves induced deterioration in the mitochondrial membrane structure and a slight disorganization in the matrix and cristae (Pastori and del Río, 1994). A decrease in mitochondrial membrane integrity could allow the leakage of H_2O_2 from the mitochondria into the cytosol during senescence. This extrusion of H_2O_2 could be favored by the decrease of APX and monodehydroascorbate reductase activities in mitochondrial membranes (Jiménez *et al.*, 1998). In parallel, the alternative respiration pathway is activated during senescence (Hiser and McIntosh, 1990; Svensson and Rasmusson, 2001; Maxwell *et al.*, 2002). An important function of the AOX is to prevent the formation of excess free oxygen radicals. AOX ensures a low reduction status of the ubiquinone pool by oxidizing ubiquinol. Thus, the electron flow is guaranteed (Millenaar and Lambers, 2003). In the fungus *Podospora anserina*, the inactivation of subunit V of the cytochrome *c* oxidase complex leads to the exclusive use of the alternative respiration pathway and a decline in ROS formation in these mutants. This inactivation of the cytochrome *c* oxidase results in an extraordinary longevity of this fungus (Dufour *et al.*, 2000). Overexpression or inactivation of AOX in the wild-type background of this fungus does not decrease ROS production and has no effect on longevity, mitochondrial stability or aging. In contrast, overexpression of AOX in a long-lived mutant deficient in cytochrome *c* oxidase considerably increased ROS production of the mutant and was able to restore senescence and mitochondrial DNA instability (Lorin *et al.*, 2001). The AOX could be identified in plants, protists, fungi and green algae (McIntosh, 1994). It acts as a chinoloxidase by transferring electrons from the reduced ubiquinone directly to molecular oxygen forming water (Siedow and Moore, 1993). The plant AOXs form homodimers (Moore *et al.*, 2002) and are encoded by small gene families. In *A. thaliana*, four genes are known, the *AOX1a*, *AOX1b*, *AOX1c* and *AOX2*, each exhibiting organ-specific expression (Saisho *et al.*, 1997).

In *Podospora anserina*, copper depletion also leads to the induction of an alternative respiratory pathway that appears to be induced by specific impairments of the copper-dependent cytochrome *c* oxidase. During senescence of the wild-type strain, copper is released from mitochondria. The involved mechanism is unknown. However, it is striking that the permeability of mitochondrial membranes in animal systems changes during apoptosis and that mitochondrial proteins with an important impact on this type of cellular death are released (Borghouts *et al.*, 2001). A copper chaperone (CCH) is up-regulated during *Arabidopsis* leaf senescence, suggesting that it mobilizes certain metal ions in leaves and transports them to other growing parts of the plants. The CCHs are also involved in defense mechanisms against oxidative stress in *Arabidopsis*, tomato and poplar. Emerging data suggest that the mechanisms regulating plant copper homeostasis could be implicated in stress and senescence signal transduction pathways (Himelblau *et al.*, 1998; Mira *et al.*, 2002; Lee *et al.*, 2005).

4.5.4 Nucleus

Since oxidative injury or DNA replication error caused by ROS is a serious problem for aerobic organisms, it is reasonable to speculate that eukaryotic cells evolved nuclear antioxidant systems distinct from the cytosolic ones. Contrasting the numerous studies on antioxidants in the cytoplasm, the nuclear antioxidant system has not been studied in much detail. However, nuclear redox states influence the activities of several transcription factors that are responsive to oxidative stress or oxidative signaling, e.g. the activity of Yap1, which induces the H_2O_2 response in yeast, is regulated by its redox status, which is modulated by some antioxidant factors (Izawa *et al.*, 1999; Delaunay *et al.*, 2002). A thiol peroxidase has been shown to function as an H_2O_2 receptor and redox transducer in this gene activation mechanism (Delaunay *et al.*, 2002). In barley, a dormancy-related peroxiredoxin antioxidant, PER1, is localized to the nucleus of embryo and aleurone cells. Stacy *et al.* (1999) have discussed the protective roles for PER1 in barley. On the other hand, it has been observed that H_2O_2 plus Cu(II) induced 8-oxo-7,8-dihydro-2′-deoxyguanosine formation in the telomere sequences more efficiently than in nontelomere sequences. In addition, oxidative damage is repaired less well in telomeric DNA than elsewhere in the chromosome, and oxidative stress accelerates telomere loss, whereas antioxidants decelerate it (Von Zglincki, 2002). In barley, telomeres shorten dramatically during development from immature embryos to mature and senescent leaves. This telomere shortening cannot be explained by the continuous loss of telomeric repeats during replication, indicating a different mechanism most likely involved in development regulation (Kilian *et al.*, 1995). In other plants, telomere length is kept constant during development and postreplicative senescence; however, in *Arabidopsis* the intranuclear localization of the telomeres at the nucleolar boundary changes with the age of the cells (Fajkus and Zentgraf, 2002) and the protein structure of the *Arabidopsis* telomeres is modified at an early stage in leaf senescence. An additional small protein is recruited to the telomeres by protein–protein interaction (Zentgraf *et al.*, 2000). Whether this has any implication on telomere function or nuclear architecture and, as a consequence, maybe also on active chromatin domains and gene-expression patterns, still has to be elucidated. In humans, oxidative DNA damage may exert deleterious effects on telomeres by disrupting the association of telomere-maintenance proteins TRF1 and TRF2 (Opresko *et al.*, 2005).

4.6 Concluding remarks

One of our future challenges will be the development of crop plants that cope better with our global climate changes. Since many different agriculturally important traits are affected by senescence, understanding senescence processes might contribute to solve these problems. ROS play an important role during leaf senescence in two different aspects: signaling and molecule degradation. Although many proteins, particularly those containing thiol groups, are known to react directly with H_2O_2, to date no complete ROS signal transduction pathways have been described (Foyer and Nocter, 2005). Obviously, plants have developed a very fine-tuned network of enzymatic and low-molecular-weight antioxidative components in different cell compartments, and different plants have different strategies to balance their redox potential and regulate their ROS status. In *Arabidopsis*, a network of at least 152 genes is involved in managing the level of ROS. This network is highly dynamic and redundant, and encodes ROS-scavenging and ROS-producing proteins. Although recent studies have unraveled some of the key players in the network, many questions related to its mode of regulation, its protective roles and its modulation of signaling networks that control growth, development and stress response remain unanswered (Mittler *et al.*, 2004). Developing *in vivo* imaging systems for different ROS to visualize local changes in ROS levels in different compartments will help to integrate our current data in a holistic view of the cells.

References

Arora, A., Sairam, R.K. and Srivastava, G.C. (2002) Oxidative stress and antioxidative system in plants. *Curr Sci* **82**, 1227–1238.

Arrigo, A.P. (1999) Gene expression and the thiol redox state. *Free Radic Biol Med* **27**, 936–944.

Baker, N.R. (1991) A possible role for photosystem II in environmental perturbations of photosynthesis. *Physiolgia Plantarum* **81**, 563–570.

Barth, C., Moeder, W., Klessig, D.F. and Conklin, P.L. (2004) The timing of senescence and response to pathogens is altered in the ascorbate-deficient *Arabidopsis* mutant vitamin c-1. *Plant Physiol* **134**, 1784–1792.

Batt, T. and Woolhouse, H.W. (1975) Changing activities during senescence and sites of synthesis of photosynthetic enzymes in leaves of labiate, *Perilla frutenscens* (L.). *Br J Exp Bot* **26**, 569–579.

Beligni, M.V. and Lamattina, L. (1999) Nitric oxide counteracts cytotoxic processes mediated by reactive oxygen species in plant tissues. *Planta* **208**(3), 337–344.

Borghouts, C., Werner, A., Elthon, T. and Osiewacz, H.D. (2001) Copper-modulated gene expression and senescence in the filamentous fungus *Podospora anserine*. *Mol Cell Biol* **21**(2), 390–399.

Bowler, C., Van Montagu, M. and Inzé, D. (1992) Superoxide dismutase and stress tolerance. *Annu Rev Plant Physiol Plant Mol Biol* **43**, 83–116.

Buchanan-Wollaston, V., Earl, S., Harrison, E., *et al.* (2003) The molecular analysis of leaf senescence – a genomics approach. *Plant Biotechol J* **1**, 3–22.

Buchanan-Wollaston, V., Page, T., Harrison, E., *et al.* (2005) Comparative transcriptome analysis reveals significant differences in gene expression and signalling pathways between developmental and dark/starvation-induced senescence in *Arabidopsis*. *Plant J* **42**(4), 567–585.

Casano, L.M., Martín, M. and Sabater, B. (1994) Sensitivity of superoxide dismutase transcript levels and activities to oxidative stress is lower in mature-senescent than in young barley leaves. *Plant Physiol* **106**, 1033–1039.

Chia, L.S., Thompson, J.E. and Dumbroff, E.B. (1981) Simulation of the effects of leaf senescence on membranes by treatment with paraquat. *Plant Physiol* **67**, 415–420.

Corpas, F.J., Barroso, J.B., Carreras, A., *et al.* (2004) Cellular and subcellular localization of endogenous nitric oxide in young and senescent pea plants. *Plant Physiol* **136**, 2722–2733.

Corpas, F.J., Barroso, J.B. and del Río, L.A. (2001) Peroxisomes as a source of reactive oxygen species and nitric oxide signal molecules in plant cells. *Trends in Plant Sci* **6**, 145–150.

Dalton, D.A., Langeberg, L. and Treneman, N.C. (1993) Correlations between the ascorbate-glutathione pathway and effectiveness in legume root nodules. *Physiologia Plantarum* **87**, 365–370.

Dat, J.F., Vandenabeele, S., Vranová, E., Van Montagu, M., Inzé, D. and Van Breusegem, F. (2000) Dual action of the active oxygen species during plant stress responses. *Cell Mol Life Sci* **57**, 779–795.

del Rio, L.A., Corpas, F.J., Sandalio, L.M., Palma, J.M. and Barroso, J.B. (2003a) Plant peroxisomes, reactive oxygen metabolism and nitric oxide. *IUBMB Life* **55**(2), 71–81.

del Río, L.A., Corpas, F.J., Sandalio, L.M., Palma, J.M., Gómez, M. and Barroso, J.B. (2002) Reactive oxygen species, antioxidant systems and nitric oxide in peroxisomes. *J Exp Bot* **53**, 1255–1272.

del Río, L.A., Pastori, G.M., Palma, J.M., *et al.* (1998) The activated oxygen role of peroxisomes in senescence. *Plant Physiol* **116**, 1195–1200.

del Rio, L.A., Sandalio, L.M., Altomare, D.A. and Zilinskas, B.A. (2003b) Mitochondrial and peroxisomal manganese superoxide dismutase: differential expression during leaf senescence. *J Exp Bot* **54**, 923–933.

Delaunay, A., Pflieger, D., Barrault, M.B., Vinh, J. and Toledano, M.B. (2002) A thiol peroxidase is an H_2O_2 receptor and redox-transducer in gene activation. *Cell* **111**, 471–481.

Dertinger, U., Schaz, U. and Schulze, E.D. (2003) Age-dependence of the antioxidative system in tobacco with enhanced glutathione reductase activity or senescence-induced production of cytokinins. *Physiologia Plantarum* **119**, 19–29.

Desikan, R., A-H-Mackerness, S., Hancock, J.T. and Neill, S.J. (2001) Regulation of the *Arabidopsis* transcriptome by oxidative stress. *Plant Physiol* **127**(1), 159–172.

Dhindsa, R.J., Dhindsa, P.P. and Thorpe, T.A. (1981) Leaf senescence: correlation with increased levels of membrane permeability and lipid peroxidation, and decreased levels of superoxide dismutase and catalase. *J Exp Bot* **32**, 93–101.

Djanaguiraman, M., Devi, D.D., Shanker, A.K., Sheeba, J.A. and Bangarusamy, U. (2005) Selenium – an antioxidative protectant in soybean during senescence. *Plant Soil* **272**(1–2), 77–86.

Dufour, E., Boulay, J., Rincheval, V. and Sainsard-Chanet, A. (2000) A causal link betweeen respiration and senescence in *Podospora anserina*. *Proc Natl Acad Sci U S A* **97**, 4138–4143.

Dufur, E. and Larsson, N.-G. (2004) Understanding aging: revealing order out of chaos. *Biochim Biophys Acta* **1658**, 122–132.

Eulgem, T., Rushton, P.J., Robatzek, S. and Somssich, I.E. (2000) The WRKY superfamily of plant transcription factors. *Trends Plant Sci* **5**, 199–206.

Evans, P.J., Gallesi, D., Mathieu, C. *et al.* (1999) Oxidative stress occurs during soybean nodule senescence. *Planta* **208**, 73–79.

Fajkus, J. and Zentgraf, U. (2002) Structure and maintenance of chromosome ends. In: *Telomeres and Telomerases: Cancer and Biology* (eds Krupp, G. and Parwaresch, R.). Bioscience, Gen-Expression, http://www.eurekah.com.

Falk, J., Krauss, N., Dahnhardt, D. and Krupinska, K. (2002) The senescence associated gene of barley encoding 4-hydroxyphenylpyruvate dioxygenase is expressed during oxidative stress. *J Plant Physiol* **159**(11), 1245–1253.

Ferri, K.F. and Kroemer, G. (2001) Organelle-specific initiation of cell death pathways. *Nat Cell Biol* **3**, E255–E263.

Foyer, C.H. (1996) Oxygen processing in photosynthesis. *Biochem Soc Trans* **24**, 427–433.

Foyer, C.H. (2004) The role of ascorbic acid in defence networks and signalling in plants. In: *Vitamin C. Functions and Biochemistry in Animals and Plants* (eds Asard, H., May, J.M. and Smirnoff, S.). Bios. Scientific, Oxon, UK, pp. 65–82.

Foyer, C.H., Lelandais, M. and Kunert, K.J. (1994) Photooxidative stress in plants. *Physiologia Plantarum* **92**, 696–717.

Foyer, C.H. and Noctor, G. (2005) Oxidant and antioxidant signalling in plants: a re-evaluation of the concept of oxidative stress in a physiological context. *Plant Cell Environ* **28**(8), 1056–1071.

Foyer, C.H., Trebst, A. and Noctor, G. (2005) Protective and signalling functions of ascorbate, glutathione and tocopherol in chloroplasts. In: *Advances in Photosynthesis and Respiration: Photoprotection, Photoinhibition, Gene Regulation, and Environment*, Vol. 19 (eds Demmig-Adams, B. and Adams, W.W.). Springer, Dordrecht, The Netherlands, pp. 241–268.

Gan, S. and Amasino, R.M. (1997) Making sense of senescence: molecular genetic regulation and manipulation of leaf senescence. *Plant Physiol* **113**, 313–319.

Gasch, A., Spellman, P., Kao, C., *et al.* (2000) Genomic expression programs in the response of yeast cells to environmental changes. *Mol Biol Cell* **11**, 4241–4257.

Green, D.R. and Reed, J.C. (1998) Mitochondria and apoptosis. *Science* **281**, 1309–1312.

Groten, K., Vanacker, H., Dutilleul, C., *et al.* (2005) The roles of redox processes in pea nodule development and senescence. *Plant Cell Environ* **28**, 1293–1304.

Guo, Y., Cai, Z. and Gan, S. (2004) Transcriptome of *Arabidopsis* leaf senescence. *Plant Cell Environ* **27**, 521–549.

Halliwell, B. and Gutteridge, J.M.C. (1989) *Free Radicals in Biology and Medicine*, 2nd edn. Oxford University Press (Clarendon), Oxford, UK.

Harman, D. (1956) Aging: a theory based on free radical and radiation chemistry. *J Gerontol* **11**, 298–300.

Harman, D. (1998) Extending functional life span. *Exp Gerontol* **33**, 95–112.

Hensel, L.L., Grbić, V., Baumgarten, D. and Bleecker, A.B. (1993) Developmental and age-related processes that influence the longevity and senescence of photosynthetic tissues in *Arabidopsis*. *Plant Cell* **5**, 553–564.

Himelblau, E., Mira, H., Lin, S.J., Culotta, V.C., Penarrubia, L. and Amasino, R.M. (1998) Identification of a functional homolog of the yeast copper homeostasis gene ATX1 from *Arabidopsis*. *Plant Physiol* **117**(4), 1227–1234.

Hinderhofer, K. and Zentgraf, U. (2001) Identification of a transcription factor specifically expressed at the onset of leaf senescence. *Planta* **213**, 469–473.

Hiser, C. and McIntosh, L. (1990) Alternative oxidase of potato is an integral membrane protein synthesized de novo during aging of tuber slices. *Plant Physiol* **93**, 312–318.

Hollander-Czytko, H., Grabowski, J., Sandorf, I., Weckermann, K. and Weiler, E.W. (2005) Tocopherol content and activities of tyrosine aminotransferase and cystine lyase in *Arabidopsis* under stress conditions. *J Plant Physiol* **162**(7), 767–770.

Izawa, S., Maeda, K., Sugiyama, K., Mano, J., Inoue, Y. and Kimura, A. (1999) Thioredoxin deficiency causes the constitutive activation of Yap1, an AP-1-like transcription factor in *Saccharomyces cerevisiae*. *J Biol Chem* **274**, 28459–28465.

Jiménez, A., Hernández, J.A., del Ríom L.A. and Sevilla, F. (1997) Evidence for the presence of the ascorbate-glutathione cycle in mitochondria and peroxisomes of pea (*Pisum sativum* L.) leaves. *Plant Physiol* **114**, 275–284.

Jiménez, A., Hernandez, J.A., Pastori, G., del Rio, L.A. and Sevilla, F. (1998) Role of the ascorbate-glutathione cycle of mitochondria and peroxisomes in the senescence of pea leaves. *Plant Physiol* **118**(4), 1327–1335.

Jones, A. (2000) Does the plant mitochondrion integrate cellular stress and regulate programmed cell death? *Trends Plant Sci* **5**, 225–230.

Kar, M. and Feierabend, J. (1984) Metabolism of activated oxygen in detached wheat and rye leaves and its relevance to the initiation of senescence. *Planta* **160**, 385–391.

Kilian, A., Stiff, C. and Kleinhofs, A. (1995) Barley telomeres shorten during differentiation but grow in callus culture. *Proc Natl Acad Sci U S A* **92**, 9555–9559.

Knox, J.P. and Dodge, A.D. (1985) Singlet oxygen and plants. *Phytochemistry* **24**, 889–896.

Kovtun, Y., Chiu, W.-L., Tena, G. and Sheen, J. (2000) Functional analysis of oxidative stress-activated mitogen-activated protein kinases cascade in plants. *Proc Natl Acad Sci U S A* **97**, 2940–2945.

Kurepa, J., Smalle, J., Van Montagu, M. and Inzé, D. (1998) Oxidative stress tolerance and longevity in *Arabidopsis*: the late flowering mutant *gigantea* is tolerant to paraquat. *Plant J* **14**, 759–764.

Kuzniak, E. and Sklodowska, M. (2005) Fungal pathogen-induced changes in the antioxidant systems of leaf peroxisomes from infected tomato plants. *Planta* **222**(1), 192–200.

Landolt, R. and Matile, P. (1990) Glyoxysome-like microbodies in senescent spinach leaves. *Plant Sci* **72**, 159–163.

Lee, H., Lee, J.S., Bae, E.K., Choi, Y.I. and Noh, E.W. (2005) Differential expression of a poplar copper chaperone gene in response to various abiotic stresses. *Tree Physiol* **25**(4), 395–401.

Lopez-Huertas, E., Charlton, W.L., Johnson, B., Graham, I.A. and Baker, A. (2000) Stress induces peroxisome biogenesis genes. *EMBO J* **19**(24), 6770–6777.

Lorin, S., Dufour, E., Boulay, J., Begel, O., Marsy, S. and Sainsard-Chanet, A. (2001) Overexpression of the alternative oxidase restores senescence and fertility in a long-lived respiration-deficient mutant of *Podospora anserine*. *Mol Microbiol* **42**(5), 1259–1267.

Martin, G.M., Austad, S.N. and Johnson, T.E. (1996) Genetic analysis of aging: role of oxidative damage and environmental stresses. *Nat Genet* **13**, 25–34.

Matile, P., Hörtensteiner, S., Thomas, H. and Kräutler, B. (1996) Chlorophyll breakdown in senescent leaves. *Plant Physiol* **112**, 1403–1409.

Maxwell, D.P., Nickels, R. and McIntosh, L. (2002) Evidence of mitochondrial involvement in the transduction of signals required for the induction of genes associated with pathogen attack and senescence. *Plant J* **29**, 269–279.

May, J.M., Qu, Z.C. and Mendiratta, S. (1998) Protection and recycling of alpha-tocopherol in human erythrocytes by intracellular ascorbic acid. *Arch Biochem Biophys* **349**, 281–289.

McIntosh, L. (1994) Molecular biology of the alternative oxidase. *Plant Physiol* **105**, 781–786.

McRae, D.G. and Thompson, J.E. (1983) Senescence dependent changes in superoxide anion production by illuminated chloroplast from bean leaves. *Planta* **158**, 185–193.

Miao, Y., Laun, T., Zimmermann, P. and Zentgraf, U. (2004) Targets of the WRKY53 transcription factor and its role during leaf senescence in *Arabidopsis*. *Plant Mol Biol* **55**, 853–867.

Miller, J.D., Arteca, R.N., Pell, E.J. (1999) Senescence-associated gene expression during ozone-induced leaf senescence. *Plant Physiol* **120**, 1015–1023.

Millenaar, F.F. and Lambers, H. (2003) The alternative oxidase: *in vivo* regulation and function. *Plant Biol* **5**, 2–15.

Mira, H., Martinez, N. and Penarrubia, L. (2002) Expression of a vegetative-storage-protein gene from *Arabidopsis* is regulated by copper, senescence and ozone. *Planta* **214**(6), 939–946.

Mittler, R., Vanderauwera, S., Gollery, M. and Van Breusegem, F. (2004) Reactive oxygen gene network of plants. *Trends Plant Sci* **9**, 490–498.

Miyake, C. and Asada, K. (1996) Inactivation mechanism of ascorbate peroxidase at low concentrations of ascorbate: hydrogen peroxide decomposes compound I of ascorbate peroxidase. *Plant Cell Physiol* **37**, 423–430.

Moore, A.L., Albury, M.S., Crichton, P.G., Affourtit, C. (2002) Function of the alternative oxidase: is it still a scavenger? *Trends Plant Sci* **7**, 478–481.

Munne-Bosch, S. and Alegre, L. (2004) Die and let live: leaf senescence contributes to plant survival under drought stress. *Funct Plant Biol* **31**(3), 203–216.

Nakagami, H., Kiegerl, S. and Hirt, H. (2004) OMTK1, a novel MAPKKK, channels oxidative stress signalling through direct MAPK interaction. *J Biol Chem* **279**, 26959–26966.

Navabpour, S., Morris, K., Allen, R., Harrison, E., Mackerness, S.A.H. and Buchanan-Wollaston, V. (2003) Expression of senescence-enhanced genes in response to oxidative stress. *J Exp Bot* **54**, 2285–2292.

Noctor, G., Veljovic-Jovanovic, S., Driscol, S., Novitskaya, L. and Foyer, C.H. (2002) Drought and oxidative load in wheat leaves: a predominant role for photorespiration? *Ann Bot* **89**, 841–850.

Noctor, G. and Foyer, C.H. (1998) Ascorbate and glutathione: keeping active oxygen under control. *Annu Rev Plant Physiol Plant Mol Biol* **49**, 249–279.

Opresko, P.L., Fan, J.H., Danzy, S., Wilson, D.M. and Bohr, V.A. (2005) Oxidative damage in telomeric DNA disrupts recognition by TRF1 and TRF2. *Nucleic Acid Res* **33**(4), 1230–1239.

Orendi, G., Zimmermann, P., Baar, C. and Zentgraf, U. (2001) Loss of stress-induced expression of catalase3 during leaf senescence in *Arabidopsis thaliana* is restricted to oxidative stress. *Plant Sci* **161**, 301–314.

Orr, W.C. and Sohal, R.S. (1994) Extension of life-span by overexpression of superoxide dismutase and catalase in *Drosophila melanogaster*. *Science* **263**, 1128–1130.

Panavas, T. and Rubinstein, B. (1998) Oxidative events during programmed cell death of daylily (*Hemerocallis* hybrid) petals. *Plant Sci* **133**, 125–138.

Panchuk, I.I., Zentgraf, U. and Volkov, R.A. (2005) Expression of the *Apx* gene family during leaf senescence of *Arabidopsis thaliana*. *Planta* **222**, 926–932.

Parthier, B. (1988) Gerontoplasts – the yellow end in the ontogenesis of chloroplasts. *Endocytobiosis Cell Res* **5**, 163–190.

Passardi, F., Cosio, C., Penel, C. and Dunand, C. (2005) Peroxidases have more functions than a Swiss army knife. *Plant Cell Rep* **24**(5), 255–265.

Pastori, G.M. and del Rio, L.A. (1994) An activated-oxygen-mediated role for peroxisomes in the mechanism of senescence of *Pisum sativum* L. leaves. *Planta* **193**, 385–391.

Pastori, G.M. and del Río, L.A. (1997) Natural senescence of pea leaves: an activated oxygen-mediated function for peroxisomes. *Plant Physiol* **113**, 411–418.

Pastori, G.M., Kiddle, G., Antoniw, J., *et al.* (2003) Leaf vitamin C contents modulate plant defense transcripts and regulate genes that control development through hormone signaling. *Plant Cell* **15**, 939–951.

Pauls, K.P. and Thompson, J.E. (1984) Evidence for the accumulation of peroxidized lipids in membranes of senescing cotyledons. *Plant Physiol* **75**, 1152–1157.

Pich, M.M., Raule, N., Catani, L., *et al.* (2004) Increased transcription of mitochondrial genes for complex I in human platelets during ageing. *FEBS Lett* **558**, 19–22.

Pistelli, L., Nieri, B., Smith, S.M., Alpi, A. and De Bellis, L. (1996) Glyoxylate cycle enzyme activities are induced in senescent pumpkin fruits. *Plant Sci* **119**, 23–29.

Polle, A. (2001) Dissecting the superoxide dismutase-ascorbate-glutathione-pathway in chloroplasts by metabolic modeling. Computer simulations as a step towards flux analysis. *Plant Physiol* **126**, 445–462.

Potters, G., Horemans, N., Bellone, S., *et al.* (2004) Dehydroascorbate influences the plant cell cycle through a glutathione-independent reduction mechanism. *Plant Physiol* **134**, 1479–1487.

Puppo, A., Groten, K., Bastian, F., *et al.* (2005) Legume nodule senescence: roles for redox and hormone signalling in the orchestration of the natural aging process. *New Phytologist* **165**(3), 683–701.

Quinn, J., Findlay, V.J., Dawson, K., *et al.* (2002) Distinct regulatory proteins control the graded transcriptional response to increasing H_2O_2 levels in fission yeast *Schizosaccharomyces pombe*. *Mol Biol Cell*, **13**, 805–816.

Richter, C. and Schweizer, M. (1997) Oxidative stress in mitochondria. In: *Oxidative Stress and the Molecular Biology of Antioxidant Defences* (ed. Scandalios, J.G.). Cold Spring Harbor Laboratory Press, New York, pp. 169–200.

Robatzek, S., and Somssich, I.E. (2002) Targets of *AtWRKY6* regulation during plant senescence and pathogen defense. *Genes Dev* **16**, 1139–1149.

Rolland, F., Moore, B. and Sheen, J. (2002) Sugar sensing and signalling in plants. *Plant Cell* **14**(Suppl.), S185–S205.

Ruis, H. and Schuller, C. (1995) Stress signaling in yeast. *BioEssays* **17**, 959–965.

Saisho, D., Nambara, E., Naito, S., Tsutsumi, N., Hirai, A. and Nakazono, M. (1997) Characterization of the gene family for alternative oxidase from *Arabidopsis thaliana*. *Plant Mol Biol* **35**, 585–596.

Scandalios, J.G. (2002) Oxidative stress responses – what have genome-scale studies taught us? *Genome Biol* **3**(7), reviews1019.1–reviews1019.6.

Siedow, J.N. and Moore, A.L. (1993) A kinetic model for the regulation of electron transfer through the cyanide-resistant pathway in plant mitochondria. *Biochim Biophys Acta* **1142**, 165–174.

Smart, C.M. (1994) Gene expression during leaf senescence. *New Phytologist* **126**, 419–448.

Smirnoff, N. and Pallanca, E. (1996) Ascorbate metabolism in relation to oxidative stress. *Biochem Soc Trans* **24**, 472–478.

Sohal, R.S. and Weindruch, R. (1996) Oxidative stress, caloric restriction, and aging. *Science* **273**, 59–63.

Srivalli, B. and Khanna-Chopra, R. (2004) The developing reproductive 'sink' induces oxidative stress to mediate nitrogen mobilization during monocarpic senescence in wheat. *Biochem Biophys Res Commun* **325**(1), 198–202.

Stacy, R.A.P., Nordeng, T.W., Culiáñez-Macià, F.A. and Aelen, R.B. (1999) The dormancy-related peroxiredoxin anti-oxidant, PER1, is localized to the nucleus of barley embryo and aleurone cells. *Plant J* **19**, 1–8.

Svensson, A.S. and Rasmusson, A.G. (2001) Light-dependent gene expression for proteins in the respiratory chain of potato leaves. *Plant J* **28**, 73–82.

Thomas, H. and Stoddart, J.L. (1980) Leaf senescence. *Annu Rev Plant Physiol* **31**, 83–111.

Thomas, H., Ougham, H., Wagstaff, C. and Stead, A.D. (2003) Defining senescence and death. *J Exp Bot* **54**, 1127–1132.

Thomas, J.C., Perron, M., LaRosa, P.C. and Smigocki, A.C. (2005) Cytokinin and the regulation of a tobacco metallothionein-like gene during copper stress. *Physiologia Plantarum* **123**(3), 262–271.

Thompson, J.E., Ledg, R.L. and Barber, R.F. (1987) The role of free radicals in senescence and wounding. *New Phytologist* **105**, 317–344.

van Dorn, W.G. and Woltering, E.J. (2004) Senescence and programmed cell death: substance or semantics. *J Exp Bot* **55**, 2147–2153.

Von Zglincki, T. (2002) Oxidative stress shortens telomeres. *Trends Biochem Sci* 27, 339–344.

Wan, J.R., Zhang, S.Q. and Stacey, G. (2004) Activation of a mitogen-activated protein kinase pathway in *Arabidopsis* by chitin. *Mol Plant Pathol* **5**, 125–135.

Willekens, H., Chamnongpol, S., Davey, M., *et al.* (1997) Catalase is a sink for H_2O_2 and is indispensable for stress defence in C_3 plants. *EMBO J* **16**, 4806–4816.

Wingler, A., Brownhill, E. and Pourtau, N. (2005) Mechanisms of the light-dependent induction of cell death in tobacco plants with delayed senescence. *J Exp Bot* **56**(421), 2897–2905.

Woo, H.R., Kim, J.H., Nam, H.G. and Lim, P.O. (2004) The delayed leaf senescence mutants of *Arabidopsis*, ore1, ore3, and ore9 are tolerant to oxidative stress. *Plant Cell Physiol* **45**(7), 923–932.

Xu, P., Rogers, S.J. and Roossinck, M.J. (2004) Expression of antiapoptotic genes bcl-xL and ced-9 in tomato enhances tolerance to viral-induced necrosis and abiotic stress. *Proc Natl Acad Sci U S A* **101**(44), 15805–15810.

Ye, Z.Z., Rodriguez, R., Tran, A., *et al.* (2000) The developmental transition to flowering represses ascorbate peroxidase activity and induces enzymatic lipid peroxidation in leaf tissue in *Arabidopsis thaliana*. *Plant Sci* **158**, 115–127.

Zapata, J.M., Guera, A., Esteban-Carrasco, A., Martin, M. and Sabater, B. (2005) Chloroplasts regulate leaf senescence: delayed senescence in transgenic ndhF-defective tobacco. *Cell Death Differ* **12**, 1277–1284.

Zavaleta-Mancera, H.A., Frnaklin, K.A., Ougham H.J., Thomas, H. and Schott, I.M. (1999a) Regreening of senescent *Nicotiana* leaves. I: Reappearance of NADPH-protophyllide oxidoreductase and light harvesting chlorophyll a/b-binding protein. *J Exp Bot* **50**, 1677–1682.

Zavaleta-Mancera, H.A., Thomas, B.J., Thomas, H. and Scott, I.M. (1999b) Regreening of *Nicotiana* leaves. II: Redifferentiation of plastids. *J Exp Bot* **50**, 1683–1689.

Zentgraf, U., Hinderhofer, K. and Kolb, D. (2000) Specific association of a small protein with the telomeric DNA-protein complex during the onset of leaf senescence in *Arabidopsis thaliana*. *Plant Mol Biol* **42**(3), 429–438.

Zentgraf, U., Jobst, J., Kolb, D. and Rentsch, D. (2004) Senescence-related gene expression profiles of rosette leaves of *Arabidopsis thaliana*: leaf versus plant age. *Plant Biol* **6**, 178–183.

Zimmermann, P., Orendi, G., Heinlein, C., Zentgraf, U. (2006) Senescence specific regulation of catalases in *Arabidopsis thaliana* (L.) Heynh. *Plant Cell Environ* **29**, 1049–1060.

5 Nutrient remobilization during leaf senescence

Andreas M. Fischer

5.1 Overview

Senescence is the last stage in the development of leaves and other plant organs. While many plants are perennial (barring adverse conditions leading to premature death), and some species even very long-lived (at least from a human perspective), senescence and death of organs such as leaves is often an annual event. Due to its importance for agriculture, the senescence of annual crops (e.g. corn, rice, wheat, barley and some legumes) has been most intensely studied (Feller and Fischer, 1994; Hayati *et al.*, 1995; Crafts-Brandner *et al.*, 1998; Yang *et al.*, 2003; Robson *et al.*, 2004; Parrott *et al.*, 2005; Weng *et al.*, 2005). Additionally, as in other areas of plant science research, *Arabidopsis* has emerged as an important model system (Diaz *et al.*, 2005; Levey and Wingler, 2005; Otegui *et al.*, 2005). These plants show monocarpic senescence, i.e. fruit set and maturation are directly associated with whole-plant senescence and death. Other types of senescence, such as top senescence (in species with bulbs, tubers, tap roots or rhizomes), deciduous senescence (in some trees and shrubs of temperate climate zones) and progressive senescence (e.g. in evergreen trees) have received less attention. In contrast to annuals, leaf (or whole-shoot) senescence is often not directly associated with seed filling in perennial plants (Feller and Fischer 1994; Noodén *et al.*, 2004). However, nutrient remobilization from senescing plant parts to surviving structures is a hallmark of the 'execution' of the senescence process in both annual plants, in which nutrients are retranslocated to the seeds, and perennial species, in which nutrients are transported to surviving structures such as bulbs and roots.

Plants need a number of elements in higher quantities or concentrations to complete their life cycle (macronutrients, including C, O, H, N, P, S, K, Mg and Ca), while a number of additional elements (micronutrients, including Fe, Mn, Zn, Cu, B, Mo, Cl and Ni) are needed in comparatively small quantities (Table 5.1) (Marschner, 1995). Some elements are essential only for specific taxonomic groups (e.g. Na, Si) and/or are considered beneficial (Marschner, 1995). Because of its quantitative importance, and of the intricacies of the biochemistry associated with its remobilization, nitrogen metabolism during leaf senescence has received most attention, and will be the primary focus of this chapter (Section 5.3).

Table 5.1 Essential mineral nutrients in plants with average tissue concentrations

Element[a]	Available form(s)	$\mu g\, g^{-1}$ dry weight (ppm)[b]	%
Nitrogen (N)	NO_3^-; NH_4^+	—	1.5
Potassium (K)	K^+	—	1.0
Calcium (Ca)	Ca^{2+}	—	0.5
Magnesium (Mg)	Mg^{2+}	—	0.2
Phosphorus (P)	$H_2PO_4^-$; HPO_4^{2-}	—	0.2
Sulfur (S)	SO_4^{2-}	—	0.1
Chlorine (Cl)	Cl^-	100	—
Boron (B)	$H_2BO_3^-$	20	—
Iron (Fe)	Fe^{2+}, Fe^{3+}	100	—
Manganese (Mn)	Mn^{2+}	50	—
Zinc (Zn)	Zn^{2+}	20	—
Copper (Cu)	Cu^+, Cu^{2+}	6	—
Nickel (Ni)	Ni^{2+}	~0.1	—
Molybdenum (Mo)	MoO_4^{2-}	0.1	—

[a]Carbon (C, from CO_2), oxygen (O, from CO_2) and hydrogen (H, from H_2O) are also essential elements, but are not, or not in ionic form, taken up from the soil.
[b]Marschner, 1995. Actual concentrations vary between tissues, species and under different physiological conditions.

5.2 Macro- and micronutrient remobilization

Developing (young) leaves constitute significant net importers ('sinks') for all nutrients, which are utilized to build the organ's cellular and molecular components. After the so-called sink–source transition (Ishimaru *et al.*, 2004; Jeong *et al.*, 2004), leaves become net exporters ('sources') of carbohydrates from photosynthesis, while import (through the xylem) and export (through the phloem) of phloem-mobile nutrients are (roughly) at an equilibrium in mature leaves (Figure 5.1) (Marschner, 1995). The onset of leaf senescence is associated with a transition to net export of 'mobile' (see below) compounds, i.e. total (per leaf) content of some nutrients starts to decrease (Figure 5.1) (Marschner, 1995). The literature often refers to this situation as 'redistribution', 'retranslocation', 'resorption' or 'remobilization' (Marschner, 1995; Killingbeck, 2004).

The main transport route from senescing leaves to nutrient sinks is the phloem (Atkins, 2000; Tilsner *et al.*, 2005). Using various approaches, including sampling and analysis of phloem sap and (radioactive) tracer studies, it has been established that macronutrients with the exception of calcium (i.e. N, P, S, K and Mg) are generally highly mobile in the phloem, while micronutrients with the exception of manganese (i.e. Fe, Zn, Cu, B, Mo, Cl and Ni) show at least moderate mobility (Marschner, 1995). As a consequence, while some mobile nutrients decrease during leaf senescence, this is not true for calcium, which continues to accumulate throughout a leaf's life span.

The molecular form, in which nutrients fulfill their biological functions, determines the biochemical steps necessary to make them phloem mobile. A certain

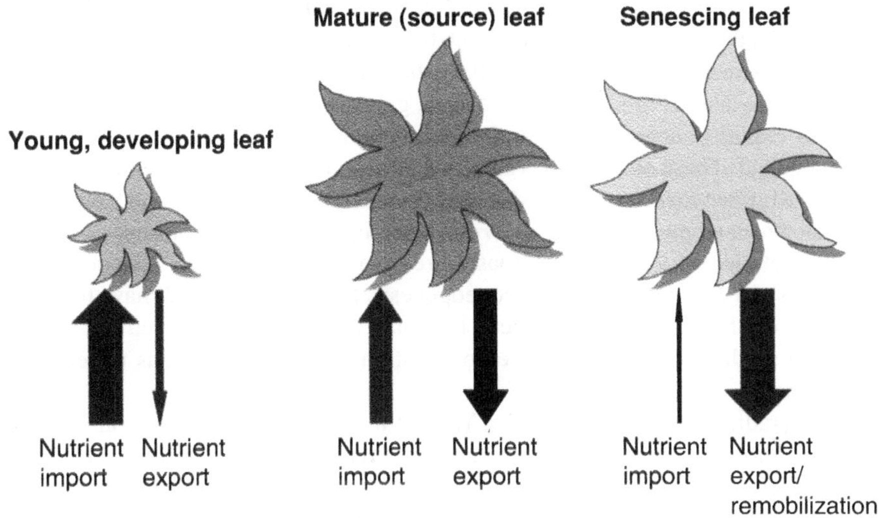

Figure 5.1 Nutrient import into young leaves, recycling between mature leaves and other plant parts and remobilization from senescing leaves.

percentage of many nutrients is biochemically inert, and cannot be remobilized (Marschner, 1995; Killingbeck, 2004). Cell wall components are a good example, and explain why fully senesced (dead) leaves are usually rich in carbon as compared to nitrogen. Some macronutrients, including carbon, nitrogen, phosphorus and sulfur, are covalently bound in myriads of both low-molecular-weight metabolites and macromolecules. Proteins and nucleic acids are important stores of nitrogen, phosphorus (nucleic acids) and sulfur (proteins); these macromolecules have to be degraded by specific hydrolases prior to phloem loading and transport. Metals (both macro- and micronutrients) can also be tightly (albeit noncovalently) bound, mostly by macromolecules, e.g. cell wall compounds or proteins. Their release is therefore often linked with the degradation of the functional complexes/macromolecules, to which they belong. The following paragraphs summarize known facts about the remobilization of carbon and some non-nitrogen mineral elements; nitrogen is discussed in detail in Section 5.3 of this chapter.

5.2.1 Carbon

Because it is taken up in gaseous form and a large amount of energy is needed for its reduction prior to its incorporation into metabolites, carbon occupies a special position in plant metabolism. Additionally, as discussed in Section 5.3, degradation of the photosynthetic apparatus is an early event during leaf senescence, leading to a decrease of photoassimilate production and export to sinks, and to an increasing dependence of senescing tissues on respiratory metabolism (Gepstein, 1988; Feller and Fischer, 1994). Metabolization and, to some degree, remobilization of reduced carbon are therefore important for senescing leaves. In this context, Gut and Matile

(1988, 1989) observed an induction of key enzymes of the glyoxylate cycle, isocitrate lyase and malate synthase, in senescent barley leaves. Based on these data, and based on low respiratory quotients (~0.6), these authors suggested a reutilization of plastidial (thylakoid) lipids via β-oxidation, glyoxylate cycle and gluconeogenesis, allowing export of at least some of the carbon 'stored' in plastidial lipids from the senescing leaf. These observations have since been confirmed and extended (Pistelli *et al.*, 1991; Graham *et al.*, 1992; McLaughlin and Smith, 1994). He and Gan (2002) have shown an essential role for an *Arabidopsis* lipase in leaf senescence; however, it is not yet clear if this or other lipases are involved in preparing substrates (free fatty acids) for β-oxidation and gluconeogenesis. Roulin *et al.* (2002) have found an induction of (1→3, 1→4)-β-D-glucan hydrolases during dark-induced senescence of barley seedlings, suggesting a remobilization of cell wall glucans under these conditions.

Using radioactive labeling studies, Yang *et al.* (2003) demonstrated considerable remobilization of pre-fixed ^{14}C from vegetative tissues to grains in senescent wheat plants. Interestingly, this process was enhanced under drought conditions, when leaf photosynthetic rates declined faster. Together, these data suggest that while C remobilization during leaf senescence has received less attention than N remobilization, it probably makes important contributions to seed development, at least in annual crops.

5.2.2 Sulfur

Besides carbon and nitrogen, sulfur is the third nutrient, which (relative to its main form of uptake, sulfate) is reduced by plants prior to its incorporation into certain metabolites and macromolecules. It is noteworthy, however, that plants also contain oxidized ('sulfated') sulfur metabolites (Crawford *et al.*, 2000). Identically to carbon and nitrogen, sulfur is an essential element of both low-molecular-weight compounds (including the protein amino acids cysteine and methionine) and macromolecules (proteins). Glutathione (γ-glutamyl-cysteinyl-glycine) represents the quantitatively most important reduced sulfur metabolite; it can reach millimolar concentrations in chloroplasts (Rennenberg and Lamoureux, 1990). Sulfur remobilization from older leaves has been shown; however, the extent of its retranslocation appears to depend on the nitrogen status, at least in some systems (Marschner, 1995). Sunarpi and Anderson (1997) demonstrated the remobilization of both soluble (nonprotein) and insoluble (protein) sulfur from senescing leaves. This study also indicated that homoglutathione (containing β-alanine instead of glycine) is the principal export form of metabolized protein sulfur from senescing soybean leaves.

5.2.3 Phosphorus

Unlike carbon dioxide, nitrate and sulfate, phosphate (main form of P uptake) is not reduced, but utilized in its oxidized form by plants (Marschner, 1995), both in low-molecular-weight metabolites and in macromolecules (nucleic acids). Studies on P remobilization from senescing leaves are scarce. Snapp and Lynch (1996) concluded that in maturing common bean plants, leaf P remobilization supplied more than half of the pod plus seed phosphorus. In contrast, Crafts-Brandner (1992) observed no net leaf P remobilization during reproductive growth of soybeans cultivated at three different P regimes. Therefore, while P is a mobile nutrient, its remobilization may be influenced by a number of exogenous and endogenous/genetic factors, making generalizations on the importance of its remobilization difficult. Nucleic acids (especially RNA) constitute a major phosphorus store but, depending on the species and growth condition investigated, considerable P amounts are also present in lipids, in esterified (organic) form, and as inorganic phosphate (Hart and Jessop, 1984; Valenzuela *et al.*, 1996). Similarly to the situation with nitrogen 'bound' in proteins, release of phosphorus from nucleic acids depends on the activities of hydrolytic enzymes. A decrease in nucleic acid levels is typical for senescing tissues, and increases in nuclease activities have also been observed (Feller and Fischer, 1994; Lers *et al.*, 2001), indicating that if P is remobilized from senescing tissues, at least part of it is derived from the degradation of RNA and DNA.

5.2.4 *Potassium*

Next to nitrogen, potassium is the mineral nutrient required in the largest amount by plants (Table 5.1). It is highly mobile within individual cells, within tissues and in long-distance transport via the xylem and phloem (Marschner, 1995). In contrast to the nutrients discussed above, potassium is not metabolized, and it forms only weak complexes, in which it is easily exchangeable. Next to the transport of carbohydrates and nitrogen compounds, potassium transport has been studied most intensely, using both physiological and molecular approaches (Kochian, 2000). Many plant genes encoding K^+ transporters have been identified, and some of them have been studied in detail in heterologous systems, such as K^+-transport-deficient yeast mutants. Similarly to the situation discussed for nitrogen transport (Section 5.3), analysis of K^+ transport is complicated by the fact that these transporters are organized in multigene families with (partially?) redundant functions (Kochian, 2000). Potassium was repeatedly reported to be remobilized in significant quantities from senescing tissues (Hill *et al.*, 1979; Scott *et al.*, 1992; Tyler, 2005). However, it has to be considered that this element easily leaches from tissues, especially senescing tissues (Tukey, 1970; Debrunner and Feller, 1995). Therefore, actually remobilized potassium quantities may be smaller than those reported in the literature.

5.2.5 *Magnesium, calcium and micronutrients*

Magnesium has not often been considered in studies on nutrient remobilization. However, despite the fact that this element is considered phloem mobile (Marschner, 1995), available results indicate a tendency of continued accumulation during leaf senescence (Killingbeck, 2004). Unsurprisingly, calcium, which is the least mobile of all macronutrients (Marschner, 1995), has repeatedly been found to increase in senescing leaves (Killingbeck, 2004).

Information on remobilization of micronutrients does not allow a generalized picture. For several of them, including Fe, Cu, Mn (which is the least phloem mobile among the micronutrients) and Zn, both remobilization from and accumulation in senescing leaves have been reported (Killingbeck, 2004, and references cited therein). Tyler (2005) gives a broad overview of the fate of numerous elements (including the micronutrients Fe, B, Mn, Zn, Cu, Mo and Ni) during senescence and decomposition of *Fagus sylvatica* leaves; however, in view of the results cited above, it is probably not possible to generalize conclusions from this study, e.g. with regard to the situation in annual crops.

5.3 Nitrogen remobilization

Quantitatively, nitrogen is the most important mineral nutrient in plants (Table 5.1) (Marschner, 1995). It is often a limiting factor for plant growth, yield and/or quality (Gastal and Lemaire, 2002; Good *et al.*, 2004). Additionally, as for carbon, the principal form in which many plants acquire nitrogen from the environment (nitrate) is more oxidized than the form in which it can be integrated into metabolites and macromolecules, demanding substantial energy input for the synthesis of nitrogen compounds. Although the biochemistry involved is different, the establishment and maintenance of a symbiosis with N_2-fixing microorganisms (e.g. in legumes) is also costly (Crawford *et al.*, 2000; Lodwig and Poole, 2003). For these reasons, efficient N remobilization increases the competitiveness of wild plants. Additionally, due to the economic and ecological (N runoff from agricultural soils) cost of N fertilization, this trait is of considerable importance to farmers.

In most plant tissues, the largest fraction of organic nitrogen, which is potentially available for remobilization during senescence, is contained in proteins. In photosynthetically active tissues of C_3 species, over 50% of this nitrogen is found in soluble (Calvin cycle) and insoluble (thylakoid) chloroplast proteins (Peoples and Dalling, 1988; Feller and Fischer, 1994). Intriguingly, ribulose-1,5-bisphosphate carboxylase/oxygenase (Rubisco) alone represents $\sim$50% of the total plastidial nitrogen. All other cellular nitrogen fractions, including cytosolic and other proteins, nucleic acids, chlorophylls and free amino acids, while not negligible, represent relatively minor stores of organic nitrogen. Efforts at understanding nitrogen remobilization during leaf senescence have therefore focused on the biochemistry of plastidial protein degradation. Mae *et al.* (1983), using elegant ^{15}N-labeling techniques, have demonstrated that the synthesis and degradation phases of Rubisco are surprisingly clearly separated during leaf development. High rates of synthesis were observed until full leaf expansion; after this point, synthesis was minimal, but degradation rates started to increase. In this context, it is well known that the photosynthetic capacity of a leaf declines early during leaf senescence, while mitochondrial integrity and respiration are maintained longer (Gepstein, 1988; Feller and Fischer, 1994). That efficient N remobilization is associated with (early) loss of CO_2 assimilation represents a formidable problem in annual crops. In this context, agronomists are well aware of the negative correlation between seed protein and yield. This situation has led to increased interest in genotypes with extended leaf duration, at least for crops in which seed protein is less critical (Robson *et al.*, 2004). However, it has to be borne in mind that photosynthetic capacity is not always associated with (visibly) extended leaf greenness. Thomas and Howarth (2000) have described five different 'ways to stay green', not all of which are associated with extended photosynthetic competence and yield.

That most of the nitrogen available for remobilization from senescing leaves is located in chloroplasts has led to two major hypotheses regarding its mobilization. While convincing experiments demonstrate that chloroplasts contain a number of proteases, and that the degradation of plastidial proteins (such as Rubisco) is performed or at least initiated within the intact organelle, a role for the highly active proteases accumulating in lytic vacuolar compartments of senescing leaves cannot be excluded (Hörtensteiner and Feller, 2002). Current knowledge about proteases present in senescing tissues, and their possible physiological functions, is discussed in the following paragraphs.

5.3.1 Protein degradation in senescing leaves

5.3.1.1 Classification of peptidases

Proteins are degraded into large fragments, oligopeptides and amino acids by the action of enzymes capable of cleaving the peptide bond between adjacent amino acids either inside the polypeptide chain, or at/near its N- or C terminus. These enzymes are known as peptide hydrolases, proteases, peptidases or simply proteolytic enzymes (Schaller, 2004). Commonly, three classification systems are used. The first system is based on the chemical mechanism of catalysis, in which serine, cysteine, aspartic, metallo- and (more recently) threonine and glutamic peptidases are distinguished (Rawlings *et al.*, 2004; Schaller, 2004; MEROPS database at http://merops.sanger.ac.uk). This system has the advantage that it immediately indicates an important common feature for all members of a catalytic type, i.e. all serine peptidases have an active-site serine residue that acts as the nucleophile at the heart of the catalytic site, and as a result many are affected by generic inhibitors of serine peptidases. A second commonly used classification system is based on the position of the (preferentially) cleaved peptide bond within the polypeptide chain and, at its most basic level, distinguishes endo- (cleaving bonds inside the chain) from exopeptidases, which attack their substrate from either the N terminus (aminopeptidases) or C terminus (carboxypeptidases) (Brouquisse *et al.*, 2001; Feller, 2004). Complete degradation of intact proteins is typically performed in collaboration of endo- and exopeptidases, with the former making additional substrates available for the latter. More recently, with the availability of sequence and structural information, peptidases are also classified by molecular structure and homology. In this system, individual peptidases are assigned to families, and families are further grouped into clans (Rawlings and Barrett, 1993). This approach is strictly followed by the MEROPS peptidase database (Rawlings *et al.*, 2004; http://merops.sanger.ac.uk/index.htm).

The availability of molecular methods has greatly facilitated the discovery and characterization of complex peptidases. The best studied and best understood proteolytic system in plants is the ubiquitin–proteasome pathway. The 'peptidase' responsible for the cleavage of ubiquitin conjugates in this pathway is the 26S proteasome complex, composed of the 20S core protease (a threonine peptidase) and the 19S regulatory particle (Smalle and Vierstra, 2004). While the proteasome has been located in the cytosol and nucleus, 'complex' proteases are also found in organelles including chloroplasts (see below). However, available data indicate that these proteolytic systems are involved in the specific degradation of damaged or rapidly turned-over (regulatory) proteins, and an involvement in bulk protein degradation during leaf senescence appears unlikely.

A surprisingly large number of genes have been found to be potentially involved in proteolytic processes. In *Arabidopsis*, besides ~1300 genes associated with the ubiquitin–proteasome pathway, ~600 additional protease genes were discovered (Schaller, 2004). This finding explains why it has been difficult to associate protease (genes) with their cellular functions, both during presenescence development, regulation and metabolism, and during senescence-associated N remobilization.

5.3.1.2 Compartmentation of peptidases

Analyses involving different plant species and tissue types indicate that peptidases are present in all cellular compartments, including the apoplast (cell walls) (Brouquisse *et al.*, 2001; Feller, 2004). In this context, it has been known for >20 years that lytic vacuolar compartments contain a number of highly active peptidases, including cysteine and serine endopeptidases and serine carboxypeptidases (Feller, 2004). Biochemically, many of the other proteases, including those present in chloroplasts, are difficult to detect. Their analysis has therefore been facilitated, or made possible, by the rigorous application of molecular and genomic tools. An exception to this 'rule' are aminopeptidases, which have been localized both in the cytosol and in plastidial compartments, and which are easy to assay using synthetic substrates such as amino acid-*p*-nitroanilides, or amino acid-β-naphthylamides (Thayer *et al.*, 1988; Fischer *et al.*, 1998).

Elegant experiments, based on chloroplast isolation, incubation under various light/dark conditions and analysis by SDS-PAGE and immunoblotting, have demonstrated the presence of proteases capable of degrading Rubisco and other stromal enzymes in pea chloroplasts (Mitsuhashi *et al.*, 1992; Mitsuhashi and Feller, 1992). These studies have suggested involvement of a metallopeptidase in stromal protein degradation (Roulin and Feller, 1998). Kokubun *et al.* (2002) demonstrated the formation of a specific 44-kDa fragment of the large subunit of Rubisco in lysates of senescing wheat chloroplasts. Determination of the N-terminal sequence of this fragment (RVSPQPGVPPEE) indicated that in these experiments, the polypeptide chain was cleaved between Phe-40 and Arg-41. Recently, efforts by several research groups have revealed the presence of a number of peptidases of prokaryotic origin in higher plant chloroplasts. These include the Clp protease (a serine peptidase) in which, similarly to the 26S proteasome, proteolytic and regulatory functions are present on different subunits in the functional complex (Clarke *et al.*, 2005). Other plastidial proteases include the FtsH (metallopeptidase), DegP, Spp and Lon proteases (all of which are serine peptidases) (Adam and Clarke, 2002). Based on results with *Arabidopsis*, these proteases are encoded by gene families, making their functional analysis quite challenging. It is known that Clp protease(s) are essential for chloroplast function, as genetic interference with some Clp proteins is detrimental to plant viability (Adam and Clarke, 2002). However, no evidence for a major role of Clp proteins in N remobilization during leaf senescence has been found to date. Similarly, while some specific functions start to be associated with some of the other mentioned plastidial proteases, they have so far not evolved as good candidates for the degradation of photosynthetic proteins at the onset of the senescence process. On the other hand, Kato *et al.* (2004) have recently reported that tobacco plants with lower levels of the DNA-binding protease CND41 had delayed leaf senescence and Rubisco degradation, giving, to the knowledge of the author of this chapter for the first time, direct evidence for the functional involvement of a defined protease gene in *in vivo* Rubisco proteolysis. Intriguingly, results in the author's laboratory have demonstrated strong upregulation of a *cnd41* gene in girdled barley leaves showing premature senescence and protein degradation (D. Parrott, K. McInnerney and A.M. Fischer, unpublished results; see Section 5.3.1.3).

It is well known that certain peptidases, including aspartic and cysteine endopeptidases and serine carboxypeptidases, are present in (lytic) vacuolar compartments; some of these enzymes accumulate in the course of the senescence process (Brouquisse *et al.*, 2001; Feller, 2004). Besides large vacuoles, certain peptidases have also been found to accumulate in smaller vesicles (Schmid *et al.*, 1998, 2001). At present, besides the involvement of plastidial proteases (such as the metallopeptidase, or CND41 discussed above), a functional involvement of proteolytic enzymes contained in such organelles remains a distinct possibility. Unfortunately, similarly to the problems encountered with plastidial peptidases, little success has been achieved in associating known peptidase genes with the degradation of distinct plastidial or even extraplastidial protein substrates. Microscopic evidence has been presented, which suggests that chloroplasts are engulfed by vacuoles, leading to the degradation of their components, including proteins, by vacuolar enzymes (Minamikawa *et al.*, 2001). While the importance of such a mechanism during leaf senescence is not generally accepted (Hörtensteiner and Feller, 2002), Chiba *et al.* (2003) demonstrated the presence of Rubisco and/or Rubisco fragments in small spherical bodies (diameter of 0.4–1.2 μm) in naturally senescing wheat leaves. Intriguingly, these bodies were found both in the cytoplasm and in vacuoles, suggesting yet another mechanism by which vacuolar peptidases could contribute to plastidial protein degradation. Advances in molecular genetics, profiting of the discovery of numerous orthologs of the yeast AUTOPHAGY (ATG) protein system in *Arabidopsis*, have recently confirmed the presence of autophagic processes in plants (Thompson and Vierstra, 2005). Additionally, indirect evidence for the involvement of vacuolar enzymes in N remobilization has been obtained using a genetic approach. Quantitative trait locus (QTL) data from this author's laboratory (Yang *et al.*, 2004) have indicated a positive correlation between carboxypeptidase activities and N remobilization at several loci. Assay conditions used favored carboxypeptidases with low pH optima; such enzymes have previously been located in vacuolar compartments (Brouquisse *et al.*, 2001). However, the same set of experiments also indicated that major (most likely vacuolar) cysteine endopeptidases were not involved in N remobilization. Clearly, both current hypotheses (plastidial vs vacuolar/extraplastidial) proteolysis need to be vigorously pursued using all available genomic, genetic and molecular techniques. Experiences with the investigation of protein processing in seeds indicate that this will be challenging, as the presence of multiple genes with similar functions and gene families can lead to considerable redundancy (Shimada *et al.*, 2003; Gruis *et al.*, 2004).

5.3.1.3 Regulation of peptidases during leaf senescence

As might be expected from the need for increased protein degradation, a number of studies have shown an upregulation of peptidases (including aspartic, cysteine and serine endopeptidases and carboxypeptidases) at the onset or during later phases of senescence, as indicated by enhanced transcript levels or increased activity or both (Feller and Fischer, 1994; Buchanan-Wollaston, 1997; Fischer *et al.*, 1998; Guo *et al.*, 2004).

Regulation of peptidases in senescing leaves has been studied using several model systems, including isolated chloroplasts, chloroplast lysates, leaf segments and plants with altered source–sink interactions (Tranbarger *et al.*, 1991; Mitsuhashi *et al.*, 1992; Fischer and Feller, 1994; Kokubun *et al.*, 2002). Blockage of phloem export from source leaves by 'steam-girdling' leads to (mostly soluble) carbohydrate accumulation and premature senescence (Figure 5.2) (Fröhlich and Feller, 1992; Feller and Fischer, 1994). This treatment can be used to induce leaf senescence in large numbers of leaves in a highly coordinated manner, and is therefore very useful to study processes occurring early during the senescence process, such as plastidial protein degradation. Furthermore, 'shift-girdling' (Figure 5.2) can be utilized to differentiate between senescence-associated processes, and signals that might be derived from wounding stress. For these reasons, leaf girdling has been utilized in the author's laboratory for the analysis of peptidases potentially involved

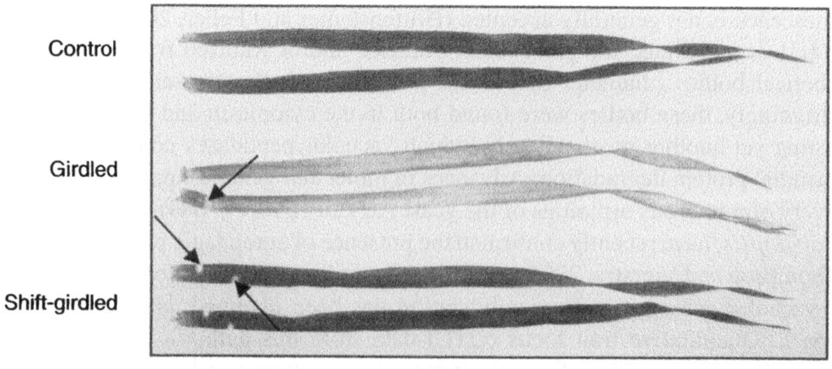

Figure 5.2 Senescence of girdled leaves with completely interrupted phloem, as compared to shift-girdled leaves (wounding control) and untreated control leaves 12 days after treatment. Leaves were excised from intact plants immediately prior to documentation. (From Parrott *et al.*, 2005, with kind permission of Springer Science and Business Media.)

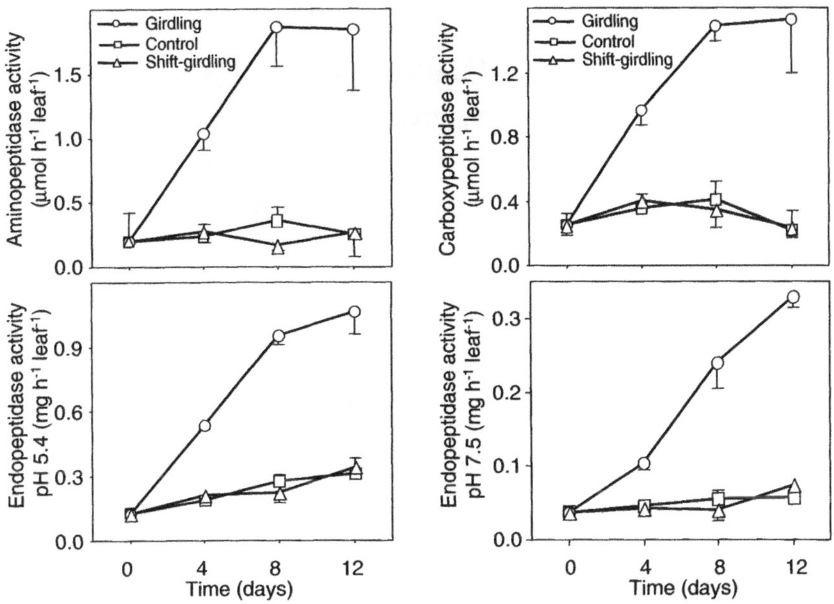

Figure 5.3 Changes in protease activities during senescence of girdled leaves. Aminopeptidases (using L-leu-*p*-nitroanilide as a substrate), carboxypeptidases (using N-CBZ-L-phe-L-ala as a substrate) and endopeptidases at pH 5.4 and 7.5 (using azocasein as a substrate) are shown. (From Parrott *et al.*, 2005, with kind permission of Springer Science and Business Media.)

in plastidial protein degradation. Biochemical methods detected a strong increase of both endo- and exoproteolytic activities in girdled leaves before the onset of chlorophyll and protein degradation (Figure 5.3) (Parrott *et al.*, 2005). Serial Northern blots indicated an upregulation of two cysteine peptidases, an aspartic peptidase, a carboxypeptidase and a proteasome subunit, while expression of two photosynthetic genes decreased rapidly in girdled leaves (Figure 5.4). Shift-girdling controls demonstrated that observed changes were not associated with wounding. Microarray analysis of girdled barley leaves, using the barley1 Affymetrix gene chip (Close *et al.*, 2004), indicated a strong upregulation of several additional peptidases, including two cysteine peptidases, a proline iminopeptidase, a gene coding for a CND41-like protein, two serine carboxypeptidases, a leucine aminopeptidase and an aspartic peptidase (Parrott *et al.*, unpublished). Conspicuously, the products of several of the listed protease genes are likely to be located in vacuolar compartments (cysteine and aspartic peptidases, serine carboxypeptidases), while others (aminopeptidase, CND41) may be located in plastids.

Besides direct regulation of peptidases at the transcriptional and/or posttranscriptional level, additional mechanisms regulating protein degradation in senescing tissues have been described. It has been demonstrated that the susceptibility of substrate proteins to degradation by active peptidases can be altered by damage, covalent modification (e.g. phosphorylation) or other changes in their three-dimensional

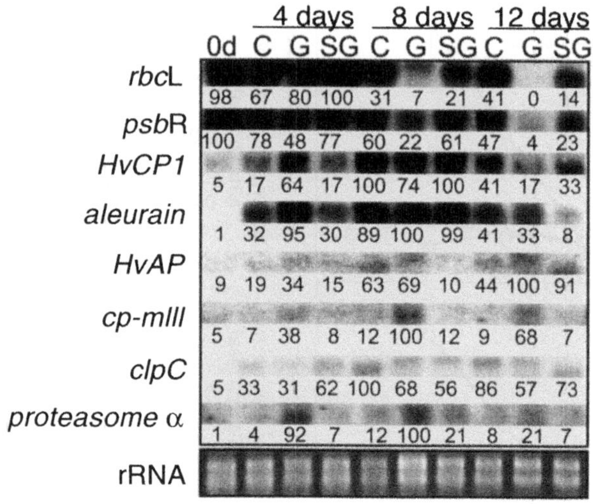

Figure 5.4 Photosynthetic and protease gene expression during senescence of girdled leaves. Transcript levels of Rubisco large subunit (*rbc*L), photosystem II 10-kDa polypeptide (*psb*R), cysteine proteinase 1 (*HvCP1*), *aleurain* (a cysteine peptidase), an aspartic proteinase (*HvAP*), carboxypeptidase III (*cp-mIII*), Clp protease ATP binding subunit (*clp*C) and proteasome α subunit (*proteasome* α) are shown. Leaves were either untreated controls (C), girdled (G; completely interrupted phloem) or shift-girdled (SG; wounding controls). Ten microgram of RNA was loaded in each lane. Numbers represent transcript levels (measured as densities) in percent of the densest band in each row. (From Parrott *et al.*, 2005, with kind permission of Springer Science and Business Media.)

structure (Feller, 2004). In this context, several enzymes have been shown to be stabilized by solutes in the presence of active peptidases (Fischer *et al.*, 1992; Feller, 2004). An especially interesting observation was made for the degradation of light-harvesting chlorophyll-binding proteins in stay-green *Lolium/Festuca* mutants with impaired chlorophyll degradation (Roca *et al.*, 2004). In mutant plants, these proteins were considerably more stable during leaf senescence, while Rubisco was degraded at the same rate as in wild-type plants. These observations suggest a stabilization of light-harvesting proteins by chlorophylls, most likely independently of the type and level of peptidases present in the senescing chloroplasts (Roca *et al.*, 2004, and references cited therein).

5.3.2 Amino acid metabolism in senescing leaves

Immediate degradation products (oligopeptides and amino acids) of plastidial and other proteins reflect the amino acid composition of these substrates. Analysis of plant transport has indicated that certain amino acids, especially the amides asparagine and glutamine, are prevalent long-distance transport forms of organic nitrogen in most higher plant species (Hayashi and Chino, 1990; Atkins, 2000; Fisher, 2000). Unsurprisingly, it has also been shown that enzymes involved in amide biosynthesis are upregulated in senescing leaves.

The best data in this context are available for glutamine synthetase (GS). In plants analyzed so far, GS isoenzymes can be separated into two classes by ion exchange chromatography. One class is localized in the cytosol (GS1), and the other in the chloroplast (GS2). GS2 is typically encoded by a single nuclear gene, while several genes are encoding GS1 subunits (Ireland and Lea, 1999; Coruzzi and Last, 2000). GS holoenzymes in plants function as octamers, and GS1 polypeptides have been found to assemble into both homo- and hetero-octamers. On a tissue level, GS2 is expressed in (photosynthetic) mesophyll cells. Cytosolic GS1 isoenzymes are expressed in vascular tissue (Kamachi *et al.*, 1992; Ireland and Lea, 1999; Coruzzi and Last, 2000), but a combination of immunocytochemical and molecular techniques has demonstrated the specific induction of a GS1 gene in mesophyll cells of senescing tobacco leaves (Brugière *et al.*, 2000). As it has been shown that GS2 is degraded early during leaf senescence, and is unstable in isolated chloroplasts (Streit and Feller, 1983; Kamachi *et al.*, 1991; Mitsuhashi and Feller, 1992), available results indicate that GS1 and not GS2 polypeptides are involved in the reassimilation of nitrogen from the metabolism of other amino acids and N compounds in senescing leaves, and possibly in preparing this nitrogen for phloem loading and export.

Similarly to GS, different isoenzymes have also been found for glutamate synthase (GOGAT). Typically, ferredoxin-dependent GOGAT is the principal form in green leaves. The NADH-dependent isoform dominates in nongreen tissues, but has also been found in leaves (Coruzzi and Last, 2000). Data obtained so far indicate that NADH-GOGAT plays an important role in N remobilization; however, this enzyme appears to be important for N metabolism in developing sink organs (young leaves and grains) rather than in senescing leaves (Obara *et al.*, 2001; Yamaya *et al.*, 2002).

Following incorporation into glutamine and glutamate through the GS–GOGAT cycle, nitrogen is distributed to the other amino acids through the action of aminotransferases (also known as transaminases) (Ireland and Lea, 1999). In this context, an upregulation of aspartate aminotransferase transcript levels has been demonstrated in senescing leaves (Yoshida *et al.*, 2001). Unsurprisingly, as asparagine is an important N transport form in plants, literature data also indicate an increase of asparagine synthetase in senescent tissues (Fujiki *et al.*, 2001; Winichayakul *et al.*, 2004). Additionally, results from the author's laboratory (D. Parrott *et al.*, unpublished), using microarray analysis of girdled barley leaves, showed strongly enhanced transcript levels of several aminotransferase genes (including an aspartate aminotransferase) at the onset of leaf senescence, suggesting a functional role for the corresponding proteins in intermediary N metabolism of senescing leaves.

5.3.3 Nitrogen transport to developing sinks

Amino acids and small peptides released from enhanced proteolysis undergo a series of transport steps between the site of protein degradation and the site of reuse or storage of remobilized nitrogen. Depending on the organelle, in which an amino acid is found after proteolysis and, possibly, metabolic conversion, membrane transport proteins are needed for export from vacuoles or chloroplasts. Within senescing mesophyll tissues, amino acids have to be transported toward the vascular bundles.

An especially interesting case exists in soybean and some related species, where a specific single-cell layer, termed the *paraveinal mesophyll* (PVM), spans the area between the palisade parenchyma, the spongy mesophyll and the minor veins (Fisher, 1967; Lansing and Franceschi, 2000). It has been hypothesized that this cell layer serves as a symplastic conduit for assimilates, including N compounds, from the photosynthetic (mesophyll) cells toward the minor veins. Furthermore, an important dataset indicates that the PVM cell layer also serves as a transitory store for N and (possibly) carbon compounds prior to their export from leaves to sinks (Tranbarger *et al.*, 1991; Staswick, 1994; Bunker *et al.*, 1995).

Loading of photoassimilates and other compounds (including amino acids) into the phloem is probably the best investigated step in higher plant transport processes. A number of studies, using microscopic, physiological and biochemical approaches have established that phloem loading occurs from the apoplast in most species, with the possible exception of plants that utilize raffinose-family oligosaccharides for long-distance carbohydrate transport (Turgeon and Medville, 2004). In consequence, metabolites have to be released from the symplasm (cytosol of leaf cells) into the cell wall area (apoplast) at some point prior to phloem loading, but little molecular detail about this step is available. The actual uptake of nitrogen (and some other) metabolites into the sieve tube/companion cell complex is mediated by specific transport proteins, which couple an electrochemical (H^+) gradient with secondary, active accumulation of amino acids (Lalonde *et al.*, 2004). Once inside the phloem, metabolites are transported from source to sink tissues. The prevalent hypothesis explaining phloem transport is based on the pressure-flow hypothesis, in which phloem loading in source and unloading in sink tissues are the forces responsible for the movement of water and soluble compounds (Fisher, 2000). An important set of data demonstrates that high concentrations of carbon, nitrogen and other compounds are present in phloem sap, and can therefore be moved toward sink tissues (Marschner, 1995; Atkins, 2000). In this context, it is noteworthy that mechanisms for xylem-to-phloem transfer of amino acids have been described in upper stem tissues (Atkins, 2000). As the transpiration of developing seeds is low, this mechanism can make additional, originally xylem-borne nitrogen available for seed import through the phloem.

While phloem unloading in sink tissues, such as developing cereal grains, has received more attention recently (Wang and Fisher, 1994a,b), its mechanism(s) is less well understood than phloem loading. The transport processes involved are complex, both at the cellular level, and with regard to the tissues involved in transporting remobilized compounds from the phloem to the actual sink tissues, such as developing endosperms or cotyledons. In developing legume seeds, sugars (and, likely, N metabolites) pass symplastically from the phloem to the inner epidermis of the seed coat, where they are unloaded to the apoplast surrounding the embryo (Weber *et al.* 1998a,b). Nutrients are then taken up by carrier-mediated transport in the outer epidermal cell layers of the embryo (McDonald *et al.*, 1996).

The microscopic and physiological approaches mentioned above were clearly able to delineate an overall picture of long-distance transport in plants. Due to the fact that membrane proteins are more resistant to biochemical characterization than (soluble) enzymes, our understanding of membrane transport has been considerably enhanced by molecular approaches, including the use of mutant yeast strains (deficient in their uptake of carbon or nitrogen metabolites) allowing the functional characterization of plant transporters. Not unlike the situation with peptidase genes and gene families, molecular methods have led to the identification of numerous transporters; the best investigated plant membrane transport proteins are involved in sucrose transport, amino acid/oligopeptide transport and potassium transport (Kochian, 2000; Lalonde *et al.*, 2004). Not including mitochondrial transporters, the *Arabidopsis* genome contains 53 putative amino acid transporter genes, and ~59 have been found in rice (Lalonde *et al.*, 2004). These genes can be grouped into three 'superfamilies', namely (a) the amino acid–polyamine-choline transporter superfamily (APC), (b) the amino acid transporter superfamily 1 (ATF1) and (c) the amino acid transporters belonging to the major facilitator superfamily (MFS) (Lalonde *et al.*, 2004). The best studied 'subfamily' (within ATF1) are amino acid permeases, which are preferentially expressed in vascular tissues and mediate the H^+-coupled uptake of a wide variety of amino acids. Roles for these genes in amino acid transport are indicated by the fact that tuber amino acid content of antisense *StAAP1* potato lines is lowered (Koch *et al.*, 2003), and that other members of this family are expressed in seeds, suggesting a role in seed supply (Lalonde *et al.*, 2004). Analysis of higher plant N transport is complicated by the presence of large numbers of peptide transporters. It is now clear that at least some of these genes are also important for the nutrition of sinks such as developing seeds. Specifically, it has been demonstrated in *Arabidopsis* that antisense repression of *AtPTR2* and T-DNA insertion in *AtOPT3* (members of two different peptide transporter families) resulted in an arrest of embryonic development (Stacey *et al.*, 2002). Unfortunately, due to the high number of amino acid combinations in even short peptides, identification of peptides preferentially transported by these proteins is difficult.

Overall, our current knowledge regarding the biochemistry, physiology and molecular biology of transport processes needed for efficient N remobilization is not unlike the situation regarding proteolysis of major photosynthetic proteins. While important insights have been gained, some gaps in our knowledge persist. In both cases, the complexity of plant genetics (dissection of functions among members of gene families) contributes to the difficulty of assigning distinct functions to key genes.

5.4 Outlook

The availability of genomic and proteomic methods has added a substantial amount of information to our knowledge on senescence processes, including nutrient remobilization from senescing leaves (Gepstein *et al.*, 2003; Guo *et al.*, 2004; Agrawal *et al.*, 2005; Buchanan-Wollaston *et al.*, 2005). One obvious problem with genomic techniques is their limitation to the detection of changes in transcript abundance. As outlined in this chapter and in the recent literature (Feller, 2004; Schaller, 2004), peptidase function is regulated at different levels, and it is at present not clear if an increase in their transcript levels is necessary or typical for all or some of the peptidases functionally involved in N remobilization from senescing tissues. The use of high-throughput proteomic methods (e.g. Carter *et al.*, 2004; Agrawal *et al.*, 2005) will be helpful, as they have the potential to complement expression-level with protein-level data. In the context of plant proteolysis, activity profiling represents an especially promising technique. In this approach, active peptidases are covalently tagged with probes derived from known peptidase inhibitors, followed by their detection and analysis with either immunological (van der Hoorn *et al.*, 2004) or mass spectrometric (Okerberg *et al.*, 2005) techniques. Mass spectrometry, in combination with the use of results of genomic or EST sequencing efforts, has the capacity to directly correlate an activity-tagged protein with the corresponding gene. Together, genomic, proteomic and activity tagging/proteomic approaches can therefore furnish information on transcript levels, protein levels and protein (peptidase) activity. The strength of these techniques is their potential to identify candidate genes and contribute to the generation of new hypotheses. Ultimately, proof of the function of peptidase genes (and other genes involved in nutrient remobilization) will be obtained only by generating or identifying (e.g. from T-DNA insertions) appropriate mutants. Experience with peptidases involved in processing seed storage proteins indicates that due to the presence of gene families with (sometimes?) redundant functions, lines with several mutations/knockouts may be needed to obtain clear phenotypes (Shimada *et al.*, 2003; Gruis *et al.*, 2004). In spite of these (and other) remaining challenges, it appears that at the beginning of the twenty-first century, the necessary tools are available to achieve a full functional understanding of nutrient remobilization and other senescence processes at the molecular and cellular level.

Acknowledgments

Research in the author's laboratory has been supported by the USDA-NRI (grants 01–01019 and 05–02022), by the US Barley Genome Project and by the Montana Board of Research and Commercialization Technology (grants 04–10 and 06–10). The author would like to thank Dr S.J. Crafts-Brandner (USDA Western Cotton Research Lab, Phoenix, AZ) and Dr U. Feller (Institute of Plant Sciences, University of Bern, Bern, Switzerland) for critically reading the manuscript, and for helpful comments.

References

Adam, Z. and Clarke, A.K. (2002) Cutting edge of chloroplast proteolysis. *Trends Plant Sci* **7**(10), 451–456.

Agrawal, G.K., Yonekura, M., Iwahashi, Y., Iwahashi, H. and Rakwal, R. (2005) Systems, trends and perspectives of proteomics in dicot plants. Part II: Proteomes of the complex developmental stages. *J Chromatogr B* **815**, 125–136.

Atkins, C. (2000) Biochemical aspects of assimilate transfers along the phloem path: N-solutes in lupins. *Aust J Plant Physiol* **27**, 531–537.

Brouquisse, R., Masclaux, C., Feller, U. and Raymond, P. (2001) Protein hydrolysis and nitrogen remobilization in plant life and senescence. In: *Plant Nitrogen* (eds Lea, P.J. and Morot-Gaudry, J.-F.). Springer, Berlin, pp. 275–293.

Brugière, N., Dubois, F., Masclaux, C., Sangwan, R.S. and Hirel, B. (2000) Immunolocalization of glutamine synthetase in senescing tobacco (*Nicotiana tabacum* L.) leaves suggests that ammonia assimilation is progressively shifted to the mesophyll cytosol. *Planta* **211**, 519–527.

Buchanan-Wollaston, V. (1997) The molecular biology of leaf senescence. *J Exp Bot* **48**(307), 181–199.

Buchanan-Wollaston, V., Page, T., Harrison, E., *et al.* (2005) Comparative transcriptome analysis reveals significant differences in gene expression and signalling pathways between developmental and dark/starvation-induced senescence in *Arabidopsis*. *Plant J* **42**, 567–585.

Bunker, T.W., Koetje, D.S., Stephenson, L.C., Creelman, R.A., Mullet, J.E. and Grimes, H.D. (1995) Sink limitation induces the expression of multiple soybean vegetative lipoxygenase mRNAs while the endogenous jasmonic acid level remains low. *Plant Cell* **7**, 1319–1331.

Carter, C., Pan, S., Zouhar, J., Avila, E.L., Girke, T. and Raikhel, N.V. (2004) The vegetative vacuole proteome of *Arabidopsis thaliana* reveals predicted and unexpected proteins. *Plant Cell* **16**, 3285–3303.

Chiba, A., Ishida, H., Nishizawa, N.K., Makino, A. and Mae, T. (2003) Exclusion of ribulose-1,5-bisphosphate carboxylase/oxygenase from chloroplasts by specific bodies in naturally senescing leaves of wheat. *Plant Cell Physiol* **44**(9), 914–921.

Clarke, A.K., MacDonald, T.M. and Sjögren, L.L.E. (2005) The ATP-dependent Clp protease in chloroplasts of higher plants. *Physiol Plant* **123**, 406–412.

Close, T.J., Wanamaker, S.I., Caldo, R.A., *et al.* (2004) A new resource for cereal genomics: 22K barley GeneChip comes of age. *Plant Physiol* **134**, 960–968.

Coruzzi, G. and Last, R. (2000) Amino acids. In: *Biochemistry and Molecular Biology of Plants* (eds Buchanan, B., Gruissem, W. and Jones, R.). American Society of Plant Physiologists, Rockville, MD, pp. 358–410.

Crafts-Brandner, S.J. (1992) Significance of leaf phosphorus remobilization in yield production in soybean. *Crop Science* **32**, 420–424.

Crafts-Brandner, S.J., Hölzer, R. and Feller, U. (1998) Influence of nitrogen deficiency on senescence and the amounts of RNA and proteins in wheat leaves. *Physiol Plant* **102**(2), 192–200.

Crawford, N.M., Kahn, M.L., Leustek, T. and Long, S.R. (2000) Nitrogen and sulfur. In *Biochemistry and Molecular Biology of Plants* (eds Buchanan, B., Gruissem, W. and Jones, R.). American Society of Plant Physiologists, Rockville, MD, pp. 786–849.

Debrunner, N. and Feller, U. (1995) Solute leakage from detached plant parts of winter wheat: influence of maturation stage and incubation temperature. *J Plant Physiol* **145**, 257–260.

Diaz, C., Purdy, S., Christ, A., Morot-Gaudry, J.-F., Wingler, A. and Masclaux-Daubresse, C. (2005) Characterization of markers to determine the extent and variability of leaf senescence in *Arabidopsis*. A metabolic profiling approach. *Plant Physiol* **138**, 898–908.

Feller, U. (2004) Proteolysis. In: *Plant Cell Death Processes* (ed. Noodén, L.D.). Elsevier Academic Press, Amsterdam, pp. 107–123.

Feller, U. and Fischer, A. (1994) Nitrogen metabolism in senescing leaves. *Crit Rev Plant Sci* **13**(3), 241–273.

Fischer, A., Brouquisse, R. and Raymond, P. (1998) Influence of senescence and carbohydrate levels on the pattern of leaf proteases in purple nutsedge (*Cyperus rotundus*). *Physiol Plant* **102**(3), 385–395.

Fischer, A. and Feller, U. (1994) Senescence and protein degradation in leaf segments of young winter wheat: influence of leaf age. *J Exp Bot* **45**(270), 103–109.

Fischer, A., Salgó, A., Hildbrand, M. and Feller, U. (1992) Cooperative protection of glucose-6-phosphate dehydrogenase by ligands in extracts from wheat grains. *Biochem Physiol Pflanz* **188**, 295–303.

Fisher, D. (1967) An unusual layer of cells in the mesophyll of the soybean leaf. *Bot Gaz* **128**(3–4), 215–218.

Fisher, D.B. (2000) Long-distance transport. In: *Biochemistry and Molecular Biology of Plants* (eds Buchanan, B., Gruissem, W. and Jones, R.). American Society of Plant Physiologists, Rockville, MD, pp. 730–784.

Fröhlich, V. and Feller, U. (1992) Effect of phloem interruption on endopeptidase and aminopeptidase activities in flag leaves of field-grown wheat. *Biochem Physiol Pflanz* **188**, 13–21.

Fujiki, Y., Yoshikawa, Y., Sato, T., *et al.* (2001) Dark-inducible genes from *Arabidopsis thaliana* are associated with leaf senescence and repressed by sugars. *Physiol Plant* **111**, 345–352.

Gastal, F. and Lemaire, G. (2002) N uptake and distribution in crops: an agronomical and ecophysiological perspective. *J ExpBot* **53**(370), 789–799.

Gepstein, S. (1988) Photosynthesis. In: *Senescence and Aging in Plants* (eds Noodén, L.D. and Leopold, A.C.). Academic Press, San Diego, CA, pp. 85–109.

Gepstein, S., Sabehi, G., Carp, M.-J., *et al.* (2003) Large-scale identification of leaf senescence-associated genes. *Plant J* **36**, 629–642.

Good, A.G., Shrawat, A.K. and Muench, D.G. (2004) Can less yield more? Is reducing nutrient input into the environment compatible with maintaining crop production? *Trends Plant Sci* **9**(12), 597–605.

Graham, I.A., Leaver, C.J. and Smith, S.M. (1992) Induction of malate synthase gene expression in senescent and detached organs of cucumber. *Plant Cell* **4**, 349–357.

Gruis, D.F., Schulze, J. and Jung, R. (2004) Storage protein accumulation in the absence of the vacuolar processing enzyme family of cysteine proteases. *Plant Cell* **16**, 270–290.

Guo, Y., Cai, Z. and Gan, S. (2004) Transcriptome of *Arabidopsis* leaf senescence. *Plant Cell Environ* **27**, 521–549.

Gut, H. and Matile, P. (1988) Apparent induction of key enzymes of the glyoxylic acid cycle in senescent barley leaves. *Planta* **176**, 548–550.

Gut, H. and Matile, P. (1989) Breakdown of galactolipids in senescent barley leaves. *Bot Acta* **102**, 31–36.

Hart, A.L. and Jessop, D. (1984) Leaf phosphorus fractionation and growth responses to phosphorus of the forage legumes *Trifolium repens*, *T. dubium* and *Lotus pedunculatus*. *Physiol Plant* **61**, 435–440.

Hayashi, H. and Chino, M. (1990) Chemical composition of phloem sap from the uppermost internode of the rice plant. *Plant Cell Physiol* **31**, 247–251.

Hayati, R., Egli, D.B. and Crafts-Brandner, S.J. (1995) Carbon and nitrogen supply during seed filling and leaf senescence in soybean. *Crop Sci* **35**, 1063–1069.

He, Y. and Gan, S. (2002) A gene encoding an acyl hydrolase is involved in leaf senescence in *Arabidopsis*. *Plant Cell* **14**, 805–815.

Hill, J., Robson, A.D. and Loneragan, J.F. (1979) The effect of copper supply on the senescence and the retranslocation of nutrients of the oldest leaf of wheat. *Ann Bot* **44**, 279–287.

Hörtensteiner, S. and Feller, U. (2002) Nitrogen metabolism and remobilization during senescence. *J Exp Bot* **53**(370), 927–937.

Ireland, R.J. and Lea, P.J. (1999) The enzymes of glutamine, glutamate, asparagine, and aspartate metabolism. In: *Plant Amino Acids. Biochemistry and Biotechnology* (ed. Singh, B.K.). Marcel Dekker, New York, pp. 49–109.

Ishimaru, K., Kosone, M., Sasaki, H. and Kashiwagi, T. (2004) Leaf contents differ depending on the position in a rice leaf sheath during sink–source transition. *Plant Physiol Biochem* **42**, 855–860.

Jeong, M.L., Jiang, H., Chen, H.-S., Tsai, C.-J. and Harding, S.A. (2004) Metabolic profiling of the sink-to-source transition in developing leaves of quaking aspen. *Plant Physiol* **136**, 3364–3375.

Kamachi, K., Yamaya, T., Hayakawa, T., Mae, T. and Ojima, K. (1992) Vascular bundle-specific localization of cytosolic glutamine synthetase in rice leaves. *Plant Physiol* **99**, 1481–1486.

Kamachi, K., Yamaya, T., Mae, T. and Ojima, K. (1991) A role for glutamine synthetase in the remobilization of leaf nitrogen during natural senescence in rice. *Plant Physiol* **96**, 411–417.

Kato, Y., Murakami, S., Yamamoto, Y., *et al.* (2004) The DNA-binding protease, CND41, and the degradation of ribulose-1,5-bisphosphate carboxylase/oxygenase in senescent leaves of tobacco. *Planta* **220**, 97–104.

Killingbeck, K.T. (2004) Nutrient resorption. In: *Plant Cell Death Processes* (ed. Noodén, L.D.). Elsevier Academic Press, Amsterdam, pp. 215–226.

Koch, W., Kwart, M., Laubner, M., *et al.* (2003) Reduced amino acid content in transgenic potato tubers due to antisense inhibition of the leaf H+/amino acid symporter StAAP1. *Plant J* **33**, 211–220.

Kochian, L.V. (2000) Molecular physiology of mineral nutrient acquisition, transport, and utilization. In: *Biochemistry and Molecular Biology of Plants* (eds Buchanan, B., Gruissem, W. and Jones, R.). American Society of Plant Physiologists, Rockville, MD, pp. 1204–1249.

Kokubun, N., Ishida, H., Makino, A. and Mae, T. (2002) The degradation of the large subunit of ribulose-1,5-bisphosphate carboxylase/oxygenase into the 44-kDa fragment in the lysates of chloroplasts incubated in darkness. *Plant Cell Physiol* **43**(11), 1390–1395.

Lalonde, S., Wipf, D. and Frommer, W.B. (2004) Transport mechanisms for organic forms of carbon and nitrogen between source and sink. *Annu Rev Plant Biol* **55**, 341–372.

Lansing, A.J. and Franceschi, V.R. (2000) The paraveinal mesophyll: a specialized path for intermediary transfer of assimilates in legume leaves. *Aust J Plant Physiol* **27**, 757–767.

Lers, A., Lomaniec, E., Burd, S. and Khalchitski, A. (2001) The characterization of LeNUC1, a nuclease associated with leaf senescence of tomato. *Physiol Plant* **112**, 176–182.

Levey, S. and Wingler, A. (2005) Natural variation in the regulation of leaf senescence and relation to other traits in *Arabidopsis*. *Plant Cell Environ* **28**, 223–231.

Lodwig, E. and Poole, P. (2003) Metabolism of *Rhizobium* bacteroids. *Crit Rev Plant Sci* **22**, 37–78.

Mae, T., Makino, A. and Ohira, K. (1983) Changes in the amounts of ribulose bisphosphate carboxylase synthesized and degraded during the life span of rice leaf (*Oryza sativa* L.). *Plant Cell Physiol* **24**(6), 1079–1086.

Marschner, H. (1995) *Mineral Nutrition of Higher Plants*. Academic Press, London.

McDonald, R., Fieuw, S. and Patrick, J.W. (1996) Sugar uptake by the dermal transfer cells of developing cotyledons of *Vicia faba* L. – experimental systems and general transport properties. *Planta* **198**, 54–63.

McLaughlin, J.C. and Smith, S.M. (1994) Metabolic regulation of glyoxylate-cycle enzyme synthesis in detached cucumber cotyledons and protoplasts. *Planta* **195**, 22–28.

Minamikawa, T., Toyooka, K., Okamoto, T., Hara-Nishimura, I. and Nishimura, M. (2001) Degradation of ribulose-bisphosphate carboxylase by vacuolar enzymes of senescing French bean leaves: immunocytochemical and ultrastructural observations. *Protoplasma* **218**, 144–153.

Mitsuhashi, W., Crafts-Brandner, S.J. and Feller, U. (1992) Ribulose-1,5-bis-phosphate carboxylase/oxygenase degradation in isolated pea chloroplasts incubated in the light or in the dark. *J Plant Physiol* **139**, 653–658.

Mitsuhashi, W. and Feller, U. (1992) Effects of light and external solutes on the catabolism of nuclear-encoded stromal proteins in intact chloroplasts isolated from pea leaves. *Plant Physiol* **100**, 2100–2105.

Noodén, L.D., Guiamét, J. L. and John, I. (2004) Whole plant senescence. In: *Plant Cell Death Processes* (ed. Noodén, L.D.). Elsevier Academic Press, Amsterdam, pp. 227–244.

Obara, M., Kajiura, M., Fukuta, Y., *et al.* (2001) Mapping of QTLs associated with cytosolic glutamine synthetase and NADH-glutamate synthase in rice (*Oryza sativa* L.). *J Exp Bot* **52**(359), 1209–1217.

Okerberg, E.S., Wu, J., Zhang, B., *et al.* (2005) High-resolution functional proteomics by active-site peptide profiling. *Proc Natl Acad Sci U S A* **102**(14), 4996–5001.

Otegui, M.S., Noh, Y.-S., Martínez, D.E., *et al.* (2005) Senescence-associated vacuoles with intense proteolytic activity develop in leaves of *Arabidopsis* and soybean. *Plant J* **41**, 831–844.

Parrott, D., Yang, L., Shama, L. and Fischer, A.M. (2005) Senescence is accelerated, and several proteases are induced by carbon 'feast' conditions in barley (*Hordeum vulgare* L.) leaves. *Planta* **222**, 989–1000.

Peoples, M.B. and Dalling, M.J. (1988) The interplay between proteolysis and amino acid metabolism during senescence and nitrogen reallocation. In: *Senescence and Aging in Plants* (eds Nooden, L.D. and Leopold, A.C.). Academic Press, San Diego, CA, pp. 181–217.

Pistelli, L., De Bellis, L. and Alpi, A. (1991) Peroxisomal enzyme activities in attached senescing leaves. *Planta* **184**, 151–153.

Rawlings, N.D. and Barrett, A.J. (1993) Evolutionary families of peptidases. *Biochem J* **290**, 205–218.

Rawlings, N.D., Tolle, D.P. and Barrett, A.J. (2004) MEROPS: the peptidase database. *Nucleic Acids Res* **32**(Database issue), D160–D164.

Rennenberg, H. and Lamoureux, G.L. (1990) Physiological processes that modulate the concentration of glutathione in plant cells. In: *Sulfur Nutrition and Sulfur Assimilation in Higher Plants* (ed. Rennenberg, H.). XPB Academic Publishers B.V., The Hague, pp. 53–65.

Robson, P.R.H., Donnison, I.S., Wang, K., *et al.* (2004) Leaf senescence is delayed in maize expressing the *Agrobacterium IPT* gene under the control of a novel maize senescence-enhanced promoter. *Plant Biotechnol J* **2**, 101–112.

Roca, M., James, C., Pružinská, A., Hörtensteiner, S., Thomas, H. and Ougham, H. (2004) Analysis of the chlorophyll catabolism pathway in leaves of an introgression senescence mutant of *Lolium temulentum*. *Phytochemistry* **65**, 1231–1238.

Roulin, S., Buchala, A.J. and Fincher, G.B. (2002) Induction of (1→3,1→4)-β-D-glucan hydrolases in leaves of dark-incubated barley seedlings. *Planta* **215**, 51–59.

Roulin, S. and Feller, U. (1998) Light-independent degradation of stromal proteins in intact chloroplasts isolated from *Pisum sativum* L. leaves: requirement for divalent cations. *Planta* **205**, 297–304.

Schaller, A. (2004) A cut above the rest: the regulatory function of plant proteases. *Planta* **220**, 183–197.

Schmid, M., Simpson, D., Kalousek, F. and Gietl, C. (1998) A cysteine endopeptidase with a C-terminal KDEL motif isolated from castor bean endosperm is a marker enzyme for the ricinosome, a putative lytic compartment. *Planta* **206**, 466–475.

Schmid, M., Simpson, D.J., Sarioglu, H., Lottspeich, F. and Gietl, C. (2001) The ricinosomes of senescing plant tissue bud from the endoplasmic reticulum. *Proc Natl Acad Sci U S A* **98**(9), 5353–5358.

Scott, D.A., Proctor, J. and Thompson, J. (1992) Ecological studies on a lowland evergreen rain forest on Maraca island, Brazil. II: Litter and nutrient recycling. *J Ecol* **80**, 705–717.

Shimada, T., Yamada, K., Kataoke, M., *et al.* (2003) Vacuolar processing enzymes are essential for proper processing of seed storage proteins in *Arabidopsis thaliana*. *J Biol Chem* **278**(34), 32292–32299.

Smalle, J. and Vierstra, R.D. (2004) The ubiquitin 26S proteasome proteolytic pathway. *Annu Rev Plant Biol* **55**, 555–590.

Snapp, S.S. and Lynch, J.P. (1996) Phosphorus distribution and remobilization in bean plants as influenced by phosphorus nutrition. *Crop Sci* **36**, 929–935.

Stacey, G., Koh, S., Granger, C. and Becker, J.M. (2002) Peptide transport in plants. *Trends Plant Sci* **7**(6), 257–263.

Staswick, P.E. (1994) Storage proteins of vegetative plant tissues. *Annu Rev Plant Physiol Plant Mol Biol* **45**, 303–322.

Streit, L. and Feller, U. (1983) Changing activities and different resistance to proteolytic activity of two forms of glutamine synthetase in wheat leaves during senescence. *Physiol Vég* **21**, 103–108.

Sunarpi and Anderson, J.W. (1997) Effect of nitrogen nutrition on remobilization of protein sulfur in the leaves of vegetative soybean and associated changes in soluble sulfur metabolites. *Plant Physiol* **115**, 1671–1680.

Thayer, S.S., Choe, H.T., Rausser, S. and Huffaker, R.C. (1988) Characterization and subcellular localization of aminopeptidases in senescing barley leaves. *Plant Physiol* **87**, 894–897.

Thomas, H. and Howarth, C.J. (2000) Five ways to stay green. *J Exp Bot* **51**, 329–337.

Thompson, A.R. and Vierstra, R.D. (2005) Autophagic recycling: lessons from yeast help define the process in plants. *Curr Opin Plant Biol* **8**, 165–173.

Tilsner, J., Kassner, N., Struck, C. and Lohaus, G. (2005) Amino acid contents and transport in oilseed rape (*Brassica napus* L.) under different nitrogen conditions. *Planta* **221**, 328–338.

Tranbarger, T.J., Franceschi, V.R., Hildebrand, D.F. and Grimes, H.D. (1991) The soybean 94-kilodalton vegetative storage protein is a lipoxygenase that is localized in paraveinal mesophyll cell vacuoles. *Plant Cell* **3**, 973–987.

Tukey, H.B. (1970) The leaching of substances from plants. *Annu Rev Plant Physiol* **21**, 305–324.

Turgeon, R. and Medville, R. (2004) Phloem loading. A reevaluation of the relationship between plasmodesmatal frequencies and loading strategies. *Plant Physiol* **136**, 3795–3803.

Tyler, G. (2005) Changes in the concentrations of major, minor and rare-earth elements during leaf senescence and decomposition in a *Fagus sylvatica* forest. *Forest Ecol Manage* **206**, 167–177.

Valenzuela, J.L., Ruíz, J.M., Belakbir, A. and Romero, L. (1996) Effects of nitrogen, phosphorus and potassium treatments on phosphorus fractions in melon plants. *Commun Soil Sci Plant Anal* **27**(5–8), 1417–1425.

van der Hoorn, R.A.L., Leeuwenburgh, M. A., Bogyo, M., Joosten, M.H.A.J. and Peck, S .C. (2004) Activity profiling of papain-like cysteine proteases in plants. *Plant Physiol* **135**, 1170–1178.

Wang, N. and Fisher, D.B. (1994a) The use of fluorescent tracers to characterize the post-phloem transport pathway in maternal tissues of developing wheat grains. *Plant Physiol* **104**, 17–27.

Wang, N. and Fisher, D.B. (1994b) Monitoring phloem unloading and post-phloem transport by microperfusion of attached wheat grains. *Plant Physiol* **104**, 7–16.

Weber, H., Borisjuk, L. and Wobus, U. (1998a) Controlling development and seed size in *Vicia faba*: a role for seed coat-associated invertases and carbohydrate status. *Plant J* **10**, 823–834.

Weber, H., Heim, U., Golombek, S., Borisjuk, L. and Wobus, U. (1998b) Assimilate uptake and the regulation of seed development. *Seed Sci Res* **8**, 331–345.

Weng, X.-Y., Xu, H.-X. and Jiang, D.-A. (2005) Characteristics of gas exchange, chlorophyll fluorescence and expression of key enzymes in photosynthesis during leaf senescence in rice plants. *J Integr Plant Biol* **47**(5), 560–566.

Winichayakul, S., Moyle, R.L., Ryan, D.J., Farnden, K.J.F., Davies, K.M. and Coupe, S.A. (2004) Distinct cis-elements in the *Asparagus officinalis* asparagine synthetase promoter respond to carbohydrate and senescence signals. *Funct Plant Biol* **31**(6), 573–582.

Yamaya, T., Obara, M., Nakajima, H., Sasaki, S., Hayakawa, T. and Sato, T. (2002) Genetic manipulation and quantitative-trait loci mapping for nitrogen recycling in rice. *J Exp Bot* **53**(370), 917–925.

Yang, J.C., Zhang, J.H., Wang, Z.Q., Zhu, Q.S. and Liu, L.J. (2003) Involvement of abscisic acid and cytokinins in the senescence and remobilization of carbon reserves in wheat subjected to water stress during grain filling. *Plant Cell Environ* **26**, 1621–1631.

Yang, L., Mickelson, S., See, D., Blake, T.K. and Fischer, A.M. (2004) Genetic analysis of the function of major leaf proteases in barley (*Hordeum vulgare* L.) nitrogen remobilization. *J Exp Bot* **55**(408), 2607–2616.

Yoshida, S., Ito, M., Nishida, I. and Watanabe, A. (2001) Isolation and RNA gel blot analysis of genes that could serve as potential molecular markers for leaf senescence in *Arabidopsis thaliana*. *Plant Cell Physiol* **42**(2), 170–178.

6 Environmental regulation of leaf senescence

Amnon Lers

6.1 Introduction

Leaf senescence is a genetically controlled degenerative process leading to cell death, which occurs at a given time, even when growth conditions are near optimal. Senescence that occurs as a part of normal development is frequently referred to as developmental or age-dependent senescence, as it is induced and controlled by endogenous factors operating during plant growth and maturation. However, senescence may be induced prematurely, via exposure to environmental stimuli, to initiate under optimal growth conditions. Many of these stimuli occur in the form of a stress for the plant. In their natural habitat, plants are frequently exposed to environmental stress conditions, which may adversely affect their growth, metabolism, development and productivity. Environmental stresses may be biotic, resulting from the interaction with other organisms, or abiotic, resulting from change in physical or chemical components in their environment compared to optimal growth conditions. It is clear that global changes endangering our future environment, such as increasing temperatures, changes in precipitation, and altered atmospheric gas composition and radiation, will have a profound effect on plant growth and development (McCarthy *et al.*, 2001b). These environmental factors have a distinct effect on the initiation and progress of plant senescence. The intricate processes that occur in the attached leaves of annual plants during senescence serve as ideal examples of the environmental regulation of senescence, and this will be the focus of this chapter.

Key environmental stresses include extremes of light or temperature, radiation, drought, nutrient deficiency, pathogen infection and the presence of toxic materials in the air, water, or soil. Developmental factors associated with senescence initiation include age, reproductive development, and phytohormone levels and interactions. A very close relationship exists between environmental conditions and plant development. Thus, it is sometimes difficult to separate the more direct effects of environmental signals on the initiation and progression of senescence from indirect effects of these signals, which instead result from attenuation of the normal plant development program. In some monocarpic plants, a correlative control of leaf senescence exists in which development of one part of the plant can influence senescence initiation or progress (Nooden, 1984; Nooden and Penney, 2001). For example, environmental stress may influence flowering, which can then alter the timing of senescence initiation. Another source of variation of the effects of environmental signals on senescence may be the development of stress adaptive mechanisms in different plants. Thus, examples demonstrating the effect of different environmental factors on the initiation of senescence may not be taken as a general trend, but must be examined for their effect on senescence in each specific plant system.

Environmental stress is frequently diverting the program of plant development from a normal and relatively slow program into a process with greater metabolic activity and accelerated progress of senescence. On the other hand, some studies suggest that a slow-down of metabolic activity throughout plant development due to specific mutations tend to result in the retardation of developmental senescence. For example, in *Arabidopsis*, the *ore4* mutation resulted in a reduced level of the plastid ribosomal small subunit protein, which was accompanied by delayed senescence (Woo *et al.*, 2002). In tobacco, a reduction of Rubisco protein levels in antisense *rbcL* plants resulted in the delay of senescence (Miller *et al.*, 2000). In contrast, when carbohydrate production is increased by growing plants in elevated CO_2 conditions, a shift in the normal stages of leaf ontogeny is observed and senescence is induced earlier in leaf development (Miller *et al.*, 1997).

Senescence initiation resulting from environmental stress may be viewed as a defense response. Although senescence leads to cell and tissue death, its occurrence is able to support plants' survival during stress by contributing to the continuity of reproduction, thus playing a role in stress resistance at the species level. For example, limited nutrient and water availability are dominant and frequent factors that affect plant development and senescence in many ecosystems. The development of accelerated leaf senescence in response to these stresses was suggested to have an adaptive significance, enabling the plant to complete its life cycle and produce viable seeds utilizing the available resources (Munne-Bosch and Alegre, 2004). It was suggested that removal of unproductive leaves as a result of accelerated senescence, accompanied by abscission, during drought stress reduces water loss through transpiration, contributing to the water balance of the intact plant (Munne-Bosch and Alegre, 2004). Organized senescence allows recycling of valuable nutrients from photosynthetically less efficient leaves to the productive organs, improving the prospect of seed production. Likewise, acceleration of senescence in leaves infected by pathogen leads to their removal, thus lowering the risk of additional pathogen spread. Thus, senescence is a developmental program that was likely selected by evolution to optimize plant survival (Granell, 1999). It is possible that plants have evolved mechanisms by which environmental stresses may induce leaf senescence to enable improved adaptation of a plant to its changing environment. In this case, the regulation of leaf senescence by environmental stimuli has an obvious adaptive value, allowing the plant to complete its life cycle even under stressful conditions.

The relationship between senescence and environmental stress is manifested by the involvement of three main plant hormones in the activation of processes related to senescence and different biotic and abiotic stress responses. In many cases, the regulatory effect of environmental stress on senescence can be mediated through these hormonal pathways. Ethylene, jasmonic acid (JA) and salicylic acid (SA) have been implicated in senescence (Grbic and Bleecker, 1995; Morris *et al.*, 2000; He *et al.*, 2002). Levels of these signaling molecules increase during senescence and induce the expression of specific genes. Comparative genomic analyses of senescence-related transcriptomes enabled groupings of associated genes into classes that are dependent on different phytohormone signaling pathways. These analyses also allowed the identification of genes that are independent of these stress-related growth regulators (Buchanan-Wollaston *et al.*, 2005).

Further support for the relationship between environmental stress and senescence is validated by the analysis of the expression profile of *Arabidopsis* transcription factor genes. This analysis suggests that biotic and abiotic stress responses and senescence may share overlapping signaling pathways (Chen *et al.*, 2002). Out of 402 distinct transcription factor genes, 43 were found to be induced during senescence. Interestingly, 28 of the 43 genes were also induced by environmental stress treatment, suggesting extensive overlap in signaling between senescence and stress response (Chen *et al.*, 2002).

Common and specific components of environmentally induced senescence and developmental senescence are illustrated by the effects of different mutations on senescence progress or on the regulation of senescence-associated genes (*SAG*s). In the *Arabidopsis ore4* mutant, developmental senescence is retarded, but not senescence induced artificially by darkness or phytohormones which sometimes are associated with environmental stress. (Woo *et al.*, 2002). On the other hand, the study of the *ore9* and *dls1* mutations, affected in the ubiquitin-dependent protein degradation pathways, revealed consequence to both age-dependent and artificially induced senescence.

The study of environmentally induced senescence has applied agricultural relevance. Photosynthetic capacity of plant leaves declines markedly with age and as senescence approaches (Gay and Thomas, 1995). The progress of the senescence program significantly impacts the ultimate contribution that a leaf makes to the plant. Delay in senescence onset can significantly increase carbon fixation in the plant (Thomas and Howarth, 2000). Agricultural crops are frequently exposed to different environmental stresses that may induce premature leaf senescence. Understanding the mechanisms by which these environmental conditions affect the senescence process may be of significant economical importance.

Early physiological and molecular studies have indicated that the regulation of leaf senescence is a highly complex process affected by both endogenous developmental signals that act independently or in concert with external environmental factors. (He *et al.*, 2001; Buchanan-Wollaston *et al.*, 2003; Yoshida, 2003). The complexity of senescence regulation is even further supported from recent genomic studies in which the expression and function of *SAG*s is investigated in a more comprehensive way (Buchanan-Wollaston *et al.*, 2003; Buchanan-Wollaston *et al.*, 2005; Guo and Gan, 2005; Lim and Nam, 2005).

The available information regarding developmental or environmental control of senescence suggests that developmental and environmental factors may regulate senescence via common internal systems or factors such as oxidative state and sugars. Environmental stimuli affect plant development in a severe manner, diverting the process from its normal course to an early and accelerated induction of senescence. Studies of the effects of different environmental stimuli on leaf senescence are reviewed with emphasis on the possible mechanisms involved and the relationship between these environmental signals and the senescence regulatory program of the plant.

6.2 Light irradiance

Light has a vital and central role in plant growth and development. It is required for photosynthesis and as a signal for normal development and interaction with the environment. This interaction is mediated through specific pigments such as the phytochrome and cryptochrome or the phototropins light receptor and its signal transduction pathways (Franklin *et al.*, 2005). In addition to light quantity and intensity, plants can measure incident light quality, direction, and periodicity, and use that information to optimize growth and development in accordance with the changing environment (Chen *et al.*, 2004). The effect of irradiance, or lack thereof, on senescence induction is complex due to the diverse effects of different light intensities and qualities on plant development. These effects may be direct or indirect, depending on the developmental stage of the plant and on the stress response it elicits. Various studies have demonstrated a direct effect of light on the initiation or advancement of senescence (Biswal and Biswal, 1984).

The influence of light on plants, and subsequently on senescence, could be mediated via several routes including the efficiency of photosynthesis, the generation of damages due to oxidative stress, signaling via interaction with light receptors, photoperiod, and the affected circadian clock. The actual effect of each of these stimuli is determined by the specific characteristics of light irradiation, which include wavelength quality and amount, determined by light intensity and duration of exposure.

6.2.1 Light intensity

6.2.1.1 Low light

Different studies demonstrate that, in general, when light intensities are either higher or lower than an optimal level, senescence is accelerated. Light intensities significantly different from optimal levels for photosynthesis may be regarded as an environmental stress leading to different physiological and biochemical consequences to the plants (Huner *et al.*, 1998; Niinemets and Valladares, 2004). Low light is known to cause the enhancement of senescence. The fraction of photosynthetically active radiation in the light perceived by the plant is an important factor in determining the initiation and advancement of senescence. For example, the shading of fully expanded cassava leaves results in accelerated senescence (Cock *et al.*, 1979). Senescence of sunflower basal leaves was found to be enhanced as they received reduced photosynthetically active light (Rousseaux *et al.*, 1996). Accordingly, increased levels of photosynthetic active light reaching the basal leaves of maize canopy can delay senescence (Ottman and Welch, 1988). In order for photosynthetically active light to retard senescence it is required to be above the photosynthetic compensation point (Veierskov, 1987). The specific mechanism mediating low-light-induced senescence is generally unknown. Shortage in energy or an effect on sugar metabolism may have an important role. The possible involvement of oxidative stress was suggested based on the observation that plant shading increases lipid peroxidation in wheat. This may result from the weakening of antioxidative protection, although low light was expected to result in reduced oxidative damage (Spundova *et al.*, 2005).

6.2.1.2 Darkness

Darkness is known to extensively enhance the rate of senescence and is frequently used for senescence studies. Senescence was demonstrated to be also induced by darkness in individually attached *Arabidopsis* leaves, but darkness inhibited the process in whole darkened plants (Weaver and Amasino, 2001). This result implies that the light status of the entire plant affects the senescence of individual leaves. One possibility is that a decrease in source strength was created in complete darkness, which led to a delay in senescence. In tobacco *rbcS* antisense mutants, which have reduced Rubisco levels, a prolongation of the senescence phase was observed which may be due to the impact of reduced source strength (Miller *et al.*, 2000). Still, dark-induced senescence in individual leaves is highly localized and is possibly cell autonomous (Weaver and Amasino, 2001).

When various *SAG*s were examined in *Arabidopsis* for their responsiveness to different hormonal and environmental treatments known to be associated with senescence, darkness was found to be most effective (Weaver *et al.*, 1998). Incubation of detached leaves in the light also had some senescence inducing effect, but was reduced when compared to darkness (Park *et al.*, 1998; Weaver *et al.*, 1998). Senescence processes, induced either naturally in attached leaves or by darkness in detached leaves, share physiological and biochemical characteristics. However, molecular analyses of the two processes show differences in the sets of induced genes (Becker and Apel, 1993; Park *et al.*, 1998). In addition, some genes have been shown to be expressed during leaf senescence regardless of whether it was induced naturally or by darkness, suggesting that the senescence processes under different conditions share common features (Oh *et al.*, 1996; Park *et al.*, 1998; Weaver *et al.*, 1998; Fujiki *et al.*, 2001). For example, in *Arabidopsis*, the *AtPaO* gene, encoding for pheophorbide an oxygenase, involved in senescence-associated chlorophyll degradation, was shown to be induced in both natural and darkness-induced senescence (Pruzinska *et al.*, 2003).

The induction of *SAG*s by darkness can occur within 3–24 h as demonstrated by the *din* genes in *Arabidopsis* (Fujiki *et al.*, 2001), but is highly dependent on the developmental stage of the leaf examined. Not much is known about the molecular regulation of darkness-induced *SAG*s. Studies involving the photosynthesis inhibitor DCMU and sucrose reveal that expression of some darkness-induced genes is related to sugar starvation in the dark, and involves different protein phosphatases and Ca^{2+}/calmodulin signaling (Fujiki *et al.*, 2000, 2001, 2005). In *Arabidopsis*, the transcript level of the *erd1*, a *clpA* protease homolog, is induced during senescence due to dehydration stress. The promoter region of *erd1* contains *cis*-acting elements that confer specifically darkness-induced expression in intact *Arabidopsis* plants (Simpson *et al.*, 2003). Other darkness-induced *SAG*s encode for various products including proteins that may have a regulatory role (Hajouj *et al.*, 2000; Guterman *et al.*, 2003). Mutations that alter natural and darkness-induced senescence support overlapping components shared by both these processes. In *Arabidopsis,* the delay of natural and dark-induced senescence was observed in the *ore1*, *ore3*, *ore9*, and *dls1* mutants (Oh *et al.*, 1997; Yoshida *et al.*, 2002a), while acceleration of both senescence types was observed due to a mutation in the *HYS1/CPR5* gene (Yoshida *et al.*, 2002b).

Analysis of the dependency of genes induced during dark-induced senescence in signaling pathways involving SA, JA, and ethylene have demonstrated that the SA pathway is not expressed in dark-induced senescence, while ethylene and JA signals are active as in natural developmental senescence (Buchanan-Wollaston *et al.*, 2005). While developmental senescence is delayed in plants defective in SA signaling, dark-induced senescence progresses normally in these plants (Buchanan-Wollaston *et al.*, 2005). The differences in regulation of dark-induced senescence is also manifested by the differential effect of the senescence-retarding mutation *ore4* for both processes in *Arabidopsis* (Woo *et al.*, 2002). In this mutant, the plastid-encoded ribosomal small subunit protein level is dramatically reduced, which has a retarding effect specifically on the age-dependent leaf senescence pathway but not on dark- or phytochrome-induced senescence. Furthermore, this mutation does not affect senescence induced by other factors such as abscisic acid (ABA), JA and ethylene, which are associated with other environmental stress factors.

6.2.1.3 High light

Prolonged exposure to high-light irradiance induces chlorophyll loss and a decrease in photosynthesis efficiency, which was sometimes referred to as senescence (Biswal and Biswal, 1984; Prochazkova and Wilhelmova, 2004). However, in early studies only chlorophyll loss and changes in photosynthetic parameters were examined and not other senescence characteristics. It is possible that in continuously highly illuminated plants, senescence-like symptoms result from photo-oxidative damage due to an excessive amount of light, resulting in chlorophyll breakdown (Choudhury and Behera, 2001). In general, the oxidative stress status of the leaf is intensified during senescence as levels of reactive oxygen species (ROS) are enhanced and antioxidant enzyme activity is reduced (del Rio *et al.*, 1998; Jimenez *et al.*, 1998; Prochazkova *et al.*, 2001; Kukavica and Jovanovic, 2004). Thus, the metabolic changes that occur during senescence may further increase susceptibility to high light-induced oxidative damage of the tissue. Senescing leaves are more sensitive to light irradiation also due to the significant decline in the photosynthetically active system. It was suggested that optical masking of the remaining chlorophyll by anthocyanins reduces risk of photo-oxidative damage to leaf cells as they senesce, which otherwise may lower the efficiency of nutrient retrieval from senescing autumn leaves (Merzlyak and Gitelson, 1995; Feild *et al.*, 2001; Hoch *et al.*, 2003). This hypothesis is supported by a study conducted in anthocyanin-deficient mutants of deciduous woody species (Hoch *et al.*, 2003). Interestingly, analysis of the reduction in photosynthetic efficiency and capacity during different stages of senescence of cotton leaves indicated no difference in the decline of photosynthesis under various light levels ranging from 15% to 100% full sun light (Sassenrath-Cole *et al.*, 1996).

When high-light stress is accompanied by an additional environmental stress, the senescence process is even further accelerated. In field conditions, high irradiance is often associated with water deficit. Examination of the combined effect of high light and water stress conditions suggests additive and possible synergistic action of both, causing an accelerated loss of pigments and proteins, compared to samples exposed to either of these stresses individually (Behera *et al.*, 2002).

6.2.2 Photoperiod

In few reports, a role for photoperiod in the control of leaf senescence was suggested mainly demonstrating a delay of leaf senescence in short days and its acceleration in long days, as is the case with temperate seasonal changes (Schwabe, 1970; Kar, 1986; Schwabe and Kulkarni, 1987; Nooden *et al.*, 1996). However, when the photoperiod effect was examined in terms of light period and dosage it was concluded that the enhancing effect was mainly a result of light dosage rather than of photoperiod (Nooden *et al.*, 1996). Prolonged accumulated light intensities during the long-day photoperiod apparently resulted in elevated stress, which in turn enhanced senescence. Natural variation in the effect of day length on leaf senescence was measured for different *Arabidopsis* ecotypes (Levey and Wingler, 2005). For most ecotypes, senescence occurred earlier in long days, and for two ecotypes it was either less pronounced or absent. However light dosage and photoperiod were not distinguished in this study.

6.2.3 Wavelength

Almost no information is available regarding the dependency of senescence on wavelength. Any effects are likely related to the effect of light on key photoreceptors such as the red/far red responsive phytochrome, the blue/UV-A responsive cryptochrome pigment, and on phototropins. An early study reported the senescence-retarding action of nonphotosynthetic light on excised wheat leaf segments using a crude action spectrum analysis (Haber *et al.*, 1969). In recent years, additional physiological and molecular studies demonstrated the importance of non-photosynthetically active light as a signal that affects senescence in plants.

6.2.3.1 Red/Far red

Phytochrome photoreceptors enable plants to sense a reduction in the ratio of red (R) to Far-Red (FR) light in their environment and change their growth or development accordingly. For example, light that has passed through the canopy of leaves has a lower ratio of R/FR due to absorbance of the red by chlorophyll of the upper leaves. The ability to sense modified R/FR light when shaded by their neighbors allows plants to avoid shading by increasing their internode extension rate (Franklin and Whitelam, 2005). It has been long recognized that the light environment determined by the density of a plant population may regulate photosynthetic characteristics as well as the timing of senescence. Reduced photosynthetically active radiation and decreased R/FR ratio are the prime senescence-triggering signals in shaded leaves of sunflower (Rousseaux *et al.*, 1996) and in the leaves of soybean grown under field conditions (Burkey and Wells, 1991). Senescence and abscission of leaves positioned in the shaded regions of the canopy were delayed by more than 4 weeks in plots where plant population densities were reduced (Burkey and Wells, 1991). In this low-density plant population, a higher ratio of R/FR light was measured in the shaded regions of the canopy. Enrichment of far-red light in field-grown sunflower accelerated senescence of individual leaves and was indicated by enhanced chlorophyll loss (Rousseaux *et al.*, 1996). The hypothesis that increasing the R/FR ratio perceived by basal leaves within canopies delays senescence was further confirmed in field-grown sunflower plants, in which enriched red light significantly delayed lower leaf senescence (Rousseaux *et al.*, 2000). The ability of red light to retard senescence was also demonstrated in the leaves of cut *Alstromeria* (van Doorn and Vanlieburg, 1993; Kappers *et al.*, 1998), while low R/FR is able to promote senescence in soybean (Guiamet *et al.*, 1989). The importance of the phytochrome in control of senescence was demonstrated for primitive plants also. In the moss *Marchantia polymorpha*, the senescence-delaying effect of white light could be reverted by FR, while red light could reverse the FR effect (De Greef *et al.*, 1971). In the fern *Nephrolepis exaltata*, senescence-accelerating effect of red light was observed which could be nullified with FR pulses (Behera and Biswal, 1990).

Leaf senescence responses to FR were found to be localized, and sensitivity to FR was also inversely correlated with the local *PHYA* phytochrome gene expression level (Rousseaux *et al.*, 1997). The localized FR response in the leaf is consistent with the localized senescence response to dark observed in *Arabidopsis* (Weaver and Amasino, 2001) and with the frequent observation that senescence can be induced locally in parts of the leaves shaded by upper leaves.

Few different molecular genetic studies further support the role of the phytochrome system in senescence control. Ectopic overexpression of an oat *PHYA* (phytochrome A gene) cDNA in tobacco under the CaMV *35S* promoter resulted in a delay in leaf senescence (Cherry *et al.*, 1991). Also in tobacco, overexpression of the oat *PHYA* gene reduced morphological responses to FR radiation and resulted in suppressed leaf-senescence responses (Rousseaux *et al.*, 1997). Transgenic potato plants constitutively expressing the *Arabidopsis PHYB* were found to have a delay in the onset of senescence under white-light irradiation (Schittenhelm *et al.*, 2004). An earlier study of these plants had claimed that the initiation of senescence in the *PHYB* overexpressing plants occurred at approximately the same time as in the wild-type, but the lifetime of the photosynthetically active transgenic plants was extended by 3–4 weeks (Thiele *et al.*, 1999). In pea a dominant mutation in *PHYA*, resulting in reduced sensitivity to FR light, had a pleiotropic effect including delayed flowering and senescence (Weller *et al.*, 1997, 2004). In some cases, the mutant pea plants grew for more than 6 months before senescence was initiated as compared to wild-type plants which grew for about 3 months (Weller *et al.*, 1997).

6.2.3.2 *Blue light*

The role of blue light in regulation of senescence is much less clear compared to that of the R/FR light and only few early studies using nonnatural senescence systems suggest such a role. Monochromatic blue light (450 nm) was suggested to retard the decline of photosynthetic activity in detached leaves induced to senescence in the dark (Choe and Thimann, 1977). Dark-stimulated chlorophyll loss was shown to be retarded by blue light pulses in senescing papaya leaf discs (Biswal and Choudhury, 1986). Delay of senescence by blue light was also demonstrated in *Hosta Tratt.* cut flowers (Rabiza-Swider and Skutnik, 2004). It was suggested that an interaction between red, FR and blue light exists to affect senescence. Such interactions and integration between the different light signaling in plants were suggested to occur throughout the development (Ni, 2005).

6.2.3.3 *Ultraviolet*

Possible increase in the level of UV radiation as a result of depletion of the stratospheric ozone layer is a major environmental concern in recent years (McKenzie *et al.*, 1999). The three UV radiation bands are UV-A (320–400 nm), UV-B (280–320 nm), and UV-C (200–320 nm). UV-C is the most damaging radiation to biological systems followed by UV-B (Biswal *et al.*, 1997); however, as sunlight passes through the atmosphere, all UV-C and approximately 90% of UV-B radiation are absorbed by ozone, water vapors, oxygen and carbon dioxide. UV-A radiation is less affected by the atmosphere. Therefore, the UV radiation reaching the earth's surface is largely composed of UV-A with a small UV-B fraction. Since UV-B is much more damaging than UV-A, more research investigating its effects on plants was performed.

Before UV-B radiation can give rise to a cellular response, it has to be perceived, which is thought to occur via a UV-B photoreceptor followed by several different signaling pathways (Brosche and Strid, 2003). These pathways include second messengers such as calcium, kinases and the signaling ROS. High levels of UV-B probably cause cellular damage and major oxidative stress; thus activating a general stress and signal transduction pathway which leads to a response similar to that which occurs after pathogen attack or other stresses (Brosche and Strid, 2003). The biological consequence of UV-B is very active when it is applied artificially at a high dose in controlled experiments. However, realistic levels of UV-B in field experiments were also shown to have physiological effect on plant growth and development as well as affect gene expression (Strid *et al.*, 1994).

The senescence-inducing effect of UV was demonstrated in various plant systems. Senescence-induced loss in pigments and proteins of detached maize leaves was significantly enhanced by UV. In both *Arabidopsis* and pea, it was found that older leaves become damaged by UV-B faster and to a greater extent than do younger leaves and an initial phase of chlorophyll loss was followed by desiccation of the tissue (Lois, 1994; Jordan *et al.*, 1998; Mackerness *et al.*, 1998; Surplus *et al.*, 1998). At the biochemical level, the rate of photosynthesis is greatly reduced in the UV-treated leaves, primarily as a result of a decline in RUBISCO protein levels and disruption to the chloroplast membranes. Some of these changes can be attributed partially to the effects of UV-B on expression of genes encoding key photosynthetic proteins (Strid *et al.*, 1994). Thus, in leaves at a certain stage of development, exposure can induce changes at the physiological, biochemical and molecular levels that resemble symptoms identified in plants undergoing senescence, including induction of SAGs as demonstrated in *Arabidopsis* (John *et al.*, 2001).

The senescence-inducing effect of UV-B radiation might be transduced by either its effect on the photosynthesis apparatus or via generation of oxidative stress (Mackerness *et al.*, 1998). UV-B exposure caused increases in JA and ethylene levels (Mackerness *et al.*, 1999), and together with the observed effects on *Arabidopsis* stress responses and induced genes, the involvement of three distinctive signal transduction associated with ROS, JA, and ethylene was suggested (Mackerness *et al.*, 1999). In the chloroplast, the thylakoid membrane seems to be much more sensitive to UV-B radiation than do the activities of the photosynthetic components within it, and a decrease in mRNA transcripts for the photosynthetic complexes and other chloroplast proteins are considered very early events of UV-B damage (Strid *et al.*, 1994). Exposure to UV-B radiation resulted in a loss of chlorophyll and an increase in lipid damage similar to that induced during natural senescence, including decline in lipids and increased lipid peroxidation indicated by rise in MDA (Dai *et al.*, 1997; John *et al.*, 2001). Difference between UV-B induced and natural senescence was found in the consequence to the maximum quantum efficiency of PS II photochemistry represented by the fluorescence marker Fv/Fm. Fv/Fm level was found to be more significantly decreased in leaves treated with UV-B (Lu and Zhang, 1998; John *et al.*, 2001).

Some of the genes identified so far as being regulated by UV-B encode proteins involved in the biosynthesis of protective pigments and antioxidative enzymes, DNA repair, photosynthetic genes, cell cycle genes, and stress genes induced by other types of stimuli (i.e. pathogenesis-related proteins) (Brosche and Strid, 2003). In few studies, the exposure of plants to UV-B resulted in up-regulation of SAGs, including that of the *SAG12* protease gene considered to be highly associated with developmentally induced senescence although to much reduced level (John *et al.*, 2001).

The effect of UV-A on senescence is not clear as only few studies were carried out. On one hand UV-A was shown, at low intensities, to be more efficient than white light in inhibiting dark-induced senescence of barley leaf segments (Cuello *et al.*, 1994) and could retard the senescence-inducing effect of UV-B in cluster bean leaves (Gartia *et al.*, 2003). On the other hand, UV-A was reported to enhance senescence in primary leaves of wheat which could be retarded by a red-light pulse (Joshi *et al.*, 1991). The effect of UV-A radiation on senescing wheat leaves over a period of days had resulted with negative impact on primary photochemistry of photosystem II (PS II) but did not show any significant effect on the level of photosynthetic pigments (Nayak *et al.*, 2003). The UV-A induced changes in PS II of chloroplasts from senescing leaves were found to be synergistically accelerated by high-temperature growth. It is possible that the specific effect of UV-A on senescence is dependent on the actual wavelength composition of the radiation used. When more enriched with longer wavelength radiation, there is a senescence-retarding effect, similar to that of blue light; however, with shorter wavelength radiation, the senescence is accelerated as with UV-B.

6.3 Ozone

Ozone (O_3) is a gas not usually emitted directly into the air, but at ground level is created by a chemical reaction between oxides of nitrogen and volatile organic compounds in the presence of sunlight. In the stratosphere, ozone has an important protective function forming a layer that protects from harmful radiation but in ground-level it has a negative effect on plants. Increasing ozone levels in urban as well as rural areas is currently viewed as a widespread and growing problem that suppresses crop productivity on a large scale (Fuhrer and Booker, 2003). The inhibitory effects of ozone on photosynthesis, plant growth, and yield have been documented recently (Ashmore, 2005; Fiscus *et al.*, 2005). Ozone injury to natural vegetation is being increasingly surveyed with growing list of species showing visible ozone-injury symptoms (Pell *et al.*, 1997).

Upon entry into the leaf intercellular space, ozone rapidly reacts with the components of the leaf apoplast to initiate a complex set of responses involving the formation of toxic metabolites and generation of plant defense responses. The interactions and perception of ozone and the immediate downstream responses and signaling were recently reviewed (Kangasjarvi *et al.*, 2005). In general, hormonal signaling has an important role for determining the outcome of ozone challenge at the cellular level, and generation of ROS by ozone degradation can cause either direct necrotic damage or induce programmed cell death (PCD). Perception of ozone or the resulting ROS in the apoplast activates several signal transduction pathways that regulate the responses of the cells to the increased oxidative stress (Kangasjarvi *et al.*, 2005).

Ozone-induced accelerated foliar senescence was demonstrated in many plant species including potato, radish, alfalfa, wheat, and hybrid poplar (Pell and Pearson, 1983; Reich, 1983; Held *et al.*, 1991; Nie *et al.*, 1993; Pell *et al.*, 1997) measured by accelerated chlorophyll and protein loss and reduced photosynthetic capacity and efficiency (Reich, 1983; Held *et al.*, 1991; Nie *et al.*, 1993). Accelerated loss of Rubisco protein (Rao *et al.*, 1995) and reduced transcript levels for *cab*, *rbcS*, and *rbcL* are also closely associated with ozone-induced senescence (Pell and Pearson, 1983; Nie *et al.*, 1993; Bahl and Kahl, 1995; Glick *et al.*, 1995; Pell *et al.*, 1997; Conklin and Barth, 2004). Even moderate elevation in ozone concentration (1.5–1.7-fold of ambient level) was found to have a senescence-accelerating effect (Yamaji *et al.*, 2003). Responses to ozone are modulated by ethylene, JA, SA and ABA, and the interactions among their signaling pathways (Pell *et al.*, 1997; Koch *et al.*, 2000; Vahala *et al.*, 2003). For example, ozone-induced ethylene production is dependent on SA, and both SA and ethylene act in concert to regulate ozone-induced cell death (Rao *et al.*, 2002). The involvement of the G-proteins pathway in plants' response to ozone was demonstrated in *Arabidopsis* based on the analysis of mutant altered in this pathway and positive correlation between the lack of a G-protein alpha subunit and decreased O_3 sensitivity (Booker *et al.*, 2004). Genomic analysis of three *Arabidopsis* lines and the related *Thellungiella halophila* suggested that the differences in sensitivity of the different lines may be related to ozone-response diversity in the signaling and transcriptional response of these plants (Li *et al.*, 2006). Ecotype WS, in particular, showed accelerated senescence in response to ozone. This ecotype also showed induction of known SAGs in ozone, while the other ecotypes showed less expressed SAGs (Li *et al.*, 2006).

6.4 Temperature

An important and relevant environmental factor to senescence is temperature, which when different from the optimal range for a given plant results in a stress and activation of response for adaptation. Temperature stress including heat, chilling or freezing is a primary cause for yield reduction in crops (Boyer, 1982). In many cases, temperature stress is accompanied with, or cannot be separated from, other environmental stresses such as drought or high light. Temperature stress frequently result in an oxidative stress (Suzuki and Mittler, 2006) and the generated ROS can injure cell membranes and proteins (Okane *et al.*, 1996; Larkindale and Knight, 2002).

Premature leaf senescence is associated with an inability of the plant to acclimate to low temperature growth conditions. Exposure of leaves to chilling temperature, level of which may be different for different plant species, in the light results in inhibition of photosynthesis. Both photosynthetic electron transport and CO_2 assimilation are down-regulated. Disruption of regulation of metabolism and insufficient antioxidant defense are postulated to cause chilling sensitivity, resulting in premature senescence (Foyer *et al.*, 2002).

High temperature was demonstrated to accelerate senescence in mature wheat flag leaves visualized by structural features such as loss of chloroplast integrity, increased thylakoid luminal volume and decreased area of appressed thylakoid membrane as well as by functional features such as decline of PS II-mediated electron transport (Harding *et al.*, 1990; Xu *et al.*, 1995). The interaction between the effect of heat on senescence progress and the developmental control of senescence was also investigated in wheat in which removal of the inflorescence slowed senescence processes but did not alter the course of high-temperature responses (Kuroyanagi and Paulsen, 1985). In addition to air temperature, soil temperature can reach high levels in summer and result in a stress to the plant. High soil temperature was found to accelerate leaf senescence in creeping bentgrass (Huang *et al.*, 2001) possibly as a result of a severe oxidative damage to leaves, limiting antioxidant activities and induction of lipid peroxidation (Huang *et al.*, 2001). Oxidative stress may have an important role in inducing the observed accelerated senescence. In some regions, grain yield of wheat is reduced by high temperature during maturation due to the acceleration of senescence resulting in diminished longevity. Applying dilute solutions of KH_2PO_4 was found to delay the senescence and give better yield (Benbella and Paulsen, 1998b) and may be related to nutrient salvage in senescence.

6.5 Drought stress

Drought stress triggers responses range from changes in gene-expression pattern to changes in plant metabolism and growth. These responses may be induced within a short period of exposure to drought, and at a slow rate such as the induction of leaf senescence. Much research regarding plant's response to drought stress, including regulation and signal transduction and defense mechanisms, has been conducted (Chaves *et al.*, 2003; Riera *et al.*, 2005). The induction of senescence as a result of drought stress is thought to have a role in plant's survival and this topic has been recently reviewed (Munne-Bosch and Alegre, 2004). Senescence can contribute to nutrient supply and perhaps more relevant is the removal of unproductive leaves as a result of accelerated senescence accompanied by abscission, which reduce water loss through transpiration and in turn contributing to water balance of the plant (Munne-Bosch and Alegre, 2004). Drought-induced senescence has been studied for different plant species including crops of economic importance grown under field conditions, such as wheat, rice or sorghum (Borrell *et al.*, 2000; Pic *et al.*, 2002; Yang *et al.*, 2003; Munne-Bosch and Alegre, 2004).

Senescence induced by mild drought in pea was shown to follow the same sequence of macroscopic, biochemical and molecular events as developmental senescence (Pic *et al.*, 2002). It was proposed that mild drought-induced senescence program was probably not a direct consequence of a water stress sensed at the cellular level, but it was triggered by an early signal occurring while leaf photosynthesis was still active, followed by a developmentally controlled senescence program (Pic *et al.*, 2002). Regulation of drought-induced senescence under field conditions was investigated in respect to the role of plant hormones in the nutrient remobilization process in crops of economical interest. Drought stress effect on endogenous levels of plant hormones has been demonstrated mainly for cytokinins and ABA (Ali *et al.*, 1999; Yang *et al.*, 2003; Munne-Bosch and Alegre, 2004). These studies revealed that drought-induced ABA was positively and significantly correlated with carbon remobilization from senescing leaves to grains in drought-stressed rice and wheat plants. Cytokinin levels were found to decrease under drought stress and show positive correlation with photosynthetic rate and chlorophyll content thus possibly preventing senescence. Drought can promote increased ethylene production in plants (Apelbaum and Yang, 1981). Loss of *ACS* expression, a gene encoding for ACC synthase, involved in ethylene biosynthesis, in transgenic maize plants resulted in delayed leaf senescence under normal growth conditions as well as inhibited drought-induced senescence. These observations suggest that ethylene determine the onset of natural senescence and mediate drought-induced senescence shown to be associated with increased stomatal conductance (Young *et al.*, 2004). This effect of ethylene may also be related to its reported inhibitory effect on ABA-mediated stomatal closure (Tanaka *et al.*, 2005), which may result in elevated stress. The interaction of the reproductive sink with drought stress and recovery was examined in cowpea plants at the seedling and pod development stages. It was observed that despite a similar difference in leaf water potential between irrigated and drought-stressed plants at both stages, the effects on senescence acceleration were more pronounced during the pod development stage (Renu Khanna *et al.*, 2000).

The oxidative stress generated in response to drought may have a central role in the induction of senescence. The relationship between drought, oxidative stress and leaf senescence was investigated in *sage* (*Salvia officinalis*), a drought-susceptible species. Drought-stressed senescing leaves showed enhanced lipid peroxidation, chlorophyll loss, reduced photosynthetic activity and reduction in membrane-bound chloroplastic antioxidant defenses, which is indicative of oxidative stress. This study demonstrated that oxidative stress in chloroplasts mediates drought-induced leaf senescence (Munne-Bosch *et al.*, 2001). Another factor that may play an important role in drought-induced senescence is an imbalance between ROS and antioxidant levels.

Genes that are induced during both drought stress and senescence may give a clue about the regulatory relationship between the processes. In *Arabidopsis*, the *erd1* gene, which encodes a protein with homology to the Clp ATP-dependent protease, was found to have in its promoter two different regulatory elements that respond to dehydration and dark-induced senescence, suggesting a role for this protease in both processes and a common molecular regulatory mechanism (Simpson *et al.*, 2003).

6.6 Flooding

Flooding is a severe constraint on plant growth and can affect wide environments due to poor drainage or excessive rainfall or irrigation. In many cases, plants that survive flooding die even after the stress is removed (Sullivan *et al.*, 2001). Many times following the flooding period senescence-associated processes are initiated in response to the original stress (Kramer, 1951). The water-submerged tissue has limited availability of oxygen, light and carbon, which can affect photosynthesis and which in turn may result in premature senescence (Mommer and Visser, 2005).

The senescence-accelerating effect of flooding in plants was associated with the two senescence-associated plant hormones, cytokinin and ethylene. Following flooding, the concentration of cytokinin in sunflower xylem sap was reported to decline sharply to a very low level (Burrows and Carr, 1969), which may lead to enhanced senescence. This reduction could result from decreased cytokinin biosynthesis or the inability of the anaerobic tissue to export cytokinin to the aerial part of the plant. The role of cytokinin in flooding-induced senescence was demonstrated in *Arabidopsis* plants in which autoregulated cytokinin was synthesized using the *SAG12:ipt* gene construct. The results indicated that endogenously produced cytokinin can regulate senescence caused by flooding stress, thereby establishing relationship between flooding tolerance and cytokinin accumulation (Zhang *et al.*, 2000; Huynh *et al.*, 2005). The question of whether or not ethylene biosynthesis is promoted by submergence is uncertain (Jackson and Ram, 2003). In rice, on the one hand, expression of the *ACC synthase* gene and its capacity to convert ACC to ethylene are up-regulated by submergence (Van der Straeten *et al.*, 2001; Zhou *et al.*, 2001). On the other hand, ethylene production underwater is probably slower than in air as indicated by work with *Rumex palustris* (Voesenek *et al.*, 1993). However, in *Rumex palustris* the expression of the *ACC synthase* gene was found to be induced by submergence (Rieu *et al.*, 2005). Although ethylene accelerated senescence of submerged *R. acetosella* plants, the process may have been caused by other factors. The slower senescence of *R. palustris* plants could not be explained by their lower ethylene concentration and it was rather due to much lower sensitivity of the senescence process to ethylene (Banga *et al.*, 1997). On the basis of induction of its synthesis, ethylene was suggested to be involved in the regulation of flooding-promoted senescence of tobacco leaves (Hurng *et al.*, 1994) and chrysanthemums cut flowers from plants subjected to flooding (Gindin *et al.*, 1989). In both cases, induction of ethylene biosynthesis was measured after 24 h. In rice, the ethylene action inhibitor, 1-methylcyclopropene (1-MCP), was used to evaluate the effect of ethylene on chlorophyll degradation and plant survival during flooding in tolerant and intolerant cultivars (Ella *et al.*, 2003). 1-MCP treatment decreased chlorophyll degradation, the activity and gene expression of chlorophyllase and improved seedling survival in the intolerant cultivar, while in the tolerant cultivar, Chlorophyllase enzyme activity and gene expression were lower even without the inhibitor treatment suggesting a link between ethylene action and flooding-induced chlorophyll degradation (Ella *et al.*, 2003).

6.7 Salinity

Salinity is becoming a growing problem in the world. Excessive soil salinity results in a major stress to plants affecting growth and development from just partial to complete inhibition of growth. Saline soils exist in many semiarid or arid regions of the world and are developed in new areas due to saline irrigation of agricultural crops with poor drainage conditions. Cellular events triggered by salinity, namely salt compartmentalization, osmotic adjustment and cell wall hardening, are connected to the whole plant responses, namely leaf senescence and necrosis, altered phenology and ultimately plant death (Volkmar *et al.*, 1998). In the first phase, salinity stress results in decreased leaf growth and in the second phase it leads to premature senescence as has been demonstrated, for example, in bean (Prisco and Oleary, 1972; Lovelli *et al.*, 2000). When excessive amounts of salt enter the plant, salt will eventually rise to toxic levels in the older transpiring leaves, causing premature senescence. Salinity, among other things, reduces the ability of plants to take up water, and this causes reductions in growth rate and metabolic changes identical to those caused by drought which, as described before, can lead to premature senescence as well (Munns, 2002).

Salt-tolerant plants may differ from salt-sensitive ones in having a low rate of Na^+ and Cl^- transport to leaves and the ability to compartmentalize these ions in vacuoles to avoid salt toxicity. When such sensitive and tolerant plants were examined for the progress of senescence under salt stress, it was found that genotype with the higher rate of Na^+ uptake showed faster leaf senescence, with injury appearing in the oldest leaf after 2 weeks (Fortmeier and Schubert, 1995; Munns, 2002). On the other hand, in sugarcane when salinity-induced senescence was compared between salinity-tolerant plants and sensitive plants, enhanced senescence and abscission was observed in the tolerant line, while in the more sensitive line senescence was less affected possibly due to pronounced growth inhibition by the salt stress, which reduced initiation of new leaves (Plaut *et al.*, 2000).

The initial reduction in shoot growth followed by accelerated senescence is probably related to hormonal signals generated in response to the stress. Increased salt concentration in the nutrient solution of tomato or cucumber resulted in increased ethylene production (Helmy *et al.*, 1994). The involvement of ethylene in salinity-induced leaf senescence is demonstrated by the effect of inhibiting the ethylene biosynthesis-related genes *ACC synthase* or *ACC oxidase* in transgenic tobacco plants (Wi and Park, 2002). When watered with various concentrations of NaCl the senescence process was inhibited in the antisense transgenic plants compared to the wild-type (Wi and Park, 2002). Two possible mechanisms were suggested for the retardation of stress-induced senescence in the antisense transgenic plants. One is based on the inhibition of stress-induced ethylene production, and the other is based on the higher polyamine content accumulated in those plants (Wi and Park, 2002).

Elevated stress resistance in transgenic or mutant plants may result also with retardation of senescence. Osmotin has been implicated in conferring tolerance to drought and salt stress in plants. Overexpression the osmotin gene in transgenic tobacco resulted with retarded leaf senescence, as well as, improved resistance to salinity and drought stress (Barthakur *et al.*, 2001). The effect on senescence is likely a result of improved stress resistance and demonstrates the indirect link between environmental stress and senescence. As is the case with other environmental stresses the combined action of two different stresses may further accelerate the senescence process as was shown for *Mesembryanthemum crystallinum* exposed to both salt and high-light stressed (Broetto *et al.*, 2002).

6.8 Environmental pollution – toxic materials

In recent years new source of environmental stress has developed due to pollution by different toxic materials and heavy metals released by the growing industrial processes such as power plants or engine exhaust or various chemicals used for different purposes in the modern world. These chemicals are accumulating in the environment and beside their negative influence on human health they have negative effects on plants development and there accumulation in the crop plants creates a risk to all herbivores including humans.

Cadmium (Cd) is a toxic trace pollutant which is highly phytotoxic causing growth inhibition and even plant death (Wagner, 1993). Different metabolic processes such as photosynthesis and cell respiration are affected by the presence of Cd (Prasad, 1995). Cd-induced acceleration of senescence was demonstrated in different plants such as pea and elodea (McCarthy *et al.*, 2001a; Dalla Vecchia *et al.*, 2005). In pea Cd was found to induces biochemical changes associated with oxidative stress which was suggest to accelerate senescence (McCarthy *et al.*, 2001a; Sandalio *et al.*, 2001; Romero-Puertas *et al.*, 2004). Senescence symptoms were visualized in leaf peroxisomes (McCarthy *et al.*, 2001a) and an increase in the level of oxidized proteins was measured, probably mediated by H_2O_2, which in turn affects important leaf antioxidative enzymes and perhaps the photosynthetic carbon assimilation (Romero-Puertas *et al.*, 2002). A study, in tomatoes, of the effects of Cd uptake on ultrastructure and lipid composition of chloroplasts suggests that Cd induces premature senescence of leaves through its effects on membrane structure and composition (Djebali *et al.*, 2005). Treatment of leaves from Cd-grown plants with different effectors and inhibitors showed that ROS production was regulated by different processes involving protein phosphatases, Ca^{2+} channels and cGMP (Romero-Puertas *et al.*, 2004). The induction of senescence by Cd is a complex process and different factors are involved, including nutritional disturbances that can affect important metabolic processes such as photosynthesis. Cd can also induce senescence by stimulating the biosynthesis of ethylene (Rascio *et al.*, 1993; Sanita di Toppi *et al.*, 1998).

Senescence-inducing effects of other metal ions have also been demonstrated, although in fewer studies. In detached wheat leaves, chromium(III) was demonstrated to enhance chlorophyll and carotenoid breakdown and increase membrane permeability and lipid peroxidation (Panda and Patra, 2000). Free radical scavengers prevented the increase in the senescence parameters indicating oxidative stress which may be involved in the accelerated senescence (Panda and Patra, 2000). Lead treatment of *Salvinia natans* promoted senescence indicated by the decrease in protein, carbohydrate and free amino acid content (Mohan and Hosetti, 1998).

6.9 Oxidative stress involvement in environmental regulation of senescence

A possible mechanism common to few different environmental stresses inducing senescence is oxidative stress, which has been correlated with and is thought to play an essential role in leaf senescence (Zimmermann and Zentgraf, 2005). Under different environmental stress conditions, the normal and balanced function of photosynthesis is damaged thus leading to photooxidation process and increased levels of free radicals. Degradation of chlorophyll and membranes also cause an increase in the production of free radicals. At the same time, the stress conditions may reduce the metabolic capabilities of the cells to produce the proper antioxidative agents required for scavenging these radicals, which results in oxidative stress (Wi and Park, 2002). Oxidative damage is generated in the plant cells following exposure to the various environmental factors discussed above, such as high light, UV radiation, ozone, drought, flooding, exposure to toxic polutants or salinity, and this oxidative stress may be involved in senescence acceleration. Temperature stress has also been associated with oxidative stress as it can have a devastating effect on plant metabolism, disrupting cellular homeostasis and uncoupling major physiological processes which in turn can enhance ROS accumulation (Suzuki and Mittler, 2006).

Redox homeostasis and antioxidant signaling has been suggested to form a metabolic interface between stress perception and physiological responses (Foyer and Noctor, 2005). Plant cells are composed of different interconnecting compartments with different antioxidant buffering capacities determined by differences in synthesis, transport and/or degradation, resulting in redox-sensitive signal transduction which may occur in specific locations (Foyer and Noctor, 2005). Recent studies have shown that ROS could play a key role in mediating important signal transduction events, and ROS, such as superoxide ($O^{2(-)}$), are produced by NADPH oxidases during abiotic stress to activate stress-response pathways (Suzuki and Mittler, 2006) and might be involved in the control of leaf senescence. Redox-related signal transduction pathways may have developed as a universal feature of aerobic life through evolution to balance information from metabolism and the environment.

The results of various studies support a function for oxidative stress and ROS in mediating environmental stress and senescence. In a study, Woo *et al.* (2004) found that in *Arabidopsis* delayed leaf senescence mutants *ore1*, *ore3* and *ore9* had increased tolerance to various types of oxidative stress and this was not due to enhanced activities of antioxidant enzymes. This study provides genetic evidence that oxidative stress tolerance is linked to control of leaf senescence in plants. The *Arabidopsis vtc1* mutant is deficient in L-ascorbic acid and was found to be hypersensitive to oxidative stress (Conklin *et al.*, 1996). In addition, *vtc1* exhibits elevated levels of some SAG transcripts indicating that reduced ascorbic acid level causes *vtc1* to enter at least some stage(s) of senescence prematurely (Barth *et al.*, 2004). The involvement of oxidative stress in control of senescence is further supported by its ability to affect certain SAGs as has been demonstrated in *Arabidopsis* (Navabpour *et al.*, 2003). Oxidative stress and ROS role in senescence could be indirect, for example affecting hormone balance. The involvement of ethylene in oxidative-stress-induced leaf senescence was demonstrated by the analysis of transgenic tobacco plants in which the ethylene level is reduced following antisense inhibition of the *ACC synthase* or *ACC oxidase* ethylene biosynthesis related genes (Wi and Park, 2002). When these plants were treated with H_2O_2 as an oxidative stress, the symptom of visible yellowing and degradation of chlorophyll were less severe in the transgenic than in the wild-type plants (Wi and Park, 2002).

6.10 Nutrient/mineral shortage

Initiation and the rate of leaf senescence depend on nutrient availability. In *Arabidopsis*, transferring rosettes from a nutrient-sufficient medium to water resulted in enhanced leaf senescence and activation of SAGs expression (Thomas and de Villiers, 1996). Nitrogen is very dominant in affecting senescence progress and when its level is sufficient the leaf senesce slowly (Ono *et al.*, 2001). Shading of plants under nitrogen limitation results in retardation of senescence, which was explained by a slower development of nitrogen deficiency due to limited growth (Ono *et al.*, 1996, 2001). Removal of nitrogen from the growth medium of wheat, just at the time of full leaf elongation, enhanced the rate of senescence (Crafts-Brandner *et al.*, 1998). It is not clear what are the differences in the senescence-inducing effect between nitrogen limitation and shortage of other nutrients, such as phosphate or sulfate. Study of hydroponically raised maize plants subjected to the deficiency of N, P, K, Ca, Mg or S has claimed that deficiency of all of these lead to premature senescence of the lower leaves which was different based on chlorophyll levels (Tewari *et al.*, 2004). However, in this study analysis of different senescence markers was not preformed; thus, it is not clear if characteristic senescence had occurred or just chlorophyll chlorosis. Few studies indicate that limited phosphate (Pi) can result in accelerated senescence. Pi status of maize leaves was suggested to be involved in senescence regulation (Usuda, 1995). In soybean, an increase of Pi in the nutrient solution delayed senescence (Graham *et al.*, 1986). Eliminating Pi from the growth medium results in enhanced leaf senescence in tomato where the senescence-accelerating effect is more pronounced when Pi removal is performed at an early developmental stage presumably due to sufficient stored Pi levels (A. Lers *et al.*, unpublished results). Foliar applications in wheat with different mineral salts demonstrated that N or Pi treatments were usually most effective, particularly in countering rapid high-temperature-induced senescence (Batten and Wardlaw, 1987; Benbella and Paulsen, 1998a; 1998b). However, a study with soybean concluded that leaf senescence was not affected by Pi deficiency (Crafts-Brandner, 1992). In corn field experiments, Pi deficiency resulted in slow growth and development and even reduced the senescence rate (Colomb *et al.*, 2000). The variation in the observed effect of Pi levels on senescence may result from differences between plants in the Pi levels stored in the plant which could be utilized.

6.11 Atmospheric CO_2

Rise in atmospheric CO_2 concentration due to fuel burning and deforestation is occurring on a global scale and is increasing carbon availability for photosynthesis (Falkowski *et al.*, 2000; Woodward, 2002; Urban, 2003). Atmospheric CO_2 concentration has increased from 270 ppm at the onset of the Industrial Revolution to a current value of 375 ppm. Elevated CO_2 directly influences plant by altering growth and development. CO_2 concentration influences the rate of carbon fixation, which subsequently influences growth processes, functioning and reproductive output (Woodward, 2002; Ward and Kelly, 2004). Upon growth under elevated CO_2 and following acclimation of photosynthetic capacity, significant reductions in N content and Rubisco content and substantial increase in sugar and starch levels have been reported (Ainsworth and Long, 2005).

High CO_2 atmosphere in levels of 400–1000 ppm in air has often been shown to accelerate leaf senescence. Since overall plants grown in elevated CO_2 are nitrogen limited, accelerated senescence would increase nitrogen availability and has been suggested to be part of the photosynthetic acclimation of plants to elevated CO_2 (Wingler *et al.*, 2006). Reports regarding the effect of elevated CO_2 on senescence are variable and majority of observations report senescence acceleration (Nie *et al.*, 1995; Miller *et al.*, 1997; Miglietta *et al.*, 1998; Fangmeier *et al.*, 2000; Ludewig and Sonnewald, 2000; Lawson *et al.*, 2001; Bindi *et al.*, 2002) some studies report no effect (Herrick and Thomas, 1999) or even retardation of senescence by increased CO_2 levels (Li *et al.*, 2000; Tricker *et al.*, 2004). The mechanism mediating the effect of CO_2 level on natural senescence is not clear and is likely to be different between different plants and under various environmental growth conditions. The dependency on the environmental conditions was demonstrated in Red Oak where elevated CO_2 accelerated senescence in sun plants and prolonged leaf function in shade plants (Cavender-Bares *et al.*, 2000). In tobacco, senescence acceleration was exhibited following down-regulation of photosynthetic genes, which was suggested to be due to temporal shift in leaf ontogeny and not due to elevated sugar levels (Ludewig and Sonnewald, 2000). In barley, CO_2-enhanced flag leaf senescence was postulated to result from increase in the nitrogen sink capacity of the grains (Fangmeier *et al.*, 2000). Interactions between high CO_2 levels and nitrogen supply were suggested to determine the effect on senescence (Wingler *et al.*, 2006). Under lower nitrogen supply conditions, CO_2 enrichment had a larger senescence-promoting effect in black cottonwood (Sigurdsson, 2001). The effect of CO_2 is possibly related also to altered hormonal balance in the plant. Elevated CO_2 was reported to increase cytokinin in cotton leaves (Yong *et al.*, 2000) which in turn may delay senescence. The possible effect of elevated CO_2 level on plants' water relation via an effect on stomatal conductance and density was also suggested to have a role in the senescence control of CO_2 (Wingler *et al.*, 2006). The involvement of sugars in the regulation of senescence by CO_2 was recently reviewed and it was concluded that presently no clear conclusions can be made (Wingler *et al.*, 2006).

Because of their different physiological effects, elevated levels of CO_2 and ozone might have interactive impacts on plants. Study of physiological, morphological and growth responses of six perennial species grown at various CO_2 and ozone concentrations suggested that ozone effects on carbon balance and growth are likely to be ameliorated by elevated concentrations of atmospheric CO_2 (Volin *et al.*, 1998).

6.12 Biotic stress

Pathogens including viruses, bacteria and fungi can induce premature senescence of the infected leaf, which results in organ removal due to enhanced abscission. This localized senescence process occurs at an accelerated rate compared to developmental senescence and may occur in relatively young leaves without affecting neighboring leaves. This induction of senescence and organ removal can be viewed as a type of plant resistance mechanism that results in decreased chances for secondary infection. On the other hand, the induction of senescence can result in an advantage to the pathogen due to a decrease in defense mechanisms in the senescing tissue. Induction of senescence by pathogens can be mediated by specific toxins such as the *Cochliobolus victoriae* victorin toxin – the causal agent of victoria blight of oats. Victorin appears to trigger accelerated senescence characterized by lipid peroxidation, proteolysis and chlorophyll loss (Duroy and Thomas, 1999). Ethylene is involved in victorin effect since ethylene inhibitors prevented victorin-induced proteolysis. Infection of bean leaves with *Pseudomonas syringae*. pv. phaseolicola causes a senescing-like region surrounding the site of inoculation which was shown to contain chlorophyll catabolites as observed in senescence (Bachmann *et al.*, 1998). This pathogen secretes phaseolotoxin, known to inhibit orinithine decarboxylase (ODC), a key enzyme involved in polyamine biosynthesis. Polyamines have been implicated in regulation of senescence (Pandey *et al.*, 2000; Walters, 2000) and it was suggested that ODC inhibition might be a contributing factor to the development of senescence-like process at the inoculation site (Chen *et al.*, 2004). In tomato *Botrytis cinerea* pathogen-induced promotion of senescence was suggested to be related to the collapse of peroxisomal antioxidant enzymes involved in the metabolism of ROS which occurred parallel to the development of the disease symptoms (Kuzniak and Sklodowska, 2005).

Another aspect of the relationship between pathogen infection and senescence is the similarity between the hypersensitive response (HR) and senescence. HR is induced by a compatible plant–pathogen interaction which involves PCD, while senescence-related cell death occurs in leaves as a normal part of the plant development or is accelerated by an environmental stress stimulus. The HR-associated resistance response involves the coordinate activation of defense systems that limit pathogen growth (Greenberg and Yao, 2004) and induces confined leaf yellowing followed by localized dry necrotic lesions, very reminiscent to the changes that a leaf is undergoing during senescence (Pontier *et al.*, 1999). As reviewed below, examples for common features to both processes exist; however, few major differences distinguish the two cell death processes. While HR occurs in a restricted site, senescence occurs on the organ level. The kinetics and the consequences of the two processes are also different. While the progress of senescence is comparatively slow, aimed for organized salvage and redistribution of metabolites, the HR response is a faster process aimed for generating a physical barrier to limit pathogen spread by executing early cell death. It is not clear whether HR involves salvage and redistribution of metabolites. In support of the occurrence of such salvage following interaction with pathogens is the observation that the two senescence-related markers GS1 (cytosolic glutamine synthetase) and GDH (glutamate dehydrogenase), thought to be involved in nitrogen mobilization in senescing, are also induced during plant defense response (Pageau *et al.*, 2006). Senescence can be reversible (Thomas and Donnison, 2000; Thomas *et al.*, 2003), with no reports for reversibility of the HR process. These differences may be related to the different functions the two processes have in plants' fitness during development and interaction with the environment: immediate defense for the HR versus efficient recycling of nutrients for senescence.

Although HR and senescence have different functions, it seems that some common features exist, such as certain regulatory or signal transduction pathways and specific genes involved in or associated with the two processes. The similarities between these two processes might be related, at least in part, to the final outcome since both of these processes ultimately result in cell death. Thus, HR and leaf senescence are forms of PCD, and common cell death execution mechanisms are possibly employed in both processes.

Significant number of SAGs were described to be also pathogenesis related (PR) or pathogen-defense related, and different SAGs were reported to be induced during plant–pathogen interactions (Hanfrey *et al.*, 1996; Butt *et al.*, 1998; Lers *et al.*, 1998; Pontier *et al.*, 1999; Quirino *et al.*, 1999, 2000; Yoshida *et al.*, 2002b).

Evidence supporting relationship in the controlling mechanisms of both processes was described. For example, the promoter of the senescence-associated metallothionein *LSC54* gene was demonstrated to activate reporter gene expression during HR at the site of infection by incompatible pathogen. When the plant was challenged with a compatible pathogen inducing necrosis, the induction of the reporter gene by LSC54 promoter was slower and associated with the development of senescence symptoms (Butt *et al.*, 1998). The SAG *SEN1* was found to be regulated by signals that link plant defense and senescence responses (Schenk *et al.*, 2005). The role of the PR or defense-related proteins in senescence is not clear. One hypothesis is that these proteins are induced to increase pathogen resistance of the senescing tissue which had become more susceptible. These defense-related genes are induced in senescence even in the absence of any pathogen infection possibly to prevent infection by opportunistic pathogens. Analysis of the regulation of defense-related genes in tobacco suggested them to be related with the mechanism of senescence rather than with protection of the plant against pathogen infection. The different examined genes are possibly controlled by slightly different sets of signals (Obregon *et al.*, 2001). In *Arabidopsis*, study of gene markers during HR and senescence indicated these markers to be relatively specific for the different cell death programs and revealed that a senescence-like process is triggered at the periphery of the HR necrotic lesion (Pontier *et al.*, 1999). It was suggested that cells committed to die during the HR release a signal able to induce senescence in the neighboring cells. It was also suggested that these defense- and senescence-related genes have a more direct role in the senescence program (Quirino *et al.*, 2000) than in other cell death-related processes.

In general, the similarities in physiological events and growth regulators involved in the control of senescence and the responses to pathogens may be related also to the induction of the defense-related genes. These include increased levels of ethylene (John *et al.*, 1995) and accumulation of H_2O_2 (Levine *et al.*, 1994; Pastori and delRio, 1997). The SA signaling pathway, known to have a central role in response to pathogens (Ryals *et al.*, 1996) was shown to have a role in the control of gene expression during developmental senescence (Quirino *et al.*, 1999). In *Arabidopsis* plants defective in the SA-signaling pathway, a number of SAGs exhibited altered expression patterns in parallel to a delay in yellowing and reduced necrosis suggesting a role for SA in the final stage of senescence. The identification of mutations that affect both senescence and responses to pathogens further supports a cross talk between the two processes. Mutations in the *HYS1* gene were found to result in early senescence (Yoshida *et al.*, 2002b) and to be allelic to mutation in the *CPR5* gene which exhibits spontaneous pathogen defense responses and resistance to virulent pathogens (Bowling *et al.*, 1997; Boch *et al.*, 1998). It was hypothesized that a common factor that can promote senescence and pathogen-defense responses exists, and the HYS1/CPR5 protein may negatively affect the activity of such common factor (Yoshida *et al.*, 2002b). The involvement of sphingolipids in HR and senescence is suggested by the induction of the serine palmitoyltransferase (*SPT*) gene, involved in sphingolipids biosynthesis in the early stages of the HR response (Birch *et al.*,

1999) and upon induction of senescence in *Arabidopsis* and broccoli (Coupe *et al.*, 2004).

Transcription factors that seem to function in both senescence and in response to pathogen infection (Chen *et al.*, 2002) were identified. In *Arabidopsis*, the WRKY6 transcription factor showed strong expression during leaf senescence and both senescence and pathogen-induced expression of the *PR1* gene depended on WRKY6 protein presence (Robatzek and Somssich, 2002). When targets of another member of the WRKY family, the senescence-associated WRKY53 transcription factor, were searched for in *Arabidopsis*, a number of transcription factors related to pathogen defense were identified (Miao *et al.*, 2004) further implying a link between the regulatory circuits of senescence and pathogen defense processes. Thus, a cross talk might exist between signaling pathways of leaf senescence and pathogen-defense responses, although this linkage remains to be elucidated.

Relationship on the level of the mechanisms involved in cell death during both senescence and HR is supported by the demonstrated involvement of caspase-like activity in both processes. The cysteine protease, VPEgamma, is associated with senescence (Rojo *et al.*, 2003) and alteration of its expression level was found to have a significant influence in the outcome of plant–pathogen interactions (Rojo *et al.*, 2004).

6.13 Concluding remarks

The observed effects of environmental stimuli on plant senescence suggest that almost any environmental stress with a negative consequence to growth conditions may result in the induction or enhancement of plant senescence. However, information is limited about the regulatory pathways involved in mediating the environmental stress signals with the observed enhancement of senescence. Recent studies of senescence in plants suggest that complex regulatory networks exist in senescence (Guo and Gan, 2005; Lim and Nam, 2005). The degree of overlap between the regulatory pathways operating during age-dependent, developmental senescence and the regulatory pathways activated by environmental stresses is not clear yet, although it seems that basic systems involving oxidative status, sugars or hormone action have an important role in both. We also do not know to what extent the regulatory pathways are kept distinct for the different senescence stimuli, and where they converge. It is likely that the convergence occurs upstream of the gene expression which is involved with the mechanisms of cellular and macromolecule breakdown leading to nutrient recycling.

What plant internal systems or factors mediate the environmental changing conditions and translate them into molecular signal transduction pathways, leading to the expression of the relevant genes? Few different internal systems or factors are suggested in the literature to be involved in the induction and control of senescence, including different growth regulators such as ethylene and cytokinin, balance between ROS and antioxidant levels, sugars, nitrogen status, photosynthesis and metabolic flux or other additional unknown age-specific factors. While each one of these systems or factors has a role in the regulation of senescence, it is possible that an integration of the overall output of these systems or the balance between some of them determine senescence initiation or rate of progress. Thus, if one of the specific systems is active to a high level it will result in senescence while moderate simultaneous change in few of these systems will also result in senescence (Figure 6.1).

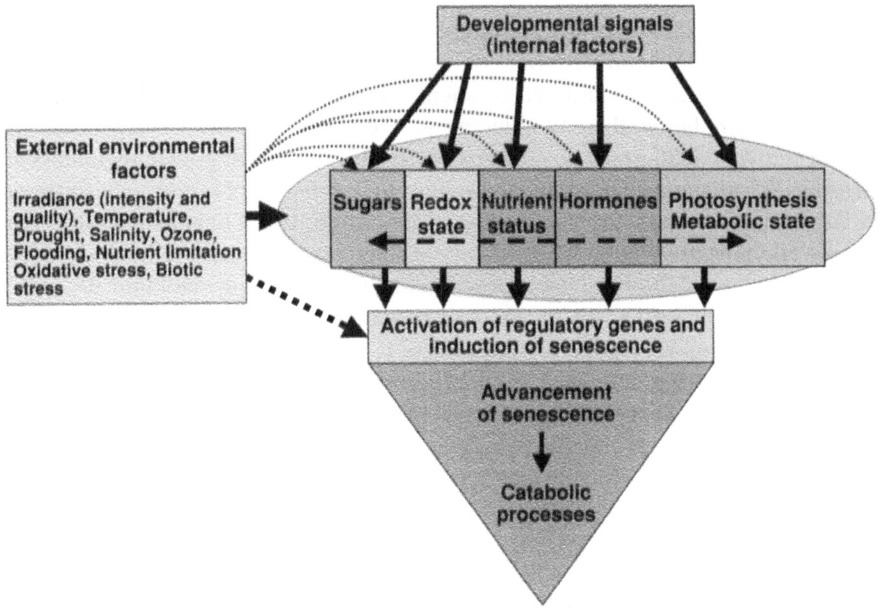

Figure 6.1 A model for regulation of senescence by environmental factors that may be mediated by the same plant internal systems/factors involved in age-dependent regulation. Few different internal systems/factors are suggested in the literature to be involved in the induction and control of senescence, including sugars, redox state (determined by the balance between ROS and antioxidant levels), nutrient status (nitrogen or other nutrient shortage), different plant hormones (ethylene, JA, SA, ABA, cytokinin and others) and photosynthesis and metabolic state. While each one of these systems/factors may mediate induction of senescence, it is possible that an integration of the overall output of these systems/factors or the balance between some of them determine senescence initiation or rate of progress. There is also an interaction between these systems/factors (illustrated by the red horizontal dashed arrow) and a change in one of them can result in a change in the others. These systems/factors are probably activating senescence-related regulatory genes. Thus, if one of the specific systems/factors is active to a high level it will result in senescence while moderate simultaneous change in few of these systems will also result in senescence. These internal systems/factors in the plant are also known to respond to environmental stimuli, which can induce senescence. During normal growth, under optimal environmental conditions, they respond or change gradually under the control of mainly the age-related factors. Changing environmental conditions, involving stress, result in more rapid and sharp changes in one or more of these internal systems/factors or the balance among them which will interfere with the age-dependent progress of senescence and lead to premature, accelerated senescence. It is also possible that the environmental signals that induce senescence activate directly regulatory genes that are involved in senescence control.

In many of the studies which investigated the effect of different environmental stresses on senescence, an important role was attributed to oxidative stress and the balance between ROS and the antioxidative system. Redox homeostasis and antioxidant signaling has been suggested to form a metabolic interface between stress perception and physiological responses (Foyer and Noctor, 2005) and it is likely to have such a role also in the regulation of senescence. Different aspects of the possible involvement of the oxidative stress in senescence have been discussed above. Sugar content and metabolism were also suggested in many studies to function as possible integrators of environmental signals during the regulation of senescence. The role of sugars was recently comprehensively reviewed (Wingler *et al.*, 2006). It was suggested that the involvement of sugars mainly involves integrating environmental signals related to nitrogen deficiency and to high-light growth which result in sugar accumulation, thereby inducing senescence via the hexose dependent signaling pathway (Wingler *et al.*, 2006). Clearly, different plant growth regulators are involved in integrating environmental signals for regulation of senescence. Different environmental stresses, such as chilling, dehydration, UV-B light and flooding, can induce ethylene production which in turn accelerates senescence (Kim *et al.*, 1998; van der Krol *et al.*, 1999). Other growth regulators including cytokinin, JA, ABA, SA and others, as well as the interactions between them, are likely to play a role in the environmental regulation of senescence. It is likely that different environmental stimuli are able to influence the photosynthetic status of the leaf which in turn results with senescence. Different hypotheses that explain how a leaf can sense its photosynthetic status within the plant were suggested (Ono *et al.*, 2001). These include source–sink balance, nitrogen status, sugar sensing or cytokinin levels among others (Ono *et al.*, 2001). Changes in different photosynthesis-related metabolites may also have a role. Interactions between some of these different systems and factors in relation to the regulation of senescence were demonstrated. For example, sugars, cytokinin, and light (Wingler *et al.*, 1998; Lara *et al.*, 2004) or ABA and sugar signaling (Pourtau *et al.*, 2004) were shown to interact during regulation of leaf senescence.

It is possible that the different systems and factors in the plant which respond to environmental stimuli are those that are involved in the age-dependent, developmental regulation of senescence as well. During normal growth, and under optimal environmental conditions, they respond or change gradually under the control of age-related factors. Changing environmental conditions involving stress result in a more rapid and pronounced change in one or more of these internal systems, or in the balance among them which interferes with the age-dependent progression of senescence and lead to premature, accelerated senescence (Figure 6.1).

The senescence accelerating effect of environmental stress may be explained in light of the possible evolutionary basis of leaf senescence discussed before (Bleecker, 1998; Jing *et al.*, 2003; Munne-Bosch and Alegre, 2004). It was suggested that the genetic programming of senescence can be explained if it is assumed that it has an adaptive aspect. For example, the activation of nutrient recycling from the mature leaves which occurs during senescence might have a strong adaptive advantage which allows plants under environmental stress to be productive and to successfully complete their life cycle determined by seed set. The presence of a complex regulatory network controlling senescence in plants may be the result of selection pressure driven by different environmental stresses for the development of senescence. It would be interesting to identify environmental stress conditions that do not lead to accelerated senescence. Comparison of such environmental stress stimuli to those that do result with senescence acceleration may supply some insights into the biology and maybe the benefit to the plant of environmentally induced senescence.

Overall it seems that different mechanisms were evolved and are involved in the translation of environmental signals into the controlling mechanisms of the senescence program. Variability due to different plant systems and different growth conditions interfere in our ability to obtain a comprehensive understanding of environmental regulation of senescence. Focus on limited number of model plant systems studied by plant senescence scientists may be required for more efficient research, and is likely to be highly relevant to agriculture as well as to our basic understanding of the senescence process in plants.

References

Ainsworth, E.A. and Long, S.P. (2005) What have we learned from 15 years of free-air CO_2 enrichment (FACE)? A meta-analytic review of the responses of photosynthesis, canopy. *New Phytologist* **165**, 351–371.

Ali, M., Jensen, C.R., Mogensen, V.O., Andersen, M.N. and Henson, I.E. (1999) Root signalling and osmotic adjustment during intermittent soil drying sustain grain yield of field grown wheat. *Field Crops Res* **62**, 35–52.

Apelbaum, A. and Yang, S.F. (1981) Biosynthesis of stress ethylene induced by water deficit. *Plant Physiol* **68**, 594–596.

Ashmore, M.R. (2005) Assessing the future global impacts of ozone on vegetation. *Plant Cell Environ* **28**, 949–964.

Bachmann, A.S., Matile, P. and Slusarenko, A.J. (1998) Inhibition of ornithine decarboxylase activity by phaseolotoxin: implications for symptom production in halo blight of French bean. *Physiol Mol Plant Pathol* **53**, 287–299.

Bahl, A. and Kahl, G. (1995) Air pollutant stress changes the steady-state transcript levels of 3 photosynthesis genes. *Environ Pollut* **88**, 57–65.

Banga, M., Bogemann, G.M., Blom, C. and Voesenek, L. (1997) Flooding resistance of *Rumex* species strongly depends on their response to ethylene: rapid shoot elongation or foliar senescence. *Physiol Plantarum* **99**, 415–422.

Barth, C., Moeder, W., Klessig, D.F. and Conklin, P.L. (2004) The timing of senescence and response to pathogens is altered in the ascorbate-deficient *Arabidopsis* mutant vitamin c-1. *Plant Physiol* **134**, 1784–1792.

Barthakur, S., Babu, V. and Bansal, K.C. (2001) Over-expression of osmotin induces proline accumulation and confers tolerance to osmotic stress in transgenic tobacco. *J Plant Biochem Biotechnol* **10**, 31–37.

Batten, G.D. and Wardlaw, I.F. (1987) Senescence of the flag leaf and grain-yield following late foliar and root applications of phosphate on plants of differing phosphorus status. *J Plant Nutr* **10**, 735–748.

Becker, W. and Apel, K. (1993) Differences in gene-expression between natural and artificially induced leaf senescence. *Planta* **189**, 74–79.

Behera, R.K., Mishra, P.C. and Choudhury, N.K. (2002) High irradiance and water stress induce alterations in pigment composition and chloroplast activities of primary wheat leaves. *J Plant Physiol* **159**, 967–973.

Behera, Y.N. and Biswal, B. (1990) Leaf senescence in fern – effect of duration, intensity and quality of light. *Environ Exp Bot* **30**, 181–186.

Benbella, M. and Paulsen, G.M. (1998a) Efficacy of treatments for delaying senescence of wheat leaves. I: Senescence under controlled conditions. *Agron J* **90**, 329–332.

Benbella, M. and Paulsen, G.M. (1998b) Efficacy of treatments for delaying senescence of wheat leaves. II:. Senescence and grain yield under field conditions. *Agron J* **90**, 332–338.

Bindi, M., Hacour, A., Vandermeiren, K. *et al.* (2002) Chlorophyll concentration of potatoes grown under elevated carbon dioxide and/or ozone concentrations. *European J Agron* **17**, 319–335.

Birch, P.R.J., Avrova, A.O., Duncan, J.M., Lyon, G.D. and Toth, R.L. (1999) Isolation of potato genes that are induced during an early stage of the hypersensitive response to *Phytophthora infestans*. *Mol Plant Microbe Interact* **12**, 356–361.

Biswal, B. and Choudhury, N.K. (1986) Photocontrol of chlorophyll loss in papaya leaf discs. *Plant Cell Physiol* **27**, 1439–1444.

Biswal, B., Joshi, P.N. and Kulandaivelu, G. (1997) Changes in leaf protein and pigment contents and photosynthetic activities during senescence of detached maize leaves: influence of different ultraviolet radiations. *Photosynthetica* **34**, 37–44.

Biswal, U.C. and Biswal, B. (1984) Photocontrol of leaf senescence. *Photochem Photobiol* **39**, 875–879.

Bleecker, A.B. (1998) The evolutionary basis of leaf senescence: method to the madness? *Curr Opin Plant Biol* **1**, 73–78.

Boch, J., Verbsky, M.L., Robertson, T.L., Larkin, J.C. and Kunkel, B.N. (1998) Analysis of resistance gene-mediated defence responses in *Arabidopsis thaliana* plants carrying a mutation in *cpr5*. *Plant Microbe Interact* **11**, 1196–1206.

Booker, F.L., Burkey, K.O., Overmyer, K. and Jones, A.M. (2004) Differential responses of G-protein *Arabidopsis thaliana* mutants to ozone. *New Phytologist* **162**, 633–641.

Borrell A.K., Hammer G.L. and Douglas A.C.L. (2000) Does maintaining green leaf area in sorghum improve yield under drought? I. Leaf growth and senescence. *Crop Sci* **40**, 1026–1037.

Bowling, S.A., Clarke, J.D., Liu, Y., Klessig, D.F. and Dong, X. (1997) The *cpr5* mutant of *Arabidopsis* expresses both NPR1-dependent and NPR1-independent resistance. *Plant Cell* **9**, 1573–1584.

Boyer, J.S. (1982) Plant productivity and environment. *Science* **218**, 443–448.

Broetto, F., Luttge, U. and Ratajczak, R. (2002) Influence of light intensity and salt-treatment on mode of photosynthesis and enzymes of the antioxidative response system of *Mesembryanthemum crystallinum*. *Funct Plant Biol* **29**, 13–23.

Brosche, M. and Strid, A. (2003) Molecular events following perception of ultraviolet-B radiation by plants. *Physiol Plantarum* **117**, 1–10.

Buchanan-Wollaston, V., Earl, S., Harrison, E. *et al.* (2003) The molecular analysis of leaf senescence – a genomics approach. *Plant Biotechnol J* **1**, 3–22.

Buchanan-Wollaston, V., Page, T., Harrison, E. *et al.* (2005) Comparative transcriptome analysis reveals significant differences in gene expression and signalling pathways between developmental and dark/starvation-induced senescence in *Arabidopsis*. *Plant J* **42**, 567–585.

Burkey, K.O. and Wells, R. (1991) Response of soybean photosynthesis and chloroplast membrane-function to canopy development and mutual shading. *Plant Physiol* **97**, 245–252.

Burrows, W.J. and Carr, D.J. (1969) Effects of flooding the root system of sunflower plants on the cytokinin content of the xylem sap. *Plant Physiol* **22**, 1105–1112.

Butt, A., Mousley, C., Morris, K. *et al.* (1998) Differential expression of a senescence-enhanced metallothionein gene in *Arabidopsis* in response to isolates of *Peronospora parasitica* and *Pseudomonas syringae*. *Plant J* **16**, 209–221.

Cavender-Bares, J., Potts, M., Zacharias, E. and Bazzaz, F.A. (2000) Consequences of CO_2 and light interactions for leaf phenology, growth, and senescence in *Quercus rubra*. *Global Change Biol* **6**, 877–887.

Chaves, M.M., Maroco, J.P. and Pereira, J.S. (2003) Understanding plant responses to drought – from genes to the whole plant. *Funct Plant Biol* **30**, 239–264.

Chen, M., Chory, J. and Fankhauser, C. (2004) Light signal transduction in higher plants. *Annu Rev Genet* **38**, 87–117.

Chen, W., Provart, N.J., Glazebrook, J. *et al.* (2002) Expression profile matrix of *Arabidopsis* transcription factor genes suggests their putative functions in response to environmental stresses. *Plant Cell* **14**, 559–574.

Cherry, J.R., Hershey, H.P. and Vierstra, R.D. (1991) Characterization of tobacco expressing functional oat phytochrome – domains responsible for the rapid degradation of Pfr are conserved between monocots and dicots. *Plant Physiol* **96**, 775–785.

Choe, H.T. and Thimann, K.V. (1977) The retention of photosynthetic activity by senescing chloroplasts of oat leaves. *Planta* **135**, 101–107.

Choudhury, N.K. and Behera, R.K. (2001) Photoinhibition of photosynthesis: role of carotenoids in photoprotection of chloroplast constituents. *Photosynthetica* **39**, 481–488.

Cock, J.H., Franklin, D., Sandoval, G. and Juri, P. (1979) The ideal cassava plant for maximum yield. *Crop Sci* **19**, 271–279.

Colomb, B., Kiniry, J.R. and Debaeke, P. (2000) Effect of soil phosphorus on leaf development and senescence dynamics of field-grown maize. *Agron J* **92**, 428–435.

Conklin, P.L. and Barth, C. (2004) Ascorbic acid, a familiar small molecule intertwined in the response of plants to ozone, pathogens, and the onset of senescence. *Plant Cell Environ* **27**, 959–970.

Conklin, P.L., Williams, E.H. and Last, R.L. (1996) Environmental stress sensitivity of an ascorbic acid-deficient *Arabidopsis* mutant. *Proc Natl Acad Sci U S A* **93**, 9970–9974.

Coupe, S.A., Watson, L.M., Ryan, D.J., Pinkney, T.T. and Eason, J.R. (2004) Molecular analysis of programmed cell death during senescence in *Arabidopsis thaliana* and *Brassica oleracea*: cloning broccoli LSD1, Bax inhibitor and serine palmitoyltransferase homologues. *J Exp Bot* **55**, 59–68.

Crafts-Brandner, S.J. (1992) Phosphorus nutrition influence on leaf senescence in soybean. *Plant Physiol* **98**, 1128–1132.

Crafts-Brandner, S.J., Holzer, R. and Feller, U. (1998) Influence of nitrogen deficiency on senescence and the amounts of RNA and proteins in wheat leaves. *Physiol Plantarum* **102**, 192–200.

Cuello, J., Sanchez, M.D. and Sabater, B. (1994) Retardation of senescence by UV-A light in barley (*Hordeum vulgare* L.) leaf segments. *Environ Exp Bot* **34**, 1–8.

Dai, Q., Yan, B., Huang, S. *et al.* (1997) Response of oxidative stress defense systems in rice (*Oryza sativa*) leaves with supplemental UV-B radiation. *Physiol Plantarum* **101**, 301–308.

Dalla Vecchia, F., Rocca, N.L., Moro, I., De Faveri, S., Andreoli, C. and Rascio, N. (2005) Morphogenetic, ultrastructural and physiological damages suffered by submerged leaves of *Elodea canadensis* exposed to cadmium. *Plant Sci* **168**, 329.

De Greef, J., Butler, W.L., Rotha, T.F. and Fredericq, H. (1971) Control of senescence in *Marchantia* by phytochrome. *Plant Physiol* **48**, 407–412.

del Rio, L.A., Pastori, G.M., Palma, J.M. *et al.* (1998) The activated oxygen role of peroxisomes in senescence. *Plant Physiol* **116**, 1195–1200.

Djebali, W., Zarrouk, M., Brouquisse, R. *et al.* (2005) Ultrastructure and lipid alterations induced by cadmium in tomato (*Lycopersicon esculentum*) chloroplast membranes. *Plant Biol* **7**, 358–368.

Duroy, A.N. and Thomas, J.W. (1999) Victorin induction of an apoptotic/senescence-like response in oats. *Plant Cell* **11**, 237–249.

Ella, E.S., Kawano, N., Yamauchi, Y., Tanaka, K. and Ismail, A.M. (2003) Blocking ethylene perception enhances flooding tolerance in rice seedlings. *Funct Plant Biol* **30**, 813–819.

Falkowski, P., Scholes, R.J., Boyle, E. *et al.* (2000) The global carbon cycle: a test of our knowledge of earth as a system. *Science* **290**, 291–296.

Fangmeier, A., Chrost, B., Hogy, P. and Krupinska, K. (2000) CO_2 enrichment enhances flag leaf senescence in barley due to greater grain nitrogen sink capacity. *Environ Exp Bot* **44**, 151–164.

Feild, T.S., Lee, D.W. and Holbrook, N.M. (2001) Why leaves turn red in autumn. The role of anthocyanins in senescing leaves of red-osier dogwood. *Plant Physiol* **127**, 566–574.

Fiscus, E.L., Booker, F.L. and Burkey, K.O. (2005) Crop responses to ozone: uptake, modes of action, carbon assimilation and partitioning. *Plant Cell Environ* **28**, 997–1011.

Fortmeier, R. and Schubert, S. (1995) Salt tolerance of maize (*Zea Mays* L) – the role of sodium exclusion. *Plant Cell Environ* **18**, 1041–1047.

Foyer, C.H. and Noctor, G. (2005) Redox homeostasis and antioxidant signaling: a metabolic interface between stress perception and physiological responses. *Plant Cell* **17**, 1866–1875.

Foyer, C.H., Vanacker, H., Gomez, L.D. and Harbinson, J. (2002) Regulation of photosynthesis and antioxidant metabolism in maize leaves at optimal and chilling temperatures: review. *Plant Physiol Biochem* **40**, 659–668.

Franklin, K.A. and Whitelam, G.C. (2005) Phytochromes and shade-avoidance responses in plants. *Ann Bot* **96**, 169–175.

Franklin, K.A., Larner, V.S. and Whitelam, G.C. (2005) The signal transducing photoreceptors of plants. *Int J Dev Biology* **49**, 653–664.

Fuhrer, J. and Booker, F. (2003) Ecological issues related to ozone: agricultural issues. *Environ Int* **29**, 141.

Fujiki, Y., Ito, M., Nishida, I. and Watanabe, A. (2000) Multiple signaling pathways in gene expression during sugar starvation. Pharmacological analysis of din gene expression in suspension-cultured cells of *Arabidopsis*. *Plant Physiol* **124**, 1139–1147.

Fujiki, Y., Nakagawa, Y., Furumoto, T. *et al.* (2005) Response to darkness of late-responsive dark-inducible genes is positively regulated by leaf age and negatively regulated by calmodulin-antagonist-sensitive signalling in *Arabidopsis thaliana*. *Plant Cell Physiol* **46**, 1741–1746.

Fujiki, Y., Yoshikawa, Y., Sato, T. *et al.* (2001) Dark-inducible genes from *Arabidopsis thaliana* are associated with leaf senescence and repressed by sugars. *Physiol Plantarum* **111**, 345–352.

Gartia, S., Pradhan, M.K., Joshi, P.N., Biswal, U.C. and Biswal, B. (2003) UV-A irradiation guards the photosynthetic apparatus against UV-B-induced damage. *Photosynthetica* **41**, 545–549.

Gay, A.P. and Thomas, H. (1995) Leaf development in *Lolium Temulentum* L. – photosynthesis in relation to growth and senescence. *New Phytologist* **130**, 159–168.

Gindin, E., Tirosh, T. and Mayak, S. (1989) Effects of flooding on ultrastructure and ethylene production in chrysanthemums. *Acta Horticulturae* **261**, 171–183.

Glick, R.E., Schlagnhaufer, C.D., Arteca, R.N. and Pell, E.J. (1995) Ozone-induced ethylene emission accelerates the loss of ribulose-1,5-bisphosphate carboxylase/oxygenase and nuclear-encoded mRNAs in senescing potato leaves. *Plant Physiol* **109**, 891–898.

Graham, L.J., Blevins, D.G. and Minor, H.C. (1986) P nutrition during seed development. Leaf senescence, pod retention, and weight of soybean. *Plant Physiol* **82**, 1008–1012.

Granell, A. (1999) Dying according to programme: occurrence in plant developmental processes and in response to environmental cues. In: *Environmental Stress And Gene Regulation* (ed. Storey, K.B.). Bios Scientific Publishers, Washington DC, pp. 159–177.

Grbic, V. and Bleecker, A.B. (1995) Ethylene regulates the timing of leaf senescence in *Arabidopsis*. *Plant J* **8**, 595–602.

Greenberg, J.T. and Yao, N. (2004) The role and regulation of programmed cell death in plant–pathogen interactions. *Cell Microbiol* **6**, 201–211.

Guiamet, J.J., Willemoes, J.G. and Montaldi, E.R. (1989) Modulation of progressive leaf senescence by the red far-red ratio of incident light. *Bot Gaz* **150**, 148–151.

Guo, Y.F. and Gan, S.S. (2005) Leaf senescence: signals, execution, and regulation. *Curr Top Dev Biol*, **71**, 83–112.

Guterman, A., Hajouj, T. and Gepstein, S. (2003) Senescence-associated mRNAs that may participate in signal transduction and protein trafficking. *Physiol Plantarum* **118**, 439–446.

Haber, A.H., Thompson, P.J., Walne, P.L. and Triplett, L.L. (1969) Nonphotosynthetic retardation of chloroplast senescence by light. *Plant Physiol* **44**, 1619–1628.

Hajouj, T., Michelis, R. and Gepstein, S. (2000) Cloning and characterization of a receptor-like protein kinase gene associated with senescence. *Plant Physiol* **124**, 1305–1314.

Hanfrey, C., Fife, M. and Buchanan-Wollston, V. (1996) Leaf senescence in *Brassica napus*: expression of genes encoding pathogenesis-related genes. *Plant Mol Biol* **30**, 597–609.

Harding, S.A., Guikema, J.A. and Paulsen, G.M. (1990) Photosynthetic decline from high temperature stress during maturation of wheat. I. Interaction with senescence processes. *Plant Physiol* **92**, 648–653.

He, Y., Fukushige, H., Hildebrand, D.F. and Gan, S. (2002) Evidence supporting a role of jasmonic acid in *Arabidopsis* leaf senescence. *Plant Physiol* **128**, 876–884.

He, Y.H., Tang, W.N., Swain, J.D., Green, A.L., Jack, T.P. and Gan, S.S. (2001) Networking senescence-regulating pathways by using *Arabidopsis* enhancer trap lines. *Plant Physiol* **126**, 707–716.

Held, A.A., Mooney, H.A. and Gorham, J.N. (1991) Acclimation to ozone stress in radish – leaf demography and photosynthesis. *New Phytologist* **118**, 417–423.

Helmy, Y.H., El-Abd, S.O., Abou-Hadid, A.F., El-Beltagy, M. and El-Beltagy, A.S. (1994) Ethylene production from tomato and cucumber plants under saline conditions. *Egyptian J Horticulture* **21**, 153–160.

Herrick, J.D. and Thomas, R.B. (1999) Effects of CO_2 enrichment on the photosynthetic light response of sun and shade leaves of canopy sweetgum trees (*Liquidambar styraciflua*) in a forest ecosystem. *Tree Physiol* **19**, 779–786.

Hoch, W.A., Singsaas, E.L. and McCown, B.H. (2003) Resorption protection. Anthocyanins facilitate nutrient recovery in autumn by shielding leaves from potentially damaging light levels. *Plant Physiol* **133**, 1296–1305.

Huang, B., Liu, X. and Xu, Q. (2001) Supraoptimal soil temperatures induced oxidative stress in leaves of creeping bentgrass cultivars differing in heat tolerance. *Crop Sci* **41**, 430–435.

Huner, N.P.A., Oquist, G. and Sarhan, F. (1998) Energy balance and acclimation to light and cold. *Trends Plant Sci* **3**, 224–230.

Hurng, W.P., Lur, H.S., Liao, C.K. and Kao, C.H. (1994) Role of abscisic acid, ethylene and polyamines in flooding-promoted senescence of tobacco leaves. *J Plant Physiol* **143**, 102–105.

Huynh, L.N., VanToai, T., Streeter, J. and Banowetz, G. (2005) Regulation of flooding tolerance of SAG12: ipt *Arabidopsis* plants by cytokinin. *J Exp Bot* **56**, 1397–1407.

Jackson, M.B. and Ram, P.C. (2003) Physiological and molecular basis of susceptibility and tolerance of rice plants to complete submergence. *Ann Bot* **91**, 227–241.

Jimenez, A., Hernandez, J.A., Pastori, G., del Rio, L.A. and Sevilla, F. (1998) Role of the ascorbate-glutathione cycle of mitochondria and peroxisomes in the senescence of pea leaves. *Plant Physiol* **118**, 1327–1335.

Jing, H.C., Hille, J. and Dijkwel, R.R. (2003) Ageing in plants: conserved strategies and novel pathways. *Plant Biol* **5**, 455–464.

John, C.F., Morris, K., Jordan, B.R., Thomas, B. and Mackerness, S.A.H. (2001) Ultraviolet-B exposure leads to up-regulation of senescence-associated genes in *Arabidopsis thaliana*. *J Exp Bot* **52**, 1367–1373.

John, I., Drake, R., Farrell, A. *et al.* (1995) Delayed leaf senescence in ethylene-deficient acc-oxidase antisense tomato plants – molecular and physiological analysis. *Plant J* **7**, 483–490.

Jordan, B.R., James, P.E. and Mackerness, S.A.H. (1998) Factors affecting UV-B-induced changes in *Arabidopsis thaliana* L. gene expression: the role of development, protective pigments and the chloroplast signal. *Plant Cell Physiol* **39**, 769–778.

Joshi, P.N., Biswal, B. and Biswal, U.C. (1991) Effect of UV-A on aging of wheat leaves and role of phytochrome. *Environ Exp Bot* **31**, 267–276.

Kangasjarvi, J., Jaspers, P. and Kollist, H. (2005) Signalling and cell death in ozone-exposed plants. *Plant Cell Environ* **28**, 1021–1036.

Kappers, I.F., Jordi, W., Maas, F.M., Stoopen, G.M. and Van Der Plas, L.H.W. (1998) Gibberellin and phytochrome control senescence in alstroemeria leaves independently. *Physiol Plantarum* **103**, 91–98.

Kar, M. (1986) The effect of photoperiod on chlorophyll loss and lipid-peroxidation in excised senescing rice leaves. *J Plant Physiol* **123**, 389–393.

Kim, Y.S., Choi, D., Lee, M.M., Lee, S.H. and Kim, W.T. (1998) Biotic and abiotic stress-related expression of 1-aminocyclopropane-1-carboxylate oxidase gene family in *Nicotiana glutinosa* L. *Plant Cell Physiol* **39**, 565–573.

Koch, J.R., Creelman, R.A., Eshita, S.M., Seskar, M., Mullet, J.E. and Davis, K.R. (2000) Ozone sensitivity in hybrid poplar correlates with insensitivity to both salicylic acid and jasmonic acid. The role of programmed cell death in lesion formation. *Plant Physiol* **123**, 487–496.

Kramer, P.J. (1951) Causes of injury to plants resulting from flooding of the soil. *Plant Physiol* **26**, 722–736.

Kukavica, B. and Jovanovic, S.V. (2004) Senescence-related changes in the antioxidant status of ginkgo and birch leaves during autumn yellowing. *Physiol Plantarum* **122**, 321–327.

Kuroyanagi, T. and Paulsen, G.M. (1985) Mode of high temperature injury to wheat. II. Comparisons of wheat and rice with and without inflorescences. *Physiol Plantarum* **65**, 203–208.

Kuzniak, E. and Sklodowska, M. (2005) Fungal pathogen-induced changes in the antioxidant systems of leaf peroxisomes from infected tomato plants. *Planta* **222**, 192–200.

Lara, M.E.B., Garcia, M.C.G., Fatima, T. *et al.* (2004) Extracellular invertase is an essential component of cytokinin-mediated delay of senescence. *Plant Cell* **16**, 1276–1287.

Larkindale, J. and Knight, M.R. (2002) Protection against heat stress-induced oxidative damage in *Arabidopsis* involves calcium, abscisic acid, ethylene, and salicylic acid. *Plant Physiol* **128**, 682–695.

Lawson, T., Craigon, J., Tulloch, A.M., Black, C.R., Colls, J.J. and Landon, G. (2001) Photosynthetic responses to elevated CO_2 and O_3 in field-grown potato (*Solanum tuberosum*). *J Plant Physiol* **158**, 309–323.

Lers, A., Jiang, W.B., Lomaniec, E. and Aharoni, N. (1998) Proteins functionally and immunogenically related to pathogenesis-related proteins are induced during parsley leaf senescence. *Physiol Plantarum* **103**, 497–502.

Levey, S. and Wingler, A. (2005) Natural variation in the regulation of leaf senescence and relation to other traits in *Arabidopsis*. *Plant Cell Environ* **28**, 223–231.

Levine, A., Tenhaken, R., Dixon, R. and Lamb, C. (1994) H_2O_2 from the oxidative burst orchestrates the plant hypersensitive disease resistance response. *Cell* **79**, 583–593.

Li, J.H., Dijkstra, P., Hymus, G.J. *et al.* (2000) Leaf senescence of *Quercus myrtifolia* as affected by long-term CO2 enrichment in its native environment. *Global Change Biol* **6**, 727–733.

Li, P., Mane, S.P., Sioson, A.A. *et al.* (2006) Effects of chronic ozone exposure on gene expression in *Arabidopsis thaliana* ecotypes and in *Thellungiella halophila*. *Plant Cell Environ* **29**, 854–868.

Lim, P.O. and Nam, H.G. (2005) The molecular and genetic control of leaf senescence and longevity in *Arabidopsis*. *Curr Top Dev Biol* **67**, 49–83.

Lois, R. (1994) Accumulation of UV-absorbing flavonoids induced by UV-B radiation in *Arabidopsis-thaliana* L.1. Mechanisms of UV-resistance in *Arabidopsis*. *Planta* **194**, 498–503.

Lovelli, S., Rivelli, A.R., Nardiello, I., Perniola, M. and Tarantino, E. (2000) Growth, leaf ion concentration, stomatal behaviour and photosynthesis of bean (*Phaseolus vulgaris* L.) irrigated with saline water. In: *Acta Horticulturae 537*, Vol. 2. International Society for Horticultural Science (ISHS), Leuven, Belgium, pp. 679–686.

Lu, C. and Zhang, J. (1998) Changes in photosystem II function during senescence of wheat leaves. *Physiol Plantarum* **104**, 239–247.

Ludewig, F. and Sonnewald, U. (2000) High CO_2-mediated down-regulation of photosynthetic gene transcripts is caused by accelerated leaf senescence rather than sugar accumulation. *FEBS Lett* **479**, 19–24.

Mackerness, S.A.H., Surplus, S.L., Blake, P. *et al.* (1999) Ultraviolet-B-induced stress and changes in gene expression in *Arabidopsis thaliana*: role of signalling pathways controlled by jasmonic acid, ethylene and reactive oxygen species. *Plant Cell Environ* **22**, 1413–1423.

Mackerness, S.A.H., Surplus, S.L., Jordan, B.R. and Thomas, B. (1998) Effects of supplementary ultraviolet-B radiation on photosynthetic transcripts at different stages of leaf development and light levels in pea (*Pisum sativum* L.): role of active oxygen species and antioxidant enzymes. *Photochem Photobiol* **68**, 88–96.

McCarthy, I., Romero-Puertas, M.C., Palma, J.M. *et al.* (2001a) Cadmium induces senescence symptoms in leaf peroxisomes of pea plants. *Plant Cell Environ* **24**, 1065–1073.

McCarthy, J.J., Canziani, O.F., Leary, N.A., Dokken, D.J. and White, K.S. (2001b) *Climate Change 2001: Impacts, Adaptation, and Vulnerability*. Cambridge University Press, New York.

McKenzie, R., Connor, B. and Bodeker, G. (1999) Increased summertime UV radiation in New Zealand in response to ozone loss. *Science* **285**, 1709–1711.

Merzlyak, M.N. and Gitelson, A. (1995) Why and what for the leaves are yellow in autumn? On the interpretation of optical spectra of senescing leaves (*Acer platanoides* L.). *J Plant Physiol* **145**, 315–320.

Miao, Y., Laun, T., Zimmermann, P. and Zentgraf, U. (2004) Targets of the WRKY53 transcription factor and its role during leaf senescence in *Arabidopsis*. *Plant Mol Biol* **55**, 853–867.

Miglietta, F., Magliulo, V., Bindi, M. *et al.* (1998) Free air CO_2 enrichment of potato (*Solanum tuberosum* L.): development, growth and yield. *Global Change Biol* **4**, 163–172.

Miller A., Schlagnhaufer C., Spalding M. and Rodermel S. (2000) Carbohydrate regulation of leaf development: prolongation of leaf senescence in Rubisco antisense mutants of tobacco. *Photosynthesis Res* **63**, 1.

Miller, A., Schlagnhaufer, C., Spalding, M. and Rodermel, S. (2000) Carbohydrate regulation of leaf development: prolongation of leaf senescence in Rubisco antisense mutants of tobacco. *Photosynthesis Res* **63**, 1–8.

Mohan, B.S. and Hosetti, B.B. (1998) Lead toxicity to Salvinia natans grown in macrophyte ponds. *J Ecotoxicol Environ Monit* **8**, 3–7.

Mommer, L. and Visser, E.J.W. (2005) Underwater photosynthesis in flooded terrestrial plants: a matter of leaf plasticity. *Ann Bot* **96**, 581–589.

Morris, K., Mackerness, S.A.H., Page, T. *et al.* (2000) Salicylic acid has a role in regulating gene expression during leaf senescence. *Plant J* **23**, 677–685.

Munne-Bosch, S. and Alegre, L. (2004) Die and let live: leaf senescence contributes to plant survival under drought stress. *Funct Plant Biol* **31**, 203–216.

Munne-Bosch, S., Jubany-Mari, T. and Alegre, L. (2001) Drought-induced senescence is characterized by a loss of antioxidant defences in chloroplasts. *Plant Cell Environ* **24**, 1319–1327.

Munns, R. (2002) Comparative physiology of salt and water stress. *Plant Cell Environ* **25**, 239–250.

Navabpour, S., Morris, K., Allen, R., Harrison, E., A-H-Mackerness, S. and Buchanan-Wollaston, V. (2003) Expression of senescence-enhanced genes in response to oxidative stress. *J Exp Bot* **54**, 2285–2292.

Nayak, L., Biswal, B., Ramaswamy, N.K., Iyer, R.K., Nair, J.S. and Biswal, U.C. (2003) Ultraviolet-A induced changes in photosystem II of thylakoids: effects of senescence and high growth temperature. *J Photochem Photobiol B-Biol* **70**, 59–65.

Ni, M. (2005) Integration of light signaling with photoperiodic flowering circadian rhythm. *Cell Res* **15**, 559–566.

Nie, G.Y., Long, S.P., Garcia, R.L. *et al.* (1995) Effects of free-air CO_2 enrichment on the development of the photosynthetic apparatus in wheat, as indicated by changes in leaf proteins. *Plant Cell Environ* **18**, 855–864.

Nie, G.Y., Tomasevic, M. and Baker, N.R. (1993) Effects of ozone on the photosynthetic apparatus and leaf proteins during leaf development in wheat. *Plant Cell Environ* **16**, 643–651.

Niinemets, U. and Valladares, F. (2004) Photosynthetic acclimation to simultaneous and interacting environmental stresses along natural light gradients: optimality and constraints. *Plant Biol* **6**, 254–268.

Nooden, L.D. (1984) Integration of soybean pod development and monocarpic senescence. *Physiol Plantarum* **62**, 273–284.

Nooden, L.D., Hillsberg, J.W. and Schneider, M.J. (1996) Induction of leaf senescence in *Arabidopsis thaliana* by long days through a light-dosage effect. *Physiol Plantarum* **96**, 491–495.

Nooden, L.D. and Penney, J.P. (2001) Correlative controls of senescence and plant death in *Arabidopsis thaliana* (Brassicaceae). *J Exp Bot* **52**, 2151–2159.

Obregon, P., Martin, R., Sanz, A. and Castresana, C. (2001) Activation of defence-related genes during senescence: a correlation between gene expression and cellular damage. *Plant Mol Biol* **46**, 67–77.

Oh, S.A., Lee, S.Y., Chung, I.K., Lee, C.H. and Nam, H.G. (1996) A senescence-associated gene of *Arabidopsis thaliana* is distinctively regulated during natural and artificially induced leaf senescence. *Plant Mol Biol* **30**, 739–754.

Oh, S.A., Park, J.H., Lee, G.I., Paek, K.H., Park, S.K. and Nam, H.G. (1997) Identification of three genetic loci controlling leaf senescence in *Arabidopsis thaliana*. *Plant J* **12**, 527–535.

Okane, D., Gill, V., Boyd, P. and Burdon, B. (1996) Chilling, oxidative stress and antioxidant responses in *Arabidopsis thaliana* callus. *Planta* **198**, 371–377.

Ono, K., Nishi, Y., Watanabe, A. and Terashima, I. (2001) Possible mechanisms of adaptive leaf senescence. *Plant Biol* **3**, 234–243.

Ono, K., Terashima, I. and Watanabe, A. (1996) Interaction between nitrogen deficit of a plant and nitrogen content in the old leaves. *Plant Cell Physiol* **37**, 1083–1089.

Ottman, M.J. and Welch, L.F. (1988) Supplemental radiation effects on senescence, plant nutrients, and yield of field-grown corn. *Agronomy J* **80**, 619–626.

Pageau, K., Reisdorf-Cren, M., Morot-Gaudry, J.F. and Masclaux-Daubresse, C. (2006) The two senescence-related markers, GS1 (cytosolic glutamine synthetase) and GDH (glutamate dehydrogenase), involved in nitrogen mobilization, are differentially regulated during pathogen attack and by stress hormones and reactive oxygen species in *Nicotiana tabacum* L. leaves. *J Exp Bot* **57**, 547–557.

Panda, S.K. and Patra, H.K. (2000) Does chromium(III) produce oxidative damage in excised wheat leaves? *J Plant Biol* **27**, 105–110.

Pandey, S., Ranade, S.A., Nagar, P.K. and Kumar, N. (2000) Role of polyamines and ethylene as modulators of plant senescence. *J Biosci* **25**, 291–299.

Park, J.H., Oh, S.A., Kim, Y.H., Woo, H.R. and Nam, H.G. (1998) Differential expression of senescence-associated mRNAs during leaf senescence induced by different senescence-inducing factors in *Arabidopsis*. *Plant Mol Biol* **37**, 445–454.

Pastori, G.M. and del Rio, L.A. (1997) Natural senescence of pea leaves – an activated oxygen-mediated function for peroxisomes. *Plant Physiol* **113**, 411–418.

Pell, E.J. and Pearson, N.S. (1983) Ozone-induced reduction in quantity of ribulose-1,5-bisphosphate carboxylase in alfalfa foliage. *Plant Physiol* **73**, 185–187.

Pell, E.J., Schlagnhaufer, C.D. and Arteca, R.N. (1997) Ozone induced oxidative stress: mechanisms of action and reaction. *Physiol Plantarum* **100**, 264–273.

Pic, E., de La Serve, B.T., Tardieu, F. and Turc, O. (2002) Leaf senescence induced by mild water deficit follows the same sequence of macroscopic, biochemical, and molecular events as monocarpic senescence in pea. *Plant Physiol* **128**, 236–246.

Plaut, Z., Meinzer, F.C. and Federman, E. (2000) Leaf development, transpiration and ion uptake and distribution in sugarcane cultivars grown under salinity. *Plant Soil* **218**, 59–69.

Pontier, D., Gan, S.S., Amasino, R.M., Roby, D. and Lam, E. (1999) Markers for hypersensitive response and senescence show distinct patterns of expression. *Plant Mol Biol* **39**, 1243–1255.

Pourtau, N., Mares, M., Purdy, S., Quentin, N., Ruel, A. and Wingler, A. (2004) Interactions of abscisic acid and sugar signalling in the regulation of leaf senescence. *Planta* **219**, 765–772.

Prasad, M.N.V. (1995) Cadmium toxicity and tolerance in vascular plants. *Environ Exp Bot* **35**, 525–545.

Prisco, J.T. and Oleary, J.W. (1972) Enhancement of intact bean leaf senescence by NaCl salinity. *Physiol Plantarum* **27**, 95–100.

Prochazkova, D., Sairam, R.K., Srivastava, G.C. and Singh, D.V. (2001) Oxidative stress and antioxidant activity as the basis of senescence in maize leaves. *Plant Sci* **161**, 765–771.

Prochazkova, D. and Wilhelmova, N. (2004) Changes in antioxidative protection in bean cotyledons during natural and continuous irradiation-accelerated senescence. *Biologia Plantarum* **48**, 33–39.

Pruzinska, A., Tanner, G., Anders, I., Roca, M. and Hortensteiner, S. (2003) Chlorophyll breakdown: pheophorbide a oxygenase is a Rieske-type iron-sulfur protein, encoded by the accelerated cell death 1 gene. *Proc Natl Acad Sci U S A* **100**, 15259–15264.

Quirino, B.F., Noh, Y.S., Himelblau, E. and Amasino, R.M. (2000) Molecular aspects of leaf senescence. *Trends Plant Sci* **5**, 278–282.

Quirino, B.F., Normanly, J. and Amasino, R.M. (1999) Diverse range of gene activity during *Arabidopsis thaliana* leaf senescence includes pathogen-independent induction of defense-related genes. *Plant Mol Biol* **40**, 267.

Rabiza-Swider, J. and Skutnik, E. (2004) Effect of light on senescence of cut leaves of Zantedeschia aethiopica Spr. and Hosta Tratt. 'Undulata Erromena'. *Folia Horticulturae* **16**, 161–166.

Rao, M.V., Lee, H.-I. and Davis, K.R. (2002) Ozone-induced ethylene production is dependent on salicylic acid, and both salicylic acid and ethylene act in concert to regulate ozone-induced cell death. *Plant J* **32**, 447–456.

Rao, M.V., Paliyath, G. and Ormrod, D.P. (1995) Differential response of photosynthetic pigments, rubisco activity and rubisco protein of *Arabidopsis thaliana* exposed to UVB and ozone. *Photochem Photobiol* **62**, 727–735.

Rascio, N., Dalla Vecchia, F., Ferretti, M., Merlo, L. and Ghisi, R. (1993) Some effects of cadmium on maize plants. *Arch Environ Contam Toxicol* **25**, 244–249.

Reich, P.B. (1983) Effects of low concentrations of O_3 on net photosynthesis, dark respiration, and chlorophyll contents in aging hybrid poplar leaves. *Plant Physiol* **73**, 291–296.

RenuKhanna, C., Reddy, P.V. and Sinha, S.K. (2000) Interaction of drought stress-induced senescence with monocarpic senescence in cowpea (*Vigna unguiculata*). I. Water relations, growth and senescence. *J Plant Biol* **27**, 171–178.

Riera, M., Valon, C., Fenzi, F., Giraudat, J. and Leung, J. (2005) The genetics of adaptive responses to drought stress: abscisic acid-dependent and abscisic acid-independent signalling components. *Physiol Plantarum* **123**, 111–119.

Rieu, I., Cristescu, S.M., Harren, F.J.M. *et al.* (2005) RP-ACS1, a flooding-induced 1-aminocyclopropane-1-carboxylate synthase gene of *Rumex palustris*, is involved in rhythmic ethylene production. *J Exp Bot* **56**, 841–849.

Robatzek, S. and Somssich, I.E. (2002) Targets of AtWRKY6 regulation during plant senescence and pathogen defense. *Genes Dev* **16**, 1139–1149.

Rojo, E., Martin, R., Carter, C. *et al.* (2004) VPE gamma exhibits a caspase-like activity that contributes to defense against pathogens. *Curr Biol* **14**, 1897–1906.

Rojo, E., Zouhar, J., Carter, C., Kovaleva, V. and Raikhel, N.V. (2003) A unique mechanism for protein processing and degradation in *Arabidopsis thaliana*. *Proc Natl Acad Sci U S A* **100**, 7389–7394.

Romero-Puertas, M.C., Palma, J.M., Gomez, M., del Rio, L.A. and Sandalio, L.M. (2002) Cadmium causes the oxidative modification of proteins in pea plants. *Plant Cell Environ* **25**, 677–686.

Romero-Puertas, M.C., Rodriguez-Serrano, M., Corpas, F.J., Gomez, M., del Rio, L.A. and Sandalio, L.M. (2004) Cadmium-induced subcellular accumulation of O-2(-) and H_2O_2 in pea leaves. *Plant Cell Environ* **27**, 1122–1134.

Rousseaux, M.C., Ballare, C.L., Jordan, E.T. and Vierstra, R.D. (1997) Directed overexpression of *PHYA* locally suppresses stem elongation and leaf senescence responses to far-red radiation. *Plant Cell Environ* **20**, 1551–1558.

Rousseaux, M.C., Hall, A.J. and Sanchez, R.A. (2000) Basal leaf senescence in a sunflower (*Helianthus annuus*) canopy: responses to increased R/FR ratio. *Physiol Plantarum* **110**, 477–482

Rousseaux, M.C., Hall, A.J. and Sanchez, R.A. (1996) Far-red enrichment and photosynthetically active radiation level influence leaf senescence in field-grown sunflower. *Physiol Plantarum* **96**, 217–224.

Ryals, J.A., Neuenschwander, U.H., Willits, M.G., Molina, A., Steiner, H.Y. and Hunt, M.D. (1996) Systemic acquired resistance. *Plant Cell* **8**, 1809–1819.

Sandalio, L.M., Dalurzo, H.C., Gomez, M., Romero-Puertas, M.C. and del Rio, L.A. (2001) Cadmium-induced changes in the growth and oxidative metabolism of pea plants. *J Exp Bot* **52**, 2115–2126.

Sanita diToppi, L., Lambardi, M., Pazzagli, L., Cappugi, G., Durante, M. and Gabbrielli, R. (1998) Response to cadmium in carrot in vitro plants and cell suspension cultures. *Plant Sci* **137**, 119.

Sassenrath-Cole, G.F., Lu, G.Y., Hodges, H.F. and McKinion, J.M. (1996) Photon flux density versus leaf senescence in determining photosynthetic efficiency and capacity of *Gossypium hirsutum* L leaves. *Environ Exp Bot* **36**, 439–446.

Schenk, P.M., Kazan, K., Rusu, A.G., Manners, J.M. and Maclean, D.J. (2005) The *SEN1* gene of *Arabidopsis* is regulated by signals that link plant defence responses and senescence. *Plant Physiol Biochem* **43**, 997–1005.

Schittenhelm, S., Menge-Hartmann, U. and Oldenburg, E. (2004) Photosynthesis, carbohydrate metabolism, and yield of phytochrome-B-overexpressing potatoes under different light regimes. *Crop Sci* **44**, 131–143.

Schwabe, W.W. (1970) The control of leaf senescence in *Kleinia articlata* by photoperiod. *Ann Bot* **34**, 43–57.

Schwabe, W.W. and Kulkarni, V.J. (1987) Senescence-associated changes during long-day-induced leaf senescence and the nature of the graft-transmissible senescence substance in *Kleinia Articulata*. *J Exp Bot* **38**, 1741–1755.

Sigurdsson, B.D. (2001) Elevated [CO_2] and nutrient status modified leaf phenology and growth rhythm of young *Populus trichocarpa* trees in a 3-year field study. *Trees Struct Funct* **15**, 403–413.

Simpson, S.D., Nakashima, K., Narusaka, Y., Seki, M., Shinozaki, K. and Yamaguchi-Shinozaki, K. (2003) Two different novel cis-acting elements of *erd1*, a *clpA* homologous *Arabidopsis* gene function in induction by dehydration stress and dark-induced senescence. *Plant J* **33**, 259–270.

Spundova, M., Sloukova, K., Hunkova, M. and Naus, J. (2005) Plant shading increases lipid peroxidation and intensifies senescence-induced changes in photosynthesis and activities of ascorbate peroxidase and glutathione reductase in wheat. *Photosynthetica* **43**, 403–409.

Strid, A., Chow, W.S. and Anderson, J.M. (1994) UV-B damage and protection at the molecular level in plants. *Photosynthesis Res* **39**, 475–489.

Sullivan, M., VanToai, T., Fausey, N., Beuerlein, J., Parkinson, R. and Soboyejo, A. (2001) Evaluating on-farm flooding impacts on soybean. *Crop Sci* **41**, 93–100.

Surplus, S.L., Jordan, B.R., Murphy, A.M., Carr, J.P., Thomas, B. and Mackerness, S.A.H. (1998) Ultraviolet-B-induced responses in *Arabidopsis thaliana*: role of salicylic acid and reactive oxygen species in the regulation of transcripts encoding photosynthetic and acidic pathogenesis-related proteins. *Plant Cell Environ* **21**, 685–694.

Suzuki, N. and Mittler, R. (2006) Reactive oxygen species and temperature stresses: a delicate balance between signaling and destruction. *Physiol Plantarum* **126**, 45–51.

Tanaka, Y., Sano, T., Tamaoki, M., Nakajima, N., Kondo, N. and Hasezawa, S. (2005) Ethylene inhibits abscisic acid-induced stomatal closure in *Arabidopsis*. *Plant Physiol* **138**, 2337–2343.

Tewari, R.K., Kumar, P., Tewari, N., Srivastava, S. and Sharma, P.N. (2004) Macronutrient deficiencies and differential antioxidant responses – influence on the activity and expression of superoxide dismutase in maize. *Plant Sci* **166**, 687.

Thiele, A., Herold, M., Lenk, I., Quail, P.H. and Gatz, C. (1999) Heterologous expression of *Arabidopsis* phytochrome B in transgenic potato influences photosynthetic performance and tuber development. *Plant Physiol* **120**, 73–81.

Thomas, H. and de Villiers, L. (1996) Gene expression in leaves of *Arabidopsis thaliana* induced to senesce by nutrient deprivation. *J Exp Bot* **47**, 1845–1852.

Thomas, H. and Donnison, I. (2000) Back from the brink: plant senescence and its reversibility. In: *Programmed Cell Death in Animals and Plants* (eds Bryant, J.A., Hughes, S.G. and Garland, J.N.). BIOS Scientific Publishers, Oxford, pp. 149–162.

Thomas, H. and Howarth, C.J. (2000) Five ways to stay green. *J Exp Bot* **51**, 329–337.

Thomas, H., Ougham, H.J., Wagstaff, C. and Stead, A.D. (2003) Defining senescence and death. *J Exp Bot* **54**, 1127–1132.

Tricker, P.J., Calfapietra, C., Kuzminsky, E. *et al.* (2004) Long-term acclimation of leaf production, development, longevity and quality following 3 yr exposure to free-air CO_2 enrichment during canopy closure in Populus. *New Phytologist* **162**, 413–426.

Urban, O. (2003) Physiological impacts of elevated CO_2 concentration ranging from molecular to whole plant responses. *Photosynthetica* **41**, 9–20.

Usuda, H. (1995) Phosphate deficiency in maize. V. Mobilization of nitrogen and phosphorus within shoots of young plants and its relationship to senescence. *Plant Cell Physiol* **36**, 1041–1049.

Vahala, J., Ruonala, R., Keinanen, M., Tuominen, H. and Kangasjarvi, J. (2003) Ethylene insensitivity modulates ozone-induced cell death in birch. *Plant Physiol* **132**, 185–195.

van der Krol, A.R., van Poecke, R.M.P., Vorst, O.F.J. *et al.* (1999) Developmental and wound-, cold-, desiccation-, ultraviolet-B-stress-induced modulations in the expression of the petunia zinc finger transcription factor gene *ZPT2–2*. *Plant Physiol* **121**, 1153–1162.

Van der Straeten, D., Zhou, Z.Y., Prinsen, E., Van Onckelen, H.A. and Van Montagu, M.C. (2001) A comparative molecular-physiological study of submergence response in lowland and deepwater rice. *Plant Physiol* **125**, 955–968.

van Doorn, W.G. and Vanlieburg, M.J. (1993) Interaction between the effects of phytochrome and gibberellic-acid on the senescence of *Alstroemeria pelegrina* leaves. *Physiol Plantarum* **89**, 182–186.

Veierskov, B. (1987) Irradiance-dependent senescence of isolated leaves. *Physiol Plantarum* **71**, 316–320.

Voesenek, L., Banga, M., Their, R.H. *et al.* (1993) Submergence-induced ethylene synthesis, entrapment, and growth in 2 plant species with contrasting flooding resistances. *Plant Physiol* **103**, 783–791.

Volin, J.C., Reich, P.B. and Givnish, T.J. (1998) Elevated carbon dioxide ameliorates the effects of ozone on photosynthesis and growth: species respond similarly regardless of photosynthetic pathway or plant functional group. *New Phytologist* **138**, 315–325.

Volkmar, K.M., Hu, Y. and Steppuhn, H. (1998) Physiological responses of plants to salinity: a review. *Can J Plant Sci* **78**, 19–27.

Wagner, G.J. (1993) Accumulation of cadmium in crop plants and its consequences to human health. *Adv Agronomy* **51**, 173–212.

Walters, D.R. (2000) Polyamines in plant–microbe interactions. *Physiol Mol Plant Path* **57**, 137–146.

Ward, J.K. and Kelly, J.K. (2004) Scaling up evolutionary responses to elevated CO_2: lessons from *Arabidopsis*. *Ecol Lett* **7**, 427–440.

Weaver, L.M. and Amasino, R.M. (2001) Senescence is induced in individually darkened *Arabidopsis* leaves but inhibited in whole darkened plants. *Plant Physiol* **127**, 876–886.

Weaver, L.M., Gan, S.S., Quirino, B. and Amasino, R.M. (1998) A comparison of the expression patterns of several senescence-associated genes in response to stress and hormone treatment. *Plant Mol Biol* **37**, 455–469.

Weller, J.L., Batge, S.L., Smith, J.J. *et al.* (2004) A dominant mutation in the pea *PHYA* gene confers enhanced responses to light and impairs the light-dependent degradation of phytochrome A. *Plant Physiol* **135**, 2186–2195.

Weller, J.L., Murfet, I.C. and Reid, J.B. (1997) Pea mutants with reduced sensitivity to far-red light define an important role for phytochrome A in day-length detection. *Plant Physiol* **114**, 1225–1236.

Wi, S.J. and Park, K.Y. (2002) Antisense expression of carnation cDNA encoding ACC synthase or ACC oxidase enhances polyamine content and abiotic stress tolerance in transgenic tobacco plants. *Mol Cells* **13**, 209–220.

Wingler, A., Purdy, S., MacLean, J.A. and Pourtau, N. (2006) The role of sugars in integrating environmental signals during the regulation of leaf senescence. *J Exp Bot* **57**, 391–399.

Wingler, A., von Schaewen, A., Leegood, R.C., Lea, P.J. and Quick, W.P. (1998) Regulation of leaf senescence by cytokinin, sugars, and light – effects on NADH-dependent hydroxypyruvate reductase. *Plant Physiol* **116**, 329–335.

Woo, H.R., Goh, C.-H. and Park, J.-H. (2002) Extended leaf longevity in the *ore4-1* mutant of *Arabidopsis* with a reduced expression of a plastid ribosomal protein gene. *Plant J* **31**, 331–340.

Woo, H.R., Kim, J.H., Nam, H.G. and Lim, P.O. (2004) The delayed leaf senescence mutants of *Arabidopsis*, *ore1*, *ore3*, and *ore9* are tolerant to oxidative stress. *Plant Cell Physiol* **45**, 923–932.

Woodward, F.I. (2002) Potential impacts of global elevated CO_2 concentrations on plants. *Curr Opin Plant Biol* **5**, 207–211.

Xu, Q., Paulsen, A.Q., Guikema, J.A. and Paulsen, G.M. (1995) Functional and ultrastructural injury to photosynthesis in wheat by high temperature during maturation. *Environ Exp Bot* **35**, 43–54.

Yamaji, K., Julkunen-Tiitto, R., Rousi, M., Freiwald, V. and Oksanen, E. (2003) Ozone exposure over two growing seasons alters root-to-shoot ratio and chemical composition of birch (*Betula pendula* Roth). *Global Change Biol* **9**, 1363–1377.

Yang, J.C., Zhang, J.H., Wang, Z.Q., Zhu, Q.S. and Liu, L.J. (2003) Involvement of abscisic acid and cytokinins in the senescence and remobilization of carbon reserves in wheat subjected to water stress during grain filling. *Plant Cell Environ* **26**, 1621–1631.

Yong, J.W.H., Wong, S.C., Letham, D.S., Hocart, C.H. and Farquhar, G.D. (2000) Effects of elevated [CO_2] and nitrogen nutrition on cytokinins in the xylem sap and leaves of cotton. *Plant Physiol* **124**, 767–780.

Yoshida, S. (2003) Molecular regulation of leaf senescence. *Curr Opin Plant Biol* **6**, 79–84.

Yoshida, S., Ito, M., Callis, J., Nishida, I. and Watanabe, A. (2002a) A delayed leaf senescence mutant is defective in arginyl-tRNA: protein arginyltransferase, a component of the N-end rule pathway in *Arabidopsis*. *Plant J* **32**, 129–137.

Yoshida, S., Ito, N., Nishida, I. and Watanabe, A. (2002b) Identification of a novel gene *HYS1/CPR5* that has a repressive role in the induction of leaf senescence and pathogen-defence responses in *Arabidopsis thaliana*. *Plant J* **29**, 427–437.

Young, T.E., Meeley, R.B. and Gallie, D.R. (2004) ACC synthase expression regulates leaf performance and drought tolerance in maize. *Plant J* **40**, 813–825.

Zhang, J., Van Toai, T., Huynh, L. and Preiszner, J. (2000) Development of flooding-tolerant *Arabidopsis thaliana* by autoregulated cytokinin production. *Mol Breeding* **6**, 135–144.

Zhou, Z.Y., Vriezen, W., Van Caeneghem, W., Van Montagu, M. and Van Der Straeten, D. (2001) Rapid induction of a novel *ACC synthase* gene in deepwater rice seedlings upon complete submergence. *Euphytica* **121**, 137–143.

Zimmermann, P. and Zentgraf, U. (2005) The correlation between oxidative stress and leaf senescence during plant development. *Cell Mol Biol Lett* **10**, 515–534.

7 Developmental and hormonal control of leaf senescence

Jos H.M. Schippers, Hai-Chun Jing, Jacques Hille and Paul P. Dijkwel

7.1 Introduction

What controls the length of life is one of the fundamental biological questions that has been puzzling scientists for centuries. Plants have many life-forms and differ greatly in the maximal life spans (Thomas, 2003). Annual and biennial plants finish life cycles in a single season or in 2 years time, respectively. An age of 4600 years has been recorded for the perennial tree bristlecone pine (*Pinus longaeva*), while some clonal plants can live over 10 000 years (Nooden, 1988). Thus, longevity is a genetically controlled life-history trait.

The phenomenon of leaf senescence can be appreciated by the colour changes among deciduous trees and in the ripening of cereal crops in late summer and autumn, which can occur at a global scale to transform the appearance of the earth from space. During leaf senescence, the sum of morphological, physiological and molecular changes is generally referred to as the senescence syndrome, which includes the visible colour changes, dismantling of chloroplasts, degradation of RNA, proteins and DNA and translocation of macro/micromolecules from senescing leaves to other parts of the plant, resulting in the death of the leaf (Bleecker and Patterson, 1997).

We propose to examine the regulation of leaf senescence from a genome optimisation perspective. We critically analyse the proposed developmental cues that are implicated in initiating leaf senescence. The prominent roles that hormones play during developmental ageing and the initiation and progression of senescence will be reviewed from a molecular point of view based partially on transcriptome data. We discuss the identified potential physiological, biochemical and molecular events during developmental senescence, although we would like to refer to the other chapters in this book for a more detailed review (Chapters 2, 3, 4 and 10).

7.2 Developmental senescence: a plant genome is optimised for early survival and reproduction

In general, a genome has evolved to contain three classes of hereditary information: (1) the basic metabolism and life maintenance programme such as photosynthesis, respiration and DNA replication and damage repair; (2) the defence programme that regulates plant responses to abiotic and biotic stresses; and (3) the growth and development programme that produces an adult plant optimised for reproduction (Gems, 2000; de Magalhaes and Church, 2005). From an evolutionary point of view, a genome is selected by the force of natural selection if it facilitates the continuous reproduction. Thus, natural selection optimises a genome for reproduction and the aforementioned three classes of genome programmes will be operating only to ensure normal plant growth and development until reproduction. After reproduction, the force of natural selection declines with age and this leads to the loss of viability and fitness of the whole plant and/or plant organs. This phenomenon is known as disposal of soma as stated in the evolutionary theory of ageing, which is developed from animal ageing studies (Rose, 1991; Kirkwood, 2005). This argues that in the genome there are no specific genetic programmes for life span and that longevity is an indirect consequence of genome optimisation for reproduction.

The annual model plant *Arabidopsis thaliana* has evolved a reproduction programme that runs in parallel with the death of the whole plant. Seeds are being produced while leaves start to senesce, and in this way the plant effectively reutilises nutrients stored in leaves for the production of seeds. Here, the evolutionary theory of ageing can explain leaf senescence if the disposal of a leaf is considered as an indirect selection for nutrient salvage (Bleecker, 1998). Indeed, selective cell death is well documented during plant development and defence responses. For instance, xylogenesis, early embryogenesis, pollen tube growth and the hypersensitive response are typical examples. Thus, a common feature of the plant body plan and architecture is that almost all the structural units are disposable for the sake of survival and reproduction. Clearly, the recruitment of nutrients from leaf tissues, which is a prominent feature of the senescence syndrome and results in the death of the leaf, is part of the genome optimisation programme. Following this line of arguments, we consider that leaf senescence, albeit genetically controlled, is a consequence of natural selection for genome reproduction. Although there are debates concerning whether ageing occurs in plants, or whether whole-plant senescence shares similarities with animal ageing (Thomas, 2002), it has been proposed that developmental ageing resembles animal ageing, especially when leaves on a plant are scaled up and viewed as equivalent of animal individuals. Leaf senescence is a typical postmitotic senescence in plants and its onset shares many similar regulatory strategies with ageing in animals (Gan, 2003; Jing *et al.*, 2003).

We consider it important to view leaf senescence from such an angle. This view helps to explain and model the molecular genetic mechanisms of leaf senescence. As predicted by the evolutionary theory of ageing, genes with early-life beneficial but late-life deleterious effects and late-acting mutations with purely deleterious effects are important for senescence regulation (Kirkwood and Austad, 2000). In the following part of the chapter, we will show that programmes important for life maintenance, stress responses and development are important for the onset and regulation of leaf senescence. We further summarise the recent progress in examining the interactions between leaf development and ethylene as an example to present the approaches we think are necessary to understand the complex process of developmental senescence.

7.3 Developmental processes that regulate leaf senescence

When a plant is grown in an environment with sufficient nutrition, away from pathogen attacks and free of abiotic stresses such as darkness, drought, extreme temperature, UV-B and ozone, leaf senescence is ultimately initiated and progresses in a leaf age dependent manner (Gan and Amasino, 1997; Quirino *et al.*, 2000). For monocarpic plants the regulation of senescence is under correlative control and the onset of whole-plant senescence is initiated by the developing reproductive sink which remobilises nutrients from the vegetative tissues (Nooden, 1988). In soybean and wheat the removal of reproductive structures usually delays leaf and whole-plant senescence (Nooden, 1984; Srivalli and Khanna-Chopra, 2004). Thus genomes of monocarpic plants are optimised for reproduction, which determines the onset of leaf and whole-plant senescence. Although whole-plant senescence in *Arabidopsis* is controlled by the reproductive structures as well (Nooden and Penney, 2001), only a weak correlation exists between the appearance of reproductive structures and the onset of leaf senescence. *Arabidopsis* leaves have a defined life span and senesce even under ideal growth conditions, which is due to the developmental programmes underlined by the genome (Hensel *et al.*, 1993; Jing *et al.*, 2002). Thus, here the onset of leaf senescence is governed mainly by age-related changes. Strategies employed in animals and humans seem to have been equally used in plants, such as hormonal modulation as discussed in the next section, reactive oxygen species (ROS), metabolic flux especially sugar and nitrogen signalling and protein degradation. Readers are advised to refer to several recent reviews for detailed discussion on them (Gan, 2003; Jing *et al.*, 2003; Lim *et al.*, 2003; Lim and Nam, 2005; also Chapters 4, 5 and 6). In this section, we will only briefly elaborate on those strategies.

7.3.1 Reactive oxygen species

Leaf senescence and the expression of various senescence-associated genes (*SAG*s) were promoted in old leaves upon exposure to UV-B, ozone or treatment with catalase inhibitors (Miller *et al.*, 1999; John *et al.*, 2001; Navabpour *et al.*, 2003). In contrast to animal ageing and plant hypersensitive responses during plant–pathogen interactions in which mitochondria are the generator of ROS (Finkel and Holbrook, 2000; Lam *et al.*, 2001; Biesalski, 2002), the main ROS source in a senescing leaf is chloroplasts (Quirino *et al.*, 2000). This is consistent with the observation that knockout of a chloroplast genome encoded *ndhF* gene, one of the components of Ndh complex involved in chlororespiratory electron transport chain, delayed leaf senescence in tobacco (Zapata *et al.*, 2005). ROS can also be generated via lipid oxidation involving membrane-associated NAD(P)H oxidases (Mittler, 2002). This is in agreement with the observed altered senescence phenotypes of *Arabidopsis* antisense-suppressed phospholipase Dα and *SAG101* plants (Fan *et al.*, 1997; He and Gan, 2002) and in plants with defects in fatty acid biosynthesis pathways (Mou *et al.*, 2000, 2002; Wellesen *et al.*, 2001). Thus, ROS generated from various sources are involved in leaf senescence. Several delayed senescence mutants exhibited enhanced tolerance to oxidative stress, indicating that the extended longevity at least in part is due to the attenuated tolerance to ROS (Woo *et al.*, 2004). Thus, the damage generated by ROS can be an important age-related change that will eventually result in the onset of leaf senescence.

7.3.2 Metabolic flux

One of the distinct features during leaf senescence is the clear metabolic shift from primary catabolism to anabolism (Smart, 1994; Buchanan-Wollaston, 1997). The number of catabolic genes highly expressed in senescing leaves is almost twofold of that of anabolic genes (Guo *et al.*, 2004). Carbon and nitrogen supplies are the two key components that reflect the control of metabolic flux on leaf senescence. An elevated CO_2 level hastened the drop in the photosynthetic activities and induced leaf senescence (Miller *et al.*, 1997; Ludewig and Sonnewald, 2000), whereas in *Rubisco* antisense tobacco plants and *Arabidopsis ore4-1* mutants, less dry weight and chlorophyll content were achieved than in the wild type at maturity, resulting in a prolonged leaf longevity (Miller *et al.*, 2000; Woo *et al.*, 2002). Thus, carbon supply achieved through photosynthesis is important for the onset of leaf senescence (Hensel *et al.*, 1993). Carbon supply may directly alter the sugar sensing and signalling, which has been shown to regulate leaf senescence as envisaged in the *gin2* mutant that has a lesion in a hexokinase gene (Moore *et al.*, 2003). The *cpr5/hys1* mutant that was originally isolated based on altered pathogen resistance was shown to have sugar hypersensitivity and early leaf senescence (Bowling *et al.*, 1997; Yoshida *et al.*, 2002b). Furthermore, glucose (carbon supply) was shown to induce early leaf senescence when combined with low, but not high nitrogen supply (Wingler *et al.*, 2004), indicating the importance of carbon–nitrogen balance. Nitrogen starvation can induce premature leaf senescence, perhaps mainly through modulating the autophagy functions (see below).

7.3.3 Protein degradation

As one of the essential activities in plant life, protein turnover involves selective and bulk removal of proteins in many processes, such as the degradation of specific regulatory gene products, the maintenance of free amino acids, the elimination of malfunctioning proteins and nutrient recycling (Smalle and Vierstra, 2004; Thompson and Vierstra, 2005). The identification of *Arabidopsis* ORE9 as an F-box protein (Woo *et al.*, 2001) and DLS1 as an arginyl-tRNA:protein arginyltransferase (ATATE1), which is involved in the N-end rule pathway (Yoshida *et al.*, 2002a), demonstrated the importance of the selective protein removal route mediated by the ubiquitin-mediated proteolysis pathway via 26S proteasome in leaf senescence. Mutations in these two genes resulted in delayed senescence, suggesting that the degraded products targeted by ORE9 and DLS1 are positive regulators of leaf senescence, or that the nondegraded products delay senescence. The bulk protein turnover is mainly achieved through vacuolar autophagy. The analyses of two autophagic mutants *apg7* and *apg9-1* demonstrated the importance of autophagy in senescence regulation (Doelling *et al.*, 2002; Hanaoka *et al.*, 2002). Many more components involved in autophagy formation, conjugation and targeting to vacuoles have been studied through mutational analyses in *Arabidopsis* (Surpin *et al.*, 2003; Yoshimoto *et al.*, 2004; Thompson *et al.*, 2005; Xiong *et al.*, 2005). In general, knockout of these components affected the survival under carbon and nitrogen starvation conditions and hastened leaf senescence under normal growth conditions. Interestingly, the mRNA and protein levels of autophagy genes are senescence enhanced, suggesting that autophagy is an important aspect of the senescence syndrome.

A major group of *SAG*s encode cysteine proteases (Bhalerao *et al.*, 2003; Guo *et al.*, 2004). For instance, RD21 remains in the vacuole as inactive aggregate and becomes active during senescence by the cleavage of its C-terminal granulin domain (Yamada *et al.*, 2001). Recently, a novel type of senescence-associated vacuole (SAV) has been observed in *Arabidopsis* and soybean which contains many proteolytic enzymes such as SAG12 (Otegui *et al.*, 2005). The development of SAVs appears to be differentially regulated from vacuole autophagy that is actively involved in leaf senescence (Doelling *et al.*, 2002; Hanaoka *et al.*, 2002). Thus, different vacuoles are functioning during senescence and play a prominent role in macromolecule degradation (Matile, 1997). Furthermore, a chloroplast nucleotide-encoded protein CND41 was shown to be responsible for the degradation of Rubisco proteins in senescent tobacco leaves (Kato *et al.*, 2004), indicative of the involvement of chloroplast genome in leaf senescence. The macromolecule degradation and nutrient recycling are prominent events during senescence. Thus, it is not surprising that protein degradation, selective or bulk, is important for senescence regulation.

7.4 Hormonal control of leaf senescence

The senescence programme is the final developmental phase of a leaf, which is influenced by several phytohormones, with cytokinin and ethylene having the most extensively documented roles in delaying or inducing leaf senescence, respectively. In addition, other hormones, such as abscisic acid (ABA), auxin, gibberellic acid (GA), jasmonic acid (JA) and salicylic acid (SA), also have effect on the senescence process. In plants, two types of senescence are evident: mitotic senescence and postmitotic senescence (Gan, 2003). Cells in leaves divide only during early development, and thus leaf senescence can be considered postmitotic.

Research on the effect of plant hormones on senescence has been started already in the late fifties. The effect of various hormones has been reported for dozens of plant species. The regulation of senescence by cytokinin and ethylene is conserved; however, the action of other hormones varies between plant species. Hormonal signalling pathways show significant overlap, which makes the study of the effect of single hormones complex. The generally used linear representation of hormonal signalling pathways controlling specific aspects of plant growth and development is too simple. In fact, hormones interact with each other and with a whole range of developmental, environmental and metabolic signals (Beaudoin *et al.*, 2000).

There are three major ways of controlling the responses to hormones in plants: regulation by hormone biosynthesis, through hormone perception and signalling pathways and downstream events leading to selective protein turnover and changes in gene expression. Although *Arabidopsis* is the model species for plant research for the last 15 years, almost all the physiological information about hormonal control of leaf senescence has been generated in other species. Evidence from hormone mutants in *Arabidopsis* strongly supports the role of several hormones in leaf senescence. Lately, exciting advances through transcriptome studies have revealed expression data for hormone biosynthesis, signalling and response genes during senescence, and a closer examination revealed a few interesting points. Combination of the physiological and genetic information will help creating a model for hormonal and developmental control of leaf senescence.

Here we try to highlight important findings of several studies that we used to present a model of hormonal regulation of leaf senescence and address remaining questions and leads for future research.

7.4.1 *Hormones that delay leaf senescence*

7.4.1.1 *Gibberellic acid*

Gibberellins are diterpenes that promote stem and leaf growth. In some species, GAs also induce seed germination and modulate flowering time and the development of flowers, fruits and seeds (Sun and Gubler, 2004). A biochemical relation between leaf senescence and GA was first reported by Fletcher and Osborne (1965) showing that GA retarded senescence of excised leaf tissue from *Taraxacum officinale* by maintaining chlorophyll levels and RNA synthesis. Another study in *Rumex* by Goldthwaite and Laetsch (1968) showed that GA could inhibit senescence in leaf disks for several days. Both protein degradation and chlorophyll degradation were delayed 4 days. Even when chlorophyll and protein loss is halfway complete, addition of GA blocks further degradation. A study performed on the leaves of romaine lettuce showed a clear age-related decline in GA levels and absence in senesced leaves. This decline in GA was caused by the conversion of free GA to a bound inactive form, probably GA glucoside (Aharoni and Richmond, 1978). Moreover, retardation of senescence by kinetin also caused a relatively high level of free GA and absence of bound GA. Mutations in genes controlling GA biosynthesis or perception have no effect on senescence. However, mutations in the F-box protein SLEEPY1 (SLY1), which result in a block of GA-responsive genes (Dill *et al.*, 2004), delay senescence when crossed to *abi1* (Richards *et al.*, 2001). Although not extensively described, several reports point to a retarding effect of GA on leaf senescence.

7.4.1.2 *Auxin*

Auxins are a group of molecules that got their name from the Greek word *auxein*, which means 'to grow'. The diversity of the auxin responses is reflected by the existence of multiple independent auxin perception mechanisms in a plant (Leyser, 2002). For soybean it has been shown that the senescence can be retarded by application of auxin (Nooden *et al.*, 1979). During abscission, auxin has been postulated to play a role in reducing the sensitivity of the cells to ethylene (Sexton and Roberts, 1982).

The endogenous auxin levels within *Coleus* leaves showed a decline with increasing age (dela Fuente and Leopold, 1968). However, a relation between endogenous auxin levels and senescence does not always seem to follow a same pattern (Nooden, 1988). The change in auxin response during ageing is the result not only of decreasing auxin levels, but also of a lower responsiveness to auxin with age (Chatterjee and Leopold, 1965). Since auxin is in general seen as a senescence-retarding compound it was a surprise that increased indoleacetic acid (IAA) levels could be detected in S3 phase leaves (Quirino *et al.*, 1999). Since leaves do not senesce uniformly, the authors suggested that auxin levels are selectively increased only in a certain population of cells corresponding to a particular senescence stage. These findings actually correlated with earlier studies that show IAA can induce the production of ethylene which opposes the senescence-retarding effect of IAA in tobacco leaf discs (Aharoni *et al.*, 1979). Interestingly, auxin effectively decreased *SAG12* expression, a marker for developmental senescence in a very short period of treatment in detached senescing leaves (Noh and Amasino, 1999).

Research performed on glucose signalling revealed that the HXK1 glucose signalling pathway interacts intimately with the auxin and cytokinin pathways. Glucose concentration and photorespiration rates are important determinants for the onset of senescence. Both cytokinin and auxin are part of a regulatory complex for nutritional status of the plant through HXK1 signalling pathway (Moore *et al.*, 2003). Thus, the role of auxin in the regulation of leaf senescence might be linked with other hormones and metabolic flux.

7.4.1.3 Cytokinins

Cytokinins have the strongest effect of all hormones on the retardation of leaf senescence. It was reported by Richmond and Lang (1957) that application of cytokinin could retard leaf senescence by preventing the chlorophyll breakdown.

While increasing cytokinin production could delay leaf senescence (Gan and Amasino, 1995; Ori *et al.*, 1999), reducing endogenous cytokinin levels resulted in accelerated senescence (Masferrer *et al.*, 2002). The drop in cytokinin levels before the onset of senescence is believed to be a key signal for the initiation (Nooden *et al.*, 1990; Gan and Amasino, 1995). Recently, exciting advances have been achieved in dissecting the components involved in cytokinin signalling (Hutchison and Kieber, 2002; Hwang *et al.*, 2002). Among the genes characterised, the receptor CKI1 (cytokinin independent 1) and the *Arabidopsis* response regulator (ARR) 2 appear to be involved in regulating leaf senescence (Hwang and Sheen, 2001). A more recent study identified an extracellular invertase whose activity is induced during cytokinin-mediated delay of senescence (Balibrea Lara *et al.*, 2004). In transgenic tobacco plants having a *SAG12–IPT* or *SAG12–KN1* construct, cytokinin biosynthesis was initiated when *SAG12* was induced resulting in a block of the senescence syndrome and delayed leaf senescence significantly (Gan and Amasino, 1995; Ori *et al.*, 1999;

also Chapter 13). When extracellular invertase activity is inhibited, cytokinin no longer can inhibit leaf senescence in transgenic *SAG12–IPT* lines (Balibrea Lara *et al.*, 2004).

Cytokinin signalling genes such as the *type-A ARR*s and biosynthesis genes show reduced transcription during leaf senescence (Buchanan-Wollaston *et al.*, 2005). Several microarray studies have been performed to reveal cytokinin-dependent gene expression (Hoth *et al.*, 2003; Rashotte *et al.*, 2003; Kiba *et al.*, 2005). Hoth *et al.* used an inducible system to assess the effects of endogenous cytokinin levels. The study identified 823 up- and 917 downregulated genes after 24 h of isopentenyl-transferase (IPT) induction. Although for these studies the seedling stage was used, this IPT system offers an attractive system to study the molecular genetics of how cytokinin can delay and/or reverse the senescence process. The study by Rashotte *et al.* (2003) showed that cytokinin-upregulated *type-A ARR*s, which were downregulated in senescing leaves (Buchanan-Wollaston *et al.*, 2005), are the primary response genes for cytokinins. Also a cytokinin oxidase (that degrades cytokinins) and several transcription factors were upregulated. Furthermore, cytokinins induce genes encoding ribosomal proteins (Crowell *et al.*, 1990) and photosynthetic genes (Mok and Mok, 2001). Application of cytokinins downregulated several peroxidases, kinases and E3 ubiquitin ligases. The regulation by cytokinin is related to auxin, light and sugar, since application of cytokinin influences the expression of genes involved in these signalling pathways.

In general it can be said that cytokinin stimulates the photosynthetic phase of a leaf. How cytokinin can maintain this phase and delay leaf senescence is still unclear. Nevertheless, the leaves of *SAG12–IPT* transgenic plants will undergo senescence, thus cytokinin action is limited to a certain developmental phase.

7.4.2 Hormones that induce leaf senescence

7.4.2.1 ABA

ABA plays a major role during processes related to seed development and germination, for instance the induction of seed dormancy, the synthesis of seed storage proteins and lipids, the acquisition of desiccation tolerance and the inhibition of the transition from embryonic to vegetative growth (Nambara and Marion-Poll, 2005). In vegetative tissue, ABA plays a role in response to drought to prevent water loss by stomatal closure and maintenance of vegetative growth by inhibiting the transition to reproductive growth. Under nonstressful conditions, ABA in plant cells is maintained at low levels, since ABA inhibits plant growth. In vegetative tissues, ABA levels increase during drought, salt and cold stress (Xiong and Zhu, 2003). Changes in gene expression during water-deficit stress are partially induced by ABA and may promote the ability of a plant to respond and survive or adapt to the stress (Bray, 2002). For long it was thought that ABA inhibits plant growth rather than maintaining plant growth. But in tomato, maize and *Arabidopsis* it has been shown that ABA maintains shoot growth by inhibiting ethylene production (Sharp, 2002). Moreover, this interaction might also play a role in early leaf senescence and leaf, flower and fruit abscission (Morgan and Drew, 1997).

Young leaves have the highest ABA levels although this is mainly produced and transported from the older leaves (Zeevaart and Creelman, 1988). During vegetative growth the ABA levels are in general very low; however, in parallel with a decline in free cytokinin and GA just before chlorophyll breakdown in lettuce leaves, an increase in ABA levels has been observed (Aharoni and Richmond, 1978). As soon as the chlorophyll breakdown is initiated, a second, more dramatic increase in endogenous ABA levels is observed. The authors suggest that lowering of GA and cytokinin levels mark the onset of leaf senescence, which results in increased ABA levels when the process has been started. Application of ethylene to the lettuce leaves resulted in a quick drop of GA in 1 day after treatment, but the ABA levels did not show any difference. This might indicate that ABA and ethylene both control different aspects of the senescence syndrome which are mediated through different but partially overlapping signalling pathways. The application of ABA to detached leaves results in a rapid senescence response; however, application to attached leaves has a less pronounced effect. Under low nitrogen conditions and high sugar the *abi5* mutant shows delayed senescence. This is consistent with a role for sugar signalling during leaf senescence. *ABI5* can be induced by glucose during later stages of development. Expression analysis of *ABI5* shows an increase during senescence (Buchanan-Wollaston *et al.*, 2005). The ABA signalling mutants *abi2-1* and *abi1-1* show signs of early leaf senescence when grown on low nitrogen with glucose and their transcripts increase during senescence (Pourtau *et al.*, 2004; Buchanan-Wollaston *et al.*, 2005). Furthermore, the enzymes controlling ABA synthesis are upregulated during senescence. This indicates that the ABA signalling and biosynthesis pathway is active during leaf senescence. Interestingly the *abi4* and *abi5* signalling mutants and the *aba1*, *aba2* and *aba3* ABA-deficient mutants all are glucose insensitive (Arenas-Huertero *et al.*, 2000). It was noted before that sugar represses photosynthesis-associated genes, which leads to a decline in photosynthesis and eventually in leaf senescence (Bleecker and Patterson, 1997). Thus the onset of leaf senescence by ABA appears to be coupled to metabolic flux changes in *Arabidopsis*.

7.4.2.2 Brassinosteroids

Brassinosteroids (BRs), polyhydroxylated steroid hormones, regulate the growth and differentiation of plants throughout their life cycle. In recent years great advances have been made in the understanding of BR signalling (Vert *et al.*, 2005). External application of BR results in premature leaf senescence for several species, but it has not been reported for *Arabidopsis*. The induction of senescence by BRs might be mediated through ROS (Clouse and Sasse, 1998). BR signal transduction takes place at the plasma-membrane-localised receptor kinase, BRI1 (Clouse *et al.*, 1996). In addition to BRI1, three homologues have been characterised. Downstream of the receptor kinases is BIN2, a negative regulator of the BR pathway. Further downstream act BES1 and BZR1 transcription factors of which BES1 promotes the expression of BR-regulated genes and BZR1 represses BR genes; both are repressed by BIN2 by targeting of BES1 and BZR1 for ubiquitination and subsequent proteasome-dependent degradation (Vert *et al.*, 2005). Interestingly, BRs can induce ethylene biosynthesis genes in mung bean (Yi *et al.*, 1999). Whether BRs also induce ethylene biosynthesis during senescence is a question that remains to be answered.

Mutants in BR biosynthesis and BR signalling do support a role for BR in senescence in *Arabidopsis*. The *det2* (de-etiolated2) mutant is defective in an early step of BR biosynthesis. When grown in light the mutant develops two times more rosette leaves than does the wild type. Chory *et al.* (1991) observed that wild-type plants showed senescence after 30 days, whereas the *det2* mutant did not show any signs of visible senescence after 49 days. One could argue that the mutant has a severe developmental defect that results in a changed senescence syndrome; however, other severely affected developmental mutants such as *ctr1* still show a normal onset of the senescence programme (Kieber *et al.*, 1993). BR mutants can also result in early leaf senescence as has been shown by the *bes1* mutant (Yin *et al.*, 2002).

Looking at the *Arabidopsis* transcriptome of leaf senescence, none of the BR signalling components are identified (Guo *et al.*, 2004). This suggests a minor role for BR during leaf senescence. Transcriptome analysis identified seven genes encoding cell-wall-associated proteins that are upregulated after BR treatment (Goda *et al.*, 2002); these genes were not identified in the transcriptome of senescing leaves (Guo *et al.*, 2004). However, one study revealed the induction of SAGs in *Arabidopsis* by BR (He *et al.*, 2001). Out of 125 enhancer-trap lines screened, 4 showed upregulation of the reporter after BR application.

Although BR mutants show an alternative onset of senescence, molecular genetic evidence of a direct role for BR is still minimal. More studies about the role of BR in senescence are necessary.

7.4.2.3 Ethylene

The gaseous plant hormone ethylene plays an important role in plant growth and development. From seed germination to organ senescence and from cell elongation to defence responses, ethylene plays its part. The diverse role that ethylene plays in growth and development suggests that ethylene action involves expression and interaction of many different genes and their products (Zhong and Burns, 2003).

Ethylene has long been seen as the key hormone in regulating the onset of leaf senescence (Zacarias and Reid, 1990). The senescence-delaying hormones like auxin and cytokinin both stimulate ethylene production in romaine lettuce leaves (*Lactuca sativa* L.), which might account for their limited stay-green properties. The author concluded that the effectiveness of exogenously applied hormones in both enhancing and retarding senescence is greatly affected by the endogenous ethylene concentration of the treated plant tissue (Aharoni, 1989). The role of the ethylene pathway in senescence is demonstrated by several studies. Both ethylene-insensitive mutants *etr1-1* and *ein2/ore3* showed increased leaf longevities (Grbić and Bleecker, 1995; Oh *et al.*, 1997), and antisense suppression of the tomato 1-aminocyclopropane-1-carboxylic acid (ACC) oxidase resulted in delayed leaf senescence (John *et al.*, 1995). In these cases, however, senescence eventually begins and progresses normally. Exogenously applied ethylene induces premature leaf senescence in *Arabidopsis*. However, constitutive application of ethylene does not change the longevity of the leaves. Both *ctr1* (constitutive triple response) mutants and

Arabidopsis plants grown in the continuous presence of exogenous ethylene did not show premature senescence (Kieber *et al.*, 1993; Grbić and Bleecker, 1995). These results suggest a dynamic regulation of the timing of leaf senescence for which the age-dependent effect of ethylene is utilised.

By making use of an ethylene-induced senescence screen, a large collection of *onset of leaf death* (*old*) mutants has been identified (Jing *et al.*, 2002, 2005). These mutants confirmed that the effect of ethylene is limited to a range of leaf ages, and that the effect of ethylene on leaf senescence increases with increase in leaf age (Grbić and Bleecker, 1995; Jing *et al.*, 2002). Another piece of evidence supporting this notion comes from a study treating *Arabidopsis* plants of an identical 24-day end age with various lengths of exogenously applied ethylene (Jing *et al.*, 2005). The results showed that increasing ethylene treatment from 3 to 12 days caused an increase in leaf senescence. Surprisingly, a drop in the number of yellow leaves occurred when a 16-day ethylene exposure was applied. Thus, varying ethylene exposure time can induce different degrees of senescence symptoms in the leaves of an identical end age, suggesting that ethylene can actively stimulate or repress age-related changes that control ethylene-induced leaf senescence. This notion is genetically supported by the altered responses of eight *old* mutants to the various ethylene treatments (Jing *et al.*, 2005). Thus, multiple genetic loci are required to regulate the action of ethylene in leaf senescence.

A transcriptome study of senescent leaves by Guo *et al.* (2004) identified three mitogen-activated protein kinases (MAPKs), three MAPKKs, nine MAPKKKs and one MAPKKKK. In the *Arabidopsis* genome, 20 MAPKs, 10 MAPKKs, 80 MAPKKKs and 10 MAPKKKKs have been identified. The few components identified of the MAPK signal cascades led the authors to the conclusion that the three MAPKs and three MAPKKs may be at the converging/cross talk points of various signal transduction pathways. One of the identified MAPKs is MPK6, which is a component of the MAPK pathway that controls ethylene signalling in plants (Ouaked *et al.*, 2003). MPK6 is upregulated during osmotic stress but also by other abiotic stresses such as low temperature, low humidity, wounding or oxidative stress, as well as by pathogens (Droillard *et al.*, 2002).

Transcriptional analyses of ethylene mutants and ethylene-treated plants revealed the molecular actions of ethylene. A study by Zhong and Burns (2003) revealed genes that are regulated by ethylene. They compared treated wild type, *etr1* and *ctr1* with untreated wild type. Ethylene treatment of 24-day-old wild-type plants for 24 h changed the expression of 184 genes. Compared to *etr1-1*, 248 genes were changed in expression level. Untreated wild type compared to *etr1-1* revealed the downregulations of nine genes and one upregulated gene in *etr1-1*. The *ctr1* mutant that did not show any signs of early senescence had 109 genes differentially expressed. Further research on these genes might help understanding the molecular regulation of ethylene-induced leaf senescence. To further assess the regulation of senescence by ethylene, expression of SAGs in *ein2* was compared with that of wild type (Buchanan-Wollaston *et al.*, 2005). Nine percent of the genes that are upregulated during senescence are at least twofold reduced in *ein2*. Seventy-seven genes are more than twofold up- or downregulated. Four genes showed upregulation, a lipid transfer protein, a heavy-metal-binding protein and a transcription factor

(HFR1). Downregulated are nine transcription factors, cell-wall-degrading proteins and nucleases. Therefore some senescence-related degradation functions may be dependent on ethylene. The generation of the transcriptome data revealed that indeed ethylene controls a subset of *SAGs* during senescence; however, the importance of the identified genes for the control of leaf senescence remains elusive.

Endogenous ethylene levels are important for the initiation of senescence. However, the age-dependent senescence induction by ethylene limits its control to a specific age range. The transcriptome studies on *SAGs* induced by ethylene, together with physiological studies, reveal extensive cross talks between ethylene and the other hormones that might be utilised to fine-tune the progression of senescence in an age-dependent way.

7.4.2.4 Jasmonic acid

Jasmonates include jasmonic acid (JA), methyl jasmonate (MeJA) and related compounds and are found in fragrant oils. This group of plant regulators is connected to plant growth and development such as germination and seedling development, flower development, tuberisation, tendril coiling, leaf senescence and fruit ripening (Wasternack and Hause, 2002). The promotional effect of MeJA on senescence was first shown by application to detached oat leaves (Ueda and Kato, 1980). Exogenously applied JA or MeJA resulted in a decreased expression of photosynthesis-related genes like *Rubisco*. Moreover, a change in the polypeptide composition in senescing tissue was observed, which shared similarity with ABA-induced senescence in detached leaves (Weidhase *et al.*, 1987). In plants two JA biosynthetic pathways have been identified; a chloroplast-localised pathway and a cytoplasm-localised pathway (Creelman and Mullet, 1995). Exogenous application of JA typically promotes senescence in attached and detached leaves of *Arabidopsis* but not in the JA-insensitive mutant *coi1*. Also the endogenous JA levels in senescing leaves increased fourfold as compared to nonsenescing leaves. Besides an increased JA level during senescence also the enzymes involved in the JA biosynthesis are differentially regulated during senescence (He *et al.*, 2002). The *coi1* mutant, which is impaired in JA signalling, did not show any altered leaf senescence. Also other JA-related mutants do not show any alterations in the senescence programme, which challenges the idea that JA plays a role in senescence. However, a study with senescence enhancer-trap lines in *Arabidopsis* showed that JA can induce GUS (β-glucuronidase) expression in 14 out of the 125 lines tested (He *et al.*, 2001). The authors developed a sensitive large-scale screening method and have screened 1300 *Arabidopsis* enhancer-trap lines, which resulted in the identification of 147 lines in which the reporter gene *GUS* is expressed in senescing leaves but not in nonsenescing ones. Application of senescence-inducing factors showed that only ethylene induced GUS expression in more lines than JA and that ABA, BR, darkness and dehydration were less effective. Based on this, JA appears to be an important senescence-promoting factor.

The identification and cloning of *coi1* resulted in the identification of an F-box protein which shows the involvement of proteasome-dependent protein degradation in JA signalling (Xie *et al.*, 1998). Interestingly, application of MeJA to *Cucurbita pepo* (zucchini) induces senescence in 7-day-old cotyledons. One of the observed effects was on the concentration of endogenous cytokinin levels, which reduced rapidly after MeJA treatment (Ananieva, 2004). A drop in cytokinin levels is necessary before senescence can be initiated; whether MeJA can directly or indirectly antagonise cytokinin levels remains to be answered.

Transcriptome analyses of MeJA-treated seedlings showed a self-activation of JA biosynthesis and cross talk with other hormones (Sasaki *et al.*, 2001). Although the *coi1* mutant does not show any visual senescence defects, a transcriptional analysis showed that 12% of the identified developmental senescence genes are not expressed during senescence of *coi1* (Buchanan-Wollaston *et al.*, 2005). In addition, certain genes that are downregulated in the *coi1* mutant also appear to be downregulated in the *ein2* or *nahG* mutants. This further demonstrates the importance of the JA pathway during leaf senescence.

7.4.2.5 Salicylic acid

SA, a phenolic compound, has been identified as a key signalling molecule in various plant responses to stress, like pathogen invasion (Glazebrook, 1999) and exposure to ozone and UV-B. The endogenous SA levels in senescing stage 2 leaves are four times higher than in nonsenescing leaves (Morris *et al.*, 2000). This is consistent with a role for SA during later stages of the senescence programme. Study of the *nahG*, *pad4* and *npr1* mutants, which are defective in the SA signalling pathway, showed an altered expression pattern of a number of *SAG*s. Furthermore, a delay in yellowing and reduced necrosis were observed in these plants (Morris *et al.*, 2000). The *pad4* mutant has a non-necrotic phenotype that has a reduced expression of *SAG12*, a well-known SAG. The authors postulated that *SAG12* may take part in a regulatory pathway leading to cell death and that it supports the transition from senescence to final cell death. Thus the senescence phenotype of *pad4* mutant suggests that SA might regulate the transition from senescence to final cell death.

Besides biochemical and physiological evidence for role of SA in senescence, genetic evidence has also been generated by a microarray approach (Buchanan-Wollaston *et al.*, 2005). Of 827 genes that were identified as senescence upregulated genes, 19% showed at least a twofold reduction in the *nahG* transgenic plants that are defective in SA signalling. Interestingly *SAG12* expression was substantially reduced compared to that in wild-type plants and SA-treated plants (Morris *et al.*, 2000, Buchanan-Wollaston *et al.*, 2003). Since *SAG12* is generally seen as a marker for developmental senescence, this further demonstrates the importance of SA in senescence.

7.5 Involvement of genome programmes in the regulation of senescence-associated genes

Developmental senescence is regulated by diverse programmes involved in plant life maintenance, defence responses and growth and development (see above). This is consistent with the evolutionary theory of senescence and the proposal of genome optimisation for reproduction, which argues that no specific genetic programmes for life span evolve. Since the expression profiles of *SAG*s are reliable markers for senescence, examining the regulation of *SAG* expression may provide evidence to support that argument. If the evolutionary theory of ageing applies to plants, we expect that many *SAG*s encode proteins with functions throughout the life cycle of the plant, and not only during developmental senescence. This notion, as a matter of fact, is well supported by the identities and expression profiles of *SAG*s. Up to now, almost all the isolated *SAG*s, including many involved in nutrient salvage, exhibit a certain basal level of expression prior to the onset of leaf senescence. This indicates that nutrient salvage is a continuous process occurring in plant cells throughout life. In this sense, leaf senescence is not different from other leaf developmental stages but is more committed to recruit the last yet important source of nutrients retained in an ageing leaf.

Recent omics techniques have allowed us to examine the genes that are upregulated during senescence on a whole-genome basis. In addition to development, leaf senescence can be induced by biotic and abiotic stresses. It is therefore possible to compare the *SAG* expression profiles of various types of senescence using currently available microarray data, which enables the better understanding of the nature of the regulation of developmental senescence. In a whole-genome transcriptome analysis, a total of 827 *SAG*s were found upregulated during developmental senescence (Buchanan-Wollaston *et al.*, 2005). However, most of those are induced by hormones (SA, JA and ethylene) or darkness as well (Buchanan-Wollaston *et al.*, 2003; Lin and Wu, 2004). Using these data, we deducted the number of *SAG*s that were enhanced by darkness-induced senescence and were downregulated during leaf senescence in *nahG*, *coi1* and *ein2* plants. The remaining *SAG*s are hence presumably regulated by other developmental cues and/or stress conditions. As shown in Table 7.1, this category under the name of 'development' includes a total of 209 *SAG*s, which interestingly spread almost in all the categories. We further dissected whether these *SAG*s are upregulated by carbon and nitrogen metabolism. For this, the array data from 'Expression patterns of genes induced by sugar accumulation during early leaf senescence' provided by Wingler's laboratory were used. Analysis was done by GENEVESTIGATOR (Zimmermann *et al.*, 2004). Nearly half of the 209 *SAG*s were upregulated after induction of senescence by glucose in combination with low nitrogen. Again, the remaining 110 *SAG*s are wide spreading in all the categories. If this list of *SAG*s is compared with the profiles of *SAG*s in senescence regulated by other developmental cues such as ROS, other hormones (cytokinins, ABA, GA, etc.) or protein degradation, it will not be surprising that perhaps nearly all the *SAG*s are be regulated by one or more of these cues. In other words, very few *SAG*s will be solely induced by developmental senescence, which is in agreement with the evolutionary theory of ageing.

The most frequently used *SAG* to monitor developmental senescence, and perhaps one of the few *SAG*s which is specifically induced by developmental senescence, is *SAG12*. *SAG12* transcripts were found to be very low or below the detection level in young and mature green leaves, contrasting to the levels of the transcripts of *SAG13* and *SAG14* (Figure 7.1; Lohman *et al.*, 1994). Unlike other *SAG*s, including

Table 7.1 Comparison of gene expression profiles of age-regulated and developmental leaf senescence

Category	Total	Dark	Cell suspension	*NahG/ coi1/ein2*	Development C/N	Development Others
Regulatory genes						
Putative transcription factors and nucleic-acid-binding proteins	96	47	38	23	9	19
Putative protein–protein interaction	9	7	3	1	1	1
Putative ubiquitination control	30	18	23	6	2	2
Protein kinase and phosphatases	66	26	20	31	6	6
Signalling	17	9	8	5	0	5
Calcium related	13	5	4	8	1	0
Hormone pathways	17	10	11	6	2	0
Macromolecule degradation and mobilisation						
Protein degradation	29	13	14	6	6	3
Amino acid degradation and N mobilisation	27	14	15	10	3	3
Nucleic acid degradation and phosphate mobilisation	14	3	8	5	1	2
Lipid degradation and mobilisation	29	18	13	5	3	5
Chlorophyll degradation	2	1	1	0	1	0
Sulphur mobilisation	2	0	0	1	1	0
Carbohydrate metabolism	63	27	22	32	6	6
Lignin synthesis	3	1	2	0	1	0
Transport	74	31	26	17	10	14
ATPases	7	1	2	3	0	1
Metal binding	10	5	4	1	3	0
Stress related						
Antioxidants	11	7	6	4	2	0
Stress and detoxification	17	7	4	5	2	4
Defence related	11	6	4	5	1	0
Secondary metabolism	1	0	0	1	0	0
Alkaloid biosynthesis	9	6	5	5	0	0
Flavonoid/anthocyanin pathway	19	2	2	5	7	5
Autophagy	5	5	3	1	0	0
Structural	4	0	0	1	1	2
Unclassified enzymes of unknown role in senescence	110	51	42	36	13	17
Unknown genes	132	65	55	34	17	15
Total	827	385	335	257	99	110

Data sources: Buchanan-Wollaston *et al.* (2005); C/N: carbon and nitrogen supply. (Expression patterns of genes induced by sugar accumulation during early leaf senescence; Zimmerman *et al.*, 2004.)

SAG13, *SEN1* and *SAG14* whose expression could be enhanced in young leaves by a range of senescence-inducing treatments such as detachment, hormonal exposure, darkness, drought, wounding and pathogen challenge, *SAG12* was only occasionally found to change its expression under these circumstances (Oh *et al.*, 1997; Park

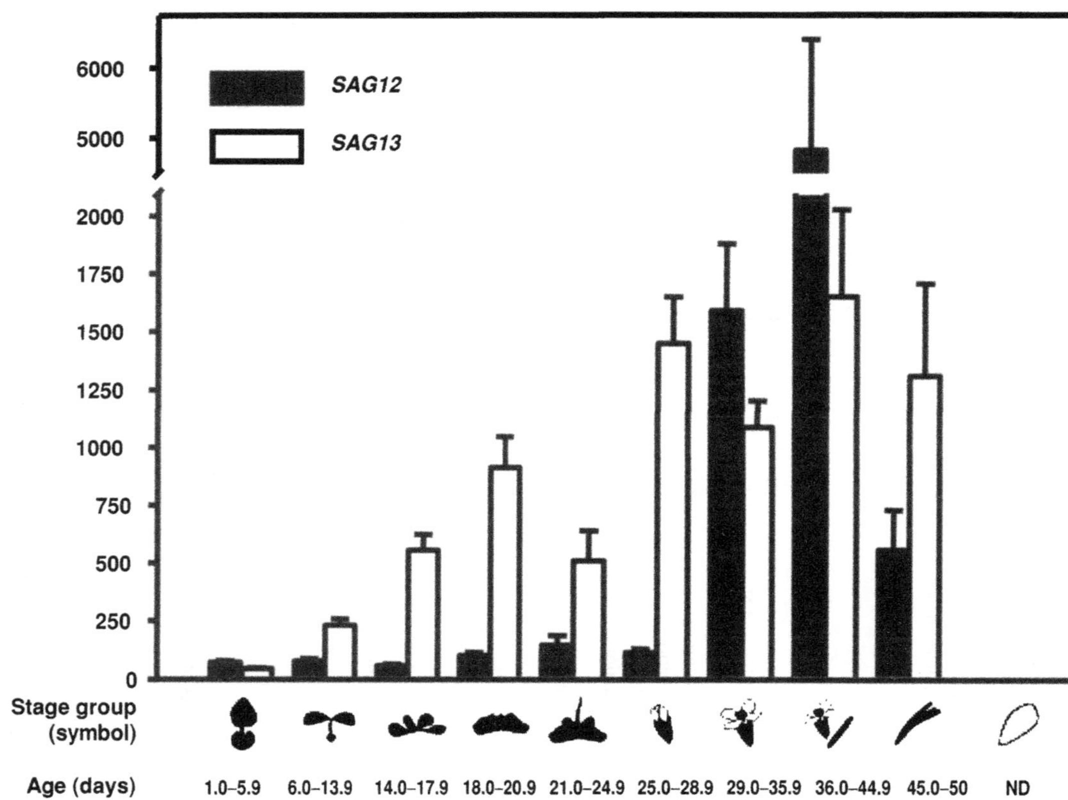

Figure 7.1 *SAG12* and *SAG13* expression during *Arabidopsis* development. Expression levels of both *SAG12* and *SAG13* are increased during senescence. In contrast to *SAG12*, basal *SAG13* expression levels are present throughout development. Data source: GENEVESTIGATOR (Zimmermann *et al.*, 2004).

et al., 1998; Weaver *et al.*, 1998; Noh and Amasino, 1999; Brodersen *et al.*, 2002). Thus, *SAG12* is considered the best marker for developmental senescence that relies on leaf age, whereas *SAG13* and *SAG14* may represent stress-induced senescence or general cell-death markers. That the *SAG12* promoter has been used for the autoregulated production of cytokinin to delay senescence in a number of species including tobacco (Gan and Amasino, 1995; Ori *et al.*, 1999), lettuce (McCabe *et al.*, 2001), petunia (Chang *et al.*, 2003) and *Arabidopsis* (Huynh *et al.*, 2005) suggests that the developmental senescence regulation of *SAG12* is conserved across species. Moreover, a conserved *cis*-element of the *SAG12* promoter was also found in the *Asparagus officinalis* asparagine synthetase promoter and was responsible for the induction of transcription of this gene by senescence (Winichayakul *et al.*, 2004). Thus, monocotyledonous and dicotyledonous plants appear to share this senescence *cis*-element, further confirming the conservation of the regulation of developmental senescence across species.

Extensive studies on the expression of *SAG*s, including *SAG12*, are presented in Chapters 9 and 10. These studies have provided exciting new insights into the developmental regulation of senescence, and future research will likely result in a better understanding of developmental senescence.

7.6 Integrating hormonal action into developmental senescence

Reproduction has specific timing and all the programmes need to be timely in place to ensure successful reproduction. The indirect consequence is that the various strategies embedded in the programmes will initiate developmental senescence in an age-dependent manner. Thus, developmental senescence is the consequence of time-specific action of genes. Understanding the timing of the various senescence strategies is a necessary step for elucidating the molecular mechanisms of developmental senescence. In this section we intend to put together the action of the hormones that control leaf senescence and thus developmental ageing in *Arabidopsis*.

Previously, we proposed a senescence window concept to explain the involvement of ethylene in leaf senescence (Jing *et al.*, 2002, 2003). Depending on whether and how senescence can be induced by ethylene, the life span of a leaf can be split into three distinct phases (Figure 7.2A). The experimental evidence supporting this view is briefly summarised as follows. (1) When plants were exposed to a short-pulse (e.g. 1–3 days) ethylene treatment, no senescence symptoms could be induced in young leaves (Grbić and Bleecker, 1995; Weaver *et al.*, 1998; Jing *et al.*, 2002). (2) Leaf senescence is not accelerated in the *ctr1* mutants (Kieber *et al.*, 1993). This indicated that there exists a never-senescence phase in which senescence cannot be induced by ethylene. (3) Furthermore, in a certain range of leaf ages, the effect of ethylene on leaf senescence increases with the increase in leaf age (Grbić and Bleecker, 1995; Weaver *et al.*, 1998; Jing *et al.*, 2002), indicative of an ethylene-dependent senescence phase. (4) Finally, beyond certain leaf ages, senescence will start even without the participation of ethylene as shown in the *etr1*

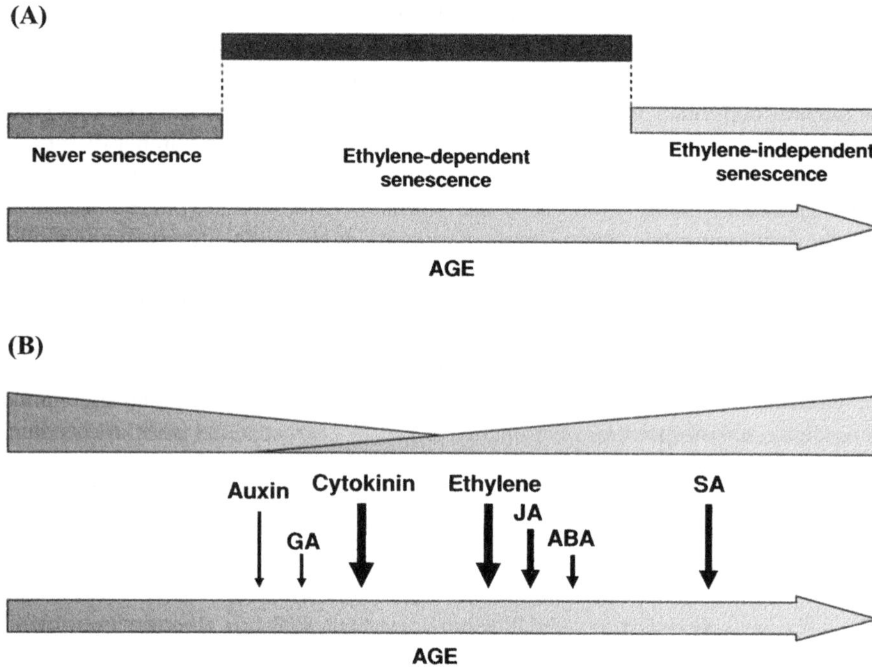

Figure 7.2 The senescence window concept. (A) The senescence window concept as deduced from the effects of ethylene on leaf senescence. At early leaf development, ethylene is not able to induce leaf senescence. This is the so-called never-senescence phase in the model. Only after a certain developmental stage, ethylene can induce leaf senescence, depending on the environmental conditions. Further development of the leaf will always result in senescence, even in the absence of ethylene. (B) Hormonal action during leaf development is age dependent. The onset of leaf senescence is modulated by the stay-green and senescence-promoting hormones. During senescence the effects of stay-green hormones (auxin, GAs, and cytokinins) diminish and the effects of senescence-promoting hormones (ethylene JA, ABA and SA) increase as indicated by the two triangles, respectively. The action of the senescence-promoting hormones is antagonistic to that of the stay-green hormones and increases with the progression of leaf senescence. Leaf age limits the action of the various plant hormones to their own specific age window.

and *ein2* mutants in which the senescence progresses normally once started (Grbić and Bleecker, 1995; Park *et al.*, 1998; Buchanan-Wollaston *et al.*, 2005), which suggests the existence of an ethylene-independent senescence phase. This senescence window concept emphasises the developmental control of leaf senescence and considers leaf age as an ultimate determinant of senescence progression. Clearly, genes that control the phase transitions of the senescence window are important for the onset of developmental senescence, and evidence suggests that many genetic loci are required (Jing *et al.*, 2002, 2005). Thus, the senescence window concept provides an explanation why the senescence-promoting effect of ethylene is variable during development.

The senescence window concept can, perhaps, integrate the action of all plant hormones involved in leaf senescence. In *Arabidopsis* the different hormones seem to control the onset and progression of senescence in an age-related manner. Figure 7.2B is an extension of the senescence window concept developed from the interaction between leaf age and ethylene and shows a tentative model illustrating the timing and action of the different hormones during developmental senescence. In this model, age-related changes, and thus development, are considered the primary regulator of leaf senescence. During ageing, developmental cues lead to the diminished action of the senescence-retarding hormones such as auxin, GA and cytokinins, as well as the concomitant strengthening of the action of senescence-enhancing hormones such as ethylene, JA, ABA and SA. The action of the different hormones during the initiation of leaf senescence does not change suddenly but gradually, allowing a gradual integration of all the hormones controlling the process. This suggests that the senescence process is partly reversible by fine-tuning hormone action and hence amenable for modulation.

The model provides a basis for the explanation of experimental data. For instance, the major senescence-retarding compound cytokinin can delay senescence when its level is maintained. However, in transgenic *SAG12–IPT* plants the senescence process will start eventually and progresses normally (Gan and Amasino, 1995; Ori *et al.*, 1999), suggesting that cytokinin action is restricted to certain developmental stages. On the other hand, cytokinin biosynthesis mutants showed a shorter leaf life span (Masferrer *et al.*, 2002). This might be explained by assuming that the effect of the senescence-promoting hormones is antagonistic to those blocking senescence; older leaves may become less sensitive for cytokinin and more sensitive for senescence-promoting hormones like JA and ABA (Weaver *et al.*, 1998). Similarly, blocking the ethylene pathway increases leaf longevity. Finally, however, the leaves go into senescence because the influence of JA, ABA and SA may increase with the age of the leaf. Thus, the age-related changes limit the action of the various hormones to their own specific window.

Taken together, although plant hormones are almost universally involved in every aspect of plant life, they may participate into developmental senescence only in very specific age windows. The proposed senescence window concept and the model for hormonal action provide a developmental view to examine the modulation of developmental senescence by hormones, which certainly requires more experimental evidence for validation.

7.7 Outlook and perspectives

Thanks to the availability of cutting-edge technology and the use of model species with known whole-genome sequences that have enabled senescence studies to be carried out at a scale that was not imaginable even 15 years ago, our knowledge on the regulation of developmental senescence has been advanced tremendously. It is clear that hormonal modulation, metabolic flux, ROS and protein degradation are the major cellular and molecular processes that are important for senescence regulation. Strikingly, these processes are embedded in the genome programmes that regulate plant life maintenance, responses to biotic and abiotic stresses, and growth and development for the sake of successful reproduction. Thus, leaf senescence can be viewed as an indirect consequence of genome optimisation for reproduction. This perspective is exciting and worthy of further exploitation, since it coincides with the evolutionary theory of senescence developed from animal ageing studies. In-depth molecular genetic studies are required to dress the evolutionary basis of leaf senescence. In particular, identification of regulatory genes with pleiotropic functions or late-life deleterious effects should be a priority for further senescence studies.

The complexity of leaf senescence is mainly due to the involvement of multiple components that exhibit overlapping effects. This is particularly true for the action of hormones. The proposed senescence window concept provides a theoretic framework to dissect the action of hormones during senescence depending on their time of action, which is important to separate the effect of hormones on senescence from their other effects. Using this concept, it is possible to study genetic components that control the action of hormones during development, which is an essential step for ultimately understanding the mode of action of hormones during development. Combined with the genetic dissection, whole-genome analysis should be employed to define the networking of various regulatory circuits.

In conclusion, senescence is one of the biological phenomena with extreme complexity. In the current postgenome era, we are provided with both opportunities and the challenge to dissect the molecular genetic mechanisms of leaf senescence. The findings in the past have enabled us to look at senescence regulation from a fresh perspective of genome optimisation. We have evolutionary and developmental theories that guard us to define the proper targets. We are also armed with cutting-edge technologies and tools. Thus, a concerted effort will eventually unveil the mystery of senescence regulation and provide a genetic basis for senescence manipulation.

References

Aharoni, N. (1989) Interrelationship between ethylene and growth regulators in the senescence of lettuce leaf discs. *J Plant Growth Regul* **8**(4), 309–317.

Aharoni, N., Anderson, J.D. and Liebermann, M. (1979) Production and action of ethylene in senescing leaf discs effect of indoleacetic acid, kinetin, silver ion, and carbon dioxide. *Plant Physiol* **64**(5), 805–809.

Aharoni, N. and Richmond, A.E. (1978) Endogenous gibberellin and abscisic acid content as related to senescence of detached lettuce leaves. *Plant Physiol* **62**(2), 224–228.

Ananieva, K., Malbeck, J., Kaminek, M. and van Staden, J. (2004) Methyl jasmonate down-regulates endogenous cytokinin levels in cotyledons of *Cucurbita pepo* (Zucchni) seedlings. *Physiol Plant* **122**(4), 496–503.

Arenas-Huertero, F., Arroyo, A., Zhou, L., Sheen, J. and Leon, P. (2000) Analysis of *Arabidopsis* glucose insensitive mutants, gin5 and gin6, reveals a central role of the plant vegetative development by sugar. *Genes Dev* **14**, 2085–2096.

Balibrea Lara, M.E.B., Garcia, M.C.G., Fatima, T., *et al.* (2004) Extracellular invertase is an essential component of cytokinin-mediated delay of senescence. *Plant Cell* **16**(5), 1276–1287.

Beaudoin, N., Serizet, C., Gosti, F. and Giraudat, J. (2000) Interactions between abscisic acid and ethylene signaling cascades. *Plant Cell* **12**(7), 1103–1115.

Bhalerao, R., Keskitalo, J., Sterky, F., *et al.* (2003) Gene expression in autumn leaves. *Plant Physiol* **131**(2), 430–442.

Biesalski, H.K. (2002) Free radical theory of aging. *Curr Opin Clin Nutr Metab Care* **5**(1), 5–10.

Bleecker, A.B. (1998) The evolutionary basis of leaf senescence: method to the madness? *Curr Opin Plant Biol* **1**(1), 73–78.

Bleecker, A.B. and Patterson, S.E. (1997) Last exit: senescence, abscission, and meristem arrest in *Arabidopsis*. *Plant Cell* **9**(7), 1169–1179.

Bowling, S.A., Clarke, J.D., Liu, Y., Klessig, D.F. and Dong, X. (1997) The cpr5 mutant of *Arabidopsis* expresses both NPR1-dependent and NPR1-independent resistance. *Plant Cell* **9**(9), 1573–1584.

Bray, E.A. (2002) Abscisic acid regulation of gene expression during water-deficit stress in the era of the *Arabidopsis* genome. *Plant cell Environ* **25**(2), 153–161.

Brodersen, P., Petersen, M., Pike, H.M., *et al.* (2002) Knockout of *Arabidopsis* accelerated-cell-death11 encoding a sphingosine transfer protein causes activation of programmed cell death and defense. *Genes Dev* **16**(4), 490–502.

Buchanan-Wollaston, V. (1997) The molecular biology of leaf senescence. *J Exp Bot* **48**(307), 181–199.

Buchanan-Wollaston, V., Earl, S., Harrison, E., *et al.* (2003) The molecular analysis of leaf senescence – a genomics approach. *Plant Biotechnol J* **1**(1), 3–22.

Buchanan-Wollaston, V., Page, T., Harrison, E., *et al.* (2005) Comparative transcriptome analysis reveals significant differences in gene expression and signaling pathways between developmental and dark/starvation-induced senescence in *Arabidopsis*. *Plant J* **42**(4), 567–585.

Chang, H.S., Jones, M.L., Banowetz, G.M. and Clark, D.G. (2003) Overproduction of cytokinins in petunia flowers transformed with P-SAG12–IPT delays corolla senescence and decreases sensitivity to ethylene. *Plant Physiol* **132**(4), 2174–2183.

Chatterjee, S.K. and Leopold, A.C. (1965) Changes in abscission processes with aging. *Plant Physiol* **40**(1), 96–101.

Chory, J., Nagpal, P. and Peto, C.A. (1991) Phenotypic and genetic analysis of det2, a new mutant that affects light-regulated seedling development in *Arabidopsis*. *Plant Cell* **3**(5), 445–459.

Clouse, S.D., Langford, M. and McMorris, T.C. (1996) A brasinosteroid-insensitive mutant in *Arabidopsis thaliana* exhibits multiple defects in growth and development. *Plant Physiol* **111**(3), 671–678.

Clouse, S.D. and Sasse, J.M. (1998) BRASSINOSTEROIDS: essential regulators of plant growth and development. *Annu Rev Plant Physiol Plant Mol Biol* **49**, 427–451.

Creelmann, R.A. and Mullet, J.E. (1995) Jasmonic acid distribution and action in plants: regulation during development and response to biotic and abiotic stress. *Proc Natl Acad Sci U S A* **92**(10), 4114–4119.

Crowell, D.N., Kadlecek, A.T., John, M.C. and Amasino, R.M. (1990) Cytokinin-induced mRNAs in cultured soybean cells. *Proc Natl Acad Sci U S A* **87**(22), 8815–8819.

de Magalhaes, J.P. and Church, G.M. (2005) Genomes optimize reproduction: aging as a consequence of the developmental program. *Physiology* **20**(4), 252–259.

dela Fuente, R.K. and Leopold, A.C. (1968) Senescence processes in leaf abscission. *Plant Physiol* **43**(9), 1496–1502.

Dill, A., Thomas, S.G., Hu, J., Steber, C.M. and Sun, T. (2004) The *Arabidopsis* F-box protein SLEEPY1 targets gibberellin signaling repressors for gibberellin-induced degradation. *Plant Cell* **16**(6), 1392–1405.

Doelling, J.H., Walker, J.M., Friedman, E.M., Thompson, A.R. and Vierstra, R.D. (2002) The APG8/12-activating enzyme APG7 is required for proper nutrient recycling and senescence in *Arabidopsis thaliana*. *J Biol Chem* **277**(36), 33105–33114.

Droillard, M., Boudsocq, M., Barbier-Brygoo, H. and Lauriere, C. (2002) Different protein kinase families are activated by osmotic stresses in *Arabidopsis thaliana* cell suspensions. Involvement of the MAP kinases AtMPK3 and AtMPK6. *FEBS Lett* **527**(1–3), 43–50.

Fan, L., Zheng, S. and Wang, X. (1997) Antisense suppression of phospholipase D alpha retards abscisic acid- and ethylene-promoted senescence of postharvest *Arabidopsis* leaves. *Plant Cell* **9**(12), 2183–2196.

Finkel, T. and Holbrook, N.J. (2000) Oxidants, oxidative stress and the biology of ageing. *Nature* **408**(6809), 239–247.

Fletcher, R.A. and Osborne, D.J. (1965) Regulation of protein and nucleic acid synthesis by gibberellin during leaf senescence. *Nature* **207**, 1176–1177.

Gan, S. (2003) Mitotic and postmitotic senescence in plants. *Sci Aging Knowledge Environ* **38**, Re7.

Gan, S. and Amasino, R.M. (1995) Inhibition of leaf senescence by autoregulated production of cytokinin. *Science* **270**(5244), 1986–1988.

Gan, S. and Amasino, R.M. (1997) Making sense of senescence (molecular genetic regulation and manipulation of leaf senescence). *Plant Physiol* **113**(2), 313–319.

Gems, D. (2000) An integrated theory of ageing in the nematode *Caenorhabditis elegans*. *J Anat* **197**(Pt 4), 521–528.

Glazebrook, J. (1999) Genes controlling expression of defense responses in *Arabidopsis*. *Curr Opin Plant Biol* **2**(4), 280–286.

Goda, H., Shimada, Y., Asami, T., Fujioka, S. and Yoshida, S. (2002) Microarray analysis of brassinosteroid-regulated genes in *Arabidopsis*. *Plant Physiol* **130**, 1319–1334.

Goldthwaite, J.J. and Laetsch, W.M. (1968) Control of senescence in *Rumex* leaf discs by gibberellic acid. *Plant Physiol* **43**(11), 1855–1858.

Grbić, V. and Bleecker, A.B. (1995) Ethylene regulates the timing of leaf senescence in *Arabidopsis*. *Plant J* **8**(4), 595–602.

Guo, Y., Cai, Z. and Gan, S. (2004) Transcriptome of *Arabidopsis* leaf senescence. *Plant Cell Environ* **27**(5), 521–549.

Hanaoka, H., Noda, T., Shirano, Y., *et al.* (2002) Leaf senescence and starvation-induced chlorosis are accelerated by the disruption of an *Arabidopsis* autophagy gene. *Plant Physiol* **129**(3), 1181–1193.

He, Y., Fukushige, H., Hildebrand, D.F. and Gan, S. (2002) Evidence supporting a role of jasmonic acid in *Arabidopsis* leaf senescence. *Plant Physiol* **128**(3), 876–884.

He, Y. and Gan, S. (2002) A gene encoding an acyl hydrolase is involved in leaf senescence in *Arabidopsis*. *Plant Cell* **14**(4), 805–815.

He, Y., Tang, W., Swain, J., Green, A., Jack, T. and Gan, S. (2001) Networking senescence-regulating pathways by using *Arabidopsis* enhancer trap lines. *Plant Physiol* **126**(2), 707–716.

Hensel, L.L., Grbic, V., Baumgarten, D.A. and Bleecker, A.B. (1993) Developmental and age-related processes that influence the longevity and senescence of photosynthetic tissues in *Arabidopsis*. *Plant Cell* **5**(5), 553–564.

Hoth, S., Ikeda, Y., Morgante, M., *et al.* (2003) Monitoring genome-wide changes in gene expression in response to endogenous cytokinin reveals targets in *Arabidopsis thaliana*. *FEBS Lett* **554**(3), 373–380.

Hutchinson, C.E. and Kieber, J.J. (2002) Cytokinin signaling in *Arabidopsis*. *Plant Cell* **14**, S47–S59.

Huynh, L.N., VanToai, T., Streeter, J. and Banowetz, G. (2005) Regulation of flooding tolerance of SAG12 : ipt *Arabidopsis* plants by cytokinin. *J Exp Bot* **56**(415), 1397–1407.

Hwang, I., Chen, H.C. and Sheen, J. (2002) Two-component signal transduction pathways in *Arabidopsis*. *Plant Physiol* **129**(2), 500–515.

Hwang, I. and Sheen, J. (2001) Two-component circuitry in *Arabidopsis* cytokinin signal transduction. *Nature* **413**(6854), 383–389.

Jing, H.C., Hille, J. and Dijkwel, P.P. (2003) Ageing in plants: conserved strategies and novel pathways. *Plant Biol* **5**(5), 455–464.

Jing, H.C., Schippers, J.H.M., Hille, J. and Dijkwel, P.P. (2005) Ethylene-induced leaf senescence depends on age-related changes and OLD genes in *Arabidopsis*. *J Exp Bot* **56**(421), 2915–2923.

Jing, H.C., Sturre, M.J., Hille, J. and Dijkwel, P.P. (2002) *Arabidopsis* onset of leaf death mutants identify a regulatory pathway controlling leaf senescence. *Plant J* **32**(1), 51–63.

John, C.F., Morris, K., Jordan, B.R., Thomas, B.S. and A-H-Mackerness, S. (2001) Ultraviolet-B exposure leads to up-regulation of senescence-associated genes in *Arabidopsis thaliana*. *J Exp Bot* **52**(359), 1367–1373.

John, I., Drake, R., Farrell, A., *et al.* (1995) Delayed leaf senescence in ethylene-deficient ACC-oxidase antisense tomato plants: molecular and physiological analysis. *Plant J* **7**, 483–490.

Kato, Y., Murakami, S., Yamamoto, Y., *et al.* (2004) The DNA-binding protease, CND41, and the degradation of ribulose-1,5-bisphosphate carboxylase/oxygenase in senescent leaves of tobacco. *Planta* **220**(1), 97–104.

Kiba, T., Naitou, T., Koizumi, N., Yamashino, T., Sakakibara, H. and Mizuno, T. (2005) Combinatorial microarray analysis revealing *Arabidopsis* genes implicated in cytokinin responses through the His→Asp phosphorelay circuitry. *Plant Cell Physiol* **46**(2), 339–355.

Kieber, J.J., Rothenberg, M., Roman, G., Feldmann, K.A. and Ecker, J.R. (1993) Ctrl, a negative regulator of the ethylene response pathway in *Arabidopsis*, encodes a member of the Raf family of protein-kinases. *Cell* **72**(3), 427–441.

Kirkwood, T.B. (2005) Understanding the odd science of aging. *Cell* **120**(4), 437–447.

Kirkwood, T.B. and Austad, S.N. (2000) Why do we age? *Nature* **408**(6809), 233–238.

Lam, E., Kato, N. and Lawton, M. (2001) Programmed cell death, mitochondria and the plant hypersensitive response. *Nature* **411**(6839), 848–853.

Leyser, O. (2002) Molecular genetics of auxin signalling. *Annu Rev Plant Biol* **53**, 377–398.

Lim, P.O. and Nam, H.G. (2005) The molecular and genetic control of leaf senescence and longevity in *Arabidopsis*. *Curr Top Dev Biol* **67**, 49–83.

Lim, P.O., Woo, H.R. and Nam, H.G. (2003) Molecular genetics of leaf senescence in *Arabidopsis*. *Trends Plant Sci* **8**(6), 272–278.

Lin, J.F. and Wu, S.H. (2004) Molecular events in senescing *Arabidopsis* leaves. *Plant J* **39**(4), 612–628.

Lohman, K.N., Gan, S.S., John, M.C. and Amasino, R.M. (1994) Molecular analysis of natural leaf senescence in *Arabidopsis thaliana*. *Physiol Plant* **92**(2), 322–328.

Ludewig, F. and Sonnewald, U. (2000) High CO2-mediated down-regulation of photosynthetic gene transcripts is caused by accelerated leaf senescence rather than sugar accumulation. *FEBS Lett* **479**(1–2), 19–24.

Masferrer, A., Arro, M., Manzano, D., *et al.* (2002) Overexpression of *Arabidopsis thaliana* farnesyl diphosphate synthase (FPS1S) in transgenic *Arabidopsis* induces a cell death/senescence-like response and reduced cytokinin levels. *Plant J* **30**(2), 123–132.

Matile, P. (1997) The vacuole and cell senescence. In: *Advances in Botanical Research: The Plant Vacuole*, Vol. 25 (eds Callow, J.A.). Academic Press, San Diego, CA, pp. 87–112.

McCabe, M.S., Garratt, L.C., Schepers, F., *et al.* (2001) Effects of P(SAG12)–IPT gene expression on development and senescence in transgenic lettuce. *Plant Physiol* **127**(2), 505–516.

Miller, A., Schlagnhaufer, C., Spalding, M. and Rodermel, S. (2000) Carbohydrate regulation of leaf development: prolongation of leaf senescence in rubisco antisense mutants of tobacco. *Photosynth Res* **63**(1), 1–8.

Miller, A., Tsai, C.H., Hemphill, D., Endres, M., Rodermel, S. and Spalding, M. (1997) Elevated CO_2 effects during leaf ontogeny (a new perspective on acclimation). *Plant Physiol* **115**(3), 1195–1200.

Miller, J.D., Arteca, R.N. and Pell, E.J. (1999) Senescence-associated gene expression during ozone-induced leaf senescence in *Arabidopsis*. *Plant Physiol* **120**(4), 1015–1024.

Mittler, R. (2002) Oxidative stress, antioxidants and stress tolerance. *Trends Plant Sci* **7**(9), 405–410.

Mok, D.W.S. and Mok, M.C. (2001) Cytokinin metabolism and action. *Annu Rev Plant Physiol Plant Mol Biol* **52**, 89–118.

Moore, B., Zhou, L., Rolland, F., *et al.* (2003) Role of the *Arabidopsis* glucose sensor HXK1 in nutrient, light, and hormonal signalling. *Science* **300**(5617), 332–336.

Morgan, P.W. and Drew, M.C. (1997) Ethylene and plant responses to stress. *Physiol Plant* **100**(3), 620–630.

Morris, K., A.-H.–Mackerness, S., Page, T., *et al.* (2000) Salicylic acid has a role in regulating gene expression during leaf senescence. *Plant J* **23**(5), 677–685.

Mou, Z., He, Y., Dai, Y., Liu, X. and Li, J. (2000) Deficiency in fatty acid synthase leads to premature cell death and dramatic alterations in plant morphology. *Plant Cell* **12**(3), 405–418.

Mou, Z., Wang, X., Fu, Z., *et al.* (2002) Silencing of phosphoethanolamine *N*-methyltransferase results in temperature-sensitive male sterility and salt hypersensitivity in *Arabidopsis*. *Plant Cell* **14**(9), 2031–2043.

Nambara, E. and Marion-Poll, A. (2005) Abscisic acid biosynthesis and catabolism. *Annu Rev Plant Biol* **56**, 165–185.

Navabpour, S., Morris, K., Allen, R., Harrison, E., A.-H.-Mackerness, S. and Buchanan-Wollaston, V. (2003) Expression of senescence-enhanced genes in response to oxidative stress. *J Exp Bot* **54**(391), 2285–2292.

Noh, Y.S. and Amasino, R.M. (1999) Identification of a promoter region responsible for the senescence-specific expression of SAG12. *Plant Mol Biol* **41**(2), 181–194.

Nooden, L.D. (1984) Integration of soybean pod development and monocarpic senescence. *Physiol Plant* **62**, 273–284.

Nooden, L.D. (1988) Whole plant senescence. In: *Senescence and Aging in Plants* (eds Nooden, L.D. and Leopold, A.C.). Academic Press, San Diego, CA, pp. 391–439.

Nooden, L.D., Kahanak, G.M. and Okatan, Y. (1979) Prevention of monocarpic senescence in soybeans with auxin and cytokinin: an antidote for self-destruction. *Science* **206**(4420), 841–843.

Nooden, L.D. and Penney, J.P. (2001) Correlative controls of senescence and plant death in *Arabidopsis thaliana* (Brassicaceae). *J Exp Bot* **52**(364), 2151–2159.

Nooden, L.D., Santokh, S. and Letham, D.S. (1990) Correlation of xylem sap cytokinin levels with monocarpic senescence in soybean. *Plant Physiol* **93**(1), 33–39.

Oh, S.A., Park, J.H., Lee, G.I., Paek, K.H., Park, S.K. and Nam, H.G. (1997) Identification of three genetic loci controlling leaf senescence in *Arabidopsis thaliana*. *Plant J* **12**(3), 527–535.

Ori, N., Juarez, M.T., Jackson, D., Yamaguchi, J., Banowetz, G.M. and Hake, S. (1999) Leaf senescence is delayed in tobacco plants expressing the maize homeobox gene knotted1 under the control of a senescence-activated promoter. *Plant Cell* **11**(6), 1073–1080.

Otegui, M.S., Noh, Y.S., Martinez, D.E., *et al.* (2005) Senescence-associated vacuoles with intense proteolytic activity develop in leaves of *Arabidopsis* and soybean. *Plant J* **41**(6), 831–844.

Ouaked, F., Rozhon, W., Lecourieux, D. and Heribert, H. (2003) A MAPK pathway mediates ethylene signaling in plants. *EMBO J* **22**(6), 1282–1288.

Park, J.H., Oh, S.A., Kim, Y.H., Woo, H.R. and Nam, H.G. (1998) Differential expression of senescence-associated mRNAs during leaf senescence induced by different senescence-inducing factors in *Arabidopsis*. *Plant Mol Biol* **37**(3), 445–454.

Pourtau, N., Mares, M., Purdy, S., Quentin, N., Ruel, A. and Wingler, A. (2004) Interactions of abscisic acid and sugar signaling in the regulation of leaf senescence. *Planta* **219**(5), 765–772.

Quirino, B.F., Noh, Y.S., Himelblau, E. and Amasino, R.M. (2000) Molecular aspects of leaf senescence. *Trends Plant Sci* **5**(7), 278–282.

Quirino, B.F., Normanly, J. and Amasino, R.M. (1999) Diverse range of gene activity during *Arabidopsis thaliana* leaf senescence includes pathogen-independent induction of defense-related genes. *Plant Mol Biol* **40**(2), 267–278.

Rashotte, A.M., Carson, S.D.B., To, J.P.C. and Kieber, J.J. (2003) Expression profiling of cytokinin action in *Arabidopsis*. *Plant Physiol* **132**(4), 1998–2011.

Richards, D.E., King, K.E., Ait-ali, T. and Harberd, N.P. (2001). How gibberellin regulates plant growth and development: a molecular genetic analysis of gibberellin signalling. *Annu Rev Plant Physiol Plant Mol Biol* **52**, 67–88.

Richmond, A.E. and Lang, A. (1957) Effect of kinetin on protein content and survival of detached *Xanthium* leaves. *Science* **125**, 650–651.

Rose, M.R. (1991) *Evolutionary Biology of Aging*. Oxford University Press, New York.

Sasaki, Y., Asamizu, E., Shibata, D., *et al.* (2001) Monitoring of methyl jasmonate-responsive genes in *Arabidopsis* by cDNA macroarray: self-activation of jasmonic acid biosynthesis and crosstalk with other phytohormone signaling pathways. *DNA Res* **8**(4), 153–161.

Sexton, R. and Roberts, J.A. (1982) Cell biology of abscission. *Annu Rev Plant Physiol* **33**, 132–162.

Sharp, R.E. (2002) Interaction with ethylene: changing views on the role of abscisic acid in root and shoot growth responses to water stress. *Plant Cell Environ* **25**(2), 211–222.

Smalle, J. and Vierstra, R.D. (2004) The ubiquitin 26S proteasome proteolytic pathway. *Annu Rev Plant Biol* **55**, 555–590.

Smart, C.M. (1994) Gene-expression during leaf senescence. *DNA Res* **126**(3), 419–448.

Srivalli, B. and Khanna-Chopra, R. (2004) The developing reproductive 'sink' induces oxidative stress to mediate nitrogen mobilization during monocarpic senescence in wheat. *Biochem Biophys Res Commun* **325**, 198–202.

Sun, T. and Gubler, F. (2004) Molecular mechanism of gibberellin signaling in plants. *Annu Rev Plant Biol* **55**, 197–223.

Surpin, M., Zheng, H., Morita, M.T., *et al.* (2003) The VTI family of SNARE proteins is necessary for plant viability and mediates different protein transport pathways. *Plant Cell* **15**(12), 2885–2899.

Thomas, H. (2002) Ageing in plants. *Mech Ageing Dev* **123**(7), 747–753.

Thomas, H. (2003) Do green plant age, and if so, how? *Top Curr Genet* **3**, 145–171.

Thompson, A.R., Doelling, J.H., Suttangkakul, A. and Vierstra, R.D. (2005) Autophagic nutrient recycling in *Arabidopsis* directed by the ATG8 and ATG12 conjugation pathways. *Plant Physiol* **138**(4), 2097–2110.

Thompson, A.R. and Vierstra, R.D. (2005) Autophagic recycling: lessons from yeast help define the process in plants. *Curr Opin Plant Biol* **8**(2), 165–173.

Ueda, J. and Kato, J. (1980) Isolation and identification of a senescence-promoting substance from wormwood (*Artemisia absinthium* L.). *Plant Physiol* **66**(2), 246–249.

Vert, G., Nemhauser, J.L., Geldner, N., Hong, F. and Chory, J. (2005) Molecular mechanisms of steroid hormone signaling in plants. *Annu Rev Cell Dev Biol* **21**, 177–201.

Wasternack, C. and Hause, B. (2002) Jasmonates and octadecanoids: signals in plant stress responses and development. *Prog Nucleic Acid Res Mol Biol* **72**, 165–221.

Weaver, L.M., Gan, S., Quirino, B. and Amasino, R.M. (1998) A comparison of the expression patterns of several senescence-associated genes in response to stress and hormone treatment. *Plant Mol Biol* **37**(3), 455–469.

Weidhase, R., Kramell, H., Lehmann, J., Liebisch, H., Lerbs, W. and Parthier, B. (1987) Methyl jasmonate-induced changes in the polypeptide pattern of senescing barley segments. *Plant Sci* **51**, 177–186.

Wellesen, K., Durst, F., Pinot, F., *et al.* (2001) Functional analysis of the LACERATA gene of *Arabidopsis* provides evidence for different roles of fatty acid omega-hydroxylation in development. *Proc Natl Acad Sci U S A* **98**(17), 9694–9699.

Wingler, A., Mares, M. and Pourtau, N. (2004) Spatial patterns and metabolic regulation of photosynthetic parameters during leaf senescence. *New Phytol* **161**(3), 781–789.

Winichayakul, S., Moyle, R.L., Coupe, S.A., Davies, K.M. and Farnden, K.J.F. (2004) Analysis of the asparagus (*Asparagus officinalis*) asparagine synthetase gene promoter identifies evolutionarily conserved cis-regulatory elements that mediate Suc-repression. *Funct Plant Biol* **31**(1), 63–72.

Woo, H.R., Chung, K.M., Park, J.H., *et al.* (2001) ORE9, an F-box protein that regulates leaf senescence in *Arabidopsis. Plant Cell* **13**(8), 1779–1790.

Woo, H.R., Goh, C.H., Park, J.H., *et al.* (2002) Extended leaf longevity in the ore4-1 mutant of *Arabidopsis* with a reduced expression of a plastid ribosomal protein gene. *Plant J* **31**(3), 331–340.

Woo, H.R., Kim, J.H., Nam, H.G. and Lim, P.O. (2004) The delayed leaf senescence mutants of *Arabidopsis, ore1, ore3*, and *ore9* are tolerant to oxidative stress. *Plant Cell Physiol* **45**(7), 923–932.

Xie, D.X., Feys, B.F., James, S., Nieto-Rostro, M., and Turner, J.G. (1998). *COI1*: an *Arabidopsis* gene required for jasmonate-regulated defense and fertility. *Science* **280**(5366), 1091–1094.

Xiong, L. and Zhu, J.K. (2003) Regulation of abscisic acid biosynthesis. *Plant Physiol* **133**(1), 29–36.

Xiong, Y., Contento, A.L. and Bassham, D.C. (2005) AtATG18a is required for the formation of autophagosomes during nutrient stress and senescence in *Arabidopsis thaliana. Plant J* **42**(4), 535–546.

Yamada, K., Matsushima, R., Nishimura, M. and Hara-Nishimura, I. (2001) A slow maturation of a cysteine protease with a granulin domain in the vacuoles of senescing *Arabidopsis* leaves. *Plant Physiol* **127**(4), 1626–1634.

Yi, H.C., Joo, S., Nam, K.H., Lee, J.S., Kang, B.G. and Kim, W.T. (1999) Auxin and brassinosteroid differentially regulate the expression of three members of the 1-aminocyclopropane-1-1carboxylate synthase gene family in mung bean (*Vigna radiata* L.). *Plant Mol Biol* **41**(4), 443–454.

Yin, Y., Wang, Z.Y., Mora-Garcia, S., *et al.* (2002) BES1 accumulates in the nucleus in response to brassinosteroids to regulate gene expression and promote stem elongation. *Cell* **109**, 181–191.

Yoshida, S., Ito, M., Callis, J., Nishida, I. and Watanabe, A. (2002a) A delayed leaf senescence mutant is defective in arginyl-tRNA:protein arginyltransferase, a component of the N-end rule pathway in *Arabidopsis. Plant J* **32**(1), 129–137.

Yoshida, S., Ito, M., Nishida, I. and Watanabe, A. (2002b) Identification of a novel gene HYS1/CPR5 that has a repressive role in the induction of leaf senescence and pathogen-defence responses in *Arabidopsis thaliana. Plant J* **29**(4), 427–437.

Yoshimoto, K., Hanaoka, H., Sato, S., *et al.* (2004) Processing of ATG8s, ubiquitin-like proteins, and their deconjugation by ATG4s are essential for plant autophagy. *Plant Cell* **16**(11), 2967–2983.

Zacarias, L. and Reid, M.S. (1990) Role of growth regulators in the senescence of *Arabidopsis thaliana* leaves. *Physiol Plant* **80**(4), 549–554.

Zapata, J.M., Guera, A., Esteban-Carrasco, A., Martin, M. and Sabater, B. (2005) Chloroplasts regulate leaf senescence: delayed senescence in transgenic ndhF-defective tobacco. *Cell Death Differ* **12**(10), 1277–1284.

Zeevaart, J.A.D. and Creelman, R.A. (1988) Metabolism and physiology of abscisic acid. *Annu Rev Plant Physiol Plant Mol Biol* **39**, 439–473.

Zhong, G. and Burns, J.K. (2003) Profiling ethylene-regulated gene expression in *Arabidopsis thaliana* by microarray analysis. *Plant Mol Biol* **53**, 117–131.

Zimmermann, P., Hirsch-Hoffmann, M., Hennig, L. and Gruissem, W. (2004) GENEVESTIGATOR. *Arabidopsis* Microarray Database and Analysis Toolbox. *Plant Physiol* **136**, 2621–2632.

8 The genetic control of senescence revealed by mapping quantitative trait loci

Helen Ougham, Ian Armstead, Catherine Howarth, Isaac Galyuon, Iain Donnison and Howard Thomas

8.1 Quantitative traits – what they are and how they are mapped

8.1.1 Genetic mapping

Linkage mapping is almost as old as the science of genetics in the modern age (Morgan, 1911). The principle is relatively simple (see Jones *et al.*, 1997): the further apart two loci are on a chromosome, the greater is the likelihood that crossing-over will occur between them at meiosis. Quantifying this likelihood in a segregating population generates an estimate of relative genetic distance. In this way, segregating loci, in the form of phenotypic characters (traits) and DNA markers, can be located relative to each other on the genetic map.

8.1.2 Major genes and QTL

'Major gene' is a term used by breeders and other applied geneticists to describe a Mendelian locus at which allelic variation gives rise to qualitatively distinct phenotypes. Mapping major traits in a segregating population is a relatively simple exercise. Jones *et al.* (1997) contrast the 'digital' nature of major gene traits with the 'analogue' characteristics of many, perhaps most, of the significant physiological, developmental and adaptive features of plants. Complex phenotypes behave as more or less continuous ranges between extremes in a segregating population. The genes underlying quantitative phenotypes are sometimes referred to as polygenes. Polygenes may be linked only in the physiological, but not the genetic sense. Quantitative trait loci (QTL; herein QTL is also used for quantitative trait locus) represent the resolution of complex characters into contributing regions of the genome by molecular marker mapping.

8.1.3 QTL mapping

Because the loci of individual polygenes cannot be identified, QTL must be mapped by a variation of the standard procedure used for molecular markers or major genes (Paterson *et al.*, 1988). The approach is to establish the statistical relationship between the inheritance of the trait and that of molecular markers whose map positions are known. The principles of QTL mapping have been outlined by Jones *et al.* (1997) (Figure 8.1). Consider a senescence-related quantitative trait, such as the greenness

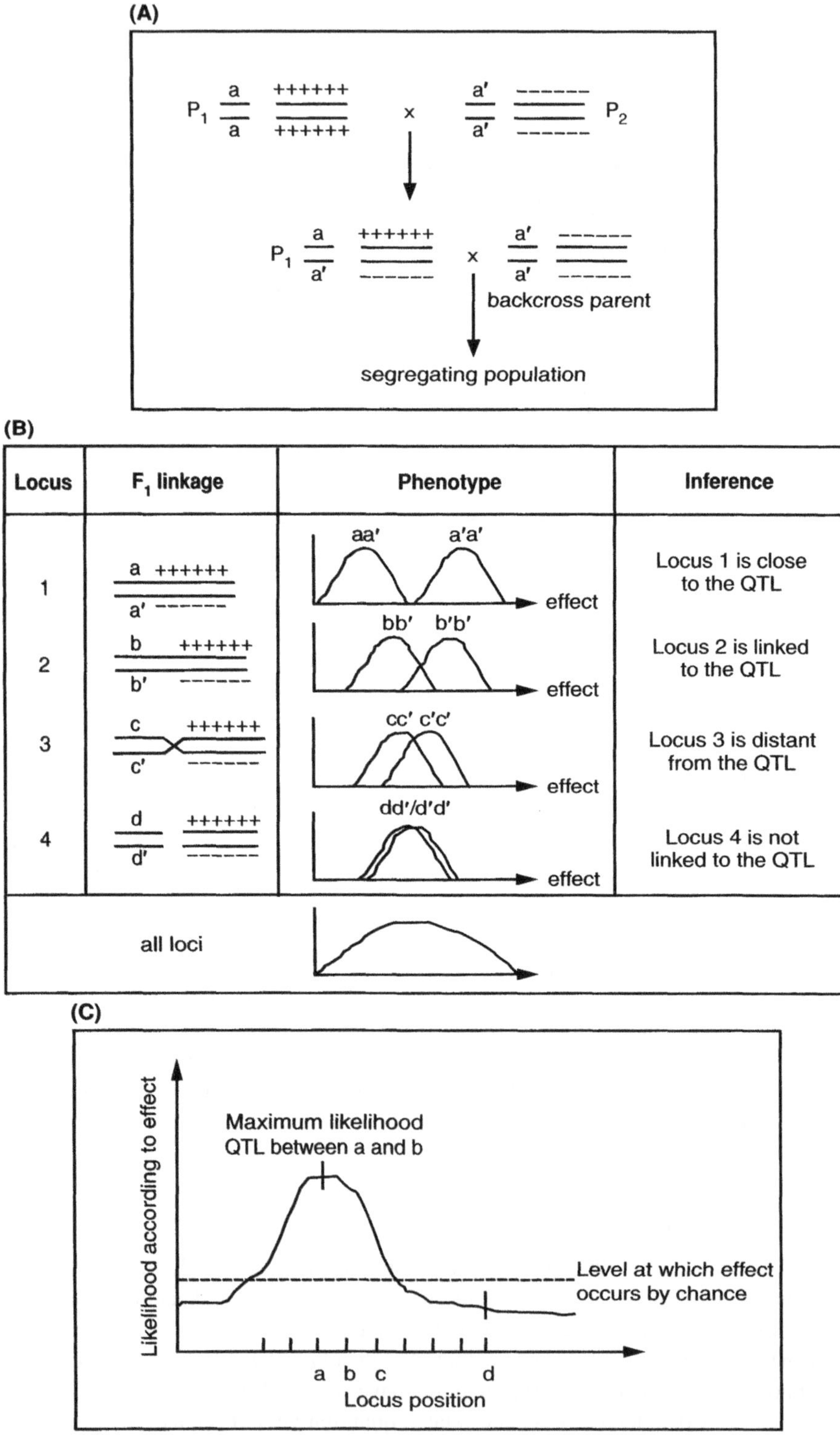
(A)
P1
a
a
++++++
++++++
x
a′
a′

P2
P1
a
a′
++++++
------
x
a′
a′

------
backcross parent
segregating population
(B)
Locus
F1 linkage
Phenotype
Inference
1
a ++++++
a′ ------
aa′
a′a′
effect
Locus 1 is close to the QTL
2
b ++++++
b′ ------
bb′
b′b′
effect
Locus 2 is linked to the QTL
3
c ++++++
c′ ------
cc′ c′c′
effect
Locus 3 is distant from the QTL
4
d
d′
++++++
------
dd′/d′d′
effect
Locus 4 is not linked to the QTL
all loci
(C)
Likelihood according to effect
Maximum likelihood QTL between a and b
Level at which effect occurs by chance
a b c d
Locus position

or yellowness of a given leaf at a given time in development of the whole plant. The mapping population is established by crossing two parent lines that are divergent in molecular markers (for example simple sequence repeats, SSRs, or restriction fragment length polymorphisms, RFLPs) and leaf colour at time *t*. RFLP or SSR alleles and QTL for colour will then segregate in the progeny. Figure 8.1 presents the simplest case of a single QTL comprising a cluster of more or less adjacent polygenes interacting to give quantitative control of colour. The possibilities are shown for this QTL in relation to four molecular markers, one nearby, one more distant, one that is remote but still linked, and an unlinked marker. All the plants with the specific allele of the molecular marker from parent 1 will display a frequency distribution for colour phenotype, as will plants with the parent 2 allele. The degree to which these individual frequency distribution curves coincide is related to the genetic distance between marker and QTL. This is quantified by statistical procedures such as maximum likelihood, and allows a QTL to be located on the molecular marker map as a score above a threshold that represents the likelihood that the effect occurs by chance.

8.1.4 'QTL for' talk

The news media have an insatiable appetite for medical breakthroughs and most familiar amongst these are announcements that scientists have discovered the gene 'for' some disease or other. Molecular biology makes the mechanistic connection between a DNA sequence and the structure of a protein, which may in turn have a definable role in phenotype X. In this case it is reasonable to speak of the region of the genome represented by that DNA sequence as a gene 'for' X. But generally, as in the case of the news story, the use of 'for' in relating genotype and phenotype is often problematical, sometimes misleading and occasionally plain wrong. Kaplan and Pigliucci (2001) have critically examined what they term *'gene for' talk* and suggest criteria – requiring the gathering of statistical, biochemical, historical and ecological information – that need to be satisfied before a gene can properly be claimed to be *for* a phenotypic trait. Awareness of the pitfalls of the little word 'for' is even more necessary when it comes to QTL analysis, which on its own is not able to distinguish between a truly causative genotype–phenotype relationship on the one hand and a correlative association on the other. Nevertheless, using the phrase 'QTL

Figure 8.1 Principle of mapping a quantitative trait locus (QTL). (A) A mapping population is established by crossing parents that are divergent for their RFLP markers at locus 1 (alleles *a* and *a′*), locus 2 (*b*, *b′*), locus 3 (*c*, *c′*), locus 4 (*d*, *d′*) and for the quantitative character concerned – for example retention (+) or loss (–) of chlorophyll during leaf senescence. The heterozygous F_1 is then backcrossed to one of the parents to give the segregating population. (B) The linkage between the QTL and various marker loci can then be ascertained by the way in which the distribution patterns of leaf pigmentation during senescence are associated with the segregation of the two alleles at each locus. (C) The map position of the QTL is determined as the maximum likelihood from the distribution of likelihood values (ratio of likelihood that the effect occurs by linkage: likelihood that the effect occurs by chance) calculated for each locus. (From Jones *et al.* 1997, reproduced with permission of the New Phytologist Trust.)

for phenotype X' avoids much circumlocution, and so will be employed on occasion in this chapter, in the full knowledge of the dangers of uncritical 'gene for' talk.

8.2 Biomarkers of the senescence process

8.2.1 Senescence is polygenic and quantitative

Senescence (here we focus the discussion specifically on green tissues) is particularly suitable for genetic analysis by the QTL approach. There is a wealth of mostly incidental or indirect evidence that a large number of genes contribute to the initiation, progress and regulation of the process (Smart, 1994; Buchanan-Wollaston, 1997; Thomas and Howarth, 2000; Buchanan-Wollaston *et al.*, 2003; Lim *et al.*, 2003; Andersson *et al.*, 2004). Although mutations in certain key loci often show inheritance of the major gene type (Thomas and Howarth, 2000), most of the variation in expression of senescence symptoms, in most species where it has been studied, is clearly quantitative, comprising a continuous range of initiation times, rates of progress and degree of connectivity between contributing sub-processes. The practical challenges confronting the researcher who wishes to analyse senescence by the QTL method are of two broad types: (a) species-specific matters, including the range of accessible variation, the generation of mapping populations and the suitability of available molecular markers; and (b) the generic question as to which constituent or function should be measured as an index of senescence. We have addressed (a) in the case studies presented in Section 8.5. The issue of how senescence should be measured as a trait is considered next.

8.2.2 Trait measurement in senescence

We need scorable characters that are directly diagnostic of senescence if we are to map the corresponding quantitative traits. The superficial features are easy to screen: greenness (Merzlyak *et al.*, 1999; Richardson *et al.*, 2002), leaf/green/photosynthetic area duration (Thomas, 1992), mobilisable N/rubisco/total protein (Hörtensteiner and Feller, 2002; Schiltz *et al.*, 2004), gas exchange (Makino *et al.*, 1985; Thomas and Howarth, 2000), chlorophyll (chl) fluorescence (Maxwell and Johnson, 2000; Wingler *et al.*, 2004) and pigment content (Roca *et al.*, 2004). There are also one or two fairly reliable biochemical and molecular markers, e.g. acid endopeptidase activity (Morris *et al.*, 1996; Masclaux *et al.*, 2001) and *SAG12* gene expression (Lim *et al.*, 2003). But all of these need to be used with caution because there are senescence-like pathological conditions (we might use the term *pseudosenescence* – Cots *et al.*, 2002) in which particular senescence biomarkers might appear to be invoked independently of the syndrome as a whole.

8.2.3 Pseudosenescence

Various kinds of bleaching responses, such as to herbicides or abiotic stresses, superficially resemble yellowing as seen in 'true' senescence (Thomas *et al.*, 2001).

QTL analysis may help to identify the degree to which such pseudosenescence processes are uncoordinated escapes from the integrated syndrome or else truly unrelated reactions of pathological origin. We are hindered by a generally poor biochemical and physiological understanding of senescence as a developmental event. Pigmentation changes are the best understood components, although even here there remain significant gaps at the molecular, chemical, enzymic and cellular levels.

8.2.4 Senescence-specific metabolism

When it comes to protein, nucleic acid, lipid and complex carbohydrate metabolism, there is a frustrating paucity of definitive enzymic, and corresponding molecular, data. For example, in spite of the scale and agroecological importance of rubisco mobilisation during senescence, neither the proteolytic activity responsible nor the controls that turn it on seem close to being identified (Hörtensteiner and Feller, 2002). Again, QTL analysis could be a useful tool (Obara *et al.* (2001) refer to it as a *post-genomic tool*), offering a way of sorting through candidates, using colocation with the map position(s) of N/protein/rubisco degradation loci as a criterion for a causal relationship. Figure 8.2 is a representation of the metabolism of the senescing green cell as currently understood. Each step in each pathway corresponds to at least one gene that may behave as a major locus and/or contribute to one or more QTL.

8.3 Correlated developmental events as second-order senescence traits

8.3.1 Remote control of senescence

Limited knowledge of the basic physiology and cell biology of foliar senescence has hindered the understanding of the contextual regulation of the process. Metabolic changes in the senescing cell, tissue and organ are sensitive to remote events during whole-plant development (Nooden and Penney, 2001). These events are themselves usually traits of a quantitative nature (for example flowering time, grain fill, the juvenility–maturity transition and so on) and frequently show up as contributing genetic factors in QTL analyses of senescence. In one sense it does not matter if a 'senescence QTL' turns out not to concern senescence, or even leaves, at all but instead underlies some remote (in spatial and ontogenetic terms) physiological process that has second-order influence on senescing foliage. If the QTL in question is responsible for modifying senescence in some way, it is by definition a senescence QTL. But the question of primary and second-order genetic associations is far from simple.

8.3.2 Allometry and QTL

Plants are modular, development is a process of module turnover and senescence represents the negative term in the turnover equation. Allometry describes the relationship between an individual module and the whole organism (Huxley, 1924).

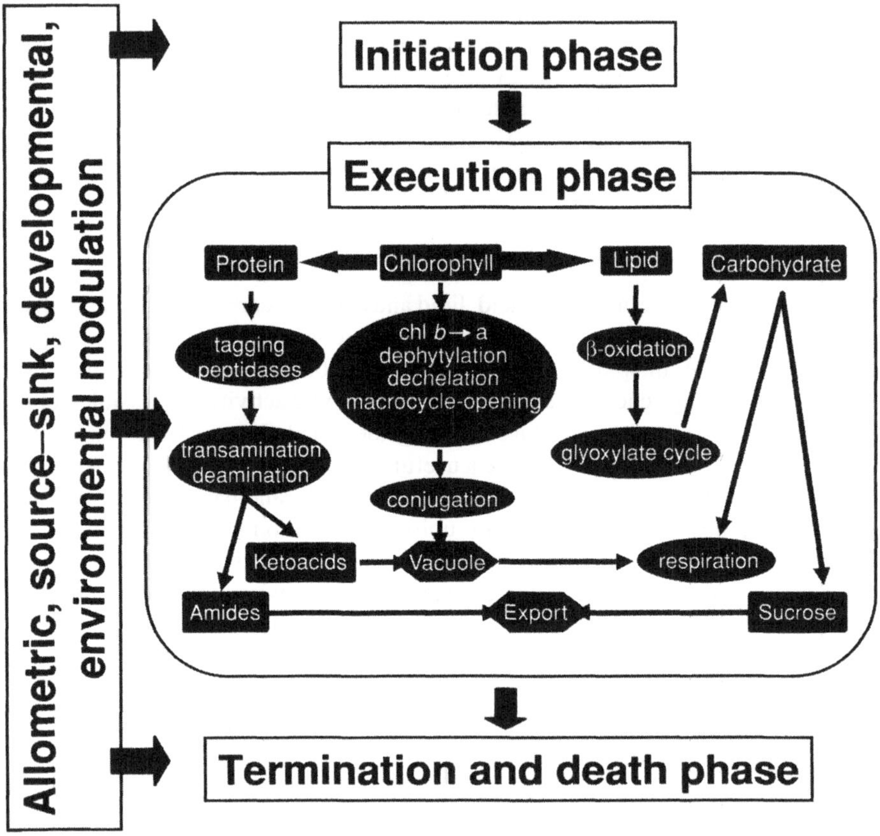

Figure 8.2 Metabolic events characteristic of the execution phase of leaf senescence and the proximal and distal influences that modify the syndrome, all of which contribute to quantitative expression of senescence-related major and poly- genes.

It follows that senescence is an allometric phenomenon, and that some senescence-related QTL may well be allometric in origin. Cheverud (1982) proposed four kinds of allometry: (1) ontogenetic allometry, describing the growth of a structural module relative to the growth of an individual organism; (2) static allometry, which refers to module-scaling relationships among individuals at a particular developmental stage; (3) plastic allometry, relating module size to different environments; and (4) evolutionary allometry, which makes comparisons across species. It is well established that the turnover of modules such as leaves shows clear ontogenetic, steady-state, acclimatory and adaptive variation (Thomas, 1992; Sachs *et al.*, 1993), and so each of Cheverud's classes of allometry would be expected to have a built-in senescence element. In general, studies of allometry focus on growth and size relationships. For example, Ma *et al.* (2002) identified QTL related to the growth of stem height and diameter in a poplar hybrid population derived from *Populus deltoides* and *Populus* × *euramericana*, and observed age-related changes in patterns of QTL expression.

Implications for the initiation and progress of senescence are apparent in some of the physiological insights emerging from these studies. QTL-associated growth curves will relate more or less directly to the timing of developmental events, such as age at first flower, age of maximum reproduction or longevity. It is easy to imagine that such growth- and size-related QTL could also emerge in a QTL study of senescence if the mapping population happens also to segregate for allometric traits.

8.3.3 QTL mapping as a tool for holistic analysis of development

The foregoing discussion illustrates particular limitations and strengths of the QTL approach. Used uncritically, it can obscure the distinction between primary, second-order and allometric controls. However, it is a powerful tool for exploring the nature of the physiological linkage between the community of developmental modules that make up an individual plant, and hence providing insights into development as a holistic process. In practical terms, it also establishes how readily a developmental correlation can be broken, which is often a major breeding objective in plant improvement.

8.4 G × E and the contribution of biotic and abiotic factors

8.4.1 Elasticity and plasticity

The discussion in Section 8.3 concerning second-order and allometric QTL is relevant to plant-environment interactions. Plant physiology and development are closely attuned to biotic and abiotic influences. Within limits, responses to environmental fluctuations are elastic and the system is self-righting, but beyond these limits plastic change happens and there is a more or less fundamental shift of state (Thomas, 1992; Sultan, 2003). Geneticists refer to the interplay between the genome and environment as the G × E ('G by E') interaction (see Wang *et al.*, 1999).

8.4.2 G × E and the now-you-see-it, now-you-don't QTL

G × E is a particular issue for QTL analysis because it is probably the rule that where a phenotype is defined by several to many QTL, some of these will be unstable and expressed in some experiments but not others. Occasionally it is possible to relate inconsistencies in the detectability of particular QTL to a particular environmental variable; in which case it can be inferred that a region of the genome significant for environmental modulation of senescence has been identified. More usually there is no clear pattern and such hide-and-seek behaviour must be put down to the statistical nature of the QTL method.

8.4.3 Implications for the design and conduct of QTL experiments

The particular significance of G × E interaction means that conclusions from QTL studies with limited or no whole-experiment replication must be drawn with caution. It is not unknown for a convincing QTL to be identified in one experiment and never to be seen again. QTL experiments need to be designed in the expectation that strong G × E interactions will occur, and data collection and reporting should include detailed records of environmental conditions to allow assessment of the robustness of loci detected. Wherever possible, scoring of traits should be carried out using an objective and internationally recognised scale of measurement, allowing information from different experiments and datasets to be drawn together and analysed meaningfully. It is therefore not unusual for QTL to be measured on different sites or over different years to determine which QTL are most robust. Moreover, different QTL are identified in different studies using different mapping families, i.e. not all QTL will be revealed in a single cross.

8.5 Case studies

8.5.1 Rice

Rice has been the subject of a number of QTL studies of leaf senescence that demonstrate not only the scientific and agricultural value, but also many of the challenges, of the approach. Ishimaru *et al.* (2001) mapped 23 physiological and agronomic characters in a backcross *japonica* × *indica* population. Senescence was determined as the difference in flag leaf chl content between 5 and 25 days after heading and mapped to four QTL, one on each of chromosomes 4, 6, 8 and 9. This study found no coincidence between the locations of these QTL and any of the measured yield components nor any coincidence with physiological or compositional characters, such as photosynthetic capacity, pre-senescence levels of chl *a* and *b* or rubisco content. The QTL on chromosome 4 mapped with one of six QTL related to intercellular CO_2 concentration. The senescence QTL on chromosome 8 was colocated with a wide region related to grain number per panicle, and the chromosome 9 QTL coincided with one of three QTL associated with shoot volume.

Abdelkhalik *et al.* (2005) also employed populations based on *indica* × *japonica* crosses to map pigmentation changes as a measure of senescence. They used a SPAD meter to determine chl content in the second youngest leaf at flowering and 25 days after flowering (DAF), and a visual score expressed as number of late-discolouring leaves per panicle at 25 DAF. SSRs were used as markers and the study was conducted on two separate mapping families, population 1, in 2001, and population 2, in 2003. Senescence measured as the loss of chl between flowering and 25 DAF mapped to a single QTL on chromosome 6 in population 2. The late-discolouring trait also mapped to chromosome 6 in this population, near to the chl loss QTL, as a broad locus comprising perhaps three peaks above the 2.5 LOD (log of the odds) threshold. None of these QTL was detected in population 1. A single QTL on chromosome 2 was determined for the late-discolouring character in population 1, but no corresponding QTL was identified in population 2. A similar lack of consistency between the two populations was apparent in the identification of QTL related to chl content at flowering (three in population 2, none in population 1).

Chlorophyll at 25 DAF was identified with a QTL on chromosome 9 in both populations.

Phenotypes in which leaves retain chl content for an extended period are collectively known as stay-green (Thomas and Smart, 1993; Thomas and Howarth, 2000). Jiang *et al.* (2004) investigated the genetic basis of stay-green in a mapping population derived from an *indica* × stay-green *japonica* cross. They took SPAD measurements of flag and second leaves on the day of heading and 30 days later. The ratio of SPAD readings at the two time points gave a measure of senescence they referred to as the relative retention of greenness. A separate estimate of senescence was made by visually scoring leaves for green area on a 1 to 5 scale. Several other yield components were measured in this study. The population, comprising 190 doubled haploid lines, was tested in replicated field trials in 2 consecutive years, and QTL were located on a genetic linkage map based on 179 SSR marker loci. Some suggestive groupings of the 46 main-effect QTL were observed. There were clusters of three or more of the six measured traits in two regions of chromosome 2 and one each on chromosomes 3, 6, 7, 8 and 10. It would be expected that the type of measurement employed would identify considerable functional and genetic commonality between the traits investigated. Pairwise examination of QTL for interaction revealed 50 epistatic associations amongst 66 loci distributed across all 12 linkage groups. Significant G × E interactions were detected for 18 main-effect QTL and 14 of the epistatic interactions, and phenotypic variations were accounted for to a large extent by epistatic effects and QTL × year interactions.

Comparing reports of rice QTL on the basis of measurements of greenness and its stability in selected leaves, some general points arise. Most of the greenness traits are poorly, or even negatively, correlated with yield components. Individually and collectively, the loci identified tend to account for a relatively small proportion of the total variation in the population. The LOD threshold needs to be set at a comparatively low level (2.5–3) for most of the loci to emerge. To read across from a QTL identified in one study to what might be a similar locus in another is not straightforward. Possible tie-ups between the maps of Jiang *et al.* (2004) and Abdelkhalik *et al.* (2005) include QTL on chromosome 2 in the region of markers RM145 and RM341, chromosome 4 near RM261, chromosome 6 in the region of RM136, and chromosome 9 between RM257 and RM215. The chromosome 9 QTL is of particular interest, since it may correspond to a locus mapped between markers C985 and RG662 on the long arm of chromosome 9 by Cha *et al.* (2002) in a study of a stay-green *japonica* generated by chemical mutagenesis. There is evidence from comparative mapping across species that an important senescence gene is located in that region of the monocot genome (see Section 8.6).

Greenness is an easily scored senescence symptom that is widely used in genetic studies, but in QTL terms it has turned out not to be a particularly useful correlate for agronomic performance. Protein breakdown is a measure of senescence that, while less convenient for screening purposes than pigmentation, is often a more direct index of resource allocation in crop species, where it has a central function in the plant's internal nitrogen economy. In a *japonica* × *indica* population, Obara *et al.* (2001) mapped QTL related to levels of cytosolic glutamine synthetase (GS1) and

NADH-glutamate synthase (NADH-GOGAT) protein in leaves. These enzymes are believed to play important roles in the recycling of N during senescence (Figure 8.2). The structural genes for NADH-GOGAT and GS1 are located on chromosomes 1 and 2, respectively. Two QTL for NADH-GOGAT content were identified on chromosome 1, near to the structural gene, but the two QTL determined for GS1 protein level on chromosome 2 were not close to the GS1 structural locus. A further three QTL related to GS1 protein content were detected on chromosome 11, and one each on chromosomes 4 and 8. NADH-GOGAT protein levels were assigned to two further QTL on chromosome 2, one on chromosome 4 and one on chromosome 7. In this study, nitrogen recycling was measured in terms of the soluble protein content of developing and senescing leaf blades. Two QTL related to protein in developing leaves were close to the NADH-GOGAT structural gene on chromosome 1. Protein content in both developing and senescing leaf blades mapped together on chromosome 2, near to one QTL each for the protein contents of GS1 and NADH-GOGAT.

In this study, pigmentation changes were determined by taking SPAD measurements from flowering to maturity and fitting a polynomial function, from which rates to half (RHD) and full (RFD) discolouration were computed. In most cases, RHD and RFD mapped together (chromosomes 2, 4, 7, 8 and 11). Interestingly, this paper does not record any QTL assignment to chromosome 9. There was also no obvious association of QTL for pigment stability with QTL for leaf protein content. A number of physiological traits were mapped and some suggestive coincidences were identified, for example between the structural gene for GS1 on chromosome 2 and a QTL region for one-spikelet weight and, also on chromosome 2, QTL related to GS1 protein content, panicle number, and panicle weight (further characterised by Obara *et al.*, 2004). In another report, Yamaya *et al.* (2002) described transgenic lines of the indica (low NADH-GOGAT) parent overexpressing NADH-GOGAT under the control of the native promoter and showed up to 80% increase in grain weight.

In spite of the familiar image of the paddy field, rice is not especially tolerant to flooding, and senescence is amongst the physiological responses to submergence. QTL studies of flooding tolerance have been reviewed by Jackson and Ram (2003) and Toojinda *et al.* (2003). Linkage mapping inheritance studies of a submergence-tolerant cultivar FR13A identified a dominant tolerance locus on chromosome 9. According to Jackson and Ram (2003), the mechanism of flooding damage consists of an ethylene-mediated supply–demand imbalance in assimilates together with strongly accelerated leaf extension and leaf senescence. FR13A and related cultivars do not display these responses. Toojinda *et al.* (2003) also identified several major QTL on chromosome 9 related to submergence responses, including plant survival, plant height, stimulation of shoot elongation, visual tolerance score and leaf senescence (measured both by visual scoring and using a SPAD meter). These QTL were robust, having been detected consistently with respect to the year when experiments were carried out and the genetic backgrounds of three separate mapping populations. To judge from the mapping interval (between markers RM41 and RG553 on the short arm of chromosome 9) this senescence QTL is located well away from the stay-green locus described by Cha *et al.* (2002) and probably represents completely different gene(s) with functions in the environmental sensitivity of senescence.

8.5.2 Sorghum and millet

The genetics of leaf senescence has been extensively studied in sorghum primarily because delayed leaf senescence or stay-green is considered a valuable agronomic trait. Water limitation during the grain development stage can cause premature leaf death and poor yield of seed and stover. Stay-green genotypes retain more green leaf area than do genotypes not possessing this trait and they also continue to fill grain normally under drought conditions (Rosenow and Clark, 1981; Borrell *et al.*, 2000). Other agronomic advantages of stay-green include higher stem carbohydrate and grain weight (Duncan *et al.*, 1981; McBee *et al.* 1983), reduced lodging (Henzell *et al.*, 1984) and good association with resistance to charcoal stem rot (Rosenow, 1984).

Conventional breeding for stay-green has been based primarily on two sources of this trait: B35 a BC_1 derivative of the Ethiopian durra sorghum IS12555; and KS19 which is derived from a Nigerian landrace (Mahalakshmi and Bidinger, 2002). Classical inheritance studies of the stay-green trait in B35 indicate that it is controlled by a major gene exhibiting varied levels of dominant gene action (and epistatic interactions) depending on the environment in which evaluations are made (Tenkouano *et al.*, 1993; Walulu *et al.*, 1994). The trait appears to be recessive in the R9188 source of stay-green, however, indicating that different genetic sources of the trait may be under different genetic control (Rosenow, 1984). Various levels of dominant gene action for stay-green had also been reported from the studies of Van Oosterom *et al.* (1996), but subsequent QTL analyses in a range of populations have indicated at least four regions of the genome associated with the trait (Table 8.1). Four of these studies used B35 as the source of stay-green (Tuinstra *et al.*, 1997; Crasta *et al.*, 1999; Subudhi *et al.*, 2000; Xu *et al.*, 2000) and one used the stay-green line QL41, which was derived partly from B35 (Tao *et al.*, 2000). The only studies using unrelated genetic sources of stay-green are those of Kebede *et al.* (2001), who used line SC56, and Haussmann *et al.* (2002), who used E36-1.

Subudhi *et al.* (2000) used markers from different sources, including cereal anchor probes, to align the various available linkage maps making it easier to compare QTL identified from different populations. More recently, Kim *et al.* (2005) have used cytological methods to align and orientate these markers relative to the ten chromosome pairs. Comparisons of all the QTL studies that have used B35 as a parent indicate that four QTL are consistent in a range of genetic backgrounds and environments (Table 8.1). These four QTL account for up to 54% of the phenotypic variance and are termed *Stg1, Stg2, Stg3* and *Stg4* (Xu *et al.*, 2000). *Stg1* and *Stg2* are both located on chromosome 3, *Stg3* is on chromosome 2 and *Stg4* is on chromosome 5. *Stg2* was found to be the most important followed by *Stg1, Stg3* and *Stg4* in order of decreasing importance (Xu *et al.*, 2000). These QTL have been confirmed in two additional post-flowering drought environments by Subudhi *et al.* (2000). Similarly, in a separate population also using B35 as the stay-green parent,

Table 8.1 Summary of recent staygreen mapping studies indicating chromosomal locations of QTL associated with staygreen identified in various studies and respective donors of the staygreen alleles

					Chromosome[a]												
					?[b]	1	2		3		4	5	6	7	8	9	10
Study	Population	Size	No. of drought environments	Trait	*stgI*	*stgF*	*Stg3*	*stgB*	*Stg1*	*Stg2*		*Stg4*					
Tuinstra *et al.* (1997)	B35 × TX7078	98	2	SG	B35	B35	B35	B35	B35	N.D.							
				Yield	B35	B35											
Crasta *et al.* (1999)	B35 × TX430	96	4	SG		B35	N.D.	B35		B35		B35				B35	
Xu *et al.* (2000)	B35 × TX7000	98	5	SG			B35		B35	B35		B35					
				SPAD			B35		B35	B35							
Subudhi *et al.* (2000)	B35 × TX7000	98	2	SG			B35		B35	B35		B35		B35			
Tao *et al.* (2000)	QL41 × QL39	152	5	SG		QL39	QL41									QL41	QL41
Kebede *et al.* (2001)	SC56 × TX7000	125	5	SG		SC56				SC56	SC56	SC56	Tx7000	SC56			Tx7000
Haussmann *et al.* (2002)	E36-1 × IS9830	226	1	%GLA		E36-1			IS9380	E36-1				E36-1	E36-1 IS9830		E36-1
	E36-1 × N13	226	1	%GLA		E36-1	N13			E36-1	N13			E36-1 N13			E36-1

SG = visual measure of staygreen at physiological maturity; SPAD = leaf chlorophyll at physiological maturity; %GLA = percent green leaf area at physiological maturity compared to that at anthesis; N.D. = region not mapped.

[a]Chromosome number as in Kim *et al.* (2005). Staygreen QTL designations from Xu *et al.* (2000) and Tuinstra *et al.* (1997).

[b] ? = Chromosome number not known.

Crasta *et al.* (1999) identified seven QTL associated with stay-green, but not all of them were stable across environments. Three of these mapped to the same genomic regions as *Stg1*, *Stg2* and *Stg4*, with minor QTL found elsewhere (Table 8.1). *Stg1* and *Stg3* were also identified in a third population based on B35 (Tuinstra *et al.*, 1997), but *Stg2* was not identified, partly due to incomplete coverage in the corresponding region of the genome. Tao *et al.* (2000) aligned their map with others and reported that one of their stay-green QTL corresponds to *Stg3*, identified by Xu *et al.* (2000), and another to a stay-green QTL on chromosome 9, also identified by Crasta *et al.* (1999). The QTL identified by Tao *et al.* (2000) in their Australian studies were partially confirmed by field trials from ICRISAT, India (Borrell *et al.*, 2001).

Tuinstra *et al.* (1997) identified two QTL with major effects on yield and stay-green under post-flowering drought. These QTL were also associated with yield under fully irrigated conditions, and so may have pleiotropic effects on yield under non-drought conditions. Further studies have shown a positive association between xylem pressure potential, on the one hand, and grain yield and stay-green on the other, indicating that the QTL for xylem pressure potential influences differences in drought tolerance by maintaining plant water status (Tuinstra *et al.*, 1998).

All the above studies employed the stay-green line B35 (or QL41, which includes B35 as a parent) as one of the parents in the mapping populations used. Two studies have used other sources of the stay-green trait, unrelated to B35. Kebede *et al.* (2001) used a cross between the stay-green SC56 (derived from a Sudanese *caudatum-nigricans* sorghum) and Tx7000, and evaluated the population in eight environments. Three major QTL were found consistently and, in all three cases, the alleles for stay-green were derived from SC56. Haussmann *et al.* (2002) identified QTL for stay-green, using two populations that included the stay-green sorghum E36-1 as a parent. Although QTL associated with stay-green were identified in which both parents contributed positive alleles, three QTL in particular were consistently found in both populations and for these the stay-green alleles originated from E36-1. Interestingly, major QTL for stay-green were contributed even by the non-stay-green parents IS 9830 and N13, suggesting that hidden genetic variability for the trait exists, which could be exploited by breeders. The congruence of some of the major QTL from the stay-green source B35 and those detected by Kebede *et al.* (2001) from SC56 is apparent on chromosomes 1 (*stgF)*, 2 (*stg3*), 3 (*stg2*) and 5 (*stg4*). *Stg2*, which was found to be the most important QTL in the B35 populations, was identified not only in SC56 but also in E36-1. Kebede *et al.* (2001) found that *stg2* showed correspondence to a maize stay-green QTL on chromosome 8 (Beavis *et al.*, 1994). Furthermore, QTL associated with drought-related traits have been identified in the syntenic region of rice chromosome 5.

Minor QTL associated with stay-green in all the seven published studies show little genomic congruency and tend to be environment specific even within the same population. Interactions among genetic, physiological and environmental factors governing expression of the trait are clearly complex (Borrell *et al.*, 2001). Nevertheless, ICRISAT has recently initiated marker-assisted backcrossing to transfer regions of the genome governing the stay-green trait from the donor parents B35 and E36-1 into a range of elite tropically adapted sorghum cultivars currently grown and preferred by resource-poor farmers in the tropics of Asia and/or Africa (Hash *et al.*, 2003). Preliminary results have indicated that using marker-assisted selection to transfer regions of the genome associated with stay-green QTL from B35 into the background of a senescent cultivar also results in the successful transfer of the functional stay-green trait. Single-QTL substitution lines for *stg1, stg2, stg3* and *stg4* in a common senescent background are also being developed, which will enable further understanding of the mechanism of stay-green in sorghum under drought conditions.

Pearl millet, a C4 cereal related to sorghum, which is often grown in drought-prone regions, has also been the subject of genetic studies into the relationship between senescence and responses to water limitations. QTL associated with drought tolerance have been identified in two independent mapping populations (Yadav *et al.*, 2002, 2004). This led to the identification, in both populations, of a genomic region on linkage group 2 (LG 2) associated with superior maintenance of grain yield performance under drought conditions. QTL on LG 1, LG 5 and LG 6 were also identified, which influenced grain yield, osmotic adjustment and leaf senescence during drought stress, and which co-mapped with QTL for maintenance of grain yield under drought conditions (R.S. Yadav *et al.*, unpublished). The allele from the drought tolerant parent (derived from *Iniadi* landrace material) was found to be associated with increased drought tolerance in both populations. This is now being studied in more detail following marker-assisted selection to produce a set of near-isogenic lines with and without putative drought-tolerance QTL.

8.5.3 Maize

Maize is a major agronomic crop with food, feed, energy and industrial uses. It is monoecious with a terminal male flower (tassel) and subsidiary female flower that subsequently emerges from one or more leaf axils and will develop into the cob. Therefore, unlike other cereals such as wheat and barley (see Section 8.5.4), the smaller terminal flag leaf is of less significance in remobilisation of nitrogen and other resources to the grain than leaves lower down the stem more closely associated with the cob. Rather than relying on chronological age of the plant or individual leaf, senescence in maize is often scored as days or weeks after pollen shed as this developmental stage frequently coincides with a significant increase in leaf senescence. Maize offers considerable classical genetic resources, including inbred lines, mapping families and detailed genetic maps, and large genetic variation between lines, including that variation which arises as a consequence of native transposon activity. In many ways therefore it is an ideal species for a QTL-based study of senescence. However, in practice, comparatively few senescence-related QTL studies have been directly made in maize. In part, this probably reflects the complex genome structure of maize, an ancient allotetraploid (Gaut and Doebly, 1997) with many duplicated genes now deleted. So although maize has twice as many chromosomes, it no longer has twice the number of functional genes. Genome organisation synteny with sorghum and millet, however, should mean that QTL in these diploid species probably exist on one or both of the two homologous chromosomes of maize. Therefore, senescence-associated QTL from these related species are often discussed in relation to maize (Crasta *et al.*, 1999; Subudhi *et al.*, 2000; Xu *et al.*, 2000; Sanchez *et al.*, 2002 – see Section 8.5.2). However, in maize, eight QTL for stay-green have also been described in progeny derived from a cross of B73 and Mo17 by Beavis *et al.* (1994). The heritability was 68–73% in the evaluated crosses and the QTL were localised on chromosomes 1, 2, 6, 8 and 9. Stay-green was defined as a visual rating of health and vigour of plants at the time of harvest on a scale of 1–9. B73 and Mo17 are not very different in their pattern of senescence and therefore it would be expected that more QTL could be identified by exploiting a specific cross to study senescence. In an earlier study, a delayed senescence line Lo876o2 was compared to B73 (Gentinetta *et al.*, 1986). The difference between lines was considerable so that at 12 weeks after pollen shed Lo876o2 had a dry matter content of only 20% compared to the 80% of B73. Inheritance studies indicated the presence of two divergent alleles at a single locus, segregating in the Lo876o2 × B73 cross, with delayed senescence dominant over senescence. However, no map position for the delayed senescence trait is available in this study.

In maize, QTL for a number of indices of nitrogen-use efficiency have also been studied and these include grain yield, activities of the cytosolic and plastidic isoforms of glutamine synthetase (GS1 and GS2, respectively) and nitrate reductase, and leaf nitrate content (Hirel *et al.*, 2001; Masclaux *et al.*, 2001). QTL for GS activity, nitrate reductase activity and nitrate content all coincided at two locations on chromosome 5, and one of these was also co-located with a gene encoding GS1, *gln4*. In another study, more QTL for traits of vegetative development (N uptake, grain yield and its components) were detected in maize grown under high-nitrogen-input conditions than that grown in low nitrogen (Gallais and Hirel, 2004). Contrastingly, more QTL for grain protein content and nitrogen-utilisation efficiency were identified in maize under low-input conditions than that under high-nitrogen input. Collectively, ten QTL for nitrogen remobilisation were identified and three coincided with QTL for kernel weight or grain yield. There were also three QTL for remobilisation which coincided with QTL for GS activity and map positions for genes encoding GS1, *gln 1*, *2* and *3*. These were on chromosomes 1, 4 and 10. A GS2 locus (gln5) on chromosome 10 coincided with QTL for senescence asparagine synthase (ASI) and nitrogen nutrition index.

As with studies in other species such as sorghum and millet, delayed foliar senescence (stay-green) has also been associated with enhanced yield or drought resistance. For example, in addition to the many indices of drought which are often employed (including flowering time, stomatal conductance, tissue abscisic acid contents, leaf water relations parameters and fluorescence characteristics, root pulling force and nodal root number), chl content per unit area or fluorescence estimates of photosynthetic efficiency have also been measured (Lebreton *et al.*, 1995). Lebreton crossed Polj17 × F-2, drought tolerant and sensitive lines, respectively, and identified QTL on chromosome 7 for F_m (maximal chlorophyll fluorescence) and on chromosomes 2 and 6 for chl content. In another study, a candidate gene for a drought tolerance QTL on chromosome 4 in maize, *Asr1*, was transgenically manipulated (Jeanneau *et al.*, 2002). *Asr1* is a putative transcription factor, and overexpressing lines exhibited an increase in foliar senescence under drought conditions. Despite such associations between drought tolerance and delayed senescence, most studies on drought tolerance measure exclusively components of, or predicted determinants of, yield and not senescence directly.

In other studies with maize, genes have been identified as senescence enhanced or associated with senescence processes, and these have been mapped. For example, senescence enhanced genes *See1* and *See2* have been mapped to homologous pairs of chromosomes 2 and 7, and 3 and 8, respectively (Griffiths *et al.*, 1997; Donnison *et al.*, unpublished). Both these genes are predicted to encode processing-type cysteine proteases: *See1* is an aleurain and *See2* a legumain. A number of other genes have also been associated with senescence in maize (Smart *et al.*, 1995; Martin *et al.*, 2005) and in the future it will be important to relate map positions for more of these genes with the integrated maize genetic maps, to identify candidate genes for QTL as has been achieved for nitrogen mobilisation.

8.5.4 Wheat and barley

The economic importance of wheat and barley as the main staple cereal grains of temperate agriculture has been the major driver for the application of molecular genetic approaches to plant breeding and improvement within these species. The prerequisite for these approaches has been the development of comprehensive and reliable genetic maps and accompanying marker systems, which have allowed for both detailed genetic analyses within the Triticeae and for wider cross-species comparisons through the use of common or 'anchor' markers (see www.gramene.org). One of the major applications of molecular genetics within the cereals has been in the development of marker-assisted breeding protocols. These are usually derived by identifying QTL for important traits within experimental breeding populations and the subsequent utilisation of the associated markers for marker-aided introgression of the desired trait into elite lines. Organ senescence per se is a trait that has received only little direct attention in terms of QTL analyses within wheat and barley. However, particularly because of the effects of leaf senescence on the partitioning of resources between vegetative and reproductive tissues within the plant, there has been considerable interest in identifying and inferring the influence of leaf senescence on QTL for grain protein content.

Grain protein content in wheat and barley is a key measure of quality and a major determinant of the economic potential of a cultivar. High grain protein content in bread and durum wheat determines nutritional quality and performance in bread and pasta making (Chee *et al.*, 2001; Khan *et al.*, 2000). In barley, high grain protein content is also considered to be a desirable trait when the grain is used as a component of animal feed. On the other hand, too high or too low protein content can reduce the malting and beer-making quality of barley developed for use in the brewing industry (See *et al.*, 2002). Masclaux *et al.* (2001) identified three major factors determining the nitrogen content of wheat grains: the nitrate availability in the soil prior to flowering, the continuous uptake of nitrogen during grain filling and the remobilisation of nitrogen from mature and senescent leaves. With respect to the latter, it has been estimated that the proportion of nitrogen accumulated by the spike from leaf nitrogen remobilisation can vary from 51 to 91% depending on the cultivar being assessed (Van Sanford and Mackown, 1987; Papakosta and Gagianas, 1991). Underlying genetic differences in the contribution of nitrogen remobilisation to grain protein content can be exploited for crop improvement and QTL detection.

A number of studies have sought to identify QTL for grain protein content in both durum and bread wheat, and factors influencing grain protein content have been identified on all chromosomes (Blanco *et al.*, 2002). For tetraploid durum wheat (*Triticum turgidum* var. *durum*), a potential source of high grain protein genes was identified in the related wild emmer (*Triticum turgidum* var. *dicoccoides*) and a number of studies showed, by the use of appropriate chromosome substitution lines and mapping populations, that there seemed to be a major QTL for grain protein content expressed in the durum wheat background, which was derived from chromosome 6B of the wild emmer relative (Steiger *et al.*, 1996; Joppa *et al.*, 1997; Chee *et al.*, 2001). Fine mapping located this QTL to a 2.6-cM region on the short arm of chromosome 6B and identified molecular markers that could be used in marker-aided introgression approaches (Olmos *et al.*, 2003; Distelfeld *et al.*, 2004). Similar analyses in hexaploid bread wheat (Mesfin *et al.*, 1999; Khan *et al.*, 2000) showed that the same segment on chromosome 6B derived from emmer wheat could also positively influence grain protein content. Uauy *et al.* (2006) reported that the major QTL for grain protein content on wheat chromosome 6B is completely linked to flag leaf chlorophyll degradation, change in peduncle colour and spike water content.

For barley, as with the *Triticum* spp., QTL that can influence grain protein content have also been identified on all chromsomes (Oziel *et al.*, 1996; Bezant *et al.*, 1997; Powell *et al.*, 1997; Mickelson *et al.*, 2003) including in a region that is likely to be homologous, in terms of conserved synteny, to that conferring high grain protein content on chromosome 6B of wheat (See *et al.*, 2002). Mickelson *et al.* (2003), using the population described by See *et al.* (2002), sought to make a direct association of nitrogen uptake, storage and remobilisation from mature and senescent leaves with various agronomic traits, including grain protein content. They identified QTL on five of the chromosomes that influenced nitrogen concentration and/or remobilisation. Two of these, on chromosomes 3H and 6H, were especially interesting as they contained overlapping QTL for total leaf nitrogen at various developmental stages. Alleles at these loci were associated with inefficient nitrogen remobilisation and depressed grain yield, related to the retention of higher levels of total or soluble organic nitrogen in flag leaves during grain filling. It was concluded that genes that directly control or regulate nitrogen remobilisation could be present on these chromosomal regions. The main QTL for overlapping nitrogen metabolism characteristics on 6H did not, however, coincide with the major QTL for grain protein concentration referred to earlier (although in one of the years of the experiment QTL for leaf nitrates and soluble organic nitrogen at mid-grain fill could be associated with this region); and, in fact, there was no overall correlation between grain protein concentration and nitrogen remobilisation. Nevertheless, strong correlations between nitrogen remobilisation, total yield and protein yield led Mickelson *et al.* (2003) to speculate that the underlying cause of the major QTL for grain protein concentration on 6H might be the presence of gene(s) controlling a different aspect of metabolite remobilisation, such as carbon or lipid. Yang *et al.* (2004) continued the study of this barley family by identifiying QTL for leaf amino-, carboxy- and endopeptidase activities at mid-grain fill and maturity and relating these to the previous work on nitrogen remobilisation and grain protein concentration and yield. Overall, their results indicated no positive association of amino- and endopeptidase activity with nitrogen remobilisation and for the aminopeptidase activity, in some cases, a negative relationship. In contrast, at a number of loci, QTL for carboxypeptidase activity were associated with measures of nitrogen remobilisation and yield in such a way as to suggest a positive role for various carboxypeptidase isoenzymes in nitrogen remobilisation and translocation from senescing leaf tissue to the developing grain.

The potential of measuring and manipulating leaf senescence to predict and improve the performance of wheat, particularly under water-stressed conditions, has also been recognised. Saulescu *et al.* (2001) investigated the relation between dark-induced senescence (in terms of SPAD measurements) in wheat seedlings and field-grown plants in a range of bread wheat cultivars and suggested that seedling response could be correlated with mature plant response and may be a good predictor of leaf senescence properties in mature plants. Similarly, studies have identified variation between cultivars in chl content and stability (Banowetz, 1997), late-onset senescence under water stress (Benbella and Paulsen, 1998; Hafsi *et al.*, 2000) and flag-leaf angle, area and duration (Simon, 1999). Verma *et al.* (2004) used this general approach to identify QTL for flag leaf senescence (measured in terms of percent green flag-leaf area remaining (%GFLA) at 14 and 35 days after anthesis) in winter wheat under optimal and drought-stressed conditions. They identified QTL for %GFLA after both time periods under drought-stressed conditions on chromosome 2D; similar, though less marked, effects were identified on chromosome 2B under optimal conditions. In both cases, the QTL for %GFLA coincided with QTL for yield and increases in %GFLA were associated with increases in yield. Verma *et al.* (2004) concluded that the timing of flag-leaf senescence, particularly under drought conditions, could affect yield and that there may be benefit in developing molecular markers to select for this QTL.

In summary, grain yield and quality in wheat and barley are, in part, determined by the timing and efficiency of remobilisation and translocation of assimilates from the leaf to the grain. Different studies reveal underlying genetic variation, which can influence the contributory physiological processes. QTL analyses have identified several regions of the genomes that have significance in this respect and have been targets for molecular marker development. As our understanding of the physiology of senescence and resource reallocation improves, it will be possible to refine the QTL analyses so as to identify, with more certainty, regions of the Triticeae genomes which determine these fundamental processes.

8.5.5 Other species

Amongst other plants in which senescence has been the subject of QTL analysis are field crops such as sunflower and various legumes, horticultural species including tomato, and non-food plants such as turf grasses. Here we selectively consider some examples. Stay-green is a valuable trait in sunflower, conferring similar advantages in this species to those identified for sorghum, notably improved plant health, resistance to drought and pests and better standability. Cukadar-Olmedo *et al.* (1997) carried out inheritance studies of the retention of green stems at maturity on parental, F1, F2 and backcross generations of two maintainer and two restorer inbred lines of sunflower. Stem greenness was estimated by a photographic image-processing method at 10- to 15-day intervals in the field between flowering and physiological maturity. Dominant gene, additive and additive $\times$ dominance epistatic effects were observed. To date stay-green has not been assigned to the sunflower genetic map, but in due course it is likely that senescence-related loci will be shown to be relevant to the higher-priority agronomic traits of disease resistance, flowering time, seed development and maturity (Bert *et al.*, 2003). Similarly, classical inheritance studies have identified a range of genes related to green tissue senescence in soybeans (Guiamet and Gianibelli, 1996) and other legumes (Thomas and Smart, 1993) but there is little information on the relations between these genes, quantitative loci and other agronomic traits on the genetic map (Zhang *et al.*, 2004). Progress in sequencing the genome of the model legume *Medicago truncatula* will open up new possibilities for analysis of senescence in legumes through comparative genomics (see Section 8.6.1).

8.6 Exploitation of QTL mapping for senescence traits

8.6.1 Model species, comparative mapping and the role of bioinformatics

In previous sections, allusion has already been made to the syntenic relationships between the genome of rice and those of other cereals and grasses. These relationships constitute a huge advantage for those working on QTL, whether for senescence or for other traits, in grass and cereal species. Rice is both a major crop and – by virtue of its small genome and diploid nature – the model for genome research in the grasses. For many of the other crops discussed here, especially those with very large and tetraploid or hexaploid genomes, it is unlikely that a complete genome sequence will be available in the near future, whereas draft sequences for both the *japonica* and the *indica* subspecies of rice were first published in 2002 (Goff *et al.*, 2002; Yu *et al.*, 2002). In addition to DNA sequencing, rice has been the subject of intensive genetic and trait mapping over the past decade. This means that where a QTL in a crop of interest is currently defined only by loosely linked genetic markers, if the same markers have also been used in mapping the corresponding region of the rice genome, the latter is a rich source of additional marker candidates that can be tested by, for example sorghum or maize specialists to dissect their QTL more precisely. The information available concerning the genes in the corresponding rice region, and their possible functions, can give additional information about the genetic basis of the QTL. Contemporary bioinformatics provides many resources and tools to facilitate the exploitation of model-species data by researchers and breeders involved with other species. In the case of rice, Gramene (www.gramene.org/; Ware *et al.*, 2002) is a major project whose aim is to bring together all available map, trait, sequence and related data for grasses and cereals and relate it to the rice genome. In September 2005, for example Gramene contained information on 178 QTL related to foliar senescence, of which 148 related to the work on an *indica* × *japonica* cross carried out by Jiang *et al.* (2004). The Gramene website allows researchers to visualise the alignment between a given rice linkage group and the syntenic regions in wheat, sorghum, maize, oats and other species, based on common genetic markers. It allows potentially comparable QTL to be displayed, the rice physical map of the region of interest to be selected and genes (or putative genes) in that region to be identified (Figure 8.3). Where reliable annotation exists, it may lead to a better understanding of how a given chromosome region contributes to a trait such as senescence. The biggest limitations to effective exploitation of comparative genomics in this way are, firstly, the relative paucity and variable quality of such annotation as regards gene function; and secondly, as discussed in Section 8.5.2, the fact that historically much genetic mapping has been carried out using marker systems such as randomly amplified polymorphic DNA, which do not transfer well between species or even, in some cases, between mapping families within the same species. The proper use of anchor markers, particularly those based on RFLPs, largely overcomes this problem, and many recent mapping projects have included such markers to facilitate comparative genomics.

The value of a model species combined with bioinformatics resources is not, of course, confined to cereals and grasses. The three other major plant models cover, between them, a large proportion of terrestrial plant groups and many important crops. *Arabidopsis thaliana*, though not itself a crop, is closely related to the brassicas, including oilseed rape (canola) as well as leaf and root vegetables, and the brassica community (see, for example www.brassica.info/) makes extensive use of the US-based TAIR (The Arabidopsis Information Resource, www.arabidopsis.org/) and the UK's Arabidopsis Ensembl (http://atensembl.arabidopsis.info/). *Arabidopsis* was the first higher plant for which a complete genome sequence was available, and resources such as insertional gene knockouts and gene expression profiles are correspondingly well developed for the species. *Arabidopsis*, as a non-crop species, has been less subjected to trait mapping studies as compared to many other plants, but there is an increasing interest in the potential of natural genetic variation within the species for identification of new loci and gene complexes controlling development and adaptation (Koornneef *et al.*, 2004). Two papers have described QTL studies of senescence-related traits in *Arabidopsis* ecotypes and recombinant inbred lines (Luquez *et al.*, 2006; Diaz *et al.* 2006). *Medicago truncatula* is not only a member of the same genus as alfalfa, but also a good model for other important food, forage and silage legumes, while poplar, by virtue of its short genome and – for a tree species – unusually short generation time, is the tree model of choice. The first draft of the poplar genome sequence was released in 2004, and *M. truncatula* is the subject of an international and rapidly progressing sequencing programme (www.medicago.org/genome/). Thus, in principle, those working on leaf senescence have access to a wide range of relevant information resources even when their own species of interest have – for logistical or financial reasons – not been the focus of much genome-level research.

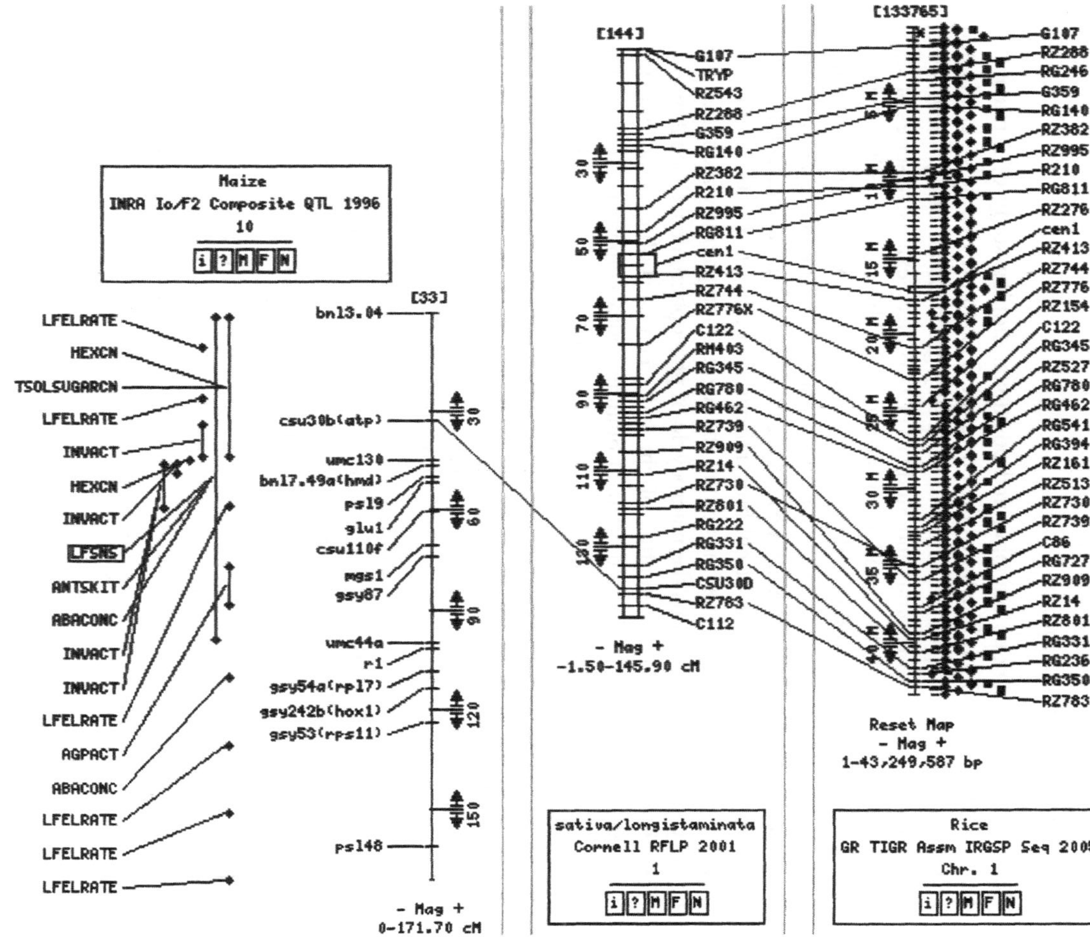

Figure 8.3 A comparative map (Cmap) view from Gramene, showing a leaf senescence QTL on maize chromosome 4, the corresponding region of rice chromosome 1 in an RFLP-based genetic map, and the rice sequence assembly for chromosome 1.

8.6.2 *Introgression landing*

Here we discuss a unique mapping and positional cloning study of a mutant of *Festuca pratensis*, which illustrates the power of QTL to provide new genetic and metabolic understanding of leaf senescence. A stay-green mutant of *F. pratensis* was originally described by Thomas and Stoddart (1975). In an extensive series of studies up to the present day, the detailed physiological and biochemical phenotype has been established. Retention of chl is not associated with extended photosynthetic capacity (Hauck *et al.*, 1997). During senescence of the mutant there is normal degradation of Calvin cycle enzymes such as rubisco (Roca *et al.*, 2004) and of other enzymes such as those of chl biosynthesis (Thomas *et al.*, 2002). Measurement of intermediates and enzymes of the chl degradation pathway identified a deficiency at the step that opens the macrocycle of phaeophorbide a (Vicentini *et al.*, 1995; Roca *et al.*, 2004). Retention of chl and green catabolites is accompanied by extended stability of thylakoid pigment-binding photosynthetic proteins (Davies *et al.*, 1990). Conventional inheritance studies establish that the mutation concerns a single recessive Mendelian locus (Thomas, 1987). The stay-green gene has been widely introgressed from *F. pratensis* into a number of *Lolium* species. Interspecific and intergeneric introgression within the *Lolium–Festuca* complex represents a powerful tool for genetic analysis in large-genome grasses. The significance of the QTL approach in this context lies in the high degree of synteny between the *L. perenne* and *F. pratensis* genetic maps on the one hand and the rice map on the other (Jones *et al.*, 2002; Alm *et al.*, 2003, in press). By reading across from the location of an introgressed locus in *Lolium–Festuca* to the equivalent position on the rice map, one can move rapidly to identify candidate genes in the corresponding region of the sequenced rice genome.

The *Festuca* stay-green locus maps to chromosome 5 (Thomas *et al.*, 1997; Moore *et al.*, 2005). Work in the International Lolium Genome Initiative (ILGI) and within a European collaboration on the *Festuca* map has established that *Lolium/Festuca* chromosome 5 is highly syntenic with rice chromosome 9 (Jones *et al.*, 2002). As described in Section 8.5.1, three independent studies have identified QTL for pigment stability in senescence in a region of rice 9. On the basis of their original paper analysing a rice stay-green (Cha *et al.*, 2002), Paek and colleagues have used sequence data from this region of the rice genome to identify a candidate sequence (accession number AY850134 in GenBank, DDBJ (DNA Data Bank of Japan) and EMBL (European Molecular Biology Laboratory's sequence database)). Fine mapping of the *Festuca* locus has narrowed it down to a limited number of sequences, which include the rice stay-green candidate (Figure 8.4). This approach to positional cloning is referred to as 'introgression landing' and represents a generally applicable method for efficiently relating major genes or QTL to specific sequences in large-genome monocots. Through a combination of sequencing, functional studies and bioinformatics, these insights into the molecular genetics of pigment metabolism are establishing for the first time the comprehensive description of a key process in plant senescence that links genotype to cellular, physiological and agroecological phenotype.

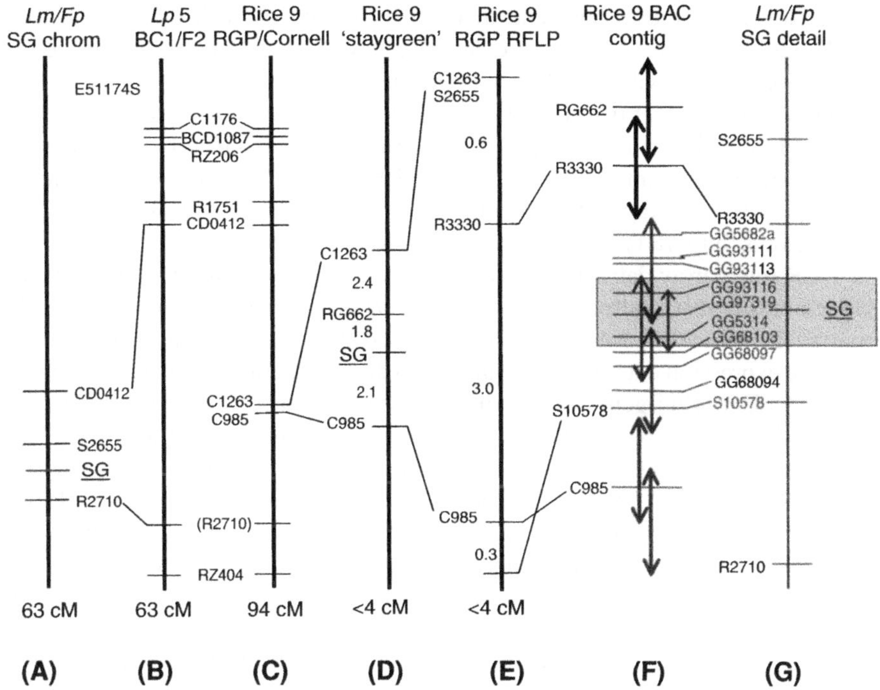

Figure 8.4 Positional cloning of a stay-green (SG) locus from *Festuca pratensis* exploiting comparative mapping with rice. (A) SG from *F. pratensis* was introgressed into a *Lolium perenne* background and markers identified. (B) The *L. perenne* reference map assigns SG to a region of chromosome 5. (C) *Lp5* is known to be syntenic with rice chromosome 9. (D) A stay-green locus has been identified in the corresponding region of rice 9. (E) This is a region of the rice 9 RFLP map that has been physically mapped with BAC contigs (see F). (G) This has allowed new markers to be associated with the SG region of *Fp/Lp5* and has identified a small number of non-recombinants (boxed) that relate to a single rice BAC.

8.6.3 Integration with omics and other technologies

Further opportunities for identifying the genes underlying QTL are opened up by the technologies developed in recent years for global profiling of gene expression: transcriptomics and proteomics. These approaches are discussed at greater length in other chapters in this volume, particularly Chapter 9. There are still relatively few published studies of the transcriptome during leaf senescence, and they mostly focus on *Arabidopsis* (Guo *et al.*, 2004; Buchanan-Wollaston *et al.*, 2005) though Andersson *et al.* (2004) investigated the transcriptional timetable of autumn senescence in poplar. Yet even the data available to date have proved important in the dissection of QTL, as exemplified by the stay-green locus described above: supporting evidence that the cloned rice gene was responsible for the character – despite its lack of homology with genes of known function in any species – came from the *Arabidopsis* transcriptome data available from public databases, which showed the gene to be strongly up-regulated during leaf and petal senescence in comparison with other tissues and treatments. In contrast with DNA sequences, which must be deposited in a public database (EMBL, GenBank or DDBJ) before most major journals will accept the corresponding papers, it is not – at the time of writing – a requirement that transcriptome data be similarly made available. However, the infrastructure is available, in the form of NCBI's Gene Expression Omnibus (www.ncbi.nlm.nih.gov/geo/) and the EBI's ArrayExpress (www.ebi.ac.uk/arrayexpress/) for array data generally, and species-specific databases such as NASCArrays (http://affymetrix.arabidopsis.info/narrays/experimentbrowse.pl) and the TAIR microarray dataset (www.arabidopsis.org/) for *Arabidopsis*, the Tomato Expression Database (http://ted.bti.cornell.edu/) and Soybean Genomics and Microarray Database (http://psi081.ba.ars.usda.gov/SGMD/default.htm), and as more and more expression data become available to the community as a whole, the probability of identifying genes underlying senescence-related QTL will continue to increase. Proteomics is still at an early stage as a practical tool for identifying the components of senescence-related QTL, but its potential is demonstrated by the work of Schiltz *et al.* (2004), who analysed the proteins abundant in mature leaves and stems of pea during the transition to nitrogen remobilisation. In addition to proteins such as rubisco and components of its remobilisation pathway, which might be expected to show changes during this period, the work identified two novel 14-3-3-like proteins, with a possible regulatory role, which showed differential expression during remobilisation. Such proteins could well be the products of genes underlying QTL. The proteomics approach is particularly appropriate for identifying genes of interest whose expression is regulated post-transcriptionally and for which the transcriptome approach may therefore be unfruitful.

8.6.4 QTL as breeding tools

Compared to their progenitors, modern crops are generally greener, and stay greener for longer in response to inputs or when subject to environmental stresses. This tells us that leaf senescence, more often incidentally than in a targeted manner, is one of the plant processes most responsive to the interventions of crop husbandry and breeding. Although the individual physiological components of the senescence syndrome are usually expressed in a well-coordinated manner, genetic and environmental interventions are able readily to prise them apart with beneficial and negative practical effects (Thomas and Howarth, 2000). QTL analysis allows this flexibility in the regulatory network of senescence to be exploited with precision, by isolating constituent processes, quantifying their contributions to agronomic performance and contributing to the design of feasible ideotypes. Introgressing single-QTL alleles into an otherwise inbred background to reduce background complexity is sometimes referred to as 'Mendelizing' the QTL (Breese and Mather, 1957) and may be achieved by new experimental and analytical methods such as sensitised analysis of polygenes (Matin and Nadeau, 2001). Coupled with efficient methods of combining a series of superior genes or loci from different parents into the same genetic background (pyramiding – Servin *et al.*, 2004), QTL dissection of senescence is a powerful crop improvement tool. This approach is already being employed successfully in crops such as sorghum (Crasta *et al.*, 1999; see Section 8.5.2) and is likely to become increasingly important as a way of channelling the advances in model and crop plant genomics into practical breeding programmes.

8.7 QTL, senescence, ageing and death

Gerontology took off as a leading area of research effort in the years after World War II, followed a couple of decades later by the rise of programmed cell death as a unifying concept in pathology, development and oncology. Plant senescence must relate somehow to these central preoccupations of the biomedical sciences, but quite how is not very clear. Attempts have been made to define the relationship between plant senescence on the one hand and ageing (Thomas, 2002, 2003) or programmed cell death (Thomas *et al.*, 2003) on the other (see Chapter 1). This may simply be a not very productive distraction concerned with semantic niceties (van Doorn and Woltering, 2004). Nevertheless, if we want to understand the mechanisms of senescence and how they are regulated, we need clarity about what these different processes entail and how much they have in common genetically and physiologically. QTL analysis can help here. For example, longevity, which in plants is expressed at least in part as perenniality and persistence, is certainly accessible to the quantitative genetic approach, and there are indications of second-order QTL commonality with senescence in some species (Thomas *et al.*, 2000). Further studies of this kind would be highly informative. It is worth remembering, however, that death is most certainly not a quantitative trait. The fact that plant senescence is definable in QTL terms is further justification for the view that it should not be classified as just another example of programmed cell death.

Acknowledgments

IGER is sponsored by the UK Biotechnology and Biological Sciences Research Council. Isaac Galyuon was supported by a studentship from the Association of Commonwealth Universities. Work on pearl millet is an output from a project funded by the UK Department for International Development (DFID) and administered by the Centre for Arid Zone Studies for the benefit of developing countries (Plant Sciences Research Programme R7375). The views expressed are not necessarily those of DFID.

References

Abdelkhalik, A.F., Shishido, R, Nomura K. and Ikehashi H. (2005) QTL-based analysis of leaf senescence in an indica/japonica hybrid in rice (*Oryza sativa* L.). *Theor Appl Genet* **110**, 1226–1235.

Alm, V., Busso, C.S., Larsen, A., Humphreys, M.W. and Rognli, O.A. Quantitative trait loci for frost tolerance, winter survival and drought tolerance in meadow fescue (*Festuca pratensis* Huds.), and comparative mapping with cereals. *Genetics* (in press).

Alm, V., Fang, C., Busso, C.S., *et al.* (2003) A linkage map of meadow fescue (*Festuca pratensis* Huds.) and comparative mapping with other Poaceae species. *Theor Appl Genet* **108**, 25–40.

Andersson, A., Keskitalo, J., Sjödin, A., *et al.* (2004) A transcriptional timetable of autumn senescence. *Genome Biol* **5**, R24.

Banowetz, G. (1977) Cultivars of hexaploid wheat of contrasting stature and chl retention differ in cytokinin content and responsiveness. *Ann Bot* **79**, 185–190.

Beavis, W.D., Smith, O.S., Grant, D. and Fincher, R. (1994) Identification of quantitative trait loci using a small sample of topcrossed and F4 progeny from maize. *Crop Sci* **34**, 882–896.

Benbella, M. and Paulsen, G.M. (1998) Efficacy of treatments for delaying senescence of wheat leaves. I: Senescence under controlled conditions. *Agronomy J* **90**, 329–332.

Bert, P.F., Jouan, I., de Labrouhe, D.T., *et al.* (2003) Comparative genetic analysis of quantitative traits in sunflower (*Helianthus annuus* L.). 2: Characterisation of QTL involved in developmental and agronomic traits. *Theor Appl Genet* **107**, 181–189.

Bezant, J.H., Laurie, D.A., Pratchett, N., Chojecki, J. and Kearsey, M.J. (1997) Mapping of QTL controlling NIR predicted hot water extract and grain nitrogen content in a spring barley cross using marker-regression. *Plant Breeding* **116**, 141–145.

Blanco, A., Pasqualone, A., Troccoli, A., Di Fonzo, N. and Simeone, R. (2002) Detection of grain protein content QTLs across environments in tetraploid wheats. *Plant Mol Biol* **48**, 615–623.

Borrell, A.K., Bidinger, F.R., Tao, Y.Z., McIntyre, C.L. and Douglas, A.C.L. (2001) Physiological and molecular aspects of stay-green in sorghum. In: *Proceedings from the Fourth Australian Sorghum Conference, Kooralbyn* (eds Borrell, A.K. and Henzell, R.G.). Range Media Pty Ltd, pp. 5–8 (CD-rom format ISBN:0–7242-2163-8).

Borrell, A.K., Hammer, G.L. and Henzell, R.G. (2000) Does maintaining green leaf area in sorghum improve yield under drought? II: Dry matter production and yield. *Crop Sci* **40**, 1037–1048.

Breese, E.L. and Mather, K. (1957) The organization of polygenic activity within a chromosome of *Drosophila*. I: Hair characters. *Heredity* **11**, 373–395.

Buchanan-Wollaston, V. (1997) The molecular biology of leaf senescence. *J Exp Bot* **48**, 181–199.

Buchanan-Wollaston, V., Earl, S., Harrison, E., *et al.* (2003) The molecular analysis of leaf senescence – a genomics approach. *Plant Biotechnol J* **1**, 3–22.

Buchanan-Wollaston, V., Page, T., Harrison, E., *et al.* (2005) Comparative transcriptome analysis reveals significant differences in gene expression and signalling pathways between developmental and dark/starvation-induced senescence in *Arabidopsis*. *Plant J* **42**, 567–585.

Cha, K.W., Lee, Y.J., Koh, H.J., Lee, B.M., Nam, Y.W. and Paek, N.C. (2002) Isolation, characterization, and mapping of the stay green mutant in rice. *Theor Appl Genet* **104**, 526–532.

Chee, P.W., Elias, E.M., Anderson, J.A. and Kianian, S.F. (2001) Evaluation of a high grain protein QTL from *Triticum turgidum* L. var. *dicoccoides* in an adapted durum wheat background. *Crop Sci* **41**, 295–301.

Cheverud, J.M. (1982) Relationships among ontogenetic, static, and evolutionary allometry. *Am J Phys Anthropol* **59**, 139–149.

Cots, J., Fargeix, C., Gindro, K. and Widmer, F. (2002) Pathogenic attack and carbon reallocation in soybean leaves (*Glycine max* L.): reinitiation of the glyoxylate cycle as a defence reaction. *J Plant Physiol* **159**, 91–96.

Crasta, O.R., Xu, W.W., Rosenow, D.T., Mullet, J. and Nguyen, H.T. (1999) Mapping of post-flowering drought resistance traits in grain sorghum: association between QTLs influencing premature senescence and maturity. *Mol Gen Genet* **262**, 579–588.

Cukadar-Olmedo, B., Miller, J.F. and Hammond, J.J. (1997) Combining ability of the stay green trait and seed moisture content in sunflower. *Crop Sci* **37**, 378–382.

Diaz C., Saliba-Colombani V., Loudet O., *et al.* (2006) Leaf yellowing and anthocyanin accumulation are two genetically independent strategies in response to nitrogen limitation in *Arabidopsis thaliana*. *Plant Cell Physiol* **47**, 74–83.

Davies, T.G.E., Thomas, H., Thomas, B.J. and Rogers, L.J. (1990) Leaf senescence in a non yellowing mutant of *Festuca pratensis*: metabolism of cytochrome f. *Plant Physiol* **93**, 588–595.

Distelfeld, A., Uauy, C., Olmos, S., Schlatter, A.R., Dubcovsky, J. and Fahima, T. (2004) Microcolinearity between a 2 cM region encompassing the grain protein content locus Gpc-6B1 on wheat chromosome 6B and a 350 kb region on rice chromosome 2. *Funct Integrat Genomics* **4**, 59–66.

Duncan, R.R., Bockholt, A.J. and Miller, F.R. (1981) Descriptive comparison of senescent and non-senescent sorghum genotypes. *Agronomy J* **73**, 849–853.

Gallais, A. and Hirel, B. (2004) An approach to the genetics of nitrogen use efficiency in maize. *J Exp Bot* **55**, 295–306.

Gaut, B.S. and Doebley, J.F. (1997) DNA sequence evidence for the segmental allotetraploid origin of maize. *Proc Natl Acad Sci U S A* **94**, 6809–6814.

Gentinetta, E., Ceppi, D., Lepori, C., Perico, G., Motto, M. and Salamini, F. (1986) A major gene for delayed senescence in maize. Pattern of photosynthates accumulation and inheritance. *Plant Breeding* **97**, 193–203.

Goff, S.A., Ricke, D., Lan T.H., *et al.* (2002) A draft sequence of the rice genome (*Oryza sativa* L. ssp. *japonica*). *Science* **296**, 92–100.

Griffiths, C.M., Hosken, S.E., Oliver, D., Chojecki, J. and Thomas, H. (1997) Sequencing, expression pattern and RFLP mapping of a senescence-enhanced cDNA from *Zea mays* with high homology to oryzain gamma and aleurain. *Plant Mol Biol* **34**, 815–821.

Guiamet, J.J. and Giannibelli, M.C. (1996) Nuclear and cytoplasmic stay-green mutations of soybean alter the loss of leaf soluble proteins during senescence. *Physiol Plantarum* **96**, 665–661.

Guo, Y., Cai, Z. and Gan, S. (2004) Transcriptome of *Arabidopsis* leaf senescence. *Plant Cell Environ* **27**, 521–549.

Hafsi, M., Mechmeche, W., Bouamama, L., Djekoune, A., Zaharieva, M. and Monneveux, P. (2000) Flag leaf senescence, as evaluated by numerical image analysis, and its relationship with yield under drought in durum wheat. *J Agronomy Crop Sci* **185**, 275–280.

Hash, C.T., Bhasker Raj, A.G., Lindup, S., *et al.* (2003) Opportunities for marker-assisted selection (MAS) to improve the feed quality of crop residues in pearl millet and sorghum. *Fleld Crops Res* **84**, 79–88.

Hauck, B., Gay, A.P., Macduff, J., Griffiths, C.M. and Thomas, H. (1997) Leaf senescence in a non-yellowing mutant of *Festuca pratensis*: implications of the stay-green mutation for photosynthesis, growth and nitrogen nutrition. *Plant Cell Environ* **20**, 1007–1018.

Haussmann, B.I.G., Mahalakshmi, V., Reddy, B.V.S., Seetharama, N., Hash, C.T. and Geiger, H.H. (2002) QTL mapping of stay-green in two sorghum recombinant inbred populations. *Theor Appl Genet* **106**, 133–142.

Henzell, R.G., Dodman, R.L., Done, A.A., Brengman, R.L., Mayers, P.E. (1984) Lodging, stalk rot, and root rot in sorghum in Australia. In: *Sorghum Root and Stalk Rots: a Critical Review* (ed. Mughogho, L.K.). ICRISAT, Patancheru, India, pp. 225–235. Proc Consultative Group Discussion of Research Needs and Strategies for Control of Sorghum Root and Stalk Diseases, Bellagio, Italy, 27 Nov–2 Dec 1983.

Hirel, B., Bertin, P., Quillere, I., *et al.* (2001) Towards a better understanding of the genetic and physiological basis for nitrogen use efficiency in maize. *Plant Physiol* **125**, 1258–1270.

Hörtensteiner, S. and Feller, U. (2002) Nitrogen metabolism and remobilization during senescence. *J Exp Bot* **53**, 927–937.

Huxley, J.S. (1924) Constant differential growth-ratios and their significance. *Nature* **114**, 895–896.

Ishimaru, K., Yano, M., Aoki, N., *et al.* (2001) Toward the mapping of physiological and agronomic characters on a rice function map: QTL analysis and comparison between QTLs and expressed sequence tags. *Theor Appl Genet* **102**, 793–800.

Jackson, M.B. and Ram, P.C. (2003) Physiological and molecular basis of susceptibility and tolerance of rice plants to complete submergence. *Ann Bot* **91**, 227–241.

Jeanneau, M., Gerentes, D., Foueillassar, X., *et al.* (2002) Improvement of drought tolerance in maize: towards the functional validation of the Zm-Asr1 gene and increase of water use efficiency by over-expressing C4-PEPC. *Biochimie* **84**, 1127–1135.

Jiang, G.H., He, Y.Q., Xu, C.G., Li, X.H. and Zhang, Q. (2004) The genetic basis of stay-green in rice analyzed in a population of doubled haploid lines derived from an *indica* by *japonica* cross. *Theor Appl Genet* **108**, 688–698.

Jones, N., Ougham, H. and Thomas, H. (1997) Markers and mapping: we are all geneticists now. *New Phytologist* **137**, 165–177.

Jones, E.S., Mahoney, N.L., Hayward, M.D., *et al.* (2002) An enhanced molecular marker based genetic map of perennial ryegrass (*Lolium perenne*) reveals comparative relationships with other Poaceae genomes. *Genome* **45**, 282–295.

Joppa, L.R., Du, C., Hart, G.E. and Hareland, G.A. (1997) Mapping a QTL for grain protein in tetraploid wheat (*Triticum turgidum* L.) using a population of recombinant inbred chromosome lines. *Crop Sci* **37**, 1586–1589.

Kaplan, J.M. and Pigliucci, M. (2001) Genes 'for' phenotypes: a modern history view. *Biol Philos* **16**, 189–213.

Kebede, H., Subudhi, P.K., Rosenow, D.T. and Nguyen, H.T. (2001) Quantitative trait loci influencing drought tolerance in grain sorghum (*Sorghum bicolor* L. Moench). *Theor Appl Genet* **103**, 266–276.

Khan, I.A., Procunier, J.D., Humphreys, D.G., *et al.* (2000) Development of PCR based markers for a high grain protein content gene from *Triticum turgidum* ssp. *dicoccoides* transferred to bread wheat. *Crop Sci* **40**, 518–524.

Kim, J.-S., Klein, P.E., Klein, R.R., Price, H.J., Mullet, J.E. and Stelly, D.M. (2005) Chromosome identification and nomenclature of *Sorghum bicolor*. *Genetics* **169**, 1169–1173.

Koornneef, M., Alonso-Blanco, C. and Vreugdenhil, D. (2004) Naturally occurring genetic variation in *Arabidopsis thaliana*. *Ann Rev Plant Biol* **55**, 141–172.

Lebreton, C., Lazicjancic, V., Steed, A., Pekic, S. and Quarrie, S.A. (1995) Identification of QTL for drought responses in maize and their use in testing causal relationships between traits. *J Exp Bot* **46**, 853–865.

Lim, P.O., Woo, H.R. and Nam, H.G. (2003) Molecular genetics of leaf senescence in *Arabidopsis*. *Trends Plant Sci* **8**, 272–278.

Luquez V.M.C., Sasal Y., Medrano M., *et al.* (2006) Quantitative trait loci analysis of leaf and plant longevity in *Arabidopsis thaliana*. *J Exp Bot* **57**, 1363–1372.

Ma, C.-X., Casella, G. and Wu, R.L. (2002) Functional mapping of quantitative trait loci underlying the character process: a theoretical framework. *Genetics* **161**, 1751–1762.

Mahalakshmi, V. and Bidinger, F.R. (2002) Evaluation of putative stay-green sorghum germplasm lines. *Crop Sci* **42**, 965–974.

Makino, A., Mae, T. and Ohira, K. (1985) Photosynthesis and ribulose-1, 5-bisphosphate carboxylase/oxygenase in rice leaves from emergence through senescence. Quantitative analysis by carboxylation/oxygenation and regeneration of ribulose 1, 5-bisphosphate. *Planta* **166**, 414–420.

Martin, A., Belastegui-Macadam, X., Quillere, I., *et al.* (2005) Nitrogen management and senescence in two maize hybrids differing in the persistence of leaf greenness: agronomic, physiological and molecular aspects. *New Phytologist* **167**, 483–492.

Masclaux, C., Quillere, I., Gallais, A. and Hirel, B. (2001) The challenge of remobilisation in plant nitrogen economy. A survey of physio-agronomic and molecular approaches. *Ann Appl Biol* **138**, 69–81.

Matin, A. and Nadeau, J.H. (2001) Sensitized polygenic trait analysis. *Trends Genet* **17**, 727–731.

Maxwell, K. and Johnson, G.N. (2000) Chlorophyll fluorescence—a practical guide. *J Exp Bot* **51**, 659–668.

McBee, G.G., Waskom, R.M., Miller, F.R. and Creelman, R.A. (1983) Effect of senescence and nonsenescence on carbohydrates in sorghum during late kernel maturity states. *Crop Sci* **23**, 372–376.

Merzlyak, M.N., Gitelson, A.A., Chivkunova, O.B. and Rakitin, V.Y.U. (1999) Non-destructive optical detection of pigment changes during leaf senescence and fruit ripening. *Physiol Plantarum* **106**, 135–141.

Mesfin, A., Frohberg, R.C. and Anderson, J.A. (1999) RFLP markers associated with high grain protein from *Trititcum turgidum* L. var. *dicoccoides* introgressed into hard red spring wheat. *Crop Sci* **39**, 508–513.

Mickelson, S., See, D., Meyer, F.D., *et al.* (2003) Mapping of QTL associated with nitrogen storage and remobilization in barley (*Hordeum vulgare* L.) leaves. *J Exp Bot* **54**, 801–812.

Moore, B.J., Donnison, I., Harper, J.A., *et al.* (2005) Molecular tagging of a senescence gene by introgression mapping of a mutant stay-green locus from *Festuca pratensis*. *New Phytologist* **165**, 801–806.

Morgan, T.H. (1911) Chromosomes and associative inheritance. *Science* (New Series) **34**, 636–638.

Morris, K., Thomas, H. and Rogers, L.J. (1996) Endopeptidases during the development and senescence of *Lolium temulentum* leaves. *Phytochemistry* **41**, 377–384.

Nooden, L.D. and Penney, J.P (2001) Correlative controls of senescence and plant death in *Arabidopsis thaliana* (Brassicaceae). *J Exp Bot* **52**, 2151–2159.

Obara, M., Kajiura, M., Fukuta, Y., *et al.* (2001) Mapping of QTLs associated with cytosolic glutamine synthetase and NADH-glutamate synthase in rice (*Oryza sativa* L.). *J Exp Bot* **52**, 1209–1217.

Obara, M., Sato, T., Sasaki, S., Kashiba, K., Nagano, A., Nakamura, I., Ebitani, T., Yano, M. and Yamaya, T. (2004) Identification and characterization of a QTL on chromosome 2 for cytosolic glutamine synthetase content and panicle number in rice. *Theor Appl Genet* **110**, 1–11.

Olmos, S., Distelfeld, A., Chicaiza, O., *et al.* (2003) Precise mapping of a locus affecting grain protein content in durum wheat. *Theor Appl Genet* **107**, 1243–1251.

Oziel, A., Hayes, P.M., Chen, F.Q. and Jones, B. (1996) Application of quantitative trait locus mapping to the development of winter-habit malting barley. *Plant Breeding* **115**, 43–51.

Papakosta, D.K. and Gagianas, A.A. (1991) Nitrogen and dry-matter accumulation, remobilization, and losses for Mediterranean wheat during grain filling. *Agronomy J* **83**, 864–870.

Paterson, A.H., Lander, E.S., Had, J.D., Paterson, S., Lincoln, S.E. and Tanksley, S.D. (1988) Resolution of quantitative traits into Mendelian factors by using a complete linkage map of restriction fragment length polymorphisms. *Nature* **335**, 721–726.

Powell, W., Thomas, W.T.B., Baird, E., *et al.* (1997) Analysis of quantitative traits in barley by the use of amplified fragment length polymorphisms. *Heredity* **79**, 48–59.

Richardson, D., Duigan, S.P. and Berlyn, G.P. (2002) An evaluation of noninvasive methods to estimate foliar chlorophyll content. *New Phytologist* **153**, 185–194.

Roca, M., James, C., Pružinská, A., Hörtensteiner, S., Thomas, H. and Ougham, H. (2004) Analysis of the chlorophyll catabolism pathway in leaves of an introgression senescence mutant of *Lolium temulentum*. *Phytochemistry* **65**, 1231–1238.

Rosenow, D.T. (1984) Breeding for resistance to root and stalk rots in Texas. In: *Sorghum Root and Stalk Rots, a Critical Review. Proceedings of Workshop held in Bellagio, Italy* (ed. Mughogho, L.K.). ICRISAT, Patancheru, AP, India, pp. 209–217.

Rosenow, D.T. and Clark, L.E. (1981) Drought tolerance in sorghum. In: *Proceedings 36th Annual Corn and Sorghum Industry Research Conference*, Chicago, IL (eds Loden, H.D., Wilkinson, D.) American Seed Trade Association, Washington DC, pp. 18–31.

Sachs, T., Novoplansky, A. and Cohen, D. (1993) Plants as competing populations of redundant organs. *Plant Cell Environ* **16**, 765–770.

Sanchez, A.C., Subudhi, P.K., Rosenow, D.T. and Nguyen, H.T. (2002) Mapping QTLs associated with drought resistance in sorghum (*Sorghum bicolor* L. Moench). *Plant Mol Biol* **48**, 713–726.

Saulescu, N.N., Ittu, G. and Mustatea, P. (2001) Dark induced senescence as a tool in breeding wheat for optimum senescence pattern. *Euphytica* **119**, 205–209.

Schiltz, S., Gallardo, K., Huart, M., Negroni, L., Sommerer, N. and Burstin, J. (2004) Proteome reference maps of vegetative tissues in pea. An investigation of nitrogen mobilization from leaves during seed filling. *Plant Physiol* **135**, 2241–2260.

See, D., Kanazin, V., Kephart, K. and Blake, T. (2002) Mapping genes controlling variation in barley grain protein concentration. *Crop Sci* **42**, 680–685.

Servin, B., Martin, O.C., Mezard, M. and Hospital, F. (2004) Toward a theory of marker-assisted gene pyramiding. *Genetics* **168**, 513–523.

Simon, M.R. (1999) Inheritance of flag-leaf angle, leaf area and flag leaf area duration in four wheat crosses. *Theor Appl Genet* **98**, 310–314.

Smart, C.M. (1994) Gene expression during leaf senescence. *New Phytologist* **126**, 419–448.

Smart, C.M., Hosken, S.E., Thomas, H., Greaves, J.A., Blair, B.G. and Schuch, W. (1995) The timing of maize leaf senescence and characterization of senescence-related cDNAs. *Physiologia Plantarum* **93**, 673–682.

Steiger, D.K., Elias, E.M. and Cantrell, R.G. (1996) Evaluation of lines derived from wild emmer chromosome substitutions. 1: Quality traits. *Crop Sci* **36**, 223–227.

Subudhi, P.K., Rosenow, D.T. and Nguyen, H.T. (2000) Quantitative trait loci for the stay green train in sorghum (*Sorghum bicolor* L. Moench): consistency across genetic backgrounds and environments. *Theor Appl Genet* **101**, 733–741.

Sultan, S.E. (2003) Phenotypic plasticity in plants: a case study in ecological development. *Evol* Dev **5**, 25–33.

Tao, Y.Z., Henzell, R.G., Jordan, D.R., Butler, D.G., Kelly, A.M. and McIntyre, C.L. (2000) Identification of genomic regions associated with stay green in sorghum by testing RILs in multiple environments. *Theor Appl Genet* **100**, 1225–1232.

Tenkouano, A., Miller, F.R., Frederiksen, R.A. and Rosenow, D.T. (1993) Genetics of non-senescence and charcoal rot resistance in sorghum. *Theor Appl Genet* **85**, 644–648.

Thomas, H. (1987) *Sid*: a Mendelian locus controlling thylakoid membrane disassembly in senescing leaves of *Festuca pratensis*. *Theor Appl Genet* **73**, 551–555.

Thomas, H. (1992) Canopy survival. In: *Crop Photosynthesis: Spatial and Temporal Determinants* (eds Baker, N. and Thomas, H.). Elsevier, Amsterdam, pp. 11–41.

Thomas, H. (2002) Ageing in plants. *Mech Ageing Dev* **123**, 747–753.

Thomas, H. (2003) Do green plants age and if so how? *Top Curr Genet* **3**, 145–171.

Thomas, H., Evans, C., Thomas, H.M., *et al.* (1997) Introgression, tagging and expression of a leaf senescence gene in *Festulolium*. *New Phytologist* **137**, 29–34.

Thomas, H. and Howarth, C.J. (2000) Five ways to stay green. *J Exp Bot* **51**, 329–337.

Thomas, H., Ougham, H., Canter, P. and Donnison, I. (2002) What stay-green mutants tell us about nitrogen remobilisation in leaf senescence. *J Exp Bot* **53**, 801–808.

Thomas, H., Ougham, H. and Hörtensteiner, S. (2001) Recent advances in the cell biology of chlorophyll catabolism. *Adv Botanical Res* **35**, 1–52.

Thomas, H., Ougham, H. and Thomas, H.M. (2000) Annuality, perenniality and cell death. *J Exp Bot* **51**, 1–8.

Thomas, H., Ougham, H., Wagstaff, C. and Stead, A.J. (2003) Defining senescence and death. *J Exp Bot* **54**, 1127–1132.

Thomas, H. and Smart, C.M. (1993) Crops that stay green. *Ann Appl Biol* **123**, 193–129.

Thomas, H. and Stoddart, J.L. (1975) Separation of chlorophyll degradation from other senescence processes in leaves of a mutant genotype of meadow fescue (*Festuca pratensis*). *Plant Physiol* **56**, 438–441.

Toojinda, T., Siangliw, M., Tragoonrung, S. and Vanavichit (2003) A molecular genetics of submergence tolerance in rice: QTL analysis of key traits. *Ann Bot* **91**, 243–253.

Tuinstra, M.R., Ejeta, G. and Goldsborough, P.B. (1998) Evaluation of near-isogenic sorghum lines for QTL markers associated with drought tolerance. *Crop Sci* **38**, 835–842.

Tuinstra, M.R., Grote, E.M., Goldsbrough, P.B. and Ejeta, G. (1997) Genetic analysis of post-flowering drought tolerance and components of grain development in *Sorghum bicolor* (L.) Moench. *Mol Breeding* **3**, 439–448.

Uauy C., Brevis J.C. and Dubcovsky J. (2006) The high grain protein content gene Gpc-B1 accelerates senescence and has pleiotropic effects on protein content in wheat. *J Exp Bot* **57**, 2785–2794.

van Doorn, W.G. and Woltering, E.J. (2004) Senescence and programmed cell death: substance or semantics? *J Exp Bot* **55**, 2147–2153.

Van Oosterom, E.J., Jayachandran, R. and Bidinger, F.R. (1996) Diallel analysis of the stay-green trait and its components in sorghum. *Crop Sci* **36**, 549–555.

Van Sandford, D.A. and Mackown, C.T. (1987) Cultivar differences in nitrogen remobilization in grain fill in soft red winter wheat. *Crop Sci* **27**, 295–300.

Verma, V., Foulkes, M.J., Worland, A.J., Sylvester-Bradley, R., Caligari, P.D.S. and Snape, J.W. (2004) Mapping quantitative trait loci for flag leaf senescence as a yield determinant in winter wheat under optimal and drought-stressed environments. *Euphytica* **135**, 255–263.

Vicentini, F., Hörtensteiner, S., Schellenberg, M., Thomas, H. and Matile, P. (1995) Chlorophyll breakdown in senescent leaves: identification of the biochemical lesion in a stay-green genotype of *Festuca pratensis* Huds. *New Phytologist* **129**, 247–252.

Walulu, R.S., Rosenow, D.T., Wester, D.B. and Nguyen, H.T. (1994) Inheritance of the stay green trait in sorghum. *Crop Sci* **34**, 970–972.

Wang, D.L., Zhu, J., Li, Z.K. and Paterson, A.H. (1999) Mapping QTLs with epistatic effects and QTL × environment interactions by mixed linear model approaches. *Theor Appl Genet* **99**, 1255–1264.

Ware, D., Jaiswal, P., Ni, J., *et al.* (2002) Gramene: a resource for comparative grass genomics. *Nucleic Acids Res* **30**, 103–105.

Wingler, A., Marès, M. and Pourtau, N. (2004) Spatial patterns and metabolic regulation of photosynthetic parameters during leaf senescence. *New Phytologist* **161**, 781–789.

Xu, W., Subudhi, P.K., Crasta, O.R., Rosenow, D.T., Mullet, J.E. and Nguyen, H.T. (2000) Molecular mapping of QTLs conferring stay-green in grain sorghum (*Sorghum bicolor* L. Moench). *Genome* **43**, 461–469.

Yadav, R.S., Hash, C.T., Bidinger, F.R., Devos, K.M. and Howarth, C.J. (2004) Genomic regions associated with grain yield and aspects of post-flowering drought tolerance in pearl millet across stress environments and testers background. *Euphytica* **136**, 265–277.

Yadav, R.S., Hash, C.T., Cavan, G.P., Bidinger, F.R. and Howarth, C.J. (2002) Quantitative trait loci associated with traits determining grain and stover yield in pearl millet under terminal drought stress conditions. *Theor Appl Genet* **104**, 67–83.

Yang, L.T., Mickelson, S., See, D., Blake, T.K. and Fischer, A.M. (2004) Genetic analysis of the function of major leaf proteases in barley (*Hordeum vulgare* L.) nitrogen remobilization. *J Exp Bot* **55**, 2607–2616.

Yamaya, T., Obara, M., Nakajima, H., Sasaki, S., Hayakawa, T. and Sato, T. (2002) Genetic manipulation and quantitative-trait loci mapping for nitrogen recycling in rice. *J Exp Bot* **53**, 917–925.

Yu, J., Hu, S., Wang, J., *et al.* (2002) A draft sequence of the rice genome (*Oryza sativa* L. ssp. *indica*). *Science* **296**, 79–92.

Zhang, W.K., Wang, Y.J., Luo, G.Z., *et al.* (2004) QTL mapping of ten agronomic traits on the soybean (*Glycine max* L. Merr.) genetic map and their association with EST markers. *Theor Appl Genet* 108, 1131–1139.

9 Genomics and proteomics of leaf senescence

Marie-Jeanne Carp and Shimon Gepstein

9.1 Introduction

Senescence represents a genetically programmed process occurring in the final stage of the development of the whole plant, organ, tissue or the cell. It is an actively ordered process that involves the synthesis of new mRNA and protein molecules leading to a dramatic metabolic shift causing cellular dismantling and dysfunction that eventually leads to death. Although senescence involves a wide range of degradation processes, selective gene expression and protein synthesis are the fundamental processes necessary for regulating and executing the senescence program (Gepstein, 2004; Lim and Nam, 2005). Research on the role of products of individual genes associated with senescence (SAGs) has indeed advanced our understanding of specific biochemical reactions and cellular events occurring during leaf senescence. During the past decade ~100 individual *SAG*s have so far been cloned from various plant species, including *Arabidopsis*, barley, *Brassica*, maize, tomato and soybean. Among these *SAG*s are genes-encoding hydrolytic enzymes such as proteinases, lipases RNases, and others like kinases, and transcription factors (Hajouj *et al.*, 2000; Buchanan-Wollaston *et al.*, 2003; Gepstein *et al.*, 2003; Guo *et al.*, 2004). The function of *SAG*s in leaf senescence has begun to be elucidated, and a few of the genes and the corresponding mutants have been shown to play functional roles during senescence. However, our understanding of the whole senescence program and the general regulatory pathways underlying this process is still very limited.

Recent technological advances enabling the analysis of hundreds to thousands of mRNAs and proteins simultaneously offer the possibility of not only identifying individual components but also dissecting signaling pathways and revealing networks by using an integrated approach. In light of the complexity of the plant senescence symptom, the new insight will lead to a better understanding of the senescence program based on molecular/biochemical criteria. Also, as discussed in this review, common and differential schemes of various senescence types, e.g. developmental and stress-induced senescence, might be established.

In this chapter, we will discuss two principal approaches, *transcriptomics* and *proteomics*. Transcriptomics is defined as the genome-wide study of mRNA expression levels, whereas proteomics is defined as the study of the full set of proteins encoded by a genome. The umbrella term for transcriptomics and proteomics is *functional genomics*, which has led to the foundation of core facilities for *functional genomics* for a wide range of processes including various developmental stages, responses to environmental (abiotic and biotic) stresses, etc.

9.2 Transcriptomics of leaf senescence

9.2.1 Technologies

Conventional methods such as Northern blot analysis offer the possibility to determine the expression of one or several genes, whereas the microarray technology (described below) offers the unique chance to analyse hundreds to thousands of transcripts simultaneously. Several other technologies such as subtractive hybridization and differential display have preceded the microarray technology.

9.2.1.1 Differential display, in situ hybridization and subtractive hybridization

In situ hybridization allows gene expression profiling at a cellular level within the tissue of interest, and Northern blot, using mRNA derived from tissue or organs of interest, displays differential transcription expression of mRNA. Yet both the approaches depend on the prior knowledge of defined gene sequences. Alternatively, techniques such as microarray and subtractive hybridization are capable of identifying known and novel genes and require only small amounts of RNA (Sagerstrom *et al.*, 1997; Byers *et al.*, 2000). The major disadvantages of differential display are the high number of false positives and the low sensitivity to detect rare species. In differential display, following reverse transcription using oligo-(dT) primers with one or two additional specific nucleotides, the cDNA sequences are amplified by PCR and the resultant products are separated on polyacrylamide gels. The bands can then be visualized, for example, by autoradiography. If present in a lane representing one sample and not the other, this band is considered to represent a differentially expressed gene and is excised and further amplified. The resultant amplification product is used to screen a library to obtain a full-length cDNA (Liang and Pardee, 1992). Subtractive hybridization is used to isolate genes that are upregulated in one cell type or tissue compared to other (Byers *et al.*, 2000). The starting material is generally cDNA obtained by the poly(A) reverse transcription polymerase chain reaction (poly(A) RT-PCR). Following first-strand cDNA synthesis for a limited time using a poly(dT) primer, PCR is used to amplify the second strand, again using a poly(dT) primer, resulting in a cDNA population of a defined size range (usually between 100 and 500 bp), which can be further amplified. To identify differentially expressed transcripts in two samples, denatured cDNA derived from the two samples are hybridized, with one sample (the driver cDNA) being biotinylated and in excess compared to the other (the tracer cDNA). After several rounds of hybridizations and removals by streptavidin precipitation, the remaining mRNA represent genes that are either unique or massively upregulated in the tracer population. A successful modified approach is the subtraction suppression hybridization. This technique has been designed specifically to identify rare mRNAs and to prevent preferred amplification of high abundant transcripts (Diatchenko *et al.*, 1999; Rebrikov *et al.*, 2000; Gepstein *et al.*, 2003).

One thing common to all of these techniques is that the genes identified as being differentially expressed need to be validated by Northern blot hybridization or qRT-PCR.

9.2.1.2 Microarrays

The development and widespread application of the gene expression array or microarray technology has been prompted due to the deciphering of the genome of an increasing number of organisms, the development of fluorochromes, miniaturization and automation. Although the general term includes a variety of distinct methodological and practical approaches, they are all high-density arrays of oligonucleotide or cDNA sequences coupled onto a solid support, most commonly a derivatized glass microscope slide or silicon wafer. The modern application of microarrays differs from classic blotting procedures in that the labeled pool of nucleic acid molecules is isolated from the tissue of interest and hybridized to a large number of immobilized DNA molecules arrayed in specific locations on a solid surface. Since each unique sequence is individually spotted onto the support, by building up thousands of such spots, an array can be made to cover the entire known expressed gene content of an organism. By hybridizing the sample to all the sequences present on the array simultaneously, a highly parallel (or high-throughput) analysis of expressed genes can be performed. RNA is extracted, labeled with a fluorescent dye and hybridized to the sequences on the array, with the signal detected from each spot being in principal proportional to the amount of that specific RNA in the original sample. By analysing each spot, a picture not only of which genes are expressed but also of the level of expression can be obtained.

In general, there are two types of arrays: oligonucleotide arrays, where defined oligonucleotides are either spotted or directly synthesized onto the support, and cDNA arrays, where selected cDNAs or PCR products are spotted. The arrayed fragments can include expressed genes obtained from various sources, including those previously identified in public databases or from expressed sequence tags (EST). ESTs are randomly selected clones constructed from mRNA derived from a specific tissue or cell type. Furthermore, there is the option of using either commercial arrays with predefined probes, which can be obtained from vendors such as Affymetrix, or custom-made arrays. Oligonucleotide arrays are composed of short 50–80 mer sequences. Compared to cDNA arrays, which use longer sequences in the range of 150–300 nucleotides, they require larger amounts of RNA to be hybridized due to a reduced sensitivity. To analyse the microarrays, mostly high-resolution laser scanners are employed, which detect the dyes incorporated into the cDNAs. The use of two dyes (e.g. Cy3 or Cy5) offers the possibility of competitive hybridization. To normalize, measure and interpret microarray data, sophisticated software has been developed. Although in the first step it is often sufficient to identify differentially transcribed genes, in the second step one would like to group genes into classes, identify patterns and interactions of genes. Eventually, by arranging differentially expressed genes into gene clusters and pathways using bioinformatics, it will be possible to define the initial genes in the pathway that lead to the downstream changes in gene expression. However, to interpret this data, the traditional hypothesis-driven approach still has validity.

9.2.2 *Altering the expression of senescence-specific genes may extend the lifespan of annual plants*

A quick review of recently published work in the field of plant senescence indicates the great advance that has been made in identifying specific genes (and gene products) that are involved and may affect the senescence processes in various organs in a variety of plant species. There are many common SAGs and molecular events occurring in various senescing organs and tissues suggesting some general metabolic and cellular features accompanying the senescence processes. Among the predominant functional classes are those of genes whose products are involved in degradation of macromolecules and remobilization of cellular constitutes. Indeed, work in different laboratories has demonstrated that lifespan extension can be affected by altering the expression (or activity) of individual gene products. These observations are highly intriguing and potentially of great importance because they indicate that it should be possible to extend the lifespan of the annual plant or of plant organs, such as leaves (Lim and Nam, 2005). Representatives of several gene functional classes have been reported to influence aging processes (Lim *et al.*, 2003; Lim and Nam, 2005). Hormonal manipulations have been carried out with genes involved in the synthesis and metabolism of cytokinins and ethylene. Genes associated with cytokinine synthesis such as the IPT (the key enzyme in cytokinins synthesis pathway) under the control of a senescence-activated promoter (Gan and Amasino, 1995), or the knotted homeobox gene under the control of the same SAG13 promoter (Ori *et al.*, 1999), significantly delayed the process of leaf senescence. In contrast to cytokinins, ethylene has been suggested to accelerate leaf senescence and this was proven by transgenic tomato plants expressing 1-aminocyclopropane-1-carboxylic-acid deaminase that exhibited significant delays in ripening and leaf deterioration (Klee *et al.*, 1991). Mutants defective in ethylene response such as the *etr* and *er* exhibited also a considerable delay in leaf senescence (Grbic and Bleecker, 1995; Oh *et al.*, 1997). These observations confirm and support the notion that ethylene plays a major role in the regulation of leaf senescence. Two independent mutations in *Arabidopsis* plants of genes associated with protein degradation have also shown alteration in leaf senescence. The *ore9* mutant with an altered activity of the ORE9, a protein containing the F-box motif which is a component of the E3 ubiquitin pathway, showed a significant delay of leaf senescence. Thus, ORE9 has been speculated to limit leaf longevity by removing target proteins that are required to delay the senescence process (Woo *et al.*, 2001). The delayed-leaf-senescence 1 (dls1) mutant, which is defective in arginyl tRNA:protein transferase (R-transferase), also showed a delay in leaf senescence. R-transferase is a component of the N-end-rule proteolytic pathway that is responsible for targeting proteins for ubiquitin-dependent proteolysis (Yoshida *et al.*, 2002). Thus, like ORE9, DLS1 might play a role in the degradation of proteins that negatively regulate leaf senescence. In both cases targeted proteolytic activity of specific regulatory proteins and not non-selective proteolysis is responsible for the creation of delayed senescence plants. Other mutations known to delay leaf senescence are those associated with the execution phases of the leaf senescence symptoms that usually act downstream in the signaling transduction pathways. These altered genes are likely to affect only a subset of senescence symptoms. A representatives of this group is the stay green mutant of *Festuca pratensis* that is defective in the chlorophyll degradation process due to a mutation in the gene encoding the enzyme pheophorbide, a oxygenase catalyzing one of the key steps in the chlorophyll degradation pathway during senescence (Roca *et al.*, 2004). Subset of genes that are functionally associated with lipid hydrolysis and metabolism were studied using a transgenic approach. Antisense suppression of the *Arabidopsis* acyl hydrolase gene (*SAG101*) delayed leaf senescence, whereas overexpression of this gene accelerated the onset of senescence (He and Gan, 2002). Transgenic plants in which senescence-induced lipase levels were reduced exhibited delayed leaf senescence (Yoshida *et al.*, 2002). A role of phospholipase D (PLD) in plant senescence was also proven using *Arabidopsis* with reduced PLDa activity. The senescence symptoms in detached leaves was delayed. However, suppression of PLDa did not influence natural senescence, suggesting that PLDa is a mediator in phytohormone-mediated senescence of detached leaves but not in age-dependent senescence (Varshavsky, 1997). Although the evidence connecting a specific gene (or a small number of genes) to the extended longevity in each case is convincing, it also appears clear that there still is no unifying explanation for these life-extending effects, let alone for the fundamental process of senescence.

9.2.3 From single to global gene expression studies of leaf senescence

In recent years, several approaches of large-scale gene expression studies have proposed the existence of various pathway networks related to several types of senescence. (He *et al.* 2001) used the enhancer-trapped libraries established in *Arabidopsis* and found 147 lines (out of 1300) that showed GUS activity exclusively in senescing leaves. Exogenously applied hormones and other factors showed specific expression of GUS and allowed the identification of genes whose expression was influenced by factors such as ABA, JA, ethylene and darkness (He *et al.*, 2001)

The comprehensive transcriptome study of Guo *et al.* has provided a large list (6200) of ESTs representing approximately 2500 SAGs in *Arabidopsis* leaves. It has been hypothesized that the ESTs found in the senescing tissues represent genes that are expressed during the senescence stage, when all transcripts present in non-senescent leaves have already been degraded in the fully senescent leaf. Since the available public database of *Arabidopsis* ESTs did not represent senescing tissues, it was not surprising that additional 100 new genes not found in the dbEST were identified in this study. The relative abundance of transcripts represented by the various categories during senescence substantially differs from other developmental stages. For example, massive degradation, a distinct feature of leaf senescence, is reflected by the high ratio (1.84) of the number of genes for primary catabolism over anabolism found in the senescence ESTdb as compared to the low ratio (0.57) for the entire *Arabidopsis* genome (Guo *et al.*, 2004).

While this EST list represents the existing gene transcripts in senescing tissues, comparative approaches such as subtractive hybridization or suppression subtractive hybridization and microarrays provide the repertoire of genes whose transcripts are either preferentially or exclusively expressed during the senescence phases (Gepstein *et al.*, 2003; Buchanan-Wollaston *et al.*, 2005).

9.2.4 Kinetics studies of gene expression define sequential changes in the pathway of the senescence program

The comparative studies of the senescence transcriptome provide global information regarding functional groups and may also suggest the kinetics of potential events occurring during leaf senescence. More importantly, clustering of genes based on their temporal expression may suggest the possible regulatory pathways, interactions between pathways, potential signaling pathways and the actual execution of biochemical and physiological events of the senescence program.

However, to establish a temporal expression pattern of SAGs one should determine first whether leaves of the same position would be sampled from plants at different ages or leaves of different ages in the same plant would be compared. Zentgraf *et al.* (2004) used the Affymetrix GeneChip to compare the two possible alternatives and found that gene expression is governed by different parameters: the age of the individual leaf and the age of the whole plant. The authors recommend appropriate designing and usage of experimental systems for gene expression studies (Zentgraf *et al.*, 2004).

Several typical temporal patterns of gene expression profiles have been described (Figure 9.1A). These include the analysis of the temporal behavior of single genes identified in the large-scale studies and were confirmed either by Northern blot or RT-PCR analyses (Smart, 1994; Buchanan-Wollaston and Ainsworth, 1997; Park *et al.*, 1998; Yoshida *et al.*, 2001; Gepstein *et al.*, 2003). The microarray approach allows depicting the temporal expression of hundreds of genes that are measured simultaneously and are clustered into various classes according to the kinetics of their expression during senescence (Figure 9.1B) (Swidzinski *et al.*, 2002; Buchanan-Wollaston *et al.*, 2003; Buchanan-Wollaston *et al.*, 2005). The cluster representing gene expression in early stages of senescence, prior to chlorophyll disappearance, is of a special interest and may suggest the involvement of upstream events in signaling pathways. Examples of individual genes are the senescence-associated receptor kinase (*SARK*) gene expressed very early in the senescence of bean leaves and may play a regulatory role in triggering the senescence symptom (Hajouj *et al.*, 2000). Other genes in this group include the MYB transcription factor, the vegetative storage protein and the dehydrin-like protein etc (Gepstein *et al.*, 2003; Buchanan-Wollaston *et al.*, 2005).

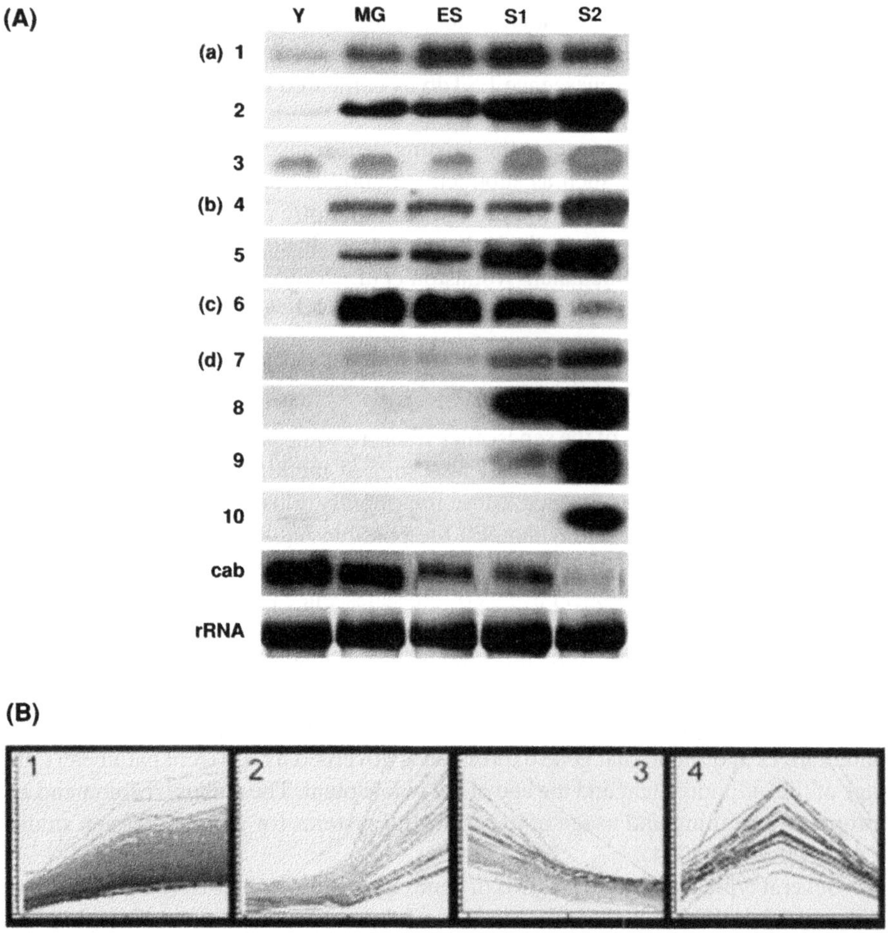

Figure 9.1 Gene expression profiles during *Arabidopsis* leaf development and senescence. (A) RNA gel blots analyses of individual SAGs, identified by subtraction hybridization. The following expression patterns are presented. (a) SAGs with basal expression in pre-senescence stages: (1) cationic amino acid transporter; (2) amino acid permease; (3) methalothionein. (b) Early expressed genes: (4) xylose isomerase; (5) RING H2 finger protein. (c) SAGs displaying transient expression: (6) lipid transfer protein. (d) Late-expressed SAGs: (7) calmodulin-like protein; (8) lipoic acid synthase; (9) lls1; (10) myo-inositol phosphate synthase: cab, chlorophyll a/b-binding protein representing a downregulated photosynthetic gene. 18S rRNA shown in the bottom panel is a loading control. The various senescence stages; Y-young leaves prior to bolting, MG-mature green leaves at the pick of photosynthetic capacity, ES-early senescent leaves with ~10% chlorophyll loss, S1-senescent leaves up to 50% chlorophyll loss, S2-late senescent leaves over 50% chlorophyll loss (from Gepstein *et al.*, 2003). (B) Expression data obtained from Affymetrix GeneChip experiments. Clustering analysis for ~1400 genes indicates four expression pattern groups: (1) genes upregulated as senescence progresses, (2) genes upregulated in late senescence, (3) genes downregulated during senescence and (4) genes with transient expression during senescence (from Buchanan-Wollaston *et al.*, 2003).

9.2.5 *Classification of the SAGs into functional classes suggests potential regulatory and biochemical pathways occurring during senescence*

The basic assumption underlying gene expression studies is that each of the genes whose expression products show increased abundance during leaf senescence plays a role in the regulation and the metabolism of the senescence program. The recent large-scale studies of gene expression during senescence have revealed, in general, similar functional gene groups as found in previous studies of individual gene expression. However, the large-scale studies have provided additional valuable indication regarding the overall perspective and proportion in the number of expressed genes and the various functional groups participating in the senescence program (Figure 9.2) (Swidzinski *et al.*, 2002; Bhalerao *et al.*, 2003; Gepstein *et al.*, 2003; Guo *et al.*, 2004; Buchanan-Wollaston *et al.*, 2005). A recent comprehensive study using the Affymetrix GeneChip arrays has indicated that 30% of the total number of *Arabidopsis* genes are expressed during developmental senescence (Swidzinski *et al.*, 2002; Bhalerao *et al.*, 2003; Gepstein *et al.*, 2003; Buchanan-Wollaston *et al.*, 2005).

However, out of total number of the ~9000 expressed genes, 800 were preferentially, or exclusively, upregulated in the senescent leaf. Another transcriptome study employing the subtraction hybridization approach (Gepstein *et al.*, 2003) presented a list of ~150 SAGs, which reflects a representing but incomplete inventory of genes. Based on their annotated putative function, each of the upregulated genes has been

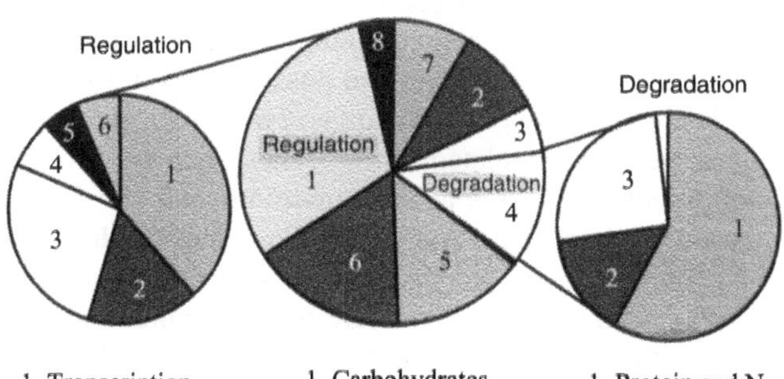

1. Transcription
2. Protein binding
3. Kinases & phosphatases
4. Signaling
5. Calcium
6. Hormones

1. Carbohydrates
2. Transport
3. Stress/defence
4. Macromolecular degradation
5. Enzymes of unknown function
6. Unknown genes
7. Regulatory genes
8. Secondary metabolism

1. Protein and N mobilization
2. Nucleic acid degradation
3. Lipid degradation

Figure 9.2 Functional classification of senescence-enhanced genes. Subdivision of the groups are illustrated in the central pie chart and the extra charts are used to divide the genes encoding putative regulatory proteins and genes involved in macromolecule degradation into smaller functional groups (from Buchanan-Wollaston *et al.*, 2005).

sorted and classified into functional groups. Prominent feature of the senescing leaf is the high ratio of genes involved in macromolecular degradation and mobilization of metabolites. A similar high ratio of genes related to catabolism during senescence was also described by Guo *et al.* (2004). Subdivision of the gene category related to degradation processes indicated the involvement of many types of proteases including cystein, serine, aspartyl and vacuolar processing enzymes (Gepstein *et al.*, 2003; Buchanan-Wollaston *et al.*, 2005). Genes related to the ubiquitin-dependent protein degradation pathway that were upregulated during senescence have been suggested to play a regulatory role in activating specific gene products rather than participating in the process of massive non-selective protein degradation occurring during leaf senescence (Woo *et al.*, 2001; Gepstein *et al.*, 2003). Massive characteristic degradation of nucleic acids by specific nucleases is predicted to supply not only the nitrogen source but also the phosphorous for recycling processes (Lers *et al.*, 2001; Buchanan-Wollaston *et al.*, 2005). The increased selective transcription of acid phosphatases in senescing leaves suggests a role for these enzymes in the storage and mobilization of phosphate released by the breakdown processes of nucleic acids (del Pozo *et al.*, 1999). The execution of the senescence-related lipid degradation processes is reflected by the increased abundance of transcripts of genes such as lipases (Thompson *et al.*, 2000; Gepstein *et al.*, 2003; Navabpour *et al.*, 2003; Lin and Wu, 2004). Chlorophyll degradation, the most conspicuous phenomenon of leaf senescence, is intensified during senescence probably due to both increased transcriptional and post-transcriptional processes of the *acd1* encoding the pheophorbide, a oxygenase gene (Buchanan-Wollaston *et al.*, 2005; Pruzinska *et al.*, 2005). The increased levels of transcripts of the early light induced protein (ELIP) during senescence may explain its role in binding the free chlorophyll molecules released from pigment protein complexes and possible prevention of the oxidative stress of phototoxic-free chlorophyll (Binyamin *et al.*, 2000; Buchanan-Wollaston *et al.*, 2005). Other genes upregulated during senescence are those classified as related to general defense processes against environmental stresses, but the most noticeable are the antioxidant genes. Activity of these gene products are crucial for protection of the cells from mass increase of the extremely toxic reactive oxygen species that otherwise would lead to premature death (Zimmermann and Zentgraf, 2005). The largest portion of the pie chart describing functional gene categories is represented by a group of genes related to regulatory events associated with senescence (Figure 9.2) (Guo *et al.*, 2004; Buchanan-Wollaston *et al.*, 2005). This general category of regulation represents genes whose products are involved in the regulation of various aspects throughout the senescence program including the final executing events associated with cell death. Subsets of this broad category include genes whose products are involved in various molecular and cellular events via the following processes: transcription, kinases and phosphatases activation and deactivation, signaling, hormone pathways, and calcium-dependent reactions. Transcription factors are the most abundant representatives in this gene list that are upregulated during senescence (Guo *et al.*, 2004; Lin and Wu, 2004; Buchanan-Wollaston *et al.*, 2005). Guo *et al.* (2004) found that ~5.0% of the total number of genes represented by ESTs in senescent leaves are genes encoding transcription factors, which is very close to the percentage of transcription factor genes in the *Arabidopsis* genome. Several super families of transcription factors have been identified in the senescence transcriptome. A total of 20 genes encoding NAC transcription factors are represented in the senescence ESTs, which is almost one fifth of all the predicted 109 members of the NAC super family in *Arabidopsis*. The second largest group of the transcription factors found both in the in the senescence EST collection and by microarray analysis of the senescence transcriptome is the WRKY super family. The *Arabidopsis* genome is predicted to encode ~100 WRKYs transcription factors. As with NACs, WRKYs are plant specific and have not been discovered in other eukaryotes.

Members of this gene family may be involved in processes including defense against pathogens, trichome development and senescence (Eulgem *et al.*, 2000). Both WRKY6 (Robatzek and Somssich, 2002) and WRKY53 (Hinderhofer and Zentgraf, 2001) have been shown to be involved in the regulation of leaf senescence. Guo *et al.* (2004) found that *WRKY75* is highly senescence-specific and that leaf senescence is significantly delayed in the related *Arabidopsis* antisense lines.

A relevant dedicated genomic study focusing on expression profiles of ~400 potential transcription factors at different developmental stages and under various biotic and abiotic stresses has shown that out of 43 transcription factor genes that were induced during senescence, 28 were also induced by stress treatments (Chen *et al.*, 2002). This data indicated high degree of overlap in the responses to stress and leaf senescence and the putative common transcription factors could be candidates for senescence controlling factors. Specific protein modifications by kinases and phosphatases that have been found to be upregulated during senescence suggest possible activation/deactivation of kinase-signalling cascade during senescence (Robatzek and Somssich, 2002). Protein kinase receptors are also appropriate candidates to regulate signaling pathways and the upregulation of the *SARK* expression was suggested to initiate the metabolic transition of the senescing leaf (Hajouj *et al.*, 2000). Upregulation in the expression of genes involved in calcium regulation during senescence suggests regulatory role for calcium and calmodulin-binding proteins in the senescence program (Gepstein *et al.*, 2003; Buchanan-Wollaston *et al.*, 2005).

The identified senescence-enhanced genes whose products are putative components of signaling pathways are indicative of the existence of various independent as well as interconnecting pathways that control genes associated with the senescence program.

9.2.6 *Stress-induced and developmental senescence can be compared by genomic studies*

Several studies have addressed the question as to what extent the molecular events occurring during developmental senescence are similar to stress-induced senescence. One of the most promising approaches to dissect and compare processes occurring in both types of senescence is to perform comparative study of gene expression profiles. Previous studies focusing on individual gene expression have clearly shown partial overlap between the two types of senescence (Weaver *et al.*, 1998; Swidzinski *et al.*, 2002; Kore-eda *et al.*, 2004). However, since the reported studies compared only limited number of randomly selected genes, the results have provided only limited information regarding specific biochemical events in the complex process of leaf senescence. Moreover, upregulation of the expression of individual gene may not necessarily indicate the simultaneous appearance of other gene products required for cooperation in the same event. Thus, global comparative gene expression studies may provide more detailed information regarding gene clusters and the extent of overlapping of predicted biochemical and cellular events occurring in the age-dependent as compared to stress-induced senescence.

Comparisons of global gene expression profiles between developmental and stress-induced senescence of detached organs are very important for future improvements of many crop species. The loss caused by post-harvest-induced senescence is one of the major economic-agricultural concerns. Manipulations aimed at improvements of shelf life of various crops require detailed information regarding molecular mechanisms including gene expression that control shelf life. Indeed, transcriptome studies of dark-induced senescence of detached *Arabidopsis* leaves have provided useful information regarding the expression of senescence-related genes (Gepstein *et al.*, 2003; Buchanan-Wollaston *et al.*, 2005).

Microarray analysis of dark-induced senescence of the *Arabidopsis* leaf was carried out to identify and characterize differentially expressed genes (Lin and Wu, 2004). This study indicated substantial alteration of metabolic pathways during the transition from a green leaf to senescing organ. Moreover, the microarray expression data was used for comparison with gene expression data obtained with similar Affymetrix GeneChip for age-dependent senescence of *Arabidopsis* leaves (Buchanan-Wollaston *et al.*, 2005). Significant differences have been found between the total number and the identities of the individual genes expressed in the two different types of senescence. Most of the genes whose transcripts were upregulated in the senescing leaf incubated in the dark could not be detected in the natural age-dependent senescing leaf. The differences in the gene expression profiles found between the two types of senescence were not restricted to a specific functional class of the expressed genes but rather found in most categories. These differences suggest the operation of different pathways related to nitrogen mobilization, lipid catabolism (β-oxidation genes), trehalose metabolism and flavonoid biosynthesis (Figure 9.3).

While the differences in gene expression between natural senescence and dark-induced senescence was quite substantial, comparison between dark-induced leaf senescence and program cell death exhibiting by starved cell suspension indicated higher degree of similarity between these two types of senescence. Independent global gene expression studies of the three types of senescence using the Arabidopsis Affymetrix Genechip were performed (Swidzinski *et al.*, 2002) and the array data of the three studies were used for combined comparative analysis (Buchanan-Wollaston *et al.*, 2005). A group of ~200 genes were found to participate in all three types of senescence, whereas a similar number of genes were exclusively upregulated in the developmental leaf senescence. More than 300 upregulated genes

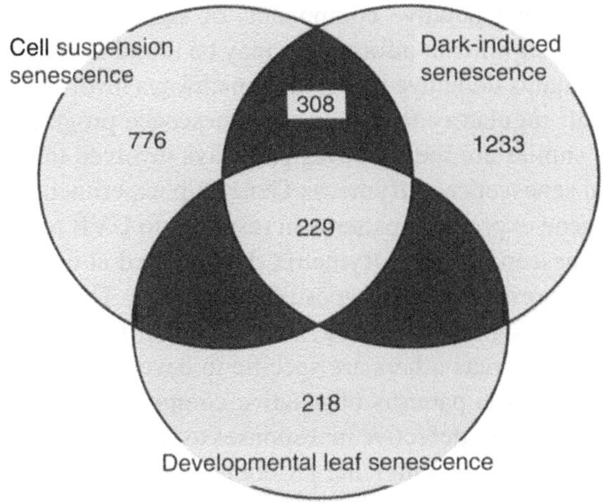

Figure 9.3 Comparative analysis of genes upregulated in three different types of senescence. The Venn diagram illustrates the number of genes upregulated in category in the three different types of senescence (from Buchanan-Wollaston *et al.*, 2005).

were common to the dark-induced leaf senescence and cell suspension senescence but not in developmental senescence. These results suggest that dark-induced leaf senescence is much closer to senescing cell suspension than to developmental leaf senescence (Figure 9.3) (Buchanan-Wollaston *et al.*, 2005). Of special interest are aspects related to the regulatory roles of plant hormones during the different types of senescence. Common upregulated genes were found to be involved in ABA biosynthesis and signaling pathways during the three types of senescence. Based on these transcriptomics results, the authors concluded that ABA is active and may play common regulatory role in the three types of the senescence programs (Buchanan-Wollaston *et al.*, 2005). Cytokinins, the key regulators in delaying leaf senescence (Gan and Amasino, 1995), were also tested and the transcriptome array data related to cytokinin-associated genes, of the three types of senescence were subjected to comparative analysis. In general, genes involved in cytokinin biosynthesis showed reduced levels of their transcripts, whereas cytokinin oxidase (involved in cytokinins degradation) showed parallel decrease in its transcripts abundance during both developmental and dark-induced senescence. However, many of the genes involved in cytokinin signaling pathways were not found to be expressed in cell culture systems, suggesting that, in contrast to both types of leaf senescence, cytokinins do not play a signaling role in cell culture senescence (Buchanan-Wollaston *et al.*, 2005).

9.2.7 Signaling pathways of the senescence program can be elucidated by global gene expression studies

One of the compelling approaches to identify regulatory elements and to dissect the signaling networks in developmental processes is to study gene expression patterns of known and unknown putative components of signaling pathways. Additional and complementary significant information may be found in comparative transcriptome studies of mutants defective in signaling pathways involving various (external as well as internal) regulatory factors in the senescence programs. To address the question of how similar are the signaling pathways involved in stress-induced and in developmental senescence, Affymetrix GeneChip experiments were designed to compare global gene expression patterns in response to UVB radiation with that of developmental senescence. The Affymetrix data pointed at common and different groups of genes expressed in both types of senescence. Therefore, it is hypothetical that some of the signaling pathways are common to stress and development related senescence, whereas others are specific to developmental senescence. Dissections of the expression patterns of putative components in signaling pathways were performed in mutants defective in responses to specific hormones. Three stress response pathways involving ethylene, jasmonic acid (JA) and salicylic acid (SA) signals are known to be associated with senescence (Grbic and Bleecker, 1995; Morris *et al.*, 2000; He *et al.*, 2002). These signaling molecules, known to increase during senescence, were also found to induce the expression of specific genes (Buchanan-Wollaston *et al.*, 2005). Buchanan-Wollaston *et al.* (2005) used Affymetrix array experiments to identify changes in the expression of senescence-related genes in the defective mutants. This study identified the downstream effects of the signalling pathways of the three stress responses.

For example, a large proportion of the genes altered by the mutants in the JA and/or ethylene pathways encode hydrolases involved in carbohydrate metabolism and others involved in nucleic acid degradation.

Some genes that depend on JA and ethylene show increased expression in the absence of the SA pathway. Examples of such genes are a chitinase, a glucanase and an osmotin-like gene (thaumatin-like) all of which have been implicated in pathogen responses. There are also a few genes that appear to be dependent on all three signalling pathways for expression. This group includes a monooxygenase, a putative subtilase and an ABC transporter. The majority of the genes that depend on JA and ethylene for expression during leaf senescence show increased abundance during both dark-induced and cell suspension senescence. This indicates that these pathways have an active signaling role in these other two types of senescence. The majority of the genes belong to the SA pathway encode kinases, transferases and hydrolases. The SA pathway has a key role in the disease resistance response and this is supported by the fact that many putative disease resistance genes are dependent on the SA pathway for expression during senescence (Buchanan-Wollaston *et al.*, 2005). SA signalling pathways was reported to be dependent on light signals and both SA-induced gene expression and the hypersensitive response were strongly reduced in tissues in darkness. The same was evident with a group of SA-dependent genes known to be associated with developmental senescence were not expressed in the dark-induced senescence. Moreover, the importance of this group of genes in senescence was assessed by comparing the progress of the two types of senescence in the NahG defective mutant in the SA pathway. While developmental senescence was delayed considerably in the NahG transgenic plants, no differences in the rates of dark-induced senescence between the wild type and the NahG transgenic plants could be detected. These results suggest an important role for SA in the induction of gene expression during developmental senescence but not in dark-induced senescence. However, in all the three types of senescence, many common genes were found to be involved in dismantling of cellular constituents and degradation of macromolecules.

9.2.8 Global gene expression studies reveal that autumn leaf senescence has much in common with the senescence in annual plants

Autumn leaf senescence in trees shares many morphological and physiological processes with monocarpic leaf senescence. It is hypothesized that the induction of autumn senescence, in most trees, is different than in annual leaf senescence and is typically dependent on short photoperiods. To decipher the molecular mechanisms involved in the regulation and the execution of autumn leaf senescence, comparative global gene expression studies have been performed in aspen tree using EST obtained from cDNA libraries and DNA microarray (Bhalerao *et al.*, 2003; Andersson *et al.*, 2004). Comparison of the list of the most abundant ESTs in the library representing autumn leaf to that of young leaf indicated considerable differences. In young leaf library, the most abundant ESTs are those encoding proteins of the photosynthetic apparatus, whereas in the autumn leaf library the prominent ones belong to clusters of ESTs representing different proteases and other hydrolases.

Since commercial microarrays have only been available for *Arabidopsis* and other few annual crops, the transcriptomic study of autumn leaf senescence in trees required large-scale EST sequencing prior to the construction of extended microarray of the aspen *(Populus)* tree. Andersson *et al.* (2004) have spotted 13 500 clones and created a DNA microarray for comparative global gene expression of natural autumn leaf senescence. Similar to monocarpic senescence, the transition from a photosynthetic leaf to a senescing organ during autumn senescence was reflected by a typical shift in gene expression. The changes in leaf pigmentation due to chlorophyll degradation and flavonoid accumulation are accompanied by the corresponding changes in the patterns of genes expression associated with carotnoids and flavonoids biosynthesis (Andersson *et al.*, 2004). Also similar to annual leaf senescence, the transition from anabolism to catabolism occurring in autumn leaf senescence is reflected by down regulation of photosynthetic genes, whereas the expression of mitochondria-related genes did not change throughout the whole period. These observations support the notion that mitochondria provide the energy required for the active processes occurring throughout the senescence period.

Microarray data indicated the upregulation of individual genes as well as functional genes clusters in autumn leaf senescence. The most relevant enzymes involved in the catabolism of macromolecules in both monocarpic and autumn leaf senescence are the proteases and the most prominent ones are the vacuolar cystein proteases. Although polyubiquitin transcripts increased during senescence, the other components of the ubiquitin-dependent degradation pathway did not show parallel increase. These results may support the notion that ubiquitin is not involved in the non selective total protein degradation but rather has a role in the degradation of specific regulatory proteins associated in leaf senescence of both systems of monocarpic senescence and autumn leaves (Bhalerao *et al.*, 2003; Gepstein *et al.*, 2003; Andersson *et al.*, 2004; Buchanan-Wollaston *et al.*, 2005).

Another group of genes whose products involved in the biosynthesis and perception of ethylene was found also to be upregulated in autumn aspen leaves (Andersson *et al.*, 2004). Methallothionein has been repeatedly reported as one of the most abundant gene families upregulated in many senescing systems and in aspen autumn leaves, five of the six identified methallothionein genes displayed a significant increase during senescence (Bhalerao *et al.*, 2003; Gepstein *et al.*, 2003; Andersson *et al.*, 2004; Buchanan-Wollaston *et al.*, 2005). ELIP transcripts were found to be upregulated in aspen trees as well as monocarpic leaf senescence. The proteins play a role in detoxifying the free chlorophyll molecules released during dismantling processes of the senescing chloroplast (Bhalerao *et al.*, 2003; Andersson *et al.*, 2004).

9.3 Proteomics of leaf senescence

Measuring gene expression at the protein level is potentially more informative than the corresponding measurement at the mRNA level. Proteins as major catalysts of biological functions contain several dimensions of information that collectively indicate the actual rather than the potential functional state as indicated by mRNA analysis. These include the abundance, state of post-translational modification, subcellular localization and association with each other (Griffin *et al.*, 2001). By definition, the proteome is the full complement of proteins produced by a particular genome of a cell, tissue or species at a given time. It represents a higher complexity than the transcriptome, and it displays a higher degree of dynamics due to post-translational modifications. The human genome for example, is estimated to include 30 000–35 000 genes; however, there are probably more than 500 000 different proteins in humans because of alternative splicing and post-translational modifications (Alam *et al.*, 1999). The number of proteins in plants is also predicted to be much higher than the number of genes due to operation of similar mechanisms of alternative splicing and modifications.

The term *proteomics* essentially refers to the large-scale analysis of many proteins simultaneously. The majority of proteomic techniques involve some form of separation procedure in order to isolate individual proteins (such as 2D PAGE and HPLC), ionization of peptides, followed by detection and identification by mass spectrometry (reviewed in Zhu *et al.*, 2003).

9.3.1 Technologies

9.3.1.1 Two-dimensional gel electrophoresis

Historically, two-dimensional (2D) gel electrophoresis provided the opportunity to determine quantitative and qualitative differences in protein composition of different samples long before gene array techniques for mRNAs were conceived.

In 2D polyacrylamide gel electrophoresis (2D PAGE), proteins are first separated according to charge and then in the second dimension by mass. This method is in principle capable of resolving up to 10 000 proteins in a single gel (O'Farrell, 1975). In the first phase of 2D PAGE proteins are separated by isoelectric focusing using immobilized pH gradients in a thin gel strip (Figure 9.4). Proteins migrate through the gel strip until their net charge is neutral (the isolelectric point, pI). Following separation in the first dimension, proteins are separated in the orthogonal direction using electrophoresis in acrylamide gels containing sodium dodecyl sulfate (SDS), where SDS imparts a net negative charge, allowing protein separation by mass (SDS-PAGE). Once proteins are separated, they can be visualized using a variety of stains. For imaging of gels, scanners are available and quantitative analysis of 2D gels can be performed by various available software programs. This can be used for analysing gels that have been stained by standard methods, as well as multi-fluorescence gels, in conjunction with the robotic equipment employed for spot picking from the gels. Recent improvements of 2D gel analysis include prior pre-fractionation to reduce the complexity of the sample, the use of fluorescent dyes with enhanced detection sensitivity

9.3.1.2 Liquid chromatography

By the 1980s, HPLC was commonly used to separate chemical compounds. This was followed by the development of micro-columns with capillary volumes ranging from 3 to 200 μm. The mobile phase acts as a carrier of the sample solution, which is injected into the mobile phase through the injector port. As the sample solution flows through the column with the mobile phase, the components of that solution migrate according to non-covalent interactions with the column. This, together with the chemical interactions of the mobile phase with the column determines the degree of migration and separation of components. The development of reversed phase columns made it possible to apply HPLC to proteins and peptides. Reversed phase operates on the basis of hydrophilic and lipophilicity. The stationary phase consists of silica-based packings with covalently bound n-alkyl chains. Using a gradient of 0.1% trifluoroacetic acid (TFA) in water to 0.1% TFA in an organic solvent, such as acetonitrile, hydrophilic compounds elute more quickly than do hydrophobic compounds. The use of size-exclusion chromatography and ion-exchange chromatography is well suited for the separation of biologically active proteins.

9.3.1.3 Mass spectrometry

Mass spectrometry (MS) (reviewed in Griffin *et al.*, 2001) is the preferred method for the identification of proteins following separation by 2D PAGE and LC (or HPLC) (Figure 9.2). It forms an essential tool in proteomic analysis, by providing structural information such as peptide mass and amino acid sequence, as well as information on protein modifications. Two major techniques were developed: matrix-assisted laser desorption ionization (MALDI) time-of-flight (TOF) MS, and electrospray ionization (ESI) MS (Hillenkamp *et al.*, 1991). The development of these relatively nondestructive, soft ionization methods made it possible to convert proteins into volatile ions. Usually, for peptide mass fingerprint (PMF) analysis, the protein to be

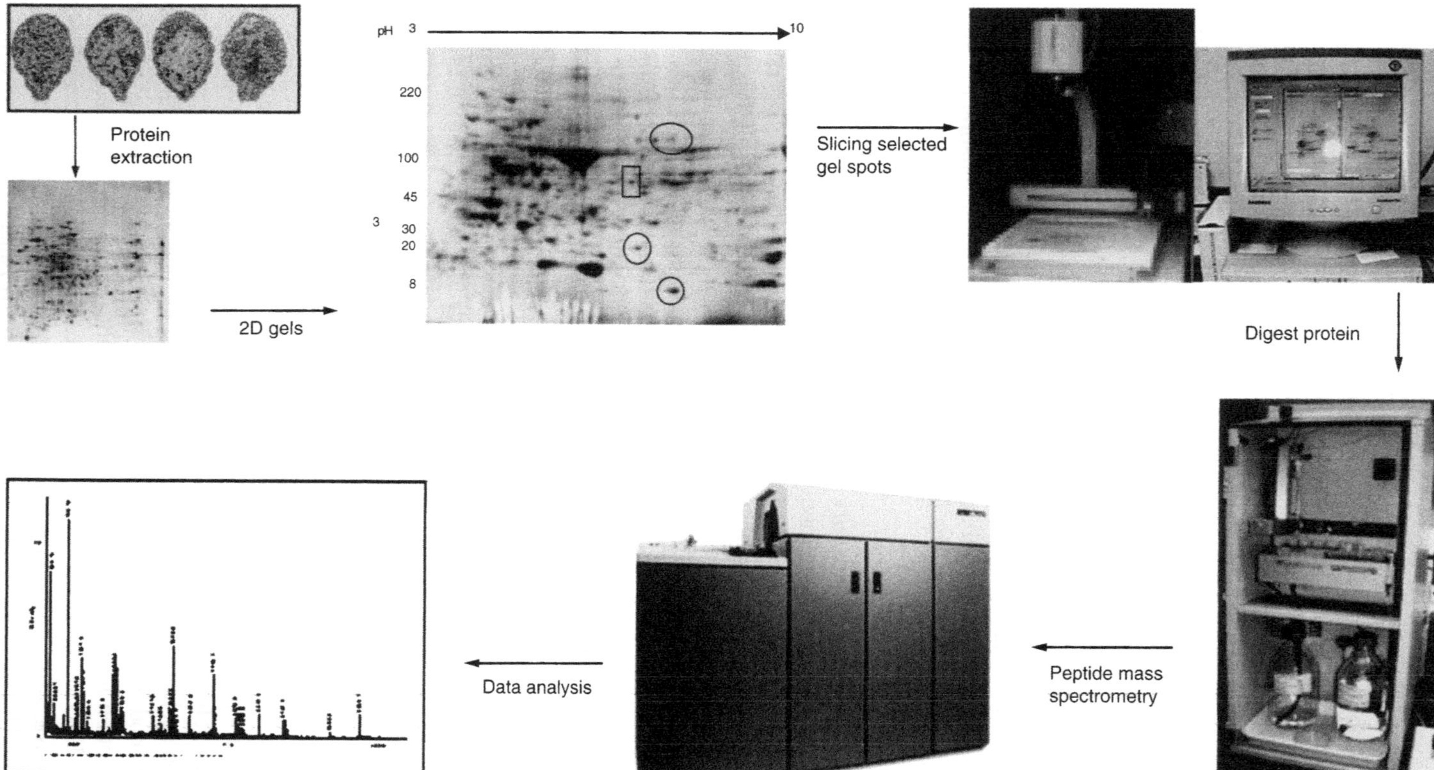

Figure 9.4 Methods for separation and identification of proteins. Proteins are extracted from plant material and separated by 2D gel electrophoresis. The gels are stained, photographed and captured. The protein spots are then sliced, digested and the resulting peptides are analysed by mass spectrometery. The mass spectrum of each peptide is searched against a protein database allowing the identification of the original protein.

analysed is purified, and digested enzymatically, generally by trypsin, to cleave the protein at specific bonds giving a reproducible pattern of digestion. For ESI–MS, the resulting peptides are injected into an HPLC column that is directly coupled to the mass spectrometer. For MALDI-MS, tryptic peptides are combined with a matrix on a MALDI target and then analysed. Gaseous ions are produced by accelerating an ionized particle, in this case a peptide, through a rarefied atmosphere to a detector. Then, the mass-to-charge (*m/z*) ratio is measured and masses are obtained, with high accuracy, for all peptides of a complex peptide mixture. This fingerprint is then compared to databases containing theoretical protein cleavage data producing a list of the closest matching proteins, using software packages such as Mascot or Sequest. Characterization of ESTs and completion of the Arabidopsis Genome Project resulted in huge databases of expressed and predicted genes. Accordingly, it is no longer necessary to identify proteins in their entirety, as it is often sufficient to simply recognize a fragment of a protein to confirm its presence even in very complex mixtures (Grant and Blackstock, 2001). Separation and identification of molecules is performed on the basis of their mass-to-charge ratio, in the case of MALDI using TOF methodologies. MALDI-MS can be combined with tandem MS/MS allowing the fragmentation of peptide species and subsequent sequence identification.

9.3.1.4 ESI mass spectrometry

Liquid chromatography (LC) coupled with MS has proven to be an important alternative to 2D gels. Of the various ionization methods developed for coupling LC to MS, ESI is the most widely used interface technique. Various LC techniques in sequence (affinity chromatography, size exclusion chromatography, reversed phase or charge) allow separation of complex samples (Wolters *et al.*, 2001). This LC–MS technique is particularly useful for membrane-associated proteins, phosphopeptides, kinases and transcription factors that are difficult to detect by 2D PAGE due to their low abundance (Link *et al.*, 1999; Washburn and Yates, 2000). In the tandem MS (MS/MS) procedure, which couples two stages of MS, a mixture of charged peptides is first separated according to their *m/z* ratios to create a list of all the detectable peptide peaks. In the second MS analysis, a specific *m/z* precursor species from this list is stored in the ion trap chamber and then directed into a collision cell to create product ions derived from the parent species. Using the appropriate collision energy, fragmentation occurs predominantly at the peptide bonds such that a ladder of fragments, each of which differs by the mass of a single amino acid, is generated. The product fragments are separated according to their *m/z*, and the sequence of the peptide can be deduced from the resulting fragments. By comparison with predicted sequences in the database, the identity of the peptide is revealed.

9.3.2 Current information on leaf senescence proteomics is limited

Searching the entrez database at NCBI using the search words "leaf-senescence proteomics" indicates less than a handful query results found. Even with the significant developments in the technologies used to quantify protein abundance over the past couple of years, protein identification and quantification still lags behind the high-throughput experimental techniques used to determine mRNA expression levels. However, when considering that proteins (a) are much more diverse than nucleic acids, (b) are much harder to isolate, (c) have limited detection sensitivity, and (d) differ in their abundance ranging over a six- to eightfold magnitude, it becomes evident that the insufficient number of leaf-senescence proteomics studies reported is due to intricacy which precludes fast progress.

The limited available studies have been performed in different biological systems; *Arabidopsis* whole leaf, *Arabidopsis* cell suspension, white clover leaves and rice leaves allowing us a glimpse at some common proteins differentially expressed during senescence and suggesting some common mechanisms. Generally, the individual protein levels were compared between leaves at their maximum photosynthetic capacity to leaves in two senescence stages: leaves with moderate loss of chlorophyll (10–20%) and with high chlorophyll loss (~50%). However, in the other two proteomic studies senescence stages were defined differently. In the cell suspension study, senescence was defined as cultures of 13–14 days old and in the study reporting changes in rice leaves old leaves were harvested during the ripening stage at the milk phase (Zhao *et al.*, 2005).

Although, it is difficult to compare data from studies of different biological systems, there are indications of common protein groups that are differentially expressed during senescence suggesting the occurrence of common mechanisms underlying the process of leaf senescence.

A recent global comparative proteomics analysis of the senescence syndrome aimed at identification of proteins and their putative functions in *Arabidopsis* leaves during senescence has been carried out by Carp and Gepstein (unpublished). The relative high resolution of the 2-D gel-electrophoresis method enabled comparing the abundance of individual protein spots during the various stages of senescence and to sample each of them for enzymatic digestion and mass spectrometry identification. Furthermore, by using the 2D-gel electrophoresis it was also possible to detect and identify some post-translational modifications such as phosphorylations (Carp and Gepstein, unpublished; Swidzinski *et al.*, 2004).

Figure 9.5 presents typical 2-D gels, each representing several hundreds of protein spots, part of which underwent changes in their abundance during senescence. Examples of down- and upregulated individual proteins have been demonstrated. The upregulated proteins identified in this study displayed several versions of distinct temporal expression profiles such as early expressed vs late expressed proteins. A proteomic study of white clover leaf senescence has also demonstrated several temporal expression patterns of proteins which correlated with the ultrastructural changes occurring during senescence, of which the most prominent are those of the chloroplasts (Wilson *et al.*, 2002). Similar to the genomics data (Figure 9.1), the temporal patterns of changes in the abundance of individual proteins may indicate the sequence of biochemical and ultrastructural and cellular events occurring during the onset and progressive stages of senescence (Carp and Gepstein unpublished; Wilson *et al.*, 2002).

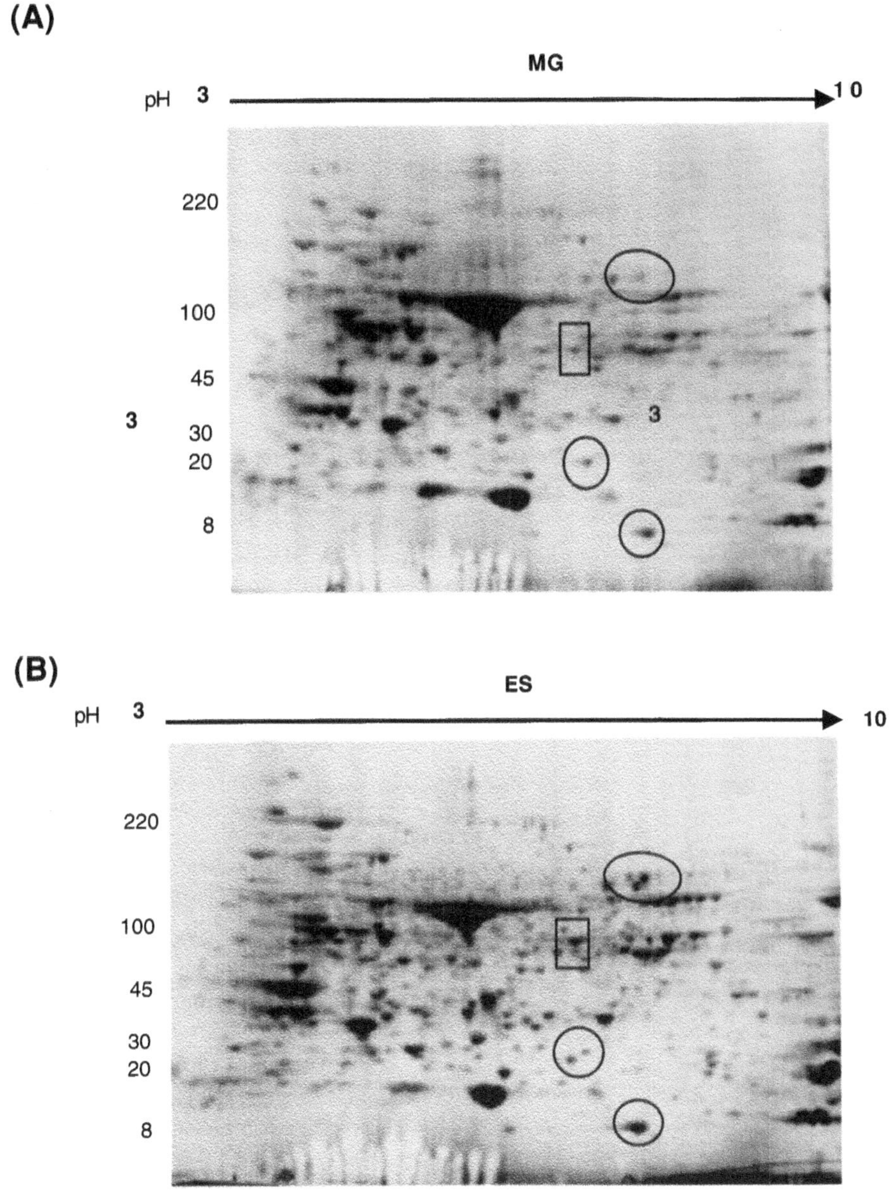

Figure 9.5 Comparative analysis of upregulated proteins during progressive stages of leaf senescence. 2D gels of proteins isolated from leaves at different developmental stages. (A) MG – mature green leaves at the pick of photosynthetic capacity; (B) ES – early senescent leaves with ~10% chlorophyll loss; (C) S1 – senescent leaves, up to 50% chlorophyll loss; (D) S2 – late senescent leaves, over 50% chlorophyll loss. Each gel presents several hundreds of protein spots, part of which underwent changes in their abundance during senescence. Examples of down- and upregulated individual proteins are circled (Carp and Gepstein, unpublished results). (*Continued overleaf*)

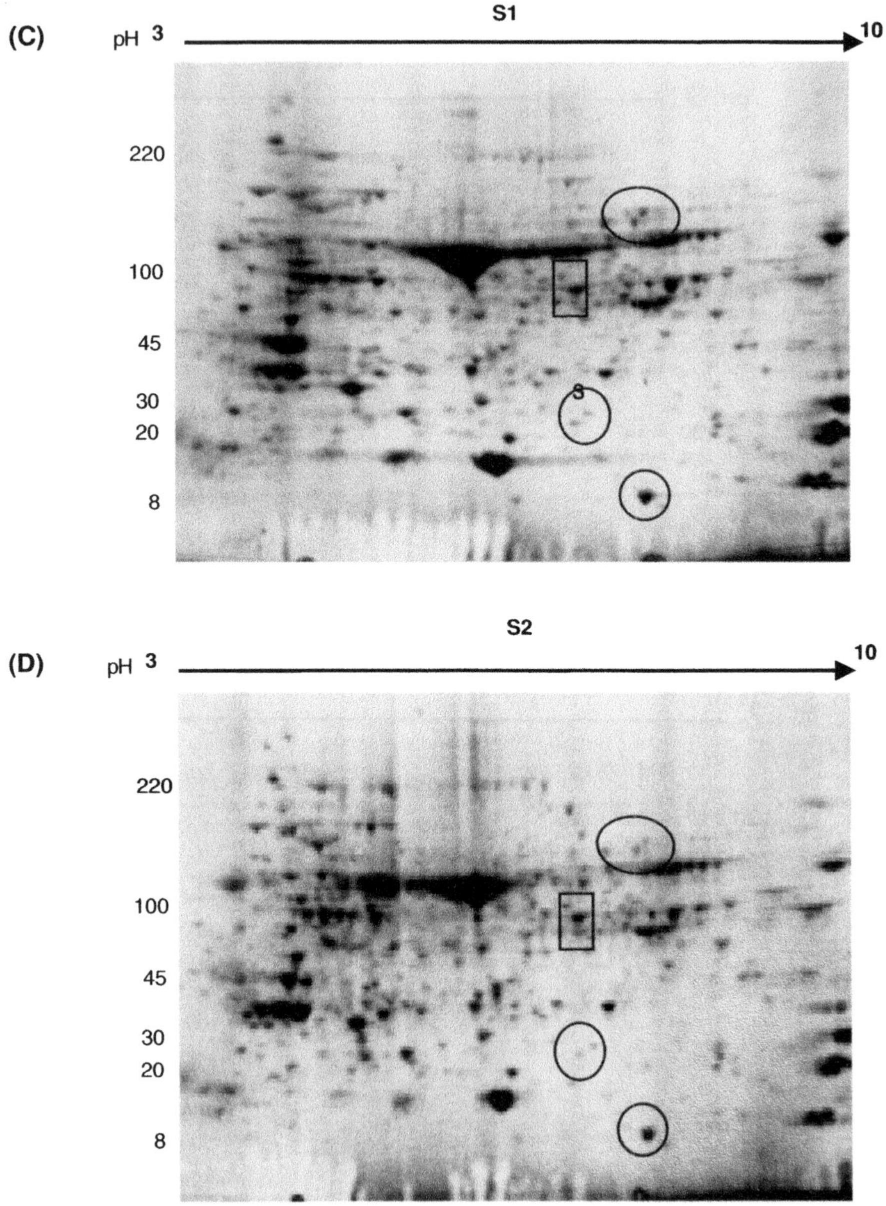

Figure 9.5 (*Continued*)

9.3.3 Functional categories of senescence-enhanced proteins

While several enzymatic activities of some of the senescence-associated proteins have been demonstrated in studies focusing on individual proteins, most of the protein functions identified in global proteomic studies have been predicted based on sequence homology. The impressive progress in the ability of protein identification by various MS methods and the availability of the sequence databases of *Arabidopsis* and other crop plants, allows unequivocal sequence identification of each of the peptides produced by digestions of the protein spots visualized on the gels.

The function of the senescence-related proteins identified falls into several categories: (1) those involved in respiration and related metabolic processes; (2) those involved in macromolecule breakdown and remobilization processes; (3) those predicted to be associated with protective responses to stress and pathogens; (4) and many others whose identity was determined by sequences homology but for which no known function in senescence is understood (Carp and Gepstein, unpublished).

9.3.4 Senescence upregulated proteins involved in respiration and various associated metabolic processes

A large proportion of the senescence-enhanced proteins are associated with respiration processes, which are known to increase during leaf senescence, fruit ripening and PCD processes. Among the proteins belonging to this category is the formate dehydrogenase (FDH) (Carp and Gepstein, unpublished). This enzyme plays an essential catalytic role in the final step of one-carbon metabolic oxidation and the generation of reducing equivalents. FDH activity is normally under metabolic control, and when sufficient reducing equivalents are available to the cell, FDH activity is inhibited and carbon assimilation is promoted from the cellular pool of formaldehyde. Under various conditions, such as during senescence, more NADH is required to fuel respiration, this enzyme is active and catalyzes dissimilation by the oxidation of formate to CO_2 (Olson *et al.*, 2000).

The behavior of the expression of this protein reflects typical upregulation of proteins involved in respiration and confirms previous observations showing that senescing tissues are energy starved due to diminishing rates of photosynthesis and active nutrient recycling processes. Moreover, the energy requirements of the senescing tissues cause carbon starvation, promote lipid to sugar gluconeogenesis and the production of sucrose. Our results indeed show enhanced expression of several enzymes involved in carbon metabolism. An example is alanine glyoxylate aminotransferase, an enzyme participating in the production of glycine and pyruvate used as sugar precursors. Additional up regulation of components of the citric acid cycle is reflected by the lipoamide dehydrogenase precursor, a component of the pyruvate dehydrogenase complex, aconitase, that catalyzes the reversible isomerization of citrate to isocitrate via *cis*-aconitate in the tricarboxylic acid cycle (Peyret *et al.*, 1995) and malate dehydrogenase, which catalyses the oxidation of malate to OAA and produces NADH, an important step in the citric acid cycle (Dangl *et al.*, 2000). The identification of upregulated proteins involved in ATP synthesis, citric acid cycle and carbohydrate metabolism, further suggests that some of the elements needed to provide energy to the leaf are enhanced during the last developmental stage of the leaf.

The dramatic upregulation of a large mass of proteins during leaf senescence suggests the operation of general complex processes including synthesis of new proteins, correct folding of their molecules and often post-translational modification (Nooden *et al.*, 1997; Yoshida, 2003). A hint of *de novo* synthesis of proteins was observed in the study of Carp and Gepstein, which showed increased expression of several proteins involved in these aspects; 50S ribosomal protein, an elongation factor and 2 different cyclophilins. All four proteins participate known to participate in processes of protein translation and post-translational modification.

9.3.5 Degradation and transport processes

A comparative study of the time course of ultrastructural and molecular changes during leaf senescence has indicated clear correlation between chloroplast dismantling and protein loss in white clover (Wilson *et al.*, 2002). Moreover, characterization of the proteomics of the senescence syndrome indicated that more than 50% of the detected soluble leaf proteins visualized in 2D gels decline in their abundance during senescence as compared to green mature leaves. The prominent chloroplast downregulated proteins with roles in photosynthesis such as large and small subunits of Rubisco and Rubisco activase declined during senescence. A typical uneven decline in the abundance of the two subunits of Rubisco indicated preferential degradation of the large subunit of Rubisco during clover leaf senescence. Surprisingly, the reported proteomic studies failed to detect up regulation of chloroplast proteases.

In contrast to many downregulated proteins during leaf senescence, many others remained stable or showed senescence-dependent increase. Similar genomics studies have shown that many genes encoding hydrolytic enzymes are upregulated during senescence. Indeed, the abundance of several enzymes involved in degradative processes was found to be increased during leaf senescence. Examples are β-glucosidase and xyloglucan endotrasnglycosylase (SEN4) associated with cell wall degradation have been found to be upregulated during *Arabidopsis* leaf senescence (Carp and Gepstein, unpublished). Since senescence is a major catabolic developmental stage and we would expect to identify a larger proportion of senescence upregulated proteins involved in degradation processes. The relatively low number of proteins reported in this functional category as compared to genomics studies can be attributed to several reasons: (1) These proteins indeed are not abundant. (2) The upregulated proteins are compartmentalized in subcellular organelles and their relative levels as compared to total cellular proteins are low. An example of by compartmentalized regulation of enzymes is presented in an EM immunolocalization study that showed that mitochondrial and peroxisomal MnSOD are regulated differently (del Rio *et al.*, 2003). Partial validation for these results is provided by Swidzinski *et al.* (2002) and Carp and Gepstein (data not published) who showed upregulation of mitochondrial MnSOD during senescence. (3) Protein activation by post-translational modifications may be an important regulatory mechanism of leaf senescence. Activation is, therefore, may not be accompanied by quantitative changes in protein levels. An example of post-translational regulation is the truncation of the protease RD21, a protein known senescence to be activated in the vacuole during senescence (Yamada *et al.*, 2001). A possible important additional activator during senescence, the γ-vacuolar processing enzyme, was also reported to be upregulated by Guterman *et al.* (2003). One of the interesting upregulated proteins is pyruvate orthophosphate dikinase (PPDK). This enzyme is well characterized in C4 plants but its function in C3 plants is unclear. However, the results of Carp and Gepstein support previous report showing that the PPDK transcript was upregulated during detached leaf senescence (Lin and Wu, 2004). It has been proposed that PPDK in C3 plants is responsible for the production of precursors for asparagine, a molecule required for nitrogen remobilization during leaf senescence (Lin and Wu, 2004).

9.3.6 Upregulated proteins related to stress and defense mechanisms

Relative increased abundance of proteins belonging to the first line of cellular enzymatic defense against oxidation damage such as catalase 3, mitochondrial MnSOD and ascorbate peroxidase are reported both in the whole leaf and in cell suspension studies of *Arabidopsis* senescence. These results are consistent with the observation that induction and execution of senescence are accompanied by oxidative stress (Swidzinski *et al.*, 2004; Buchanan-Wollaston *et al.*, 2005) (Carp and Gepstein, data not published). The identification of senescence-associated proteins that have been identified as participating in various plant stress responses supports and compliments the concept of the existence of a common plant response to stress. The following proteins provide such examples; the enzyme myrosinase which degrades glucosinolates due to wounding and the products found to be actively involved in pest defense, the cold tolerant dehydrin, COR47 involved in freeze stress tolerance (Puhakainen *et al.*, 2004) and the enzyme allene oxide cyclase, a key component of the JA biosynthesis pathway. Jasmonic acid was found to be induced both by various stress treatments and senescence (He *et al.*, 2002). In the context of defense against oxygen free radicals, two forms of catalase 3 were upregulated during leaf senescence. MS analysis of the two protein spots separated by 2D gel electrophoresis suggested post-translational phosphorylation of catalase 3 and the phsoporylated form was found to be preferentially expressed during the late stages of senescence (Carp and Gepstein, unpublished)

9.3.7 Comparison between pattern of changes in mRNA and protein levels during senescence indicates partial correlation

With all the excitement about the high throughput data of functional genomics of the senescence syndrome, however, it is also widely recognized that the correlation between mRNA concentrations and the identities and biological activities of the corresponding proteins is often not very strong. Events critical to the function of gene products are frequently determined post-transcriptionally through alternate splicing, post-translational protein modifications, protein–protein interactions or protein trafficking, and these post-transcriptional events can hardly be learned from gene expression patterns alone.

Because of post-transcriptional processing, a typical cell often harbors many more functional gene products than genes. Finally, protein–protein interactions and networking, which do not show up in the genomic information, are extremely important, and their alteration can feature in development and aging. From the discussion above, it is clear that in order to understand complex, multigene-controlled, biological phenomena such as leaf senescence more completely, we need to be able to determine, in real time, the concentrations of all gene products in the cell, the nature of any modifications in them, their cellular localization and interactions, and how the latter parameters affect their functional properties. Primarily because of limited availability of precise technologies to identify and measure protein abundance, gene expression studies have used mRNA levels to study leaf senescence. Therefore, analyses of mRNAs and proteins are both necessary and complementary for understanding how the senescing cell functions. The limited number of reports on the correlation between mRNA and protein expression level are most notable in yeast and human (Gygi *et al.*, 1999; Greenbaum *et al.*, 2003; Kadota *et al.*, 2003). These studies suggest that mRNA abundance is not necessarily a predictor of protein abundance (Gygi *et al.*, 1999; Greenbaum *et al.*, 2003; Swidzinski *et al.*, 2004; Tian *et al.*, 2004). In this context, leaf senescence is not exceptional. In a recent proteomic study performed in our laboratory, we have compared the temporal expression of proteins and their corresponding mRNA and found that ~60% of the studied upregulated proteins were accompanied by an increase in the steady state levels of the corresponding transcripts. However, the remainder of the proteins displaying differential expression during both detached and natural leaf senescence, exhibited opposite trends in their changes as compared to their mRNA levels. A prominent example of such inconsistency is the PPDK protein that was not detected during natural senescence but still exhibited significant mRNA expression (Carp and Gepstein, unpublished). Many other proteins associated with the senescence of both attached and detached leaves showed opposite or inconsistent trends in the pattern of changes of their corresponding mRNA levels.

There are presumably at least three reasons for the poor correlations between the level of mRNA and the level of protein, and these may not be mutually exclusive. First, there are many complicated and varied post-transcriptional mechanisms involved in translating mRNA into protein that are not yet sufficiently defined, to be able to compute protein concentrations from mRNA. Second, proteins may differ substantially in their in vivo half lives. Third, to date, there is still a significant amount of error and noise in both protein and mRNA experiments that limit our ability to get a clear picture.

An additional method to determine the proteome state during leaf senescence would be the analysis of microarray experiments comparing free versus polysome-bound mRNAs (translation state array analysis – TSAA). These experiments can be indicative of the protein levels providing a simultaneous comparison for the level of transcript and protein in the cell. Moreover, the method has no bias against proteins known to be difficult to isolate and identify as the integral membrane proteins.

However, although the finding of inconsistency between mRNA and protein levels during senescence is not surprising, yet the information regarding the changes in actual levels of proteins is important not only for the elucidation of the pathways and events involved in this process but also for biotechnological purposes. Identification of major proteins playing role in the senescence program, which are not necessarily detected at the mRNA levels, is much more relevant for further agrobiotechnological improvement especially in aspects of shelf life of vegetables and crop yield improvements.

9.4 Conclusions

A major goal of the biological sciences is to generate predictive organismal models. In the specific case of aging and senescence, we would like to construct a model that can predict all the changes occurring during the transition leading to the senescence syndrome. In our nascent post-genome era, we already have the potential to address higher levels of molecular organization than has previously been possible. Clearly, a deeper understanding of developmental and physiological processes can be attained through high-throughput studies rather than reactions of individual gene products or by the mere identification of the genes present in this organism. Global gene expression studies have indicated similarities and differences between the different types of leaf senescence. Common processes participating in developmental and stress-induced senescence include dismantling of cellular constituents and breakdown of the major macromolecules, proteins, lipids and nucleic acids. Rapid development in recent year of microarray technologies enables researchers to follow expression patterns of complete genomes, a feat that only a few years ago looked far-fetched. A primary role of functional genomic and proteomics is to generate the data and the tools (both analytical and computational) needed to build these models. At present, we clearly are at the very beginning of the proteomics era. The few studies related to proteomics of leaf senescence and other developmental processes indicate the potential of this tool to provide new knowledge at a level of detail that was impossible to envision just a few years ago. However, it is also clear that major technological and computational developments need to take place before the full potential of proteomics can be tapped. As the field of proteomics evolves, there is a need for a shift from the simple functional analyses of global protein concentrations to the identification of the roles of proteins in signaling networks and cellular pathways.

To achieve the above-stated goal, however, we need to be able to compare proteomes obtained at different time points in a reliable, accurate, sensitive and reproducible way and to develop detailed protein interaction maps.

References

Alam, S.L., Atkins, J.F. and Gesteland, R.F. (1999) Programmed ribosomal frameshifting: much ado about knotting. *Proc Natl Acad Sci USA* **96**, 14177.

Andersson, A., Keskitalo, J., Sjodin, A., *et al.* (2004) A transcriptional timetable of autumn senescence. *Genome Biol* **5**, R24.

Bhalerao, R., Keskitalo, J., Sterky, F., *et al.* (2003) Gene expression in autumn leaves. *Plant Physiol* **131**, 430–442.

Binyamin, L., Falah, M., Portnoy, V., Soudri, E. and Gepstein, S. (2000) The early light-induced protein is also produced during leaf senescence of *Nicotiana tabacum*. *Planta*, **211**, 591–597.

Buchanan-Wollaston, V. and Ainsworth, C. (1997) Leaf senescence in *Brassica napus*: cloning of senescence related genes by subtractive hybridisation. *Plant Mol Biol* **33**, 821–834.

Buchanan-Wollaston, V., Earl, S., Harrison, E., *et al.* (2003) The molecular analysis of leaf senescence – a genomics approach. *Plant Biotechnol J* **1**, 3–22.

Buchanan-Wollaston, V., Page, T., Harrison, E., *et al.* (2005) Comparative transcriptome analysis reveals significant differences in gene expression and signalling pathways between developmental and dark/starvation-induced senescence in *Arabidopsis*. *Plant J* **42**, 567–585.

Byers, R.J., Hoyland, J.A., Dixon, J. and Freemont, A.J. (2000) Subtractive hybridization – genetic takeaways and the search for meaning. *Int J Exp Pathol* **81**, 391.

Chen, W., Provart, N.J., Glazebrook, J., *et al.* (2002) Expression profile matrix of Arabidopsis transcription factor genes suggests their putative functions in response to environmental stresses. *Plant Cell* **14**, 559–574.

Dangl, J., Dietrich, R. and Thomas, H. (2000) Senescence and programmed cell death. In: *Biochemistry and Molecular Biology of plants* (eds B. Buchanan, W. Gruissem, and R. Jones). American Society of Plant Phsiologists, Rockville, MD, pp. 1044–1100.

del Pozo, J.C., Allona, I., Rubio, V., *et al.* (1999) A type 5 acid phosphatase gene from *Arabidopsis thaliana* is induced by phosphate starvation and by some other types of phosphate mobilising/oxidative stress conditions. *Plant J* **19**, 579–589.

del Rio, L.A., Sandalio, L.M., Altomare, D.A. and Zilinskas, B.A. (2003) Mitochondrial and peroxisomal manganese superoxide dismutase: differential expression during leaf senescence. *J Exp Bot* **54**, 923–933.

Diatchenko, L., Lukyanov, S., Lau, Y.F. and Siebert, P.D. (1999) Suppression substractive hybridization: a verstile method for identifying differentially expressed genes. *Methods Enzymol* **303**, 349–380.

Eulgem, T., Rushton, P.J., Robatzek, S. and Somssich, I.E. (2000) The WRKY superfamily of plant transcription factors. *Trends Plant Sci* **5**, 199–206.

Gan, S. and Amasino, R.M. (1995) Inhibition of leaf senescence by autoregulated production of cytokinin. *Science* **270**, 1986–1988.

Gepstein, S. (2004) Leaf senescence – not just a 'wear and tear' phenomenon. *Genome Biol* **5**, 212.

Gepstein, S., Sabehi, G., Carp, M.-J., *et al.* (2003) Large-scale identification of leaf senescence-associated genes. *Plant J* **36**, 629–642.

Grant, S.G.N. and Blackstock, W.P. (2001) Proteomics in neuroscience: from protein to network. *J Neurosci* **21**, 8315.

Grbic, V. and Bleecker, A.B. (1995) Ethylene regulates the timing of leaf senscence in *Arabidopsis*. *Plant J* **8**, 595–602.

Greenbaum, D., Colangelo, C., Williams, K. and Gerstein, M. (2003) Comparing protein abundance and mRNA expression levels on a genomic scale. *Genome Biol* **4**, 117.

Griffin, T.J., Goodlett, D.R. and Aebersold, R. (2001) Advances in proteome analysis by mass spectrometry. *Curr Opin Biotech* **12**, 607.

Guo, Y., Cai, Z. and Gan, S. (2004) Transcriptome of *Arabidopsis* leaf senescence. *Plant Cell Environ* **27**, 521–549.

Guterman, A., Hajouj, T. and Gepstein, S. (2003) Senescence-associated mRNAs that may participate in signal transduction and protein trafficking. *Physiol Plant* **118**, 439–446.

Gygi, S.P., Rochon, Y., Franza, B.R. and Aebersold, R. (1999) Correlation between protein and mRNA abundance in yeast. *Mol Cell Biol* **19**, 1720–1730.

Hajouj, T., Michelis, R. and Gepstein, S. (2000) Cloning and characterization of a receptor-like protein kinase gene associated with senescence. *Plant Physiol* **124**, 1305–1314.

He, Y., Fukushige, H., Hildebrand, D.F. and Gan, S. (2002) Evidence supporting a role of jasmonic acid in *Arabidopsis* leaf senescence. *Plant Physiol* **128**, 876–884.

He, Y. and Gan, S. (2002) A gene encoding an acyl hydrolase is involved in leaf senescence in *Arabidopsis*. *Plant Cell* **14**, 805–815.

He, Y., Tang, W., Swain, J.D., Green, A.L., Jack, T.P. and Gan, S. (2001) Networking senescence-regulating pathways by using *Arabidopsis* enhancer trap lines. *Plant Physiol* **126**, 707–716.

Hillenkamp, F., Karas, M., Beavis, R.C. and Chait, B.T. (1991) Matrix-assisted laser desorption/ionization mass spectrometry of biopolymers. *Anal Chem* **63**, 1193A–1203A.

Hinderhofer, K. and Zentgraf, U. (2001) Identification of a transcription factor specifically expressed at the onset of leaf senescence. *Planta* **213**, 469–473.

Kadota, K., Tominaga, D., Asai, R. and Takahashi, K. (2003) Correlation analysis of mRNA and protein abundances in human tissues. *Genome Lett* **2**, 139–148.

Klee, H.J., Hayford, M.B., Kretzmer, K.A., Barry, G.F. and Kishore, G.M. (1991) Control of ethylene synthesis by expression of a bacterial enzyme in transgenic tomato plants. *Plant Cell* **3**, 1187–1193.

Kore-eda, S., Cushman, M.A., Akselrod, I., *et al.* (2004) Transcript profiling of salinity stress responses by large-scale expressed sequence tag analysis in *Mesembryanthemum crystallinum*. *Gene* **341**, 83.

Lers, A., Lomaniec, E., Burd, S. and Khalchitski, A. (2001) The characterization of LeNUC1, a nuclease associated with leaf senescence of tomato. *Physiol Plantarum*, **112**, 176–182.

Liang, P. and Pardee, A.B. (1992) Differential display of eukaryotic messenger RNA by means of the polymerase chain reaction. *Science* **257**, 967–971.

Lim, P.O. and Nam, H.G. (2005) The molecular and genetic control of leaf senescence and longevity in *Arabidopsis*. In: *Current Topics in Developmental Biology*. Academic Press, New York, pp. 49.

Lim, P.O., Woo, H.R. and Nam, H.G. (2003) Molecular genetics of leaf senescence in *Arabidopsis*. *Trends in Plant Sci* **8**, 272.

Lin, J.F. and Wu, S.H. (2004) Molecular events in senescing *Arabidopsis* leaves. *Plant J* **39**, 612–628.

Link, A.J., Eng, J., Schieltz, D.M., *et al.* (1999) Direct analysis of protein complexes using mass spectrometry. *Nature Biotechnol* **17**, 676–682.

Morris, K., MacKerness, S.A., Page, T., *et al.* (2000) Salicylic acid has a role in regulating gene expression during leaf senescence. *Plant J* **23**, 677–685.

Navabpour, S., Morris, K., Allen, R., Harrison, E., A-H-Mackerness, S. and Buchanan-Wollaston, V. (2003) Expression of senescence-enhanced genes in response to oxidative stress. *J Exp Bot* **54**, 2285–2292.

Nooden, L.D., Guiamet, J.J. and John, I. (1997) Senescence mechanisms. *Physiol.Plant* **101**, 746–753.

O'Farrell, P.H. (1975) High resolution two-dimensional electrophoresis of proteins. *J. Biol. Chem.* **250**, 4007–4021.

Oh, S.A., Park, J.H., Lee, G.I., Paek, K.H., Park, S.K. and Nam, H.G. (1997) Identification of three genetic loci controlling leaf senescence in Arabidopsis thaliana. *Plant J* **12**, 527–535.

Olson, B.J.S.C., Skavdahl, M., Ramberg, H., Osterman, J.C. and Markwell, J. (2000) Formate dehydrogenase in *Arabidopsis thaliana*: characterization and possible targeting to the chloroplast. *Plant Sci* **159**, 205–212.

Ori, N., Juarez, M.T., Jackson, D., Yamaguchi, J., Banowetz, G.M. and Hake, S. (1999) Leaf senescence is delayed in tobacco plants expressing the maize homeobox gene knotted1 under the control of a senescence-activated promoter. *Plant Cell* **11**, 1073–1080.

Park, J.H., Oh, S.A., Kim, Y.H., Woo, H.R. and Nam, H.G. (1998) Differential expression of senescence-associated mRNAs during leaf senescence induced by different senescence-inducing factors in *Arabidopsis*. *Plant Mol Biol* **37**, 445–454.

Peyret, P., Perez, P. and Alric, M. (1995) Structure, genomic organization, and expression of the *Arabidopsis thaliana* Aconitase gene. *J. Biol. Chem.* **270**, 8131–8137.

Pruzinska, A., Tanner, G., Aubry, S., *et al.* (2005) Chlorophyll breakdown in senescent Arabidopsis leaves. Characterization of chlorophyll catabolites and of chlorophyll catabolic enzymes involved in the degreening reaction. *Plant Physiol.* **139**, 52–63.

Puhakainen, T., Hess, M.W., Makela, P., Svensson, J., Heino, P. and Palva, E.T. (2004) Overexpression of multiple dehydrin genes enhances tolerance to freezing stress in *Arabidopsis*. *Plant Mol Biol* **54**, 743–753.

Rebrikov, D.V., Britanova, O.V., Gurskaya, N., Lukyanov, K., Trarabykin, V.S. and Lukyanov, S.A. (2000) Mirror orientation selection (MOS): a method for eliminating false positive clones from libraries generated by supression subtractiv hybridization. *Nucleic Acids Res* **28**, e90.

Robatzek, S. and Somssich, I.E. (2002) Targets of AtWRKY6 regulation during plant senescence and pathogen defense. *Genes Dev.* **16**, 1139–1149.

Roca, M., James, C., Pruzinska, A., Hortensteiner, S., Thomas, H. and Ougham, H. (2004) Analysis of the chlorophyll catabolism pathway in leaves of an introgression senescence mutant of *Lolium temulentum*. *Phytochem* **65**, 1231–1238.

Sagerstrom, C.G., Sun, B.I. and Sive, H.L. (1997) Subtractive cloning: Past, present, and future. *Annu Rev Biochem* **66**, 751.

Smart, C.M. (1994) Gene expression during leaf senescence. *New Phytology* **126**, 419–448.

Swidzinski, J.A., Leaver, C.J. and Sweetlove, L.J. (2004) A proteomic analysis of plant programmed cell death. *Phytochem* **65**, 1829.

Swidzinski, J.A., Sweetlove, L.J. and Leaver, C.J. (2002) A custom microarray analysis of gene expression during programmed cell death in *Arabidopsis thaliana*. *Plant J* **30**, 431–446.

Thompson, J., Taylor, C. and Wang, T.W. (2000) Altered membrane lipase expression delays leaf senescence. *Biochem Soc Trans* **28**, 775–777.

Tian, Q., Stepaniants, S.B., Mao, M., *et al.* (2004) Integrated genomic and proteomic analyses of gene expression in mammalian cells. *Mol Cell Proteomics* **3**, 960–969.

Varshavsky, A. (1997) The N-end rule pathway of protein degradation. *Genes Cells* **2**, 13–28.

Washburn, M.P. and Yates, J.R., III (2000) Analysis of the microbial proteome. *Curr Opin Microbiol* **3**, 292–297.

Weaver, L.M., Gan, S., Quirino, B. and Amasino, R.M. (1998) A comparison of the expression patterns of several senescence-associated genes in response to stress and hormone treatment. *Plant Mol Biol* **37**, 455–469.

Wilson, K.A., McManus, M.T., Gordon, M.E. and Jordan, T.W. (2002) The proteomics of senescence in leaves of white clover, *Trifolium repens* (L.). *Proteomics* **2**, 1114–1122.

Wolters, D.A., Washburn, M.P. and Yates, J.R., III (2001) An automated multidimensional protein identification technology for shotgun proteomics. *Anal Chem* **73**, 5683–5690.

Woo, H.R., Chung, K.M., Park, J.H., *et al.* (2001) ORE9, an F-Box protein that regulates leaf senescence in *Arabidopsis*. *Plant Cell* **13**, 1779–1790.

Yamada, K., Matsushima, R., Nishimura, M. and Hara-Nishimura, I. (2001) A slow maturation of a cysteine protease with a granulin domain in the vacuoles of senescing *Arabidopsis* leaves. *Plant Physiol.* **127**, 1626–1634.

Yoshida, K. (2003) Molecular regulation of leaf senescence. *Curr Opin Plant Biol* **6**, 79–84.

Yoshida, S., Ito, M., Callis, J., Nishida, I. and Watanabe, A. (2002) A delayed leaf senescence mutant is defective in arginyl-tRNA:protein arginyltransferase, a component of the N-end rule pathway in *Arabidopsis*. *The Plant J* **32**, 129–137.

Yoshida, S., Ito, M., Nishida, I. and Watanabe, A. (2001) Isolation and RNA gel blot analysis of genes that could serve as potential molecular markers for leaf senescence in *Arabidopsis thaliana*. *Plant Cell Physiol* **42**, 170–178.

Zentgraf, U., Jobst, J., Kolb, D. and Rentsch, D. (2004) Senescence-related gene expression profiles of rosette leaves of *Arabidopsis thaliana*: leaf age versus plant age. *Plant Biol (Stuttg)* **6**, 178–183.

Zhao, C., Wang, J., Cao, M., *et al.* (2005) Proteomic changes in rice leaves during development of field-grown rice plants. *Proteomics* **5**, 961–972.

Zhu, H., Bilgin, M. and Snyder, M. (2003) Proteomics. *Ann Rev Biochem* **72**, 783.

Zimmermann, P. and Zentgraf, U. (2005) The correlation between oxidative stress and leaf senescence during plant development. *Cell Mol Biol Lett* **10**, 515–534.

10 Molecular regulation of leaf senescence

Hyo Jung Kim, Pyung Ok Lim and Hong Gil Nam

10.1 Introduction

10.1.1 Leaf senescence

Leaf senescence is a highly regulated degenerative process that involves a series of concerted biochemical reactions (Nam, 1997; Quirino *et al.*, 2000; Buchanan-Wollaston *et al.*, 2003; Lim and Nam, 2005). Chloroplast degradation and concomitant attenuation of anabolic activities such as photosynthesis and overall protein synthesis begin as a leaf surpasses its peak of assimilatory capacity. In contrast, catabolic activities such as the macromolecular breakdown of nucleic acid, proteins and lipids are increased through the induction of a number of hydrolytic enzymes. The main purpose of the cellular activities of leaf senescence is to convert accumulated materials into exportable nutrients that can then be transported and supplied to developing seeds or other actively growing tissues. Thus, leaf senescence is a type of programmed cell death (PCD) that occurs via a genetic program as a developmental process (Nam, 1997). Cell death during natural senescence occurs relatively slower than that of many other PCDs. The slow degeneration of cells during leaf senescence is in part to ensure efficient remobilization of nutrients that are generated by macromolecular hydrolysis during senescence (Buchanan-Wollaston *et al.*, 2003; Lim *et al.*, 2003). Accordingly, the temporal control of leaf senescence is critical to a plant's fitness to ensure optimal production of their offspring and better survival of plants.

10.1.2 Senescence-associated genes

Consistent with the notion that leaf senescence is a genetically controlled developmental process with critical physiological changes from a photo-assimilatory organ to a degenerating one, there occurs a massive change of gene expression. The transcriptome analysis performed with a senescing leaf of *Arabidopsis* revealed that a large fraction of gene expression in a green leaf is downregulated, while there is also an upregulation of the expression of up to 2500 genes (Guo *et al.*, 2004). In general, the downregulated genes are involved in anabolic activities. Upregulated genes, generally referred to as senescence-associated genes, or *SAG*s, are mostly catabolic ones. In this section, we will focus on the nature, function and regulatory modes and mechanisms of *SAG*s.

10.2 Isolation and classification of *SAGs*

10.2.1 Isolation of SAGs

Over the last decade, fairly extensive research efforts have been conducted to reveal the underlying molecular mechanisms of leaf senescence by identifying and analysing *SAG*s. Identification of *SAG*s has been aided by various methodologies studying differential gene expression (Lohman *et al.*, 1994; Buchanan-Wollaston and Ainsworth, 1997; John *et al.*, 1997; Park *et al.*, 1998; Weaver *et al.*, 1998; Quirino *et al.*, 1999; Guterman *et al.*, 2003). The initial collections of *SAG*s in *Arabidopsis* were identified by the differential screening methods (Lohman *et al.*, 1994; Park *et al.*, 1998). Lately, suppression subtractive hybridization was used to add more *SAG*s collection in *Arabidopsis* (Quirino *et al.*, 1999; Gepstein *et al.*, 2003). We have also identified several putative regulatory genes of senescence in *Arabidopsis* by a similar method (Lim *et al.*, unpublished data). Generation of expressed sequence tags has been used for identifying genes preferentially expressed in autumn leaves of field-grown aspen (*Populus tremula*) (Bhalerao *et al.*, 2003) and for identifying *SAG*s from senescing leaves of *Arabidopsis* (Guo *et al.*, 2004).

Recent global analyses of *SAG*s with DNA microarrays are revealing an unprecedented level of the nature and expression patterns of *SAG*s at the genome-wide level. For example, 677 *SAG*s of aspen leaves during autumn senescence were analysed by a DNA microarray (Andersson *et al.*, 2004). Buchanan-Wollaston *et al.* (2005) identified more than 2200 *SAG*s of *Arabidopsis* leaves, which show a reproducible increase in transcript abundance during developmental ageing. Lin and Wu (2004) surveyed expression changes of transcription factor genes, establishing an apparent association of the transcription factor families with leaf senescence. Chen *et al.* (2002) analysed expression profiles of 402 potential transcription factor genes during leaf developmental stages and also during various biotic and abiotic stress conditions.

One approach that allows identification of *SAG*s and their *in planta* function simultaneously is to screen enhancer or promoter trap lines. In this approach, *SAG*s are identified by examining the expression of a reporter gene. He *et al.* (2001) screened 1300 enhancer trap lines of *Arabidopsis* and identified 147 lines that show enhanced expression of the *GUS* reporter gene in senescing leaves. We have developed a new gene-trap system carrying promoterless GUS-luciferase fusion reporter genes that provides a non-destructive monitoring of gene expression via a luminescence assay and at the same time allows an anatomical examination of gene expression by GUS staining. During initial screenings, it was possible to isolate a few gene-trap lines, in which the reporter activities were enhanced in yellowing areas of senescing leaves (Koo *et al.*, unpublished data).

Although the amount of total cellular RNA and the major portions of mRNA species decrease with the progression of senescence, it is obvious that a large number of mRNA species are newly synthesized and experience an increase in abundance. These studies provide a strong molecular foundation for the notion that leaf senescence is an active and genetically programmed process of high complexity.

10.2.2 *Functional classification of SAGs*

The functional characterization of *SAG*s should lead to the molecular understanding of leaf senescence. Although the function of many *SAG*s is unknown, the functionally assignable spectrum of *SAG*s is generally consistent with known biochemical and physiological aspects and has provided new insight into the molecular regulation of leaf senescence. Based on predicted physiological functions, *SAG*s may be grouped into the following functional categories.

10.2.2.1 *Macromolecule degradation*

Protein degradation. Many genes involved in protein degradation are induced during leaf senescence. These include the genes encoding cysteine proteases, cathepsin B-like cystein proteases, aspartic proteases, vacuolar processing enzymes (see Chapter 5) and components of the novel autophagic (APG) pathway (Kinoshita *et al.*, 1999; Shikanai *et al.*, 2001; Thompson and Vierstra, 2005). It is notable that although chloroplasts contain up to 75% of leaf protein, subcellular fractionation studies indicate that most of the proteolytic activity in senescing leaf cells is located elsewhere, particularly in the vacuole. Among chloroplast stroma-localized Clp protease family members, ClpD/ERD1 and ClpC1 are induced in darkness-treated leaves (Nakashima *et al.*, 1997; Nakabayashi *et al.*, 1999). This implies that these two Clp proteases may play major regulatory roles in senescing leaves. Expression of several genes encoding components of autophagy is also increased during senescence (Doelling *et al.*, 2002). Autophagy is a key process for vacuolar bulk degradation of cytoplasmic components, either to remove damaged or unnecessary proteins or to recycle nutrients. As observed in yeast, autophagy may contribute to maintaining cell viability during leaf senescence since disruption of *Arabidopsis* autophagy genes, *AtAPG7* and *AtAPG9*, by T-DNA insertion resulted in accelerated leaf senescence (Doelling *et al.*, 2002; Hanaoka *et al.*, 2002). It has also been shown that *AtAPG18a,* a member of the *AtAPG18* gene family, is required for a proper process of leaf senescence since the reduced expression of *AtAPG18a* resulted in an inability to produce autophagosomes, resulting in premature senescence (Xiong *et al.*, 2005).

Lipid degradation. During leaf senescence, a decline in the structural and functional integrity of cellular membranes is clearly observed at an ultrastructural level, which is the result of the hydrolysis and metabolism of membrane lipids (see Chapter 3).

Chlorophyll degradation. During plant senescence, chlorophyll is degraded to non-fluorescent chlorophyll catabolites through a multi-step pathway (Matile *et al.*, 1999). This reaction is catalyzed by the two essential enzymes, pheophorbide *a* oxygenase (PaO) and red chlorophyll catabolite reductase (RCCR) (Hörtensteiner *et al.*, 1998). More detailed discussions are given in Chapter 2.

Nucleic acid degradation. Nucleic acids are important sources of nitrogen and phosphorous. *SAG*s encoding nucleases are expressed differentially at various stages of leaf senescence. The overall decrease in RNA levels in a cell is accompanied by the upregulation of *SAG*s for several RNases including BFN1. BFN1 is a bifunctional nuclease I enzyme that degrades both RNA and DNA (Pérez-Amador *et al.*, 2000). Furthermore, upregulation of acid phosphatase genes in senescing leaves supports the notion that nucleic acid breakdown is important for salvageing phosphorus (Buchanan-Wollaston *et al.*, 2005).

10.2.2.2 Nutrient salvage and translocation

The observed changes in metabolism and cell structure in senescent leaves are closely associated with a nutrient recycling program. This nutrient re-mobilization includes a general loss of protein within the senescing leaves and thus is dependent on developmentally regulated degradation of endogenous proteins for nitrogen salvage (see Chapter 5).

10.2.2.3 Defence and detoxification genes

A number of *SAG*s encode proteins related to pathogenesis and defence (Quirino *et al.*, 1999; Gepstein *et al.*, 2003). Some of the *PR* genes, which are expressed in response to pathogens, are also expressed during the senescence of healthy leaves (Hörtensteiner *et al.*, 1998). This indicates that *PR* genes play a role in leaf senescence in addition to their role in pathogen infection.

Many stress-inducible genes are upregulated during senescence, implying that the cells are under stress conditions during senescence (Blein *et al.*, 2002; Eriksson *et al.*, 2002). These *SAG*s include genes for lipid transfer proteins and myrosinase-binding proteins, which have roles in plant resistance to biotic and abiotic stresses. Other stress-related *SAG*s include genes for metallothionein and ferritin, which may be involved in the chelation of metal ions released during cellular degradation and in the transport of these ions into developing organs. Increased levels of reactive oxygen species (ROS) are commonly observed with different stress responses. During leaf senescence, macromolecule degradation also increases levels of ROS (del Rio *et al.*, 1998). In addition, levels of lipid peroxidation products increase in senescing *Arabidopsis* leaves (Ye *et al.*, 2000; John *et al.*, 2001). *SAG*s accordingly include the genes involved in the oxidative stress response such as the expression of genes for Fe^{2+}-ascorbate oxidase, anionic peroxidase, glutathione S-transferase and a blue copper-binding protein.

10.2.2.4 Regulatory genes

It is clear that the global and complex gene expression patterns and metabolic pathways during leaf senescence involve a sophisticated regulatory system. This regulatory system includes genes that encode various transcription factors and signalling components that perceive or distribute senescence signals.

Transcription factors. Members of the various transcription factor families including WRKY, NAC, bZIP, C2H2, EREBP and MYB show increased expression during senescence. In *Arabidopsis*, the WRKY family includes plant-specific transcription factors that regulate several plant processes such as pathogen defence and senescence. In particular, expression of *AtWRKY6* is influenced by several external and internal signals associated with senescence (Robatzek and Somssich, 2001). The potential targets of AtWRKY6 include genes encoding senescence-associated protein 1 (SEN1), a protease, the jasmonic acid (JA) regulatory protein NAC2 and a glutathione transferase. Furthermore, AtWRKY6 and AtWRKY42 act upstream of the senescence-induced receptor kinase gene, *SIRK*, in the process of leaf senescence (Robatzek and Somssich, 2002). This implies that regulation of *SIRK* promoter activity involves functionally redundant members of the WRKY family. Another *WRKY* gene, *AtWRKY53*, is specifically expressed at a very early stage of leaf senescence (Hinderhofer and Zentgraf, 2001), and putative target genes of WRKY53 include various *SAG*s, PR genes, stress-related genes and transcription factors including other WRKY factors (Miao *et al.*, 2004). This indicates that *AtWRKY53* may play a regulatory role in the early events of leaf senescence.

NAC proteins make one of the largest families of plant-specific transcription factors in *Arabidopsis* with more than 100 members. (Guo *et al.*, 2004; Olsen *et al.*, 2005). Several members of the NAC domain proteins showed enhanced expression during natural senescence and dark-induced senescence (John *et al.*, 1997). *AtNAP*, an *Arabidopsis* NAC family transcription factor, has been shown to play an important role in regulating leaf senescence in *Arabidopsis* and possibly in other crops too (Guo and Gan, 2006). The tomato *SAG, SENU5*, also encodes a protein that belongs to the NAC domain family. Recently we have isolated a delayed leaf senescence mutant, which is due to nonsense mutation in one of NAC transcription factors (Kim *et al.*, unpublished data). It is likely that at least a few of the NAC domain proteins play a regulatory role in leaf senescence.

In plants, basic leucine zipper motif (bZIP) transcription factors regulate diverse biological processes such as pathogen defence, light and stress signalling, seed maturation and flower development (Jakoby *et al.*, 2002). Interestingly, almost half of the bZIP transcription factors are induced in dark-treated leaves of *Arabidopsis*. It remains to be seen how these transcription factors control leaf senescence. The two genes belonging to the bZIP gene family in tobacco, *TBZF* and *TBZ17*, are both expressed specifically in the guard cells of senescing leaves (Yang *et al.*, 2001). The stomata of senescing leaves remain operable until the very last stage of leaf senescence. These two proteins might function in senescing guard cells as regulators maintaining the cellular functions.

Many potential transcription factors have been identified as *SAG*s through DNA microarray analysis, some of which are likely involved in leaf senescence, although *in vivo* functions and target genes remain to be elucidated.

Signalling molecules. Senescence is associated with the induction of various genes that are potentially involved in signal perception and transduction, including protein kinases and phosphatases. For example, receptor kinases may trigger the transduction cascade of senescence signals via protein phosphorylation. Senescence-associated receptor-like kinase of bean (*SARK*) is exclusively expressed during senescence, especially prior to the loss of chlorophyll and the decrease in *CAB* mRNA (Hajouj *et al.*, 2000). Treatment with light or cytokinin delayed the initial appearance of the *SARK* transcript, while darkness or ethylene accelerated it. Functional assignment of this gene may shed light on the phosphorylation cascades involved in leaf senescence. Another receptor-like kinase gene, which is gradually accumulated during progress of leaf senescence, was isolated in our laboratory (Koo *et al.*, unpublished data). Inducible expression of this receptor kinase gene hastened the onset of natural leaf senescence as well as leaf senescence accelerated by ABA among various senescence-inducing factors.

MAPK (mitogen-activated protein kinase) cascades are responsible for several aspects of plant growth and development (Bögre *et al.*, 1999; Wilson and Heberle-Bors, 2000). They are activated in response to a variety of stress stimuli, such as wounding and pathogen attack. In *Arabidopsis*, MAP kinase 4 (MPK4) functions as a regulator of pathogen defence responses (Petersen *et al.*, 2000). Senescence-associated gene *SAG101* showed increased expression in the *mpk4* knockout mutant, indicating that MAPK components may play a role in coordinating the senescence process. One of MAPK genes, *ZmMPK5*, was also isolated as a *SAG* in maize (Berberich *et al.*, 1999).

The involvement of calcium in many types of cell death has been demonstrated (Jones, 2001), and elevation of cytoplasmic calcium concentration has been observed during dark-induced leaf senescence in detached parsley (*Petroselinum crispum*) leaves (Huang *et al.*, 1997). DNA microarray analysis revealed that expression of several genes involved in calcium regulation is increased during senescence. They include calcium- and calmodulin-binding proteins and a specific calcium-dependent protein kinase. Calcium ions may be second messengers in some aspects of leaf senescence and these proteins may function in the calcium-mediated processes.

10.2.3 Comparison of SAGs in various plant species

While most *SAG* collections are derived from *Arabidopsis*, there are several reports on *SAG*s from other plant species, which reveal many characteristic differences at the molecular level. While this aspect has not been rigorously studied yet, we describe a few efforts on the study of *SAG*s in plant species other than *Arabidopsis*.

A number of different cDNA clones representing *SAG*s were isolated by subtractive hybridization in *Brassica napus* (Buchanan-Wollaston and Ainsworth, 1997). They mostly overlap in sequence with *Arabidopsis SAG*s. For example, *Brassica SAG*s include genes encoding proteases and enzymes involved in amino acid biosynthesis such as glutamine synthetase and ATP sulphurylase, which function in the degradation of cellular components and mobilization of essential nutrients. The genes that have protective roles during senescence, such as genes encoding catalase, metallothionein and ferritin, are also included in *Brassica*. The expression of *SAG*s with similar functions is similarly upregulated in both *Arabidopsis* and *Brassica* during senescence. The *Arabidopsis SAG12* gene is induced during senescence in an age-specific manner. The two *B. napus* orthologues of the *Arabidopsis SAG12* gene

BnSAG12-1 and *BnSAG12-2* also display upregulation in senescing leaves (Noh and Amasino, 1999a). Furthermore, expression of these *Brassica* genes is regulated by an *Arabidopsis* protein that binds to the promoter region of these genes, indicating that regulatory mechanisms for these genes are conserved between the two species (Noh and Amasino, 1999a).

*SAG*s of aspen were investigated through microarray analysis to understand molecular genetic basis of senescence of autumn leaves. Autumn leaf senescence in aspen is characteristically induced by the shortening of the photoperiod (Bhalerao *et al.*, 2003), while *Arabidopsis* leaves senesce largely in an age-dependent manner. This analysis revealed that many *SAG*s of autumn leaf senescence functionally overlap with those in *Arabidopsis*. For example, genes encoding metallothionein, cellular proteolysis proteins and components of the ubiquitin degradation pathway are abundant in autumn leaves. Genes for ferritin, PR1, phospholipase D, Asn synthetase, and ATP sulphulyase were also enriched in autumn leaves as in senescing *Arabidopsis* leaves. The gene encoding early light inducible protein, a light-dependent and stress-induced chlorophyll a/b-binding protein, also showed a pronounced increase in expression during senescence of both *Arabidopsis* and aspen. A pigment carrier, early light inducible protein may play a role in the protection from oxidative stress by binding to phototoxic-free chlorophyll at the initial stage of senescence (Binyamin *et al.*, 2001). The high degree of functional overlapping between *SAG*s of *Arabidopsis* and those of autumn leaves strongly suggest that much of the underlying molecular regulation of leaf senescence is similar in a broad range of plant species. It is notable that among the genes involved in hormone biosynthesis and signalling, genes involved in ethylene metabolism and perception showed upregulation, while genes involved in cytokinin, auxin or gibberellin metabolism and perception did not show any clear pattern (Andersson *et al.*, 2004). This contrasts to the case in *Arabidopsis*, whose genes involved in cytokinin biosynthesis show reduced transcript levels during senescence.

10.3 Regulatory modes of *SAGs*

The regulatory modes of *SAG*s should reveal the ways in which initiation and progression of leaf senescence are regulated. The regulatory modes of *SAG*s may be understood in terms of expression levels, temporal and spatial expression patterns, duration of expression and regulation by various senescence factors. The expression level and duration of *SAG*s would provide clues to the degree and to the length of their activities required for senescence. Temporal expression patterns provide clues to the timing and to the order of their necessity during senescence. Spatial expression patterns provide insight into which tissues they are functioning within during senescence. The regulation patterns by various senescence factors reveal how various senescence factors are integrated to control leaf senescence. In this section, we describe the temporal regulation of *SAG*s as well as the complex regulatory network of *SAG*s by various senescence factors.

10.3.1 Temporal regulation of SAGs during senescence

Leaf senescence is a highly regulated process that involves orderly and sequential changes in cellular physiology and biochemistry. The biochemical and physiological changes are most easily understandable from the viewpoint of nutrient salvage, e.g. the hydrolysis of macromolecules and their subsequent remobilization, which require the coordination of a complex array of metabolic pathways (Himelblau and Amasino, 2001). These changes also reflect the ultrastructural changes observed in senescing leaves. The order of physiological, biochemical and structural changes occurring during leaf senescence should be reflected in the expression timing of *SAG*s.

The first visible phenotypic change at the onset of leaf senescence is the colour change due to the preferential breakdown of chlorophyll molecules (Woolhouse, 1984; Gut *et al.*, 1987). Accordingly, genes for chlorophyll degradation are upregulated early in leaf senescence. Downregulation of photosynthetic-related genes has been proposed to be the possible signal for the initiation of leaf senescence and for the related upregulation of *SAG*s for macromolecular degradation and transport. Interestingly, many of the *SAG*s encoding hydrolytic enzymes are not specific to senescence. For example, many genes encoding enzymes that participate in the degradation of macromolecules for remobilization of reserved nutrients during seed germination also function during senescence. These include tomato cysteine protease genes, *SENU2* and *SENU3*, and genes for *Arabidopsis* aleurain-like protein, which has a high homology with the barley enzyme aleuain and with γ-oryzain. Thus, these *SAG*s may have been conserved to function in senescence by being regulated by senescence associated regulatory factors.

A massive decrease in nucleic acids occurs at an early stage of leaf senescence: total RNA levels are rapidly reduced with the progression of senescence (Taylor *et al.*, 1993). The initial decrease in the RNA levels consists largely of chloroplast rRNAs and of cytoplasmic rRNAs. This decrease of the amount of rRNAs is followed by that of the cytoplasmic mRNA and tRNA and is accompanied by the upregulation of *SAG*s for various nucleases. The lytic changes in the cell do not extend initially to the mitochondria or nucleus, which are essential for gene expression and energy production, respectively, and remain intact until the last stages of senescence. Accordingly, the nuclear and mitochondrial DNAs are degraded at the later stages of senescence as well. This reflects the senescing cells need to be functional for the complete progression of senescence, and possibly for the efficient reutilization of cellular materials. Interestingly, telomerases and chromosome fragmentation appear to be involved in mitotic senescence of plant cells, although the telomere lengths do not significantly change during plant development (Riha *et al.*, 1998; Fitzgerald *et al.*, 1999). This balance between telomere elongation and degradation mechanisms may be important for plant cells because, unlike animal cells, plant cells are totipotent.

Protein degradation is a major biochemical event during leaf senescence. The activity of various proteases such as cysteine proteases are specifically expressed in senescing leaves. A senescence-associated RD21 cysteine protease accumulates in the vacuole as an inactive form and slowly matures to produce a soluble active enzyme at a later stage of leaf senescence (Yamada *et al.*, 2001), implying that RD21 may play a role in this later stage. Protein degradation for efficient nutrient remobilization should occur in a highly regulated manner according to plant age; however, detailed knowledge concerning this process is insufficient.

The loss of integrity of the plasma membrane is the final stage of senescence, ending the life of the cell. Genes involved in disintegration of the plasma and vacuolar membranes are upregulated at a later phase of senescence.

Possible regulatory genes are also expressed in a time-dependent manner. This temporal order of expression may represent the temporal regulatory cascades that control the orderly process of leaf senescence.

10.3.2 Regulation of SAGs by various endogenous and external factors

Senescence is influenced by developmental age and also by various internal and external factors. The apparent senescence symptoms influenced by various factors appear similar but the *SAG*s expressed under different induction conditions may be different. In other words, the molecular states of leaf senescence caused by various senescence factors are different, using different subsets of *SAG*s thereby executing a characteristic senescence process coping with the internal and external conditions experienced by the plants.

This initial observation was confirmed in analyses of enhancer trap lines (He *et al.*, 2001). In this analysis, regulation of *SAG*s by the six senescence-promoting factors, ABA, ethylene, JA, brassinosteroids, darkness and dehydration, was analysed in the 125 senescence-upregulated enhancer trap lines, resulting in a regulatory network of leaf senescence initiating from the different senescence promoting factors. Recently, microarray experiments have been used to compare gene expression patterns during natural age-dependent leaf senescence with those during dark-induced and sucrose starvation-induced senescence (Buchanan-Wollaston *et al.*, 2005). These three types of senescence involved the induction of expression of common genes for the disruption of cellular constituents and of macromolecule degradation. However, this study also showed that the pathways for essential metabolic processes such as nitrogen mobilization and lipid catabolism were used variably among the different senescence conditions (Hörtensteiner and Feller, 2002). Genes involved in ammonia assimilation such as cytosolic glutamine synthetase genes are upregulated during the developmental senescence (Buchanan-Wollaston *et al.*, 2005). In contrast, genes encoding glutamate dehydrogenase, aspartate amino transferase and asparagines synthase are upregulated during dark-induced leaf senescence and sucrose-starved senescence of suspension cells (Lin and Wu, 2004). Many lipid catabolic genes and sucrose degradation genes are also upregulated during dark-induced leaf senescence and sucrose starvation-induced senescence of suspension cells. Interestingly, the gene expression pattern in the sucrose starvation-induced senescence of suspension cell culture was more similar to those in dark-induced senesce, indicating the dark induced senescence may be mostly due to reduced sugar level in a non-photosynthetic environment.

Ethylene, JA and salicylic acid (SA) signals affect leaf senescence. The role and importance of these signals in senescence was investigated by examining expression of *SAGs* in mutant plants with a lesion in the signalling pathways of these hormones (Buchanan-Wollaston *et al.*, 2005). The analysis revealed that all the three pathways are involved in regulating many *SAGs* during age-dependent senescence, providing a detailed molecular picture for the role of each signal in senescence. A remarkable discovery was that, in contrast to the JA and ethylene pathways, the SA pathway is not involved in dark- or starvation-induced senescence. The importance of the SA pathway in leaf senescence was further supported by the finding that the age-dependent, but not dark-induced leaf senescence, is delayed in plants defective in the SA pathway.

Many defence-related genes such as *ELI3-2, NIT2, SAG26, SAG29* and *AtOSM34* are also induced during leaf senescence in pathogen-free conditions (Quirino *et al.*, 1999). The expression of these genes is induced by SA, but still increased during senescence in *NahG* overexpressor transgenic lines that do not accumulate SA. These suggest that induction of defence-related genes during leaf senescence is independent of pathogen and SA accumulation is not essential for its induction. The *hypersenescence 1* (*hys1*) mutation, which exhibited an early-senescence phenotype, was assumed to have a lesion in a potential negative regulator of leaf senescence (Yoshida *et al.*, 2002b). This mutation turned out to be allelic to the *constitutive expresser of pathogenesis-related gene 5* (*cpr5*) mutation showing spontaneous pathogen-defence responses. These results suggested some degree of molecular overlaps between leaf senescence and pathogen responses. This may be understandable given that leaf senescence and pathogen defence response show common physiological symptoms such as accumulation of H_2O_2 (Levine *et al.*, 1994), increased ethylene and SA (John *et al.*, 1995; Morris *et al.*, 2000) and accumulation of PR genes (Hanfrey *et al.*, 1996; Pontier *et al.*, 1999; Yoshida *et al.*, 2001), despite senescence and defence having no immediate functional relation.

10.3.3 Cis-acting regulatory elements of SAGs

The complex regulatory modes of *SAGs* should be reflected in their *cis*-acting regulatory elements. Little progress has been made in the identification of *cis*-acting regulatory elements. However, the promoter sequence of *SAG12*, a gene encoding a cysteine protease, appears to contain a highly conserved region that is responsible for senescence-specific expression (Noh and Amasino, 1999b). This promoter region is not similar to any other *cis*-element sequences for known DNA-binding factors, suggesting novel transcription factors are involved in the regulation of senescence. The promoter of 12-oxo-phytodienoic acid-10,11-reductase (OPR1) gene was identified from an enhancer trap line (He and Gan, 2001). The *GUS* reporter gene driven by the promoter of the *OPR1* gene was upregulated both during age-dependent senescence and by JA, indicating modulation of *OPR1* gene during senescence and by JA shares common molecular mechanisms. Two regulatory *cis*-elements, JASE1 and JASE2, were identified by promoter deletion assay. However, no conserved sequences were recognized among JASE1, JASE2 and the regulatory sequence of *SAG12*, indicating

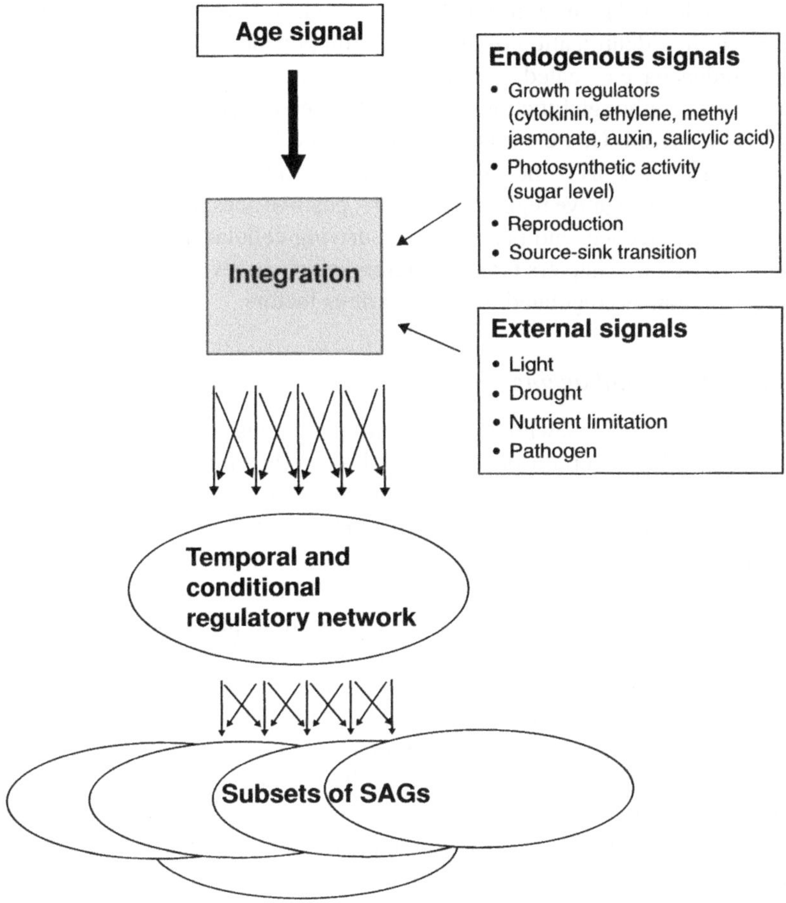

Figure 10.1 A model for regulatory pathways in leaf senescence. Leaf senescence is considered as a complex process, in which the effects of various endogenous and external signals are integrated into the developmental age-dependent senescence pathways. Multiple pathways that respond to various factors are possibly interconnected to form a regulatory network. These regulatory pathways activate distinct sets of senescence-associated genes (SAGs); some genes are shared by these pathways, whereas others are unique to specific pathways.

that different molecular mechanisms may be employed to regulate *SAG12* and *OPR1* during senescence.

10.4 Molecular regulatory mechanisms of leaf senescence

Leaf senescence is regulated through an interconnected network of multiple regulatory pathways responding to various internal and external factors (Figure 10.1). Thus, the regulatory network should involve numerous regulatory factors. Genetic strategies have been highly effective in identifying regulatory components of senescence (Table 10.1). In addition, functional analysis of potential regulatory *SAGs* led to further identification of regulatory components of leaf senescence.

Recent studies have revealed some new insight into the regulatory mechanisms in leaf senescence, although functions of most of the potential regulatory genes have not been characterized. One clear conclusion derived from the study of the regulatory molecules is that, as is the case with other biological processes, senescence involves positive and negative regulatory factors, positive ones driving initiation and progression of senescence, and negative ones driving cellular maintenance. These have been reviewed in Chapters 6 and 7. Herein we try to review these in terms of gene expression that are regulated by these various factors.

10.4.1 Developmental ageing

Leaf senescence is associated with the developmental ageing process in *Arabidopsis*, and thus occurs after a certain developmental age. Therefore, there should be a cellular mechanism(s) that measure age of a cell, tissue, organ or whole body for the initiation and/or progression of senescence. However, there is no formal report on genes that alter senescence by controlling developmental ageing, and the nature of genes that control ageing in plants is still not known.

Metabolic rate may regulate developmental ageing in plants, as has been suggested for animals (Ewbank *et al.*, 1997; Kimura *et al.*, 1997). Studies of the *ore4-1* mutant in *Arabidopsis* have shown that the mutation causes a delay in leaf senescence during natural age-dependent senescence, but not in hormone or dark-induced senescence (Woo *et al.*, 2002). The *ore4-1* mutant has a partial lesion in chloroplast functions, including photosynthesis, which resulted from reduced expression of the plastid ribosomal protein small subunit 17 (*PRPS17*) gene. The delayed leaf senescence phenotype is likely due to reduced metabolic rate, in turn due to the chloroplasts which are only partially functional in the mutant. This observation suggested that metabolic rate might be one of the key mechanisms involved in developmental leaf senescence.

Intracellular ROS levels may be a developmental ageing factor (Munne-Bosch and Alegre, 2002). It is known that one intracellular ROS source is the mitochondria, which is thought to be an integral component of PCD regulation in animals (Jones, 2000). However, mitochondrial involvement has not been identified in the PCD of leaf senescence, suggesting that processes such as those involving ROS are controlled by chloroplasts. Analysis of Ndh (a plastid complex with NADH dehydrogenase activity) and its subunits in thylakoid membranes has indicated that the topologies of the Ndh complex are similar to the mitochondria respiratory complex I (ND5/NuoL/NQO12), suggesting that these complexes are structurally homologous (Casano *et al.*, 2004). Supporting data were obtained from transgenic tobacco plants having the plastid *ndhF* gene knocked-out and thus low levels of the plastid Ndh complex (Zapata *et al.*, 2005). The transgenic lines lacking the *NdhF* plastid exhibited delayed leaf senescence compared to wild type. When the level of MDA (malondialdehyde), a decomposition product of lipid peroxidation, was measured, the transgenic lines showed much lower level than the control plants. Based on these

Table 10.1 Genes involved in regulation of leaf senescence

Gene	Molecular nature	Effects of mutations on senescence phenotype or characteristics	Reference
Genes affecting developmental ageing factor			
ORE4	Plastid ribosomal protein subunit 17	*ore4* delays leaf senescence only in an age-dependent manner and shows a reduced leaf growth rate. Delayed senescence phenotype might be due to slower metabolic rate	Woo *et al.*, 2002
NdhF	A pastid respiratory complex with NADH dehydrogenase activity	The transgenic lines lacking plastid *NdhF* exhibited delayed leaf senescence phenotype, possibly due to reduced endogenous level of ROS	Zapata *et al.*, 2005
BOP1	BTB/POZ domain protein with ankyrin repeats	The leaves of bop1-1 mutant exhibited a prolonged lifespan by modulating meristematic activity	Ha *et al.*, 2003
Genes affecting senescence through modulating endogenous factors (hormones or sugar)			
EIN2 (ORE3)	Putative metal ion transporter	*ein2* shows ethylene-insensitive phenotype; *ein2* delays leaf senescence	Oh *et al.*, 1997
ETR1	Ethylene receptor	*etr1* shows ethylene-insensitive phenotype; *etr1* delays leaf senescence	Grbic and Bleecker, 1995
OLD1	Not identified	*old1* accelerates leaf senescence in an age-dependent manner as well as ethylene-induced condition. *OLD1* might function as a repressor for integrating ethylene action into leaf senescence	Jing *et al.*, 2002
Cin1	Extracellular invertase	Transgenic tobacco expressing *Cin1* under the control of SAG12 promoter-induced delayed leaf senescence through action on the cytokinin control	Lara *et al.*, 2004
KN1	Homeobox protein	Transgenic tobacco expressing *KN1* under the control of SAG12 promoter exhibited delayed leaf senescence phenotype by increasing the cytokinin level	Ori *et al.*, 1999
GIN2	Hexokinase	*gin2* delays leaf senescence. Overexpression of *HXK* induces early senescence	Dai *et al.*, 1999; Jang *et al.*, 1997
HYS1	Unknown protein	*hys1* accelerates leaf senescence; allelic to *cpr5*; constitutive expression of pathogenesis-related genes; enhanced response to sugar	Yoshida *et al.*, 2002b
Genes affecting in response to environmental factors			
PHYA	Phytochrome A	Delays chlorophyll and protein loss when overexpressed	Cherry *et al.*, 1991
PHYB	Phytochrome B	Delays leaf senescence when overexpressed	Thiele *et al.*, 1999
Genes for transcription factors		Genes encoding potential transcription factors, which are induced by environmental stresses as well as senescence	Chen *et al.*, 2002

(Continued)

Table 10.1 Genes involved in regulation of leaf senescence (*Continued*)

Gene	Molecular nature	Effects of mutations on senescence phenotype or characteristics	Reference
Genes involved in perceiving senescence signals and transducing into senescence execution programs			
AtWRKY6	WRKY transcription factor	Induced by pathogen as well as during leaf senescence. The WRKY6 controls senescence through regulating the senescence-induced receptor kinase gene	Robatzek and Somssich, 2001, 2002
WRKY53	WRKY transcription factor	Senescence-induced. Overexpressor and knockout lines showed accelerated and delayed senescence phenotypes, respectively	Miao *et al.*, 2004
SIRKa (At2g19190)	Receptor-like kinase	Senescence-induced	Robatzek and Somssich, 2002
Genes involved in the degradation of senescence regulatory protein			
ORE9	F-box protein	*ore9* mutation delays leaf senescence. The ORE9 might function to limit leaf longevity through regulating degradation process of senescence negative factors	Woo *et al.*, 2001
DSL1	Arginyl t-RNA transferase	*dls1* mutation delays leaf senescence. The *DSL1* might play a role in degradation of proteins that negatively regulate leaf senescence	Yoshida *et al.*, 2002a
Genes for senescence execution			
ACD2	Red chlorophyll catabolite reductase	*acd2* accelerates cell death. The ACD2 detoxifies phytotoxic chlorophyll product during senescence process	Mach *et al.*, 2001
PLD	Phospholipase Dα	Retardation of ABA- and ethylene-promoted senescence, when expression is blocked	Thompson *et al.*, 2000
SAG101	Acyl hydrolase	Delays leaf senescence when expression is blocked	He and Gan, 2002
AtAPG7	Autophagy gene	Early leaf senescence in the knockout line. Autophagy gene is required for nutrient recycling	Doelling *et al.*, 2002
AtAPG9	Autophagy gene	Early leaf senescence in the knockout line	Hanaoka *et al.*, 2002
AtATG18a	Autophagy gene	Early leaf senescence in the RNA interference line	Xiong *et al.*, 2005

This table illustrates possible senescence-regulatory genes. Genes that alter leaf senescence phenotype or potential regulatory genes, such as transcription factors or receptor-like kinase, that are induced during leaf senescence were included. This is only a representation of published results and many genes are not included.

data, it was suggested that the chloroplast Ndh complex is functionally homologous to mitochondrial respiration complex I and would be involved in increasing the reduction level of electron transport and generation of ROS. Accordingly, low levels of plastid Ndh complex in transgenic plants would lead to a delayed senescence phenotype, possibly due to reduced level of ROS. The control of leaf senescence by chloroplast in plant provided a new mechanism of PCD.

Leaf senescence should be intimately related to the previous developmental stages of leaf, such as leaf initiation, growth and maturation. Thus, it is possible that genes controlling these processes including meristematic activity could influence age-dependent senescence. Transgenic tobacco plants that express the maize homeobox gene *knotted1* under the control of the *SAG12* promoter exhibit delayed leaf senescence (Ori *et al.*, 1999). The delayed leaf senescence in the transgenic lines was accompanied by increased levels of cytokinin in the older leaves. Thus, the effect of *KN1* on senescence was suggested to be mediated through changes in cytokinin levels. However, expression of *KN1* in the transgenic lines may keep the leaf cells at a young stage because *KN1* controls the meristematic activity. We also observed that the leaves of *bop1-1* mutants showing enhanced meristematic activity in leaves exhibited a prolonged life span (Ha *et al.*, 2003). Exact mechanisms by which these genes regulate leaf senescence need to be investigated.

10.4.2 Internal factors

Although the leaf senescence occurs in an age-dependent manner, it is also finely tuned by other endogenous developmental factors. Several lines of evidence using mutants or transgenic lines have revealed that factors regulating developmental processes such as phytohormones, sugars and other developmental regulators are involved in regulation of leaf senescence. Studies have also revealed complex and extensive interactions between these factors, as shown in other developmental processes.

10.4.2.1 Phytohormones

Cytokinins are known to have a major effect on leaf senescence (Richmond and Lang, 1957). A striking example of the senescence-suppressing effect of cytokinins was observed in various transgenic plants expressing *IPT*, an *Agrobacterium*-originated cytokinin biosynthesis gene, under the control of the senescence-specific *SAG12* promoter (see Chapter 13). These transgenic plants demonstrated markedly delayed leaf senescence without noticeable pleiotropic phenotypes. Recently, an interesting link between the anti-senescence effect of cytokinins and primary metabolism was suggested based on the finding that cytokinin-mediated delay of senescence is correlated with the activity of extracellular invertase, the enzyme functionally linked to an apoplastic phloem unloading pathway. When the extracellular invertase activity was inhibited, senescence was no longer delayed by cytokinins, demonstrating one possible role of extracellular invertase in this process. Hence, it was proposed that nutrient mobilization via an extracellular invertase might be an important component of the underlying regulatory mechanism (Lara *et al.*, 2004). It is notable that regulation of leaf senescence is related to changes in source-sink relations, and that a link between cytokinin action and primary metabolism exists. Our recent observation that a mutation in a cytokinin receptor of *Arabidopsis* causes delayed leaf senescence through a constitutive cytokinin response adds exciting insight into the molecular mechanism of cytokinin action in controlling senescence (Kim *et al.*, 2006).

Ethylene has long been known as a senescence-accelerating phytohormone with its pronounced effect on fruit ripening, and flower and leaf senescence. The importance of endogenous ethylene in senescence was clearly demonstrated in ethylene insensitive mutants and transgenic plants with altered ethylene synthesis. Ethylene-insensitive mutants exhibited a measurable delay in the initiation of leaf senescence, but the mutation has little effect on progression of senescence (Grbic and Bleecker, 1995; Oh *et al.*, 1997). This observation, thus, suggests that ethylene plays a role in coordinating the timely transition of the leaf to senescence state but it is not essential to the process itself. It has been indicated that ethylene-mediated pathways leading to leaf senescence in *Arabidopsis* depend on age-dependent factors; thus ethylene can only induce senescence after leaves reach a certain developmental stage (Bleecker and Patterson, 1997). A potential regulator that may be involved in integrating ethylene signalling into age-dependent pathways has been recently reported. The *onset of leaf death1* (*old1-1*) mutant displayed a phenotype with earlier onset of senescence in an age-dependent manner. The early senescence phenotype was further accelerated by ethylene exposure, suggesting that *old1* mutation generates alternation in two sets of pathways: age-regulated leaf senescence and ethylene signalling. When the senescence phenotype was examined in *old1etr1* double mutant where ethylene perception was blocked, earlier onset of leaf senescence still occurred, but was not exaggerated by ethylene treatment. These data led to a suggestion that *OLD1* might function as a repressor for integrating ethylene action into leaf senescence (Jing *et al.*, 2002).

SA signalling pathway has a role in the control of gene expression during developmental senescence (Morris *et al.*, 2000). Expression of *SAG*s such as *PR1a*, *chitinase,* and *SAG12* is considerably reduced or undetectable in *Arabidopsis* plants defective in the SA signalling pathway. Involvement of SA in the regulation of leaf senescence was also demonstrated in *NahG* overexpressor transgenic lines that do not accumulate SA. The transgenic lines exhibited a delayed leaf senescence phenotype during age-dependent senescence, supporting the microarray data (Buchanan-Wollaston *et al.*, 2005). This is a clear evidence that the SA pathway has an important and specific role to play in developmental senescence, quite possibly in its final death phase.

Exogenous treatment of methyl jasmonate (MJ) induced leaf senescence by activating a subset of *SAG*s, suggesting that MJ is a senescence-promoting hormone (He *et al.*, 2002). This notion is consistent with the observation that expression of *SAG*s was reduced in JA insensitive mutant, *coronative insensitive1* (*coi1*), although the *coi1* mutant did not show any visible phenotype, presumably due to functional redundancy. Recently, *cos1* (*coi1 suppressor*) that restores the *coi1*-related phenotypes was identified. *COS1* encodes lumazine synthase, a key component in riboflavin synthesis pathways (Xiao *et al.*, 2004). More studies are required to understand the role of riboflavin pathways in controlling JA-mediated leaf senescence.

Auxin regulates many aspects of plant growth and development and was also suggested to affect senescence (Sexton and Roberts, 1982). It was recently reported that the disruption of *AUXIN RESPONSE FACTOR2* (*ARF2*) by T-DNA insertion causes a delay in leaf senescence as well as in floral organ abscission (Ellis *et al.*, 2005; Okushima *et al.*, 2005). Furthermore, the expression of *ARF2* is induced in senescing leaves. Diverse phenotype in *arf2* mutant plants suggests that ARF2 functions at multiple stages and are regulated by different factors during plant development.

It should be noted that plant hormones interact or cross talk with one another, constituting a complex network of regulation, and is likely to be the case in controlling plant senescence. However, caution must be taken when these results were interpreted because many phytohormone mutants are characterized at the seedling stage, and hormone signalling at the seedling stage may differ from that at the senescence stage. It should be also noted that these plant hormones may be involved in correlative control of senescence, although it is not the case in *Arabidopsis.*

10.4.2.2 Sugar signalling

Source-sink balance, which can affect the partitioning of sugar in plants, may be important in regulating senescence (Ono *et al.*, 2001; Paul and Foyer, 2001; Yoshida, 2003). Young leaves serve as sink organs, whereas old leaves are source organs that provide sugars. When young leaves mature, they develop photosynthetic machinery, which leads to the elevated level of sugars, but sugars are immediately used for new growth. However, when full expansion is attained, sugars accumulate at high levels. Based on the notion that elevated sugars can repress the expression of photosynthesis-associated genes, it was proposed that the accumulation of sugar in mature leaves will lead to a decline of photosynthetic activity and to the reduction of photosynthetic activity at a certain threshold level, which may act as a senescence signal (Hensel *et al.*, 1993).

Accordingly, sugar signalling has emerged as an important regulator of leaf senescence (Rolland *et al.*, 2002). Studies using hexokinase overexpressors show that increased hexokinase levels simulate a rise in sugar, and is also associated with reduced photosynthetic activity (Jang *et al.*, 1997; Dai *et al.*, 1999). One notable phenotype found in these transgenic plants is accelerated leaf senescence, supporting the notion that reduced photosynthetic activity may be related to premature leaf senescence via hexokinase. Moreover, glucose-insensitive *Arabidopsis* mutant (*gin2*) with a lesion in one of the hexokinases showed delayed senescence (Moore *et al.*, 2003). However, the senescence phenotypes observed in transgenic and mutant lines were not thoroughly examined and require further analysis. The *hys1* mutant has an increased sensitivity to exogenously applied sugars as well as an accelerated leaf senescence phenotype (Yoshida *et al.*, 2002b). This observation led to a suggestion that an enhanced sugar signal in the mutant causes reduced photosynthetic activity and induces premature senescence likely via hexokinase.

It has been known that sugar-signalling pathways interact intimately with the signalling pathways regulated by hormones such as auxin, cytokinin, or ABA during plant development. It is likely the case in regulating *Arabidopsis* leaf senescence. The control of senescence by sugar signalling is also likely affected by other factors such as nitrogen status and developmental stage. Integration of these factors into the senescence program may be important in properly regulating the onset and progression of leaf senescence.

10.4.3 External factors

Like many other plant growth and developmental processes, senescence is also highly influenced by environmental conditions such as pathogen infection, nutrient stress, water stress, or oxidative stresses induced by ozone or UV-B. Thus, senescence should be an integrated response of plants to external environmental factors as well as to endogenous developmental signals.

Light is perceived by a variety of photoreceptors and affects plant developmental processes throughout its life span (Cherry *et al.*, 1991; Thiele *et al.*, 1999). Transgenic plants overexpressing phytochrome A (*PhyA*) or phytochrome B (*PhyB*), were shown to have greater longevity. In the case of *PhyB* overexpressor, the initiation of leaf senescence was similar to that of wild type, however the progression of senescence appeared to be retarded.

Exposure to environmental factors can also trigger leaf yellowing. Components of signalling pathways that are associated with environmental stresses may regulate leaf senescence. Recently, expression profiles of 402 potential stress-related genes that encode known or putative transcription factors from *Arabidopsis* were monitored in various organs at different developmental stages and under various biotic and abiotic stresses (Chen *et al.*, 2002). Among the 43 transcription factor genes that are induced during senescence, 28 genes are also induced by stress treatment, suggesting extensive overlap responses to these stresses. Downstream genes of senescence-enhanced transcription factors might play a role in executing leaf senescence or in protecting the cellular function required for proper progression or completion of leaf senescence.

10.4.4 Regulatory role of protein degradation

Specific control of protein degradation and stability has emerged as a pivotal mechanism that regulates the growth and development of eukaryotic organisms. Recent genetic studies in *Arabidopsis* demonstrated that protein degradation is also involved in controlling leaf senescence (Woo *et al.*, 2001; Yoshida *et al.*, 2002a). The ORE9, a protein containing an F-box motif, which is a component of the ubiquitin E3 ligase complex, was identified as a positive regulator of leaf senescence (Woo *et al.*, 2001). The SCF complexes are known to ubiquitinate specific target substrates (Patton *et al.*, 1998). Hence, it is likely that the ORE9 functions via ubiquitin-dependent proteolysis to limit leaf longevity by degrading target proteins that are required to suppress the leaf senescence program in *Arabidopsis*. However, target substrates of ORE9 are yet to be identified.

Proteolysis by the N-end rule ubiquitin pathway also appears to be a mechanism involved in regulating leaf senescence in *Arabidopsis*. The *delayed-leaf-senescence1* (*dls1*) mutant, which is defective in arginyl tRNA:protein transferase (R-transferase), showed delayed development of leaf senescence symptoms (Yoshida *et al.*, 2002a). R-transferase transfers arginine to the amino-terminus of proteins with amino terminal glutamyl or aspartyl residues and thereby targets the proteins for ubiquitin-dependent proteolysis (Varshavsky, 1997). Thus, DLS1 may play a role in the degradation of proteins that negatively regulate leaf senescence. Nonetheless, the *ore9* and *dls1* mutations may have different roles in leaf senescence. Unlike R-transferase, the SCF complex is not involved in the N-end rule-dependent pathway. Furthermore, the *dls1* mutation delays not only initiation of leaf senescence but also its progression, whereas the *ore9* mutation mainly affects the initiation of leaf senescence.

While specific protein degradation mechanisms appear to have regulatory roles in leaf senescence, much more needs to be revealed to understand their exact roles including identification of the specific substrates.

10.5 Conclusions and future challenges

Leaf senescence is an integrated response to age, developmental status, and environmental conditions. Thus, plants need to incorporate all these factors when executing a proper senescence progression to maximize its fitness. This control may be achieved via the delicate regulation of gene expression. In the last decade, efforts have focused on identifying senescence regulators and elucidating regulatory mechanisms for leaf senescence through genomics and genetic approaches.

One future challenge is to elucidate the roles of potential regulatory genes in senescence. Utilization of large collections of T-DNA insertion lines or TILLING approach in *Arabidopsis* will allow the functional analysis of genes identified from microarray data. Senescence phenotypes of mutants may not be obvious due to gene redundancy or to functional redundancy caused by various pathways leading to senescence. In this case, it may be necessary to generate double, triple, and even higher order combination of mutants with these genes or in redundant functional pathways. Transgenic approaches using inducible overexpression, minigene, or RNA interference could also complement the limit of loss of function mutants (Gan and Amasino, 1999; Hinderhofer and Zentgraf, 2001). Recent genomics technologies such as microarray, proteomics, or metabolomics may reveal distinct molecular phenotypes, even in the case that no visible phenotype was observed, as shown in the *wrky6* mutant.

Genetic approaches have also led to discoveries of many regulatory components. Surprisingly, some of senescence-delayed mutants we have recently isolated are allelic to known mutants. Considering the nature of senescence, there should be many more mutants that can be identified by a well-designed screening scheme. Thus, novel genetic screening schemes of senescence mutants may identify new regulatory genes. Alternatively, diverse genetic resources, which are available in *Arabidopsis*, could be used. For example, an activation tagging approach may be useful in identifying a negative regulator of leaf senescence. Also, generation of a T-DNA pool in which the inducible promoters are inserted in the genome would be a useful resource for screening negative elements by inducing a gene at a maturation or senescence stage, followed by observing the phenotypic effect. Future analyses of these mutants with global gene expression analysis through use of a DNA chip, in combination with proteomic approaches, the altered subcellular localization, signalling, or metabolic functions and of *in vivo* protein-protein interactions are likely to reveal molecular mechanism of leaf senescence mediated by these regulatory factors.

Like other biological systems, senescence should be understood at a systems biology level. Many components of senescence are likely to show complex interactions whether it is direct or indirect. These complex interactions are likely to yield emergent properties that cannot be explained from one simple extrapolation of function of each component. These emergent properties may occur at the gene expression level, at the metabolite level, or at the cellular level. Likewise, these interactions are not static but should be highly dynamic. During age-dependent senescence or senescence affected by other endogenous and exogenous senescence signals, the pattern of these interactions will continuously change to reflect the progressive stage of senescence and to reflect the senescence state affected by various senescence signals. For a better understanding of senescence, it is necessary to observe the process at a systems level and with a dynamic view. Technologies are rapidly improving these types of analysis and therefore we are close to a breakthrough in understanding the molecular mechanisms of senescence. Knowledge of senescence will also lead to the generation of transgenic agronomic, horticultural, and vegetable crop plants with improved yield, better post-harvest quality, and increased shelf life.

Acknowledgment

We dedicate this review to the memory of Dr. Anthony Bleecker who greatly contributed to this field with far-reaching insights. We apologize to all our colleagues whose work could not be properly reviewed here because of space limitation. The work by Nam was supported in part by MOST (KOSEF) through National Core Research Center for Systems Bio-Dynamics (R15-2004-033-03002-0) and in part by the Crop Functional Genomics Research Program (CG1312). Kim was supported in part by the Program for the Training of Graduate Students from Ministry of Commerce, Industry and Energy of Korea. The work by Lim was supported by the Korea Research Foundation Grant funded by Korea Government (MOEHRD, Basic Research Promotion Fund, KRF-2005-C00075).

References

Andersson, A., Keskitalo, J., Sjodin, A. *et al.* (2004) A transcriptional timetable of autumn senescence. *Genome Biol* **5**, R24.

Berberich, T., Sano, H. and Kusano, T. (1999) Involvement of a MAP kinase, ZmMPK5, in senescence and recovery from low-temperature stress in maize. *Mol Gen Genet* **262**, 534–542.

Bhalerao, R., Keskitalo, J., Sterky, F. *et al.* (2003) Gene expression in autumn leaves. *Plant Physiol* **131**, 430–442.

Binyamin, L., Falah, M., Portnoy, V., Soudry, E. and Gepstein, S. (2001) The early light-induced protein is also produced during leaf senescence of *Nicotiana tabacum*. *Planta* **212**, 591–597.

Bleecker, A.B. and Patterson, S.E. (1997) Last exit: senescence, abscission, and meristem arrest in *Arabidopsis*. *Plant Cell* **9**, 1169–1179.

Blein, J.P., Coutos-Thevenot, P., Marion, D. and Ponchet, M. (2002) From elicitins to lipid-transfer proteins: a new insight in cell signaling involved in plant defence mechanisms. *Trends Plant Sci* **7**, 293–296.

Bögre, L., Calderini, O., Binarova, P. *et al.* (1999) A MAP kinase is activated late in plant mitosis and becomes localized to the plane of cell division. *Plant Cell* 11, 101–114.

Buchanan-Wollaston, V. and Ainsworth, C. (1997) Leaf senescence in *Brassica napus*: cloning of senescence related genes by subtractive hybridization. *Plant Mol Biol* **33**, 821–834.

Buchanan-Wollaston, V., Earl, S., Harrison, E. *et al.* (2003) The molecular analysis of leaf senescence – a genomics approach. *Plant Biotechnol J* **1**, 3–22.

Buchanan-Wollaston, V., Page, T., Harrison, E. *et al.* (2005) Comparative transcriptome analysis reveals significant differences in gene expression and signalling pathways between developmental and dark/starvation-induced senescence in *Arabidopsis*. *Plant J* **42**, 567–585.

Casano, L.M., Lascano, H.R., Martin, M. and Sabater B. (2004) Topology of the plastid Ndh complex and its NDH-F subunit in thylakoid membranes. *Biochem J* **382**, 145–155.

Chen, W., Provart, N.J., Glazebrook, J. *et al.* (2002) Expression profile matrix of *Arabidopsis* transcription factor genes suggests their putative functions in response to environmental stresses. *Plant Cell* **14**, 559–574.

Cherry, J.R., Hershey, H.P. and Vierstra, R.D. (1991) Characterization of tobacco expressing functional oat phytochrome. *Plant Physiol* **96**, 775–785.

Dai, N., Schaffer, A., Petreikov, M. *et al.* (1999) Overexpression of *Arabidopsis* hexokinase in tomato plants inhibits growth, reduces photosynthesis, and induces rapid senescence. *Plant Cell* 11, 1253–1266.

del Rio, L.A., Pastori, G.M., Palma, J.M. *et al.* (1998) The activated oxygen role of peroxisomes in senescence. *Plant Physiol* **116**, 1195–1200.

Doelling, J.H., Walker, J.M., Friedman, E.M., Thompson, A.R. and Vierstra, R.D. (2002) The APG8/12-activating enzyme APG7 is required for proper nutrient recycling and senescence in *Arabidopsis thaliana*. J Biol Chem **277**, 33105–331141.

Ellis, C.M., Nagpal, P, Young, J.C., Hagen, G., Guilfoyle, T.J. and Reed, J.W. (2005) Auxin response factor1 and auxin response factor2 regulate senescence and floral organ abscission in *Arabidopsis thaliana*. *Development* **132**, 4563–4574.

Eriksson, S., Andreasson, E., Ekbom, B., Graner, G. *et al.* (2002) Complex formation of myrosinase isoenzymes in oilseed rape seeds are dependent on the presence of myrosinase-binding proteins. *Plant Physiol* **129**, 1592–1599.

Ewbank, J.J., Barnes, T.M., Lakowski, B., Lussier, M., Bussey, H. and Hekimi S. (1997) Structural and functional conservation of the *Caenorhabditis elegans* timing gene *clk-1*. *Science* **275**, 980–983.

Fan, L., Zheng, S. and Wang, X. (1997) Antisense suppression of phospholipase D alpha retards abscisic acid- and ethylene-promoted senescence of postharvest *Arabidopsis* leaves. *Plant Cell* **9**, 2183–2196.

Fitzgerald, M.S., Riha, K., Gao, F., Ren, S., McKnight, T.D. and Shippen, D.E. (1999) Disruption of the telomerase catalytic subunit gene from *Arabidopsis* inactivates telomerase and leads to a slow loss of telomeric DNA. *Proc Natl Acad Sci USA* **96**, 14813–14818.

Gan, S. and Amasino, R.M. (1999) Developmental targeting of gene expression by the use of a senescence-specific promoter. In: *Inducible Gene Expression in Plants* (ed. Reynolds, P.H.S.). CAB International, New York, pp. 169–186.

Gepstein, S., Sabehi, G, Carp, M.J. *et al.* (2003) Large-scale identification of leaf senescence-associated genes. *Plant* **36**, 629–642.

Grbic, V. and Bleecker, A.B. (1995) Ethylene regulates the timing of leaf senescence in *Arabidopsis*. *Plant J* **8**, 595–602.

Guo,Y., Cai, Z. and Gan, S. (2004) Transcriptome of *Arabidopsis* leaf senescence. *Plant Cell Environ* **27**, 521–549.

Guo,Y. and Gan, S. (2006) AtNAP, a NAC family transcription factor, has an important role in leaf senescence. *Plant J* **46**, 601–612.

Gut, H., Ruts, C., Matile, P. and Thomas H. (1987) Leaf senescence in a non-yellowing mutant of *Festuca pratensis*: Degradation of carotenoids. *Physiol Plant* **70**, 659–663.

Guterman, A., Hajouj, T. and Gepstein, S. (2003) Senescence-associated mRNAs that may participate in signal transduction and protein trafficking. *Physiol Plant* **118**, 439–446.

Ha, C.M., Kim, G.T., Kim, B.C. *et al.* (2003) The *BLADE-ON-PETIOLE 1* gene controls leaf pattern formation through the modulation of meristematic activity in *Arabidopsis*. *Development* **130**, 161–172.

Hajouj, T., Michelis, R. and Gepstein, S. (2000) Cloning and characterization of a receptor-like protein kinase gene associated with senescence. *Plant Physiol* **124**, 1305–1314.

Hanaoka, H., Noda, T., Shirano, Y. *et al.* (2002) Leaf senescence and starvation-induced chlorosis are accelerated by the disruption of an *Arabidopsis* autophagy gene. *Plant Physiol* **129**, 1181–1193.

Hanfrey, C., Fife, M. and Buchanan-Wollaston, V. (1996) Leaf senescence in *Brassica napus*: expression of genes encoding pathogenesis-related proteins. *Plant Mol Biol* **30**, 597–609.

He, Y., Fukushige, H., Hildebrand, D.F. and Gan, S. (2002) Evidence supporting a role of jasmonic acid in *Arabidopsis* leaf senescence. *Plant Physiol* 128: 876–884.

He, Y. and Gan, S. (2001) Identical promoter elements are involved in regulation of the OPR1 gene by senescence and jasmonic acid in *Arabidopsis*. *Plant Mol Biol* **47**, 595–605.

He, Y. and Gan, S. (2002) A gene encoding an acyl hydrolase is involved in leaf senescence in *Arabidopsis*. *Plant Cell* **14**, 805–815.

He, Y., Tang, W., Swain, J.D., Green, A.L., Jack, T.P. and Gan, S. (2001) Networking senescence-regulating pathways by using *Arabidopsis* enhancer trap lines. *Plant Physiol* **126**, 707–716.

Hensel, L., Grbic, V., Baumgarten, D.A. and Bleecker, A.B. (1993) Developmental and age-related processes that influence the longevity and senescence of photosynthetic tissues in *Arabidopsis*. *Plant Cell* **5**, 553–564.

Himelblau, E. and Amasino, R.M. (2001) Nutrients mobilized from leaves of *Arabidopsis thaliana* during leaf senescence. *J Plant Physiol* **158**, 1317–1323.

Hinderhofer, K. and Zentgraf, U. (2001) Identification of a transcription factor specifically expressed at the onset of leaf senescence. *Planta* **213**, 469–473.

Hörtensteiner, S., Wüthrich, K.L., Matile, P., Ongania, K.-H. and Kräutler, B. (1998) The key step in chlorophyll breakdown in higher plants. Cleavage of pheophorbide a macrocycle by a monooxygenase. *J Biol Chem* **271**, 15335–15339.

Hörtensteiner, S. and Feller, U. (2002) Nitrogen metabolism and remobilization during senescence. *J Exp Bot* **53**, 927–937.

Huang, F.Y., Philosoph-Hadas, S., Meir, S. *et al.* (1997) Increases in cytosolic Ca^{2+} in parsley mesophyll cells correlate with leaf senescence. *Plant Physiol* **115**, 51–60.

Jakoby, M., Weisshaar, B., Droge-Laser, W. *et al.* (2002) bZIP transcription factors in *Arabidopsis*. *Trends Plant Sci* **7**, 106–111.

Jang, J.C., Leon, P., Zhou, L. and Sheen, J. (1997) Hexokinase as a sugar sensor in higher plants. *Plant Cell* 9, 5–19.

Jing, H.-C., Sturre, M.J.G., Hille, J. and Dijkwel P.P. (2002) *Arabidopsis* onset of leaf death mutants identify a regulatory pathway controlling leaf senescence. *Plant J* **32**, 51–63.

John, C.F., Morris, K., Jordan, B.R., Thomas, B. and Mackerness, S. (2001) Ultraviolet-B exposure leads to upregulation of senescence-associated genes in *Arabidopsis thaliana*. *J Exp Bot* **52**, 1367–1373.

John, I., Drake, R., Farrell, A. *et al.* (1995) Delayed leaf senescence in ethylene-deficient ACC-oxidase antisense tomato plants: molecular and physiological analysis. *Plant J* **7**, 483–490.

John, I., Hackett, R., Cooper, W., Drake, R., Farrell, A. and Grierson, D. (1997) Cloning and characterization of tomato leaf senescence-related cDNAs. *Plant Mol Biol* **33**, 641–651.

Jones, A. (2000) Does the plant mitochondrion integrate cellular stress and regulate programmed cell death? *Trends Plant Sci* **5**, 225–230.

Jones, A.M. (2001) Programmed cell death in development and defense. *Plant Physiol* **125**, 94–97.

Kim, H.J., Ryu, H., Hong, S.H. *et al.* (2006) Cytokinin-mediated control of leaf longevity by AHK3 through phosphorylation of ARR2 in *Arabidopsis*. *Proc Natl Acad Sci USA* **103**, 814–819.

Kimura, K.D., Tissenbaum, H.A., Liu, Y. and Ruvkun, G. (1997) *daf-2*, an insulin receptor-like gene that regulates longevity and diapause in *Caenorhabditis elegans*. *Science* **277**, 942–946.

Kinoshita, T., Yamada, K., Hiraiwa, N., Kondo, M., Nishimura, M., and Hara-Nishimura, I. (1999) Vacuolar processing enzyme is up-regulated in the lytic vacuoles of vegetative tissues during senescence and under various stressed conditions. *Plant J* **19**, 43–53.

Lara, M.E.B., Garcia, M.C.G., Fatima, T. *et al.* (2004) Extracellular invertase is an essential component of cytokinin-mediated delay of senescence. *Plant Cell* **16**, 1276–1287.

Levine, A., Tenhaken, R., Dixon, R. and Lamb, C. (1994) H_2O_2 from the oxidative burst orchestrates the plant hypersensitive disease resistance response. *Cell* **79**, 583–593.

Lim, P.O., Woo, H.R. and Nam, H.G. (2003) Molecular genetics of leaf senescence in *Arabidopsis*. *Trends Plant Sci* **8**, 272–278.

Lin, J.F. and Wu, S.H. (2004) Molecular events in senescing *Arabidopsis* leaves. *Plant J* **39**, 612–628.

Lohman, K.N., Gan, S., John, M.C. and Amasino, R.M. (1994) Molecular analysis of natural leaf senescence in *Arabidopsis thaliana*. *Physiol Plant* **92**, 322–328.

Mach, J.M., Castillo, A.R., Hoogstraten, R. and Greenberg, J.T. (2001) The *Arabidopsis*-accelerated cell death gene *ACD2* encodes red chlorophyll catabolite reductase and suppresses the spread of disease symptoms. *Proc Natl Acad Sci USA* **98**, 771–776.

Matile, P., Hörtensteiner S. and Thomas, H. (1999) Chlorophyll degradation. *Annu Rev Plant Physiol Plant Mol Biol* **50**, 67–95.

Miao, Y., Laun, T., Zimmermann, P. and Zentgraf, U. (2004) Targets of the WRKY53 transcription factor and its role during leaf senescence in *Arabidopsis*. *Plant Mol Biol* **55**, 853–867.

Moore, B., Zhou, L., Rolland, F. *et al.* (2003) Role of the *Arabidopsis* glucose sensor HXK1 in nutrient, light, and hormonal signaling. *Science* **300**, 332–336.

Morris, K., Mackerness, S.A., Page, T. *et al.* (2000) Salicylic acid has a role in regulating gene expression during senescence. *Plant J* **23**, 677–685.

Munne-Bosch, S. and Alegre, L. (2002) Plant ageing increases oxidative stress in chloroplasts. *Planta* **214**, 608–615.

Nakabayashi, K., Ito, M., Kiyosue, T., Shinozaki, K. and Watanabe, A. (1999) Identification of *clp* genes expressed in senescing *Arabidopsis* leaves. *Plant Cell Physiol* **40**, 504–514.

Nakashima, K., Kiyosue, T., Yamaguchi-Shinozaki, K. and Shinozaki, K. (1997) A nuclear gene, erd1, encoding a chloroplast-targeted Clp protease regulatory subunit homolog is not only induced by water stress but also developmentally up-regulated during senescence in *Arabidopsis thaliana*. *Plant J* **12**, 851–861.

Nam, H.G. (1997) Molecular genetic analysis of leaf senescence. *Curr Opin Biotech* **8**, 200–207.

Noh, Y.S. and Amasino, R. (1999a) Regulation of developmental senescence in conserved between *Arabidopsis* and *Brassica napus*. *Plant Mol Biol* **41**, 195–206.

Noh, Y.S. and Amasino, R. (1999b) Identification of a promoter region responsible for the senescence-specific expression of *SAG12*. *Plant Mol Biol* **41**, 181–194.

Oh, S.A., Park, J.H., Lee, G.I., Paek, K.H., Park, S.K. and Nam, H.G. (1997) Identification of three genetic loci controlling leaf senescence in *Arabidopsis thaliana*. *Plant J* **12**, 527–535.

Okushima, Y., Mitina, I., Quach, H.L. and Theologis, A. (2005) AUXIN RESPONSE FACTOR 2 (ARF2): a pleiotropic developmental regulator. *Plant J* **43**, 29–46.

Olsen, A.N., Ernst, H.A., Leggio, L.L. and Skriver, K. (2005) NAC transcription factors: structurally distinct, functionally diverse. *Trends Plant Sci* **10**, 79–87.

Ono, K., Nishi, Y., Watanabe, A. and Terashima, I. (2001) Possible mechanisms of adaptive leaf senescence. *Plant Biol* **3**, 234–243.

Ori, N., Juarez, M.T., Jackson, D., Yamaguchi, J., Banowetz, G.M. and Hake, S. (1999) Leaf senescence is delayed in tobacco plants expressing the maize homobox gene knotted1 under the control of a senescence-activated promoter. *Plant Cell* **11**, 1073–1080.

Park, J.H., Oh, S.A., Kim, Y.H., Woo, H.R. and Nam, H.G. (1998) Differential expression of senescence-associated mRNAs during leaf senescence induced by different senescence-inducing factors in *Arabidopsis*. *Plant Mol Biol* **37**, 445–454.

Patton, E.E., Willems, A.R. and Tyers, M. (1998) Combinational control in ubiquitin-dependent proteolysis: don't Skp the F-box hypothesis. *Trends Genet* **14**, 236–243.

Paul, M.J. and Foyer, C.H. (2001) Sink regulation of photosynthesis. *J Exp Bot* **52**, 1383–1400.

Pérez-Amador, M.A., Abler, M.L., De Rocher, E.J. *et al.* (2000) Identification of BFN1, a bifunctional nuclease induced during leaf and stem senescence in *Arabidopsis*. *Plant Physiol* **122**, 169–180.

Petersen, M., Brodersen, P. and Naested, H. *et al.* (2000) *Arabidopsis* map kinase 4 negatively regulates systemic acquired resistance. *Cell* **103**, 1111–1120.

Pontier, D., Gan, S., Amasino, R.M., Roby, D. and Lam, E. (1999) Markers for hypersensitive response and senescence show distinct patterns of expression. *Plant Mol Biol* **39**, 1243–1255.

Pruzinska, A., Tanner, G. Anders, I., Roca, M. and Hortensteiner, S. (2003) Chlorophyll breakdown: pheophorbide a oxygenase is a Rieske-type iron-sulfur protein, encoded by the accelerated cell death 1 gene. *Proc Natl Acad Sci USA* **100**, 15259–15264.

Quirino, B.F., Noh, Y.S., Himelblau, E. and Amasino, R.M. (2000) Molecular aspects of leaf senescence. *Trends Plant Sci* **5**, 278–282.

Quirino, B.F., Normanly, J. and Amasino, R.M. (1999) Diverse range of gene activity during *Arabidopsis thaliana* leaf senescence includes pathogen-independent induction of defense-related genes. *Plant Mol Biol* **40**, 267–278.

Richmond, A.E. and Lang, A. (1957) Effect of kinetin on protein content and survival of detached *Xanthium* leaves. *Science* **125**, 650–651.

Riha, K., Fajkus, J., Siroky, J. and Vyskot, B. (1998) Developmental control of telomere lengths and telomerase activity in plants. *Plant Cell* **10**, 1691–1698.

Robatzek, S. and Somssich, I.E. (2001) A new member of the *Arabidopsis* WRKY transcription factor family, AtWRKY6, is associated with both senescence- and defence-related processes. *Plant J* **28**, 123–133.

Robatzek, S. and Somssich, I.E. (2002) Targets of AtWRKY6 regulation during plant senescence and pathogen defense. *Genes Dev* **16**, 1139–1149.

Rolland, F., Moore, B. and Sheen, J. (2002) Sugar sensing and signalling in plants. *Plant Cell Suppl* **14**, S185–205.

Sexton, R. and Roberts, J.A. (1982). Cell biology of abscission. *Ann Rev Plant Physiol* **33**, 133–162.

Shikanai, T., Shimizu, K., Ueda, K., Nishimura, Y., Kuroiwaa, T. and Hashimoto, T. (2001) The chloroplast clpP gene, encoding for a proteolytic subunit of ATP-dependent protease, is indispensable for chloroplast development in tobacco. *Plant Cell Physiol* **42**, 264–273.

Sullivan, J.A., Shirasu, K. and Deng, X.W. (2003) The diverse roles of ubiquitin and the 26S proteasome in the life of plants. *Nat Rev Genet* **4**, 948–958.

Taylor, C.B., Bariola, P.A., delCardayre, S.B., Raines, R.T. and Green, P.J. (1993) RNS2: a senescence-associated RNase of *Arabidopsis* that diverged from the S-RNase before speciation. *Proc Natl Acad Sci USA* **90**, 5118–5122.

Thiele, A., Herold, M., Lenk, I., Quail, P.H. and Gatz, C. (1999) Heterologous expression of *Arabidopsis* phytochrome B in transgenic potato influences photosynthetic performance and tuber development. *Plant Physiol* **120**, 73–81.

Thompson, A.R. and Vierstra, R.D. (2005) Autophagic recycling: lessons from yeast help define the process in plants. *Curr Opin Plant Biol* **8**, 165–173.

Thompson, J., Taylor, C. and Wang, T.W. (2000) Altered membrane lipase expression delays leaf senescence. *Biochem Soc Trans* **28**, 775–777.

Varshavsky, A. (1997) The N-end rule pathway of protein degradation. *Genes Cells* 2, 13–28.

Wilson, C. and Heberle-Bors, E. (2000) MAP kinases in pollen. *Results Probl Cell Differ* **27**, 39–51.

Weaver, L.M., Gan, S., Quirino, B. and Amasino, R.M. (1998) A comparison of the expression patterns of several senescence-associated genes in response to stress and hormone treatment. *Plant Mol Biol* **37**, 455–469.

Woo, H.R., Chung, K.M., Park, J.H. *et al.* (2001) ORE9, an F-box protein that regulates leaf senescence in *Arabidopsis*. *Plant Cell* **13**, 1779–1790.

Woo, H.R., Goh, C.H., Park, J.H. *et al.* (2002) Extended leaf longevity in the *ore4-1* mutant of *Arabidopsis* with a reduced expression of a plastid ribosomal protein gene. *Plant J* **31**, 331–340.

Woolhouse, H.W. (1984) The biochemistry and regulation of senescence in chloroplasts. *Can J Bot* **62**, 2934–2942.

Xiao, S., Dai, L., Liu, F., Wang, Z., Peng, W. and Xie, D. (2004) COS1: An *Arabidopsis* coronatine insensitive1 suppressor essential for regulation of Jasmonate-mediated plant defense and senescence. *Plant Cell* **16**, 1132–1142.

Xiong, Y., Contento, A.L. and Bassham, D.C. (2005) AtATG18a is required for the formation of autophagosomes during nutrient stress and senescence in *Arabidopsis thaliana*. *Plant J* **42**, 535–546.

Yamada, K., Matsushima, R., Nishimura, M. and Ikuko, H. (2001) A Slow maturation of a cysteine protease with a granulin Domain in the vacuoles of senescing *Arabidopsis* leaves. *Plant Physiol* **127**, 1626–1634.

Yang, S.H., Berberich, T., Sano, H. and Kusano, T. (2001) Specific association of transcripts of tbzF and tbz17, tobacco genes encoding basic region leucine zipper-type transcriptional activators, with guard cells of senescing leaves and/or flowers. *Plant Physiol* **127**, 23–32.

Ye, Z., Rodriguez, R., Tran, A. *et al.* (2000) The developmental transition to flowering repressed ascorbate peroxidase activity and induces enzymatic lipid peroxidation in leaf tissue in *Arabidopsis thaliana*. *Plant Sci* **158**, 115–127.

Yoshida, S. (2003) Molecular regulation of leaf senescence. *Curr Opin Plant Biol* **6**, 79–84.

Yoshida, S., Ito, M., Callis, J., Nishida, I. and Watanabe, A. (2002a) A delayed leaf senescence mutant is defective in arginyl-tRNA:protein arginyltransferase, a component of the N-end rule pathway in *Arabidopsis*. *Plant J* **32**, 129–137.

Yoshida, S., Ito, M., Nishida, I. and Watanabe, A. (2002b) Identification of a novel gene HYS1/CPR5 that has a repressive role in the induction of leaf senescence and pathogen-defence responses in *Arabidopsis thaliana*. *Plant J* **29**, 427–437.

Yoshida, S., Ito, M., Nishida, I. and Watanabe, A. (2001) Isolation and RNA gel blot analysis of genes that could serve as potential molecular markers for leaf senescence in *Arabidopsis thaliana*. *Plant Cell Physiol* **42**, 170–178.

Zapata, J.M., Guera, A., Esteban-Carrasco, A., Martin, M. and Sabater, B. (2005) Chloroplasts regulate leaf senescence: delayed senescence in transgenic ndhF-defective tobacco. *Cell Death Differ* **12**, 1277–1284.

11 Flower senescence

Michael S. Reid and Jen-Chih Chen

11.1 Introduction

The wilting of flower petals is perhaps the most deliberate and precisely controlled senescence event in plants. Frequently rapid, and often very accurately programmed in response to environmental or physiological cues, petal senescence (and/or flower abscission) evolved so as to optimize pollination and husband resources. Studies with different model systems suggest that much of the biochemistry of disassembly of the petal cells is shared, and quite similar to the events that have been demonstrated in leaves and other organs. On the other hand, the control mechanisms are diverse, both in terms of environmental or endogenous cues and of the response cascade, where an ethylene-dependent or ethylene-independent pathway may be invoked. The importance to commercial floriculture of extending flower life has led to studies that have generated a wealth of information on environmental and hormonal cues that may play a role in the flower senescence signal, and recent reports from several model systems have illuminated the changes in gene transcription associated with the onset and progress of the process. This chapter provides an opportunity to meld these findings with past physiological and biochemical studies and suggests future directions in the study and manipulation of this fascinating process.

11.2 Flower opening and senescence

Among the characteristics that distinguished the angiosperms from their ancestors was the evolution of a remarkable symbiosis – the elaboration of floral organs that attracted specific pollinators by providing nectar or pollen. The petals, billboards advertising the availability of these resources, were large and brightly colored, reflecting the limited vision and high color sensitivity of the birds and insects that are the predominant pollinators. An interesting demonstration of this evolution comes from the flora of New Zealand. In that isolated ecosystem, there is an unusually high proportion of white flowers (Godley, 1979). Heine (1938) proposed that the dearth of red, blue and purple flowers in the New Zealand flora '... is due to the absence of native long-tongued bees ... yellow and white flowers are best adapted for pollination by short-tongued bees and flies, which are the chief pollinators in New Zealand'.

The life of a flower is an elegantly choreographed sequence of anatomical and physiological changes designed to optimize the opportunity for outcrossing and fertilization. The rapid growth of petals and the operation of the different mechanisms that are involved in flower opening have to be coordinated with the development of the ovary and the pollen grains so that the open flower is receptive to pollen and able to shed pollen during pollinator visits. During angiosperm evolution, the selective advantage of organs that attracted pollinators by scent, size, form or color was enhanced by the evolution of systems that shut down the advertising system once pollination has been accomplished or the gynoecium was no longer receptive – cessation of scent production, changes in petal color, petal senescence and/or petal abscission. The removal of the floral organs once they have accomplished their task also has a resource implication. Ashman and Schoen (1994) showed a strong positive correlation between flower longevity and floral maintenance and construction cost. In this chapter, it is our particular interest to focus on the events of petal senescence, recognizing that this is but one event in the program that results in cross fertilization and the formation of seed.

11.3 Model systems

The senescence behavior of flowers can be categorized in several ways (Woltering and van Doorn, 1988). A primary differentiation is between plants where the corolla is shed once the flower is pollinated and those where the petals wilt (and sometimes change color) while attached to the plant. Flowers of the latter type have been the primary focus of research on flower senescence, since wilting in abscising-type flowers is likely to be primarily due to separation from water supply.

The precise regulation and rapidity of petal wilting has attracted both applied and basic researchers to use flowers in their studies of senescence. Initial studies of flower senescence were engendered by commercial interest in increasing the life of cut flowers and were focused primarily on the major flowers of commerce at the time, namely roses, carnations and chrysanthemums. Many early studies focused on carnations, whose short life was a major commercial problem. Researchers recognized that the predictable and rapid senescence of carnation petals made them a useful model system for studying the events of senescence. Indeed, isolated single petals of carnation flowers make excellent replicates for physiological and biological studies, and senesce in synchrony with those on an intact flower of the same age (Mor and Reid, 1981). The identification of ethylene as the triggering factor in carnation senescence (Nichols, 1996) led to the development of the silver thiosulfate (STS) complex as a commercial treatment to extend the life of these and other flowers (Veen and van der Geijn, 1978). Petunias, another ethylene-sensitive flower, are easily produced throughout the year, and continue to be a favored tool for physiological and molecular studies of petal wilting (Whitehead *et al.*, 1984; Chen *et al.*, 2004).

Studies of physiology and biochemistry are enhanced by systems that senesce predictably, and preferably rapidly, and Matile and his colleagues made use of the flowers of morning glory, which senesce in a single day, apparently triggered by a burst of ethylene biosynthesis (Matile and Winkenbach, 1971). The repeatable and rapid 'blushing' of the orchid labellum that follows pollination or simply removal of the pollinia in some orchid species has made these a useful system for a number of researchers (Arditti and Knauft, 1969; O'Neill *et al.*, 1993).

For studies of ethylene-independent flower senescence, research has focused on daylily (Valpuesta *et al.*, 1995) and iris (van Doorn *et al.*, 2003). Useful information has also been obtained in studies of other taxa, including both ethylene-dependent (lathyrus, gypsophila) and ethylene-independent (daffodil, gladiolus, four o'clock and alstroemeria) species. In daffodil, curiously, the senescence pathway can be invoked by ethylene (applied exogenously, or synthesized endogenously following pollination), but natural senescence cannot be delayed by application of inhibitors of ethylene synthesis and action (Hunter *et al.*, 2004b). We term flowers with this behavior pattern 'mixed'. Recently we have been using another 'mixed' flower, the four o'clock (*Mirabilis jalapa*), as a model system (Gookin *et al.*, 2003).

In cut roses and chrysanthemums, petal wilting is normally the result of adverse water relationships, frequently associated with vascular occlusion. Because of their commercial importance, these flowers have also been a frequent target for senescence studies, but since their senescence is of the 'absciscing' type, most or these studies have addressed water relations. On the plant, rose petals usually abscise while fully turgid, and the same is true for the ligules of chrysanthemums. Unfortunately, this same behavior is characteristic of the flowers of *Arabidopsis thaliana*, where the fully turgid petals abscise from the flower. The powerful molecular genetic tools available in this quintessential model species for dissecting the regulation of growth, development and senescence are thus of limited direct value in the analysis of floral senescence.

11.4 Hormonal regulation of flower senescence

Plant hormones and other signal molecules have long been implicated in the regulation of flower senescence. Early studies of petal senescence using carnation and morning glory revealed a climacteric-like pattern, where a substantial increase in ethylene production triggered an increase in respiration and in-rolling of the petals. Inhibitors of ethylene action, including high CO_2 and, famously, silver ion, resulted in a dramatic extension of life of the petals, as did inhibitors of ethylene biosynthesis, such as amino-ethoxyvinyl glycine (AVG) and amino-oxyacetic acid (AOA) (Reid and Wu, 1992). Studies with a range of plant hormones suggested interaction among them in the initiation of senescence. In particular, cytokinins seemed to delay petal senescence, apparently by reducing petal sensitivity to ethylene (Chang *et al.*, 2003). In contrast, auxin typically accelerates flower wilting. The blushing of the orchid labellum was suggested to be a response to the high auxin content of orchid pollen (Burg and Dijkman, 1967).

11.4.1 Ethylene

Ethylene is the primary regulator of flower senescence in a number of species where petal wilting is accompanied by a burst of ethylene production and accelerated by exogenous ethylene. In many other taxa, flowers or petals abscise without wilting, and this is also an ethylene response. Ethylene sensitivity shows strong taxonomic conservation and tends to be characteristic of certain families (Woltering and van Doorn, 1988; van Doorn, 2001). Application of ethylene biosynthesis inhibitors, such as AOA, or action inhibitors, such as STS and 1-methylcyclopropene (1-MCP), extends the longevity of ethylene-sensitive flowers (Lovell *et al.*, 1987; Serek *et al.*, 1994). Some flowers, including daffodil, campanula and four o'clock, show an intermediate pattern of senescence (Hunter *et al.*, 2004). Although these flowers wilt earlier when exposed to exogenous ethylene, their normal senescence is not retarded by inhibitors either of ethylene biosynthesis or of ethylene action. Furthermore, without pollination, no climacteric rise of in ethylene production is detected during senescence. These findings suggest the probability that ethylene-dependent and ethylene-independent petal senescence differ largely at the level of the regulators that initiate the senescence cascade.

Aminocyclopropane-1-carboxylic (ACC) synthase (ACS) and ACC oxidase (ACO) catalyze the last two reactions in the biosynthesis of ethylene (Wang *et al.*, 2002). Both enzymes are encoded by small gene families whose members are differentially expressed in response to different stimuli (Wang *et al.*, 2002). The induction of *ACS* and *ACO* transcripts and the activity of both enzymes are correlated with ethylene production in senescing and ethylene-treated petunias and carnations (Tang *et al.*, 1994; ten Have and Woltering, 1997). Transgenic carnations expressing an antisense fragment of *ACO* have a longer shelf life than untransformed flowers (Savin *et al.*, 1995). Likewise, virus-induced gene silencing (VIGS) of ACO in petunia resulted in extended flower life (Chen *et al.*, 2004). Research is still needed to clarify the regulation of *ACS/ACO* expression and to identify the regulatory components (Wang *et al.*, 2002).

Ethylene's role in flower senescence has also been studied by genetically altering the ethylene-signaling pathway. Flowers from transgenic petunias expressing *etr1-1*, the mutated ethylene receptor gene, from *Arabidopsis*, which confers dominant ethylene insensitivity, exhibit ethylene insensitivity and long shelf life (Wilkinson *et al.*, 1997). Transgenic petunias expressing an antisense fragment of petunia *EIN2*, a positive regulator of ethylene action, show similar ethylene insensitivity and delayed flower senescence (Shibuya *et al.*, 2004).

11.4.2 Abscisic acid

Recent studies suggest that abscisic acid (ABA) may also play an important role in the regulation of flower senescence. Onoue *et al.* (2000) showed that ABA content began to increase after the harvest of cut carnation flowers and exogenous ABA treatment triggered endogenous ABA production and flower senescence. Ethylene production increased 2 days after the start of the ABA increase, and exogenous ABA greatly stimulated ethylene production. Inhibition of ethylene action by application of STS prevented the ABA-mediated stimulation of senescence. These results suggest that if ABA is involved in the regulation of carnation senescence, it is mediated through an increase in ethylene production (Onoue *et al.*, 2000).

In daylily and cacao flowers, it appears that ABA has a direct effect on flower senescence (Panavas *et al.*, 1998; Aneja *et al.*, 1999). Exogenous application of ABA, but not ethylene, accelerated flower senescence in both these taxa. The endogenous ABA concentration also increased dramatically in these flowers before any visible signs of senescence. In daylily, moreover, the RNA-AFLP profile in ABA-treated flowers was similar to that in naturally senescing petals. Application of fluridone, an inhibitor of ABA biosynthesis, reduced ABA levels and extended longevity of cacao flowers. In these species, therefore, ABA seems to be directly involved in the regulation of flower senescence.

It has not been determined how common the ABA-regulated pattern of flower senescence is in ethylene independent flower species. Hunter *et al.* (2004) demonstrated that ABA would stimulate senescence in daffodils, a flower that responds to exogenous ethylene but whose natural senescence is ethylene independent. As in carnations, however, their data indicate that the ABA effect is mediated through stimulated ethylene production. As had been shown in carnations, they also demonstrated a rise in ABA concentration in petals before the onset of senescence, but since the increase in senescence-associated genes (SAGs) commenced before the rise in ABA content, they concluded that the change in ABA concentration was not the primary regulator of senescence.

11.4.3 Cytokinins

While ethylene and ABA act to promote flower senescence, cytokinins, which are well known to delay leaf senescence, have been reported also to delay flower senescence in a number of important ornamental species, including carnations, petunias, daffodils and roses. Cytokinin content was found to be higher in young flowers than in old flowers in some rose and carnation cultivars (Mayak *et al.*, 1972; van Staden and Dimalla, 1980), and the longevity of flowers of different rose cultivars was found to be correlated with their cytokinin content (Mayak and Halevy, 1970). Application of cytokinin can delay flower senescence but the effect depends on the type of cytokinin, the flower species and the developmental stage at which treatment occurs (Taverner *et al.*, 1999). Recently, Chang *et al.* (2003) demonstrated that the longevity of petunia flowers was greatly extended in plants expressing *IPT*, a bacterial gene providing precursors to cytokinin biosynthesis, under the control of the senescence-associated SAG12 promoter.

It has been suggested that the effect of cytokinins in retarding flower senescence may be associated with changes in ethylene signaling. Application of cytokinins to petals of both carnationss and petunias changed their ethylene sensitivity and biosythesis (Cook *et al.*, 1985; Chang *et al.*, 2003). The interplay between ethylene and cytokinins is also indicated by the finding that showed that ethylene promoted cytokinin degradation during floral senescence in petunias (Taverner *et al.*, 1999).

11.4.4 Gibberellic acid

Like cytokinins, gibberellins (GA) have been found to delay flower senescence in a number of ornamental crops, including carnations, daffodils, roses and *Sandersonia*

(Saks and van Staden, 1993; Sultan and Farooq, 1999; Eason, 2002, Hunter *et al.*, 2004). GA is a natural antagonist of ABA action in many physiological events, such as seed germination, programmed cell death (PCD) in aleurone layer cells and anthocyanin biosynthesis in petunia flowers (Weiss *et al.*, 1995; Fath *et al.*, 2000). In regulation of flower senescence, it also appears that GA may counteract the promotive effects of ABA action and thereby retard flower senescence (Hunter *et al.*, 2004). However, there is little evidence to suggest a direct involvement of GA in the regulation of flower senescence.

11.4.5 Auxin

Auxin has long been suspected of involvement in the onset of flower senescence. It is known that auxin can stimulate ethylene production and thereby accelerate flower senescence in some ethylene-sensitive flowers, such as orchids, and carnations (Stead, 1992; O'Neill, 1997). It has also been shown that pollen contains auxin. It has been proposed that auxin serves as a primary signaling molecule in pollination, triggering ethylene production and promoting flower senescence (O'Neill, 1997). However, in daylily, application of auxin delayed petal senescence (Rubinstein, 2000). This may suggest that auxin plays a different role in ethylene insensitive flowers.

11.4.6 Jasmonic acid

Jasmonic acid is an important regulator of plant defense responses, and there is evidence that increased synthesis of this hormone plays a role in leaf senescence (He *et al.*, 2002). However, little is known about its role in the regulation of flower senescence. Exogenous jasmonic acid accelerated flower senescence in two orchid species and in petunia by promoting the production of 1-ACC acid and ethylene (Porat *et al.*, 1993). Linolenic acid, a precursor of jasmonic acid and a substrate for lipoxygenase (LOX), was also found to stimulate ethylene production and thereby accelerate flower senescence in orchid flowers (Porat *et al.*, 1995). However, this study suggested that there was no direct involvement of jasmonic acid in the regulation of flower senescence. Inhibitors of LOX did not retard the senescence of pollinated orchids, and there was no change in jasmonic acid content of the flowers following pollination (Porat *et al.*, 1995).

11.4.7 Polyamines

The fact that polyamine and ethylene biosyntheses use the same precursor, S-adenosylmethionine (SAM), has long intrigued physiologists, given that these regulators commonly have antagonistic effects (Pandey *et al.*, 2000). Application of polyamines commonly delays petal wilting (Rubinstein, 2000). The extension of flower longevity in cut carnation flowers treated with spermine was correlated with a reduction in ethylene production, ACC content and the activities of ACS and ACO in the petals (Lee *et al.*, 1997). Application of methylglyoxal bis-(guanylhydrazone) (MGBG), an inhibitor of polyamine biosynthesis, increased ethylene production and accelerated petal wilting (Lee *et al.*, 1997).

Contrasting results have also been reported; exogenous polyamines stimulated ethylene production in pea leaves (Pennazio and Roggero, 1990), and a higher level of free putrescsine was detected in senescing carnation flowers (Roberts *et al.* 1984). Moreover, though spermidine and putrescine content declined during pea flower senescence, the level of spermine and the conjugated polyamine, N4-hexanoylspermidine (hexanoyl-spd), increased (Perez-Amador *et al.*, 1996), suggesting that different polyamines may have different biological functions.

Polyamines have also been implicated in plant defense responses. Application of spermine induced pathogenesis-related (PR) protein expression and conferred tobacco mosaic virus (TMV) resistance in tobacco, implicating spermine in the induction of hypersensitive response (HR) cell death (Yamakawa *et al.*, 1998). Harpin (pathogenic bacteria-secreted proteinaceous elicitor)-induced 1 (HIN1) and two closely related proteins, HIN9 and HIN18, were later identified as potential spermine-signaling factors (Takahashi *et al.*, 2004). The abundance of *HIN1* transcripts increased with flower aging (Takahashi *et al.*, 2004). These results conflict with the notion that polyamines are general retardants of flower senescence. It would be interesting to examine polyamine content changes during wilting of tobacco corollas and the effect of different polyamines on the longevity of tobacco flowers.

11.4.8 Sugars

Carbohydrates are not only a primary energy source but also potential signal molecules (Gibson, 2004). Changing source–sink relations are a common event during plant development and have been suggested to be a factor in leaf senescence (Roitsch and Gonzalez, 2004). Thimann *et al.* (1977) noted that a reduction in photosynthesis rate accelerates leaf senescence and suggested that sugar starvation may be a cause of leaf senescence. Bieleski *et al.* (1992) demonstrated that sugar depletion was the cause of 'leaf blackening', necrotic death of the cells in leaves of cut *Protea* flowers. High sugar content has also been shown to promote leaf senescence (van Doorn, 2004). Moreover, plants overexpressing hexokinase, a potential sugar sensor, showed early leaf senescence while those with down-regulated hexokinase showed delayed senescence (Moore *et al.*, 2003).

Exogenous sugar has frequently been shown to extend cut flower longevity, suggesting that sugar starvation may also lead to petal wilting. However, this effect may be due to reduced ethylene sensitivity and/or increased osmotic potential that improves water uptake and thereby delays senescence (van Doorn, 2004). In addition to carbon supply and osmotic effects, applied sugars may also interact with hormonal signals. Mayak and Dilley (1976) demonstrated that sugar application reduced the sensitivity of carnation petals to ethylene, but in a recent review, van Doorn (2004) concluded that more research is needed to elucidate the mechanism by which sugar application delays petal wilting (van Doorn, 2004).

11.5 Flower senescence and remobilization of resources

Flower senescence is an active process that not only increases pollination efficiency but also allows for nutrient recycling. In gladiolus spikes, for example, removal of the older florets before they senesced resulted in a decrease in the dry weight and opening of younger florets (Waithaka *et al.*, 2001). Nutrient remobilization from petal cells occurs through export of simple metabolites; the senescence of lower florets on a gladiolus spike, for example, was associated with disappearance of most of the soluble sugars (Waithaka *et al.*, 2001). The dry matter content of these same florets falls to less than half, presumably due to the degradation of macromolecules, including nucleic acids, cell walls and other complex carbohydrates, and lipids. Genes encoding hydrolytic enzymes are prominent in the cDNA libraries that have been isolated from senescing flowers.

11.5.1 Protein degradation

By analogy with animal systems, where proteases play an important signaling role in the onset of PCD, it is suggested that the strong up-regulation of a variety of proteases that occurs during flower senescence may play a regulatory role. However, protein degradation is also an important part of the remobilization process during flower senescence. A dramatic reduction in petal protein content during flower senescence (Lay-Yee *et al.*, 1992) is accompanied by increased protease activity (Wagstaff *et al.*, 2002; Xu and Hanson, 2000). Cysteine proteases appear to play a particularly important role in flower senescence. Vacuolar processing enzyme (VPE) and papain-type cysteine proteases are consistently identified in populations of SAGs isolated from both ethylene-sensitive and ethylene-insensitive flowers (Jones *et al.*, 1995; Valpuesta *et al.*, 1995; Hunter *et al.*, 2002; Wagstaff *et al.*, 2002; van Doorn *et al.*, 2003).

It has been demonstrated that VPEs are required to initiate HR cell death (Hatsugai *et al.*, 2004). Although they have not yet been proved to be important in the regulation of flower senescence, the importance of vacuoles during flower senescence may suggest that they may play a significant role.

Increased transcription of genes encoding papain-type cysteine proteases is commonly associated with flower senescence. This group of proteases appears to regulate plant PCD and is perhaps required for the process (Beers *et al.*, 2000). Papain has an optimum activity at an acidic pH, a broad substrate range and a requirement for the presence of a reducing agent for optimum activity *in vitro* (Beers *et al.*, 2000). In plant cells, papain-type cysteine proteases are targeted to different cellular compartments, including the endoplasmic reticulum (ER), vacuole and apoplast (Beers *et al.*, 2000). Because of their acidic preference, they are expected to be active in the vacuole and cell wall, but not in the ER. Like VPEs, they may be involved in vacuole processing and thereby important for plant PCD. Inhibitor studies support the involvement of papain-type proteases in the regulation of PCD regulation. Ectopic expression of cystatin, a natural inhibitor of papain, blocked H_2O_2-induced PCD (Beers *et al.*, 2000). Cystatin was also identified in carnation petals and showed a significant negative correlation with petal protease activity during carnation petal senescence (Sugawara *et al.*, 2002).

11.5.2 Nucleic acid degradation

DNA fragmentation is a hallmark of animal apoptosis. This event is also found in some plant PCDs, including flower senescence (Xu and Hanson, 2000). Like protein degradation, degradation of nucleic acids is an important facet of nutrient recycling. Increased nuclease activity is followed by the loss of nucleic acid content during flower senescence (Orzáez and Granell, 1997; Rubinstein, 2000; Xu and Hanson, 2000; Hunter *et al.*, 2002). Two groups of nucleases were found to be responsible for the degradation of nucleic acids. One was up-regulated during flower senescence and the other was constitutively expressed. The activity of both nucleases was substantially promoted by Ca^{2+} or Mg^{2+} (Xu and Hanson, 2000).

cDNA fragments that are homologous with S1-type nucleases and highly associated with flower senescence were identified in cDNA libraries from senescing daylily and daffodil (Panavas *et al.*, 1999; Hunter *et al.*, 2002). Recently, this type of nuclease was demonstrated to be a primary nuclease in nuclear DNA degradation in zinnia cells during the PCD that accompanies the development of tracheary elements (Ito and Fukuda, 2002). Down-regulation of this nuclease in zinnia cells using antisense technology prevented nuclear DNA degradation but did not prevent vacuolar collapse in cells undergoing PCD (Ito and Fukuda, 2002). It will be interesting to determine whether the S1-type nucleases that are associated with flower senescence play a similar role.

11.5.3 Membrane degradation

During petal wilting, cellular membranes progressively lose their integrity, resulting in leakage of pigments, nutrients and electrolytes from the petal cells (Rubinstein, 2000). During wilting of carnation petals, some loss of membrane integrity was observed even before the climacteric rise in ethylene production (Smith *et al.*, 1992). Moreover, changes in lipid content and membrane structure were demonstrated before the start of electrolyte leakage (Thompson *et al.*, 1998; Leverentz *et al.*, 2002). Generally, membrane degradation is characterized by replacement of phospholipids with neutral lipids. This change may be a result of increasing activity of phospholipasess and acyl hydrolases (Paliyath and Droillard, 1992). The saturation–unsaturation index of the membrane fatty acids also increases. This oxidative event is due to either increased LOX activity or autoxidation. Increased LOX activity prior to obvious electrolyte leakage has been shown in a numbers of flowers, including carnation, daylily and rose (Fobel *et al.*, 1987; Panavas and Rubinstein, 1998; Fukuchi-Mizutani *et al.*, 2000). However, no such increase was observed in *Alstroemeria peruviana* (Leverentz *et al.*, 2002), suggesting a LOX-independent lipid oxidation pathway in that species.

Membrane peroxidation is thought to increase reactive oxygen species (ROS) production (Rubinstein, 2000). ROS haves long been suggested to play an important role in animal aging, apoptosis and plant PCD. Indeed, H_2O_2 content starts increasing before flower opening in daylily flowers. Moreover, the expression of two genes encoding ROS scavenge scavenging enzymes, catalase (CAT) and ascorbate peroxidase (APX), decreases during daylily petal wilting (Panavas and Rubinstein, 1998). Sodium benzoate, an antioxidant that has been shown to delay petal wilting in other species, also reduced H_2O_2 production and delayed petal wilting in this species. Although these results strongly suggest that ROS is an important player in the regulation of flower senescence, CAT and APX activity remained unchanged during carnation petal wilting (Bartoli *et al.*, 1995). Patterns of ROS production during carnation petal wilting have not yet been determined. It is possible that reduced activity of other ROS-scavenging enzymes and continued ROS generation results in high ROS content even though the activities of CAT and APX remain unchanged. It is also possible that the role of ROS in senescence regulation is entirely different between the daylily (ethylene-independent) and carnation (ethylene-dependent) flowers.

ROS can also be generated from the mitochondrial respiratory chain, and declining respiratory efficiency leading to increased ROS production is a common event in aging animal cells. Mitochondria are also very sensitive to ROS, whose production may lead to mitochondrial permeability transition (MPT) and mitochondrial membrane deterioration. Recently, MPT has been implicated as a common event in plant PCD (Yao *et al.*, 2004). MPT allows the death effectors and activators, such as cytochrome c and nucleases, to be released from the mitochondria and trigger PCD in animals (Jiang and Wang, 2004). Cytochrome c release has also been evidenced in several plant PCD events (Curtis and Wolpert, 2002; Tiwari *et al.*, 2002). However, Xu and Hanson (2000) did not find MPT in pollination-mediated petunia flower senescence, indicating that more research is needed to investigate the role of ROS and MPT in petal senescence.

11.5.4 Cell wall changes

Ultrastructural studies showed loosening of wall fibrillar structure and the appearance of intracellular cytoplasmic debris in postclimacteric carnation tissues (Smith *et al.*, 1992), indicating dramatic cell wall changes in the later stages of flower senescence. Increased activity of wall-based hydrolytic enzymes in senescing daylily and *Sandersonia* flowers is a further evidence of cell-wall degradation in senescing flower tissues (Panavas *et al.*, 1998; O'Donoghue *et al.*, 2002). Polysaccharides are released from the cell wall during cell-wall degradation, and there is accumulating evidence that polysaccharides may serve as signal molecules that regulate plant development and defense mechanisms (Pilling and Hofte, 2003). However, there is no evidence of a role for a polysaccharide signal in the regulation of flower senescence.

11.6 Petal senescence as programmed cell death

Senescence is more precisely and deliberately controlled in flowers than in any other plant organ. It contrasts strikingly with leaves and fruits, where senescence is often a final phase of an aging process that starts with reduced photosynthetic capacity (in the case of leaves) or with ripening (in the case of fruits). The difference between senescence in leaves and petals is particularly intriguing. In leaves, the presence of stomates results in diurnal water stress, and the photosynthetic system constantly generates damaging free radicals that must be detoxified. By contrast, the stomates in most petals, if present, are rudimentary and/or nonfunctional so that petals are only exposed to water stress as a result of water stress generated by the leafy parts of the plant. Moreover, most petals are not photosynthetic, which obviates the damaging effects of free radicals, and many are pigmented with known antioxidants, anthocyanins and carotenes. Senescence patterns do not reflect the difference in stresses endured by these different organs. Despite the daily stresses to which leaves are subjected, they commonly remain highly functional for many months, and even years. In contrast, the longest lasting flowers senesce within months, and most fade within days. In ephemeral species, the whole process of opening and senescence may take place in a matter of a few hours. This contrast clearly implies that the life and death of the petals is strictly regulated, and this surely reflects the selective advantage of directing pollinators away from flowers that are already pollinated or whose stigmas are no longer receptive (Stead and Reid, 1990; Stead, 1992).

The precise coordination of petal senescence in many flowers suggests that this is a strictly programmed event. The program clearly involves changes in gene expression, since application of inhibitors of transcription and translation can dramatically delay the onset of petal senescence. Application of cycloheximide, a metabolic poison that inhibits protein synthesis by eukaryotic ribosomes, can extend the life of flower petals by many days (Jones *et al.*, 1994). Marked reduction in protein synthesis does not result in premature cell death, as might have been predicted, and it would be interesting to examine how the petals are able to persist without the ability to replace essential proteins lost to the protein turnover mechanism.

The programmed nature of petal senescence has invited comparison with programmed cell death in animal systems, and the more specifically characterized apoptosis. Programmed cell death in animal cells (often referred to as apoptosis) is thought to be initiated by release of mitochondrial cytochrome c and activity of specific proteases, the caspases. The process is defined by the ultrastructural changes, including chromatin condensation, cytoplasm shrinkage, membrane blebbing and appearance of fragmented, membrane-bounded, apoptotic bodies (Kerr, 2002). Many of the characteristics of apoptosis have also been reported in plant PCD, including flower senescence. DNA fragmentation, the most stringent hallmark of animal apoptosis, as well as increased nuclease and protease activity, has been reported in plant PCD (Wang *et al.*, 1996; O'Brien *et al.*, 1998; Solomon *et al.*, 1999; Xu and Hanson, 2000). More importantly, there is indirect evidence suggesting the possible existence of caspase-like proteases, key players in apoptosis, in plant PCD. First, extracts from plant cells undergoing death contain enzyme activities capable of cleaving synthetic caspase substrates (Woltering *et al.*, 2002). Second, plant PCD can be blocked by caspase inhibitors, such as IAP, Op-IAP and p35 (Woltering *et al.*, 2002; Lincoln *et al.*, 2002).

Although this evidence suggests that a conserved pathway similar to apoptosis is involved in plant PCD, a lively debate continues as to the validity of such a comparison. The absence from the *Arabidopsis* genome of any sequences with homology to animal caspases (the proteases that is considered to be the trigger and the hallmark of apoptosis) has been cited as evidence that senescence in plants, even when programmed, is not strictly homologous with animal apoptosis. Others point out that other types of cysteine proteases do exist in the plant genome and some are clearly linked with PCD. For example, it has been demonstrated that a VPE, a type of cysteine protease, is required for apoptosis-like virus-induced hypersensitive cell death, a type of plant PCD, in tobacco cells (Hatsugai *et al.*, 2004). Metacaspases are another type of cysteine proteases showing caspase-like activity (Woltering *et al.*, 2002; Woltering, 2004), and metacaspases isolated from *Arabidopsis* were shown to induce apoptosis-like cell death in yeast (Watanabe and Lam, 2005).

One of the fundamental features of plant cells that differentiate them from animal cells, the presence of a rigid cell wall, precludes the phagocytosis that is the endpoint of apoptosis. van Doorn and Woltering (2005) noted that in addition to apoptosis, many cell death events in animals can be categorized as autophagic. In reviewing the evidence, they concluded that petal cells (and plant cells in general) do not senesce by an apoptotic mechanism but utilize autophagy in recovering and redistributing nutrients to the rest of the plant. Three types of autophagy, micro-, macro- and mega-autophagy, have been reported. All of them involve lysosomal vacuoles where molecules or organelles from the cytoplasm are digested. Their review suggested that plant PCD was most properly categorized as the mega-autophagic type where hydrolases released from an enlarged vacuole through permeabilization of the vacuolar membrane degrade all the cell contents. Autophagy has long been suggested to be a primary mechanism of nutrient recycling in eukaryotes (Thompson and Vierstra, 2005), and autophagy has been implicated in the cell death accompanying the development of tracheary elements, formation of aerenchyma, the formation of root cap cells, leaf senescence and the senescence of iris petals (Gahan, 1982; Drew *et al.*, 2000; Obara *et al.*, 2001; van Doorn *et al.*, 2003).

Autophagy is partially reversible, and the chloroplast remodeling that is an important feature of leaf senescence can be reversed by hormonal and other treatments. Although regreening is rare in petals, it does happen naturally in certain orchids where pollination is followed by a regreening of the petals, which are then capable of photosynthesis and a leaf-like life span (Thomas *et al.*, 2003). The fact that a vacuolar protease, the VPE, is important in plant PCD lends support to the hypothesis that plant PCD is autophagic in nature, and may provide insight into the regulation of plant senescence (Hara-Nishimura *et al.*, 2005).

11.7 Molecular biology of petal senescence

11.7.1 Senescence-associated genes

The fact that α-aminitin – an RNA polymerase inhibitor – and cycloheximide – a protein synthesis inhibitor – retard flower senescence suggests that the process is under tight genetic control and depends on *de novo* protein synthesis (Bieleski and Reid, 1992). Many researchers have attempted to understand the control of flower senescence by identifying genes involved in this process. It is no surprise that genes encoding hydrolytic and ethylene biosynthesis-related enzymes have been identified from a number of species (Rubinstein, 2000). It is perhaps more interesting that genes encoding transcription factors, kinases and proteins with unknown function also are commonly identified (Hunter *et al.*, 2002; Leverentz *et al.*, 2002; van Doorn *et al.*, 2003).

Two recent major studies employed a microarray approach to examining changes in the transcriptome during petal senescence. Leverentz *et al.* (2002) determined changes in transcript abundance during senescence in *Alstroemeria*, and van Doorn *et al.* (2003) conducted a similar study in *Iris*. Expressed sequence tags (EST) libraries, constructed following subtractive enrichment to highlight genes whose abundance was affected by petal senescence, were used to create microarrays that were then probed with mRNA from flowers harvested at different stages of senescence. The transcriptome in both species changed dramatically, with genes in the microarray showing up-regulation, down-regulation or constitutive expression through the senescence program. In their analysis, van Doorn *et al.* (2003) correlated the changes in gene abundance with a morphological and anatomical anaylsis that took account of differential senescence of cells in different parts of the petal. One of the early events that they identified was closing of the plasmodesmata, which occurred 2 days before opening in mesophyll cells and at opening in epidermal cells. The onset of visible senescence in these two different tissues was similarly separated by 2 days. Genes related to the final stages of senescence included genes putatively involved in signal transduction and the remobilization of phospholipids, proteins and cell-wall components. Among the genes that they found particularly intriguing were an ion channel protein and some putative regulators of transcription and translation, including a MADS (MCMI of yeast, AGMOUS of Arabidopsis, DEFICIENS of snapdragon and SRF of human) domain factor.

In a study of four o'clock, an ephemeral flower that we have been using as a model system, we have found changes in numerous genes that are closely correlated with the timing of the commitment to senescence (X. Xu, T. Gookin, M.S. Reid, and C. Jiang, unpublished data). In some cases, the changes are dramatic; transcripts of a gene homologous to a ring zinc finger protein, for example, increased in abundance, as measured by real-time PCR, by more than 30 000 fold!

11.7.2 Functional analysis of SAGs

Evaluating the importance of the many genes whose abundance has been associated with flower senescence depends on functional studies where the abundance of the transcript or its encoded protein is altered during the onset of senescence. Because *Arabidopsis* shows an abscission phenotype during flower senescence, the powerful tools that are available for analysis of the role of different genes in physiological events have not proved very useful for studies of flower senescence. Moreover, many of the model species that have been used for physiological studies of flower senescence are unsuited to analysis of gene function because transformation, regeneration and phenotype characterization are difficult and/or time consuming. Petunia is a striking exception to this generalization; it is readily transformed and regenerated, has a relatively short generation time, and has been the subject of substantial genetic studies over the years.

11.7.2.1 Ethylene-dependent senescence

Senescence of petunia flowers is ethylene dependent, and as might be anticipated, down-regulation of ethylene biosynthetic genes (Chen *et al.*, 2004) and ethylene response cascade (Shibuya *et al.*, 2004) extends flower life. We have also been utilizing an RNAi approach to testing gene function using VIGS, a powerful and rapid technique for analysis of gene function in plant growth and development (Baulcombe, 1999; Ratcliffe *et al.*, 2001). The tobacco rattle virus (TRV) silencing vector has several interesting properties for studies of the function of candidate genes in floral senescence; it infects floral tissues, has only mild symptoms and has a broad host range. We tested the effect of TRV containing a fragment of the phytoene desaturase (PDS) gene on a range of host plants (Chen *et al.*, 2005a). In most cases, infection resulted in no apparent phenotype, or local inoculation effects. In petunia, after a lag of several weeks, the characteristic photo bleaching of upper portions of the plant indicated movement and silencing of the virus and PDS. In tests with silencing chalcone synthase (CHS) with a TRV/CHS construct, the typical effects of posttranscriptional gene silencing of CHS in petunia – symmetrical and asymmetrical white patches on the dark purple petals – indicated the ability to use the system in petunia studies. We subsequently demonstrated that we could simultaneously silence both CHS and PDS by incorporating fragments of both genes in the viral vector (Chen *et al.*, 2004). This suggested that we could use CHS silencing, in purple petunias, as a silencing indicator, and we demonstrated this strategy by infecting purple petunias with a viral vector containing fragments of CHS and ACO, the terminal step in ethylene biosynthesis. Infection with this combined construct resulted in white sectors on purple flowers, and completely white flowers. As anticipated, the white flowers had reduced ethylene production and their senescence was significantly delayed. Although petunia has several ACO homologs, there was evidently sufficient homology between them that the fragment we used (of ACO4) resulted in silencing of the homologous genes as well. Another intriguing finding with these flowers was the observation that after pollination, white segments on sectored flowers lasted longer than the purple sectors. This finding directly tests the hypothesis that intracellular ethylene might be the 'pollination' signal (Woltering *et al.*, 1996). The differential senescence of silenced and unsilenced segments on the same flower suggests, at least in petunia, that it is local ethylene production in response to a translocated signal, rather than intercellular ethylene itself that is responsible for the accelerated senescence resulting from pollination.

11.7.2.2 Ethylene-independent senescence

Lack of a suitable system for transformation and regeneration, as well as the extended period required for expression of any transgenic phenotype, has so far limited functional analysis of the genes associated with senescence in ethylene-independent flowers (like daffodil, iris and tulip). We have been exploring the use of the flowers of four o'clock as a model system. The large nocturnal flowers of this member of the

Nyctaginaceae open in the afternoon (hence the common name), are fully open at midnight and senesce shortly after dawn. Although senescence of *Mirabilis* flowers is associated with a small peak of ethylene production (Li *et al.*, 1994) and can be accelerated by exogenous ethylene (Gookin *et al.*, 2001), natural senescence is little affected by the ethylene action inhibitors STS or 1-MCP (Gookin *et al.*, 2003). *M. jalapa* therefore exhibits what a 'mixed' pattern of floral senescence is, similar to that seen in daffodils, in which senescence can proceed via the ethylene-sensitive or the ethylene-insensitive senescence pathway. *M. jalapa* can go from seed to seed in as little as 3 months. Being dicotyledonous, we anticipate that it will readily be transformed (Xu *et al.*, 2005), allowing relatively rapid examination of the function of genes associated with senescence of its ephemeral flowers.

11.7.3 Regulation of petal senescence – a regulatory network?

It has been elegantly demonstrated that the control of senescence in leaves is the result of interactions among a network of regulators (He *et al.*, 2001). We hypothesize that senescence of petals probably involves a comparable and perhaps even homologous network. Although a considerable body of information is now available describing the early steps in the ethylene signal transduction cascade, the way in which this pathway interacts with the other parts of the senescence network is not yet understood. The probable complexity of the petal senescence network is attested to by the numerous genes that have already been shown to change in transcript abundance during senescence. A number of those genes are putative transcription factors, and kinases, all potential nodes in a senescence network.

Researchers at the University of Florida have been developing a number of floral-specific libraries containing a large number of ESTs from petunia (David Clark, personal communication). The database includes more than 6000 independent ESTs, including many homologs of genes that have been previously identified in flower and leaf senescence. Among the flower SAGs in the database are a number of transcription factors, including those shown in Table 11.1. These transcription factors are potential candidates for regulators of floral senescence, and members of the WRKY, NAC and MADS-box families are of particular interest.

WRKY proteins are a new family of zinc-finger type transcription factors found exclusively in plants. These proteins have been shown to bind specifically to W Box-type [(C/T)TGAC (C/T)] DNA sequence elements both *in vitro* and *in vivo* (Eulgem *et al.*, 2000). In *Arabidopsis thaliana*, the WRKY gene family has 72 members (Riechmann *et al.*, 2000), and AtWRKY53 is expressed very early in leaf senescence, suggesting that this gene may play a regulatory role (Hinderhofer and Zentgraf, 2001). The high representation of WRKY homologs in the petunia floral EST collection (Table 11.1) and the fact that the *Arabidopsis* homologs appear to be a central part of the network regulating leaf senescence (Guo *et al.*, 2004) suggest these genes as prime candidates in the regulation of petal senescence.

The NAC family is another group of transcription factors unique to plants, which share a highly conserved N-terminal 150 amino acids – the NAC domain – and appear to play regulatory roles in plant growth and development (Xie *et al.*, 1999).

Table 11.1 DNA-binding factors identified in the University of Florida petunia floral EST databases

Transcription factor family	Number of EST clones
Zinc finger	35
Aux/IAA	12
AP2/EREBP	11
MYB	10
MADS	10
HB	8
BHLH	5
ARF	5
GRAS	4
WRKY	3
BZIP	3
NAC	2
YABBY	2
TCP	1
SBP	1
Nin-like	1
Others	18
Total	131

SENU5, a gene encoding a NAC domain protein, is up-regulated during tomato leaf senescence (John *et al.*, 1997), and *SAG107* is up-regulated during the senescence of *Arabidopsis* rosette leaves. The gene encoding the tomato *nor* mutation which strongly retards tomato fruit ripening has recently been identified as a NAC type transcription factor (J.J. Giovannoni, personal communication). Two NAC-type genes are also present in the petunia EST collection, suggesting the possibility that these genes are also involved in the regulation of petal senescence.

MADS-box transcription factors share the MADS domain, a 56-amino acid region involved in dimerization and DNA binding. These genes include the transcription factors involved in the determination of floral meristem and organ identity (Shore and Sharrocks, 1995; Riechmann and Meyerowitz, 1997). A MADS-box protein (Vrebalov *et al.*, 2002) has been shown to be involved in the control of fruit ripening, and MADS-box transcription factors have also been implicated in flower senescence. Thus, overexpression of AGL15, a member of the MADS-box family, delayed flower senescence in *Arabidopsis* (Fang and Fernandez, 2002). Although MADS-box transcription factors play major roles in flower development, it was surprising that none of the 82 members of the MADS-box genes in the *Arabidopsis* genome was found to be expressed in the senescent leaves (Guo *et al.*, 2004). However, we showed up-regulation of a MADS-box clone in senescing daffodil flowers (Hunter *et al.*, 2002), and our analysis identified 10 MADS-box genes in the petunia floral EST database (Table 11.1). Since silencing of a petunia MADS-box clone with VIGS extends the life of petunia flowers (J. Chen, C. Jiang, and M.S. Reid, unpublished data), it seems highly probable that MADS-box genes play a role in the regulation of floral senescence.

Our studies of the function of these genes using VIGS suggest that many of the intriguing correlations that we and other workers have seen between abundance of regulatory genes and the onset or progression of flower senescence may not have any functional meaning. Silencing many of these genes has no effect on petal senescence (Chen *et al.*, unpublished data). These results could be the result of functional redundancy, although one of the attractions of VIGS is that it silences any genes with high homology to the target gene. We are continuing to test the effect of silencing more than 100 genes with regulatory or unknown function that have been shown to change in abundance in concert with the onset of senescence in petunia and other flowers.

11.7.4 New frontier: prohibitins – mitochondrial proteins with a possible role in floral senescence

In the case of one particular group of proteins we did find a striking effect of silencing. Prohibitins, highly conserved mitochondrial proteins, have been shown to play important roles in cell cycling and senescence in animals and yeast. Sequences with high homology to prohibitins have been identified in a number of plant species, but their function had not previously been demonstrated. The deduced amino acid sequence of *PhPHB1,* a sequence that we isolated from a petunia floral EST database, shows high homology with that of other reported prohibitins from plants, animals and yeast (Chen et al., 2005b). When we down-regulated expression of *PhPHB1* using VIGS, we observed plants with smaller and distorted leaves and flowers. Cells in the silenced flowers were *larger* than in control flowers, indicating a substantial reduction in the number of cell divisions that took place during corolla development. The life of the silenced flowers was shorter than that of the controls, whether on the plant or detached. The respiration of the silenced flowers was higher than that of the controls. We observed a marked increase in the abundance of transcripts of CAT in the silenced flowers, which could suggest increased production of ROS by the impaired mitochondria, which may initiate premature senescence. Our data indicate that, as in animal systems, prohibitins play a key role in maintaining mitochondrial function. Since the abundance of transcripts encoding prohibitin genes decreases in the early stages of floral senescence, it is possible that reduced levels of this protein, and consequent malfunction of mitochondria, could be, in floral tissues as in animal systems, an inducer of programmed cell death.

References

Aneja, M., Gianfagna, T. and Ng, E. (1999) The roles of abscisic acid and ethylene in the abscission and senescence of cocoa flowers, *Plant Growth Regul* **27**, 149–155.

Arditti, J.H. and Knauft, R.L. (1969). The effects of auxin, actinomycin D, ethionine, and puromycin on post-pollination behaviour in orchid flowers. *Am J Bot* **56**, 620–628.

Ashman, T.-L. and Schoen, D.J. (1994) How long should flowers live? *Nature* **371**, 788–791.

Bartoli, C.G., Simontacchi, M., Guiamet, J., Montaldi E. and Puntarulo, S. (1995) Antioxidant enzymes and lipid peroxidation during aging of Chrysanthemum morifolium RAM petals. *Plant Sci* **104**, 161–168.

Baulcombe, D.C. (1999) Fast forward genetics based on virus-induced gene silencing. *Curr Opin Plant Biol* **2**, 109–113.

Beers, E.P., Woffenden, B.J. and, Zhao C. (2000) Plant proteolytic enzymes: possible roles during programmed cell death. *Plant Mol Biol* **44**, 399–415.

Bieleski, R.L. and Reid, M.S. (1992) Physiological changes accompanying senescence in the ephemeral daylily flower. *Plant Physiol* **98**, 1042–1049.

Bieleski, R.L., Ripperda, J., Newman, J.P. and Reid, M.S. (1992) Carbohydrate changes and leaf blackening in cut flower stems of Protea eximia. *J Am Soc Hort Sci* **117**, 124–127.

Burg, S.P. and Dijkman, M.J. (1967) Ethylene and auxin participation in pollen induced fading of *Vanda* orchid blossoms. *Plant Physiol* **42**, 1648–1650.

Chang, H., Jones, M.L., Banowetz, G.M. and Clark, D.G. (2003) Overproduction of cytokinins in petunia flowers transformed with P(SAG12)-IPT delays corolla senescence and decreases sensitivity to ethylene. *Plant Physiol* **132**, 2174–2183.

Chen, J.C., Jiang, C.Z., Gookin, T.E., Hunter, D.A, Clark, D.G. and Reid, M.S. (2005a) Chalcone synthase as a reporter in virus-induced gene silencing studies of flower senescence. *Plant Mol Biol* **55**, 521–530.

Chen, J.C., Jiang, C.Z. and Reid, M.S. (2005b) Silencing a prohibitin alters plant development and senescence. *Plant J* **44**, 16–24.

Chen, J.-C., Johnson, F., Clark, D.G., Gookin, T., and Reid, M.S. (2003) Potential application of virus-induced gene silencing (VIGS) in flower senescence studies. *Acta Horticulturae* **669**, 147–152.

Cook, D., Rasche, M. and Eisinger, W. (1985) Regulation of ethylene biosynthesis and action in cut carnation flower senescence by cytokinins. *J Am Soc Hort Sci* **110**, 24–27.

Curtis, M.J. and Wolpert, T.J. (2002) The oat mitochondrial permeability transition and its implication in victorin binding and induced cell death. *Plant J* **29**, 295–312.

Drew, M.C., He, C.J. and Morgan, P.W. (2000) Programmed cell death and aerenchyma formation in roots. *Trends Plant Sci* **5**, 123–126.

Eason, J.R. (2002) *Sandersonia aurantiaca*: An evaluation of postharvest pulsing solutions to maximise cut flower quality. *N Z J Crop Hort Sci* **30**, 273–279.

Eulgem, T., Rushton, P.J., Robatzek, S., and Somssich, I.E. (2000) The WRKY superfamily of plant transcription factors. *Trends Plant Sci* **5**, 199–206.

Fang, S.-C., and Fernandez, D.E. (2002) Effect of regulated overexpression of the MADS domain factor AGL15 on flower senescence and fruit maturation. *Plant Physiol* **130**, 78–89.

Fath, A., Bethke, P., Lonsdale, J., Meza-Romero, R. and Jones, R. (2000) Programmed cell death in cereal aleurone. *Plant Mol Biol* **44**, 255–66.

Fobel, M., Lynch, D.V. and Thompson, J.E. (1987) Membrane deterioration in senescing carnation flowers. *Plant Physiol* **85**, 204–211.

Fukuchi-Mizutani, M., Ishiguro, K., Nakayama, T. *et al.* (2000) Molecular and functional characterization of a rose lipoxygenase cDNA related to flower senescence. *Plant Sci* **160**, 129–137.

Gahan, P.B. (1982) Cytochemical and ultrastructural changes in cell senescence and death. In: *Growth Regulators in Plant Senescence* (eds Michael Jackson, Brain Grout, and Ian Mackenzie). British Plant Growth Regulator Group, Wantage (UK). Monograph 8, pp . 47–57.

Gibson, S.I. (2004) Sugar and phytohormone response pathways: navigating a signalling network. *J Exp Bot* **55**, 253–264.

Godley, E.J. (1979) Flower biology in New Zealand. *N Z J Bot* **17**, 441–466.

Gookin, T.E., Hunter, D.A. and Reid, M.S. (2003) Temporal analysis of alpha- and beta-expansin expression during floral opening and senescence. *J Plant Sci* **164**, 769–781.

Gookin, T., Hunter, D., Steele, B. and Reid, M.S. (2001) Evulation of *Mirabilis jalapa* as a model system for studying floral senescence. *HortScience* **36**, 599.

Guo, Y., Cai, Z. and Gan, S. (2004) Transcriptome of *Arabidopsis* leaf senescence. *Plant Cell Environ* **27**, 521–549.

Hara-Nishimura, I., Hatsugai, N., Nakaune, S., Kuroyanagi, M. and Nishimura, M. (2005) Vacuolar processing enzyme: an executor of plant cell death. *Curr Opin Plant Biol* **8**, 404–408.

Hatsugai, N., Kuroyanagi, M., Yamada, K. *et al.* (2004) A plant vacuolar protease, VPE, mediates virus-induced hypersensitive cell death. *Science* **305**, 855–858.

He, Y., Fukushige, H., Hildebrand, D.F. and Gan, S. (2002) Evidence supporting a role of jasmonic acid in *Arabidopsis* leaf senescence. *Plant Physiol* **128**, 876–884.

He, Y., Tang W, Swain, J.D., Green, A.L., Jack, J.P. and Gan, S. (2001) Networking senescence-regulating pathways by using Arabidopsis enhancer trap lines. *Plant Physiol* **126**, 707–716.

Heine, E.M. (1938) Observations on the pollination of New Zealand flowering plants. *Trans R Soc N Z* **67**, 133–148.

Hinderhofer, K. and Zentgraf, U. (2001). Identification of a transcription factor specifically expressed at the onset of leaf senescence. *Plant* **213**, 469–473.

Hunter, D.A., Ferrante, A., Vernieri, P. and Reid, M.S. (2004a) Role of abscisic acid in perianth senescence of daffodil (*Narcissus pseudonarcissus* 'Dutch Master'). *Physiol Plantarum* **121**, 313–321.

Hunter, D.A., Steele, B.C. and Reid, M.S. (2002) Identification of genes associated with perianth senescence in daffodil (*Narcissus pseudonarcissus* L. 'Dutch Master'). *Plant Sci* **163**, 13–21.

Hunter, D.A., Yi, M.F., Xu, X. and Reid, M.S. (2004b) Role of ethylene in perianth senescence of Daffodil (*Narcissus pseudonarcissus* L. 'Dutch Master'). *Postharvest Biol Technol* **32**, 269–280.

Ito, J. and Fukuda, H. (2002) ZEN1 is a key enzyme in the degradation of nuclear DNA during programmed cell death of tracheary elements. *Plant Cell* **14**, 3201–3211.

Jiang, X. and Wang, X. (2004) Cytochrome C-mediated apoptosis. *Annu Rev Biochem* **73**, 87–106.

John, I., Hackett, R., Cooper, W., Drake, R., Farrell, A. and Grierson, D. (1997) Cloning and characterization of tomato leaf senescence-related cDNAs. *Plant Mol Biolog* **33**, 641–651.

Jones, R.B., Serek, M., Kuo, C-L. and Reid, M.S. (1994) The effect of protein synthesis inhibition on petal senescence in cut bulb flowers. *J Am Soc HortSci* **119**, 1243–1247.

Kerr, J.F. (2002) History of the events leading to the formulation of the apoptosis concept. *Toxicology* **181–182**, 471–474.

Lay-Yee, M., Stead, A.D. and Reid, M.S. (1992) Flower senescence in daylily (*Hemerocallis*). *Physiol Plantarum* **86**, 308–314.

Lee, M.M., Lee, S.H. and Park, K.Y. (1997) Effects of spermine on ethylene biosynthesis in cut carnation (*Dianthus caryophyllus* L.) flowers during senescence. *J Plant Physiol* **151**, 68–73.

Leverentz, M.K., Wagstaff, C., Rogers, H.J. *et al.* (2002) Characterization of a novel lipoxygenase-independent senescence mechanism in *Alstroemeria peruviana* floral tissue. *Plant Physiol* **130**, 273–283.

Li, X.Z., Hou, J.Z., Shao, L.M. and Yin, W.Y. (1994) Physiological, biochemical and cytological changes in senescing corolla of four o'clock flower. *Acta Botanica Sinica* **36**, 116–122.

Lincoln, J.E., Richael, C., Overduin, B., Smith, K., Bostock, R. and Gilchrist, D.G. (2002) Expression of the antiapoptotic baculovirus p35 gene in tomato blocks programmed cell death and provides broad-spectrum resistance to disease. *Proc Natl Acad Sci U S A* **99**, 15217–15221.

Lovell, P.J., Lovell, P.H. and Nichols, R. (1987) The control of flower senescence in petunia (*Petunia hybrida*). *Ann Bot* **60**, 49–60.

Matile, P. and Winkenbach, F. (1971). Function of lysosomes and lysosomal enzymes in the senescing corolla of the Morning Glory (*Ipomoea purpurea*). *J Exp Bot* **22**, 759–771.

Mayak, S. and Dilley, D.R. (1976) Effect of sucrose on the response of cut carnation to kinetin, ethylene and abscisic acid. *J Am Soc HortSci* **101**, 583–585.

Mayak, S. and Halevy, A.H. (1970) Cytokinin activity in rose petals and its relation to senescence. *Plant Physiol* **46**, 497–499.

Mayak, S., Halevy, A.H. and Katz, M. (1972) Correlative changes in phytohormones in relation to senescence processes in rose petals. *Physiol Plantarum* **27**, 1–4.

Moore, B., Zhou, L., Rolland, F. *et al.* (2003) Role of the *Arabidopsis* glucose sensor HXK1 in nutrient, light, and hormonal signaling. *Science* **11**, 332–336.

Mor, Y., and Reid, M.S. (1981) Isolated petals – a useful system for studying flower senescence. *Acta Horticulturæ* **113**, 19–25.

Nichols, R. (1996) Ethylene production during senescence of flowers. *J Hort Sci* **41**, 279–290.

Obara, K., Kuriyama, H. and Fukuda, H. (2001) Direct evidence of active and rapid nuclear degradation triggered by vacuole rupture during programmed cell death in *Zinnia*. *Plant Physiol* **125**, 615–626.

O'Brien, I.E.W., Baguley, B.C., Murray, B.G., Morris, B.A.M. and Ferguson, I.B. (1998) Early stages of the apoptotic pathway in plant cells are reversible. *Plant J* **13**, 803–814.

O'Donoghue, E.M., Somerfield, S.D. and Heyes, J.A. (2002) Organization of cell walls in *Sandersonia aurantiaca* floral tissue, *J Exp Bot* **53**, 513–23.

O'Neill, S.D. (1997) Pollination regulation of flower development. *Ann Rev Plant Physiol* **48**, 547–574.

O'Neill, S.D., Nadeau, J.A., Zhang, X.S., Bui, A.Q. and Halevy, A.H. (1993) Interorgan regulation of ethylene biosynthetic genes by pollination. *Plant Cell* **5**, 419–432.

Onoue, T., Mikami, M., Yoshioka, T., Hashiba, T. and Satoh, S. (2000) Characteristics of the inhibitory action of 1,1-dimethyl-4-(phenylsulfonyl)semicarbazide (DPSS) on ethylene production in carnation (*Dianthus caryophyllus* L.) flowers. *Plant Growth Regul* **30**, 201–207.

Orzáez, D. and Granell, A. (1997) The plant homologue of the defender against apoptotic death gene is down-regulated during senescence of flower petals. FEBS *Letters* **404**, 275–278.

Paliyath, G. and Droillard, M.J. (1992) The mechanisms of membrane deterioration and disassembly during senescence. *Plant Physiol Biochem* **30**, 789–812.

Panavas, T., Pikula, A., Reid, P.D., Rubinstein, B. and Walker, E.L. (1999) Identification of senescence-associated genes from daylily petals. *Plant Mol Biol* **40**, 237–48.

Panavas, T., Reid, P.D. and Rubinstein, B. (1998) Programmed cell death of daylily petals: activities of wall-based enzymes and effects of heat shock. *Plant Physiol Biochem* **36**, 379–388.

Panavas, T. and Rubinstein, B. (1998) Oxidative events during programmed cell death of daylily (*Hemerocallis* hybrid) petals. *Plant Sci* **133**, 125–138.

Pandey, S., Ranade, S.A., Nagar, P.K. and Kumar, N. (2000) Role of polyamines and ethylene as modulators of plant senescence. *J Biosci* **25**, 291–299.

Pennazio, S. and Roggero, P. (1990) Exogenous polyamines stimulate ethylene synthesis by soybean leaf tissue. *Ann Bot* **65**, 45–50.

Perez-Amador, M.A., Carbonell, J., Navarro, J.L. *et al.* (1996) N4-hexanoylspermidine, a new polyamine-related compound that accumulates during ovary and petal senescence in pea. *Plant Physiol* **110**, 1177–1186.

Pilling, E. and Hofte, H. (2003) Feedback from the wall. *Curr Opin Plant Biol* **6**, 611–616.

Porat, R., Borochov, A. and Halevy, A.H. (1993) Enhancement of petunia and dendrobium flower senescence by jasmonic acid methyl ester is via the promotion of ethylene production. *Plant Growth Regul* **13**, 297–301.

Porat, R., Reiss, N., Atzorn, R., Halevy, A.H. and Borochov, A. (1995) Examination of the possible involvement of lipoxygenase and jasmonates in pollination-induced senescence of *Phalaenopsis* and *Dendrobium* orchid flowers. *Physiol Plantarum* **94**, 205–210.

Ratcliffe, F., Martin-Hernendez, A.M. and Baulcombe, C.D. (2001) Tobacco rattle virus as a vector for analysis of gene function by silencing. *Plant J* **25**, 237–245.

Reid, M.S. and Wu, M-J. (1992) Ethylene and flower senescence. *Plant Growth Regul* **11**, 37–43.

Riechmann, J.L., Heard, J., Martin, G. *et al.* (2000) *Arabidopsis* transcription factors: Genome-wide comparative analysis among eukaryotes. *Science* **290**, 2105–2110.

Riechmann, J.L. and Meyerowitz, E.M. (1997). MADS domain proteins in plant development. *Biol Chem* **378**, 1079–1101.

Roberts, D.R., Walker, M.A., Thompson, J.E. and Dumbroff, E.B. (1984) The effect of inhibitors of polyamine and ethylene biosynthesis on senescence, ethylene production and polyamine levels in cut carnation flowers, *Plant Cell Physiol* **25**, 315–322.

Roitsch, T. and Gonzalez, M.C. (2004) Function and regulation of plant invertases: sweet sensations. *Trends Plant Sci* **9**, 606–613.

Rubinstein, B. (2000) Regulation of cell death in flower petals. *Plant Mol Biol* **44**, 303–318.

Saks, Y. and van Staden, J. (1993) Evidence for the involvement of gibberellins in developmental phenomena associated with carnation flower senescence. *Plant Growth Regul* **12**, 105–110.

Savin, K.W., Baudinette, S.C., Graham, M.W. *et al.* (1995) Antisense ACC oxidase RNA delays carnation petal senescence. *HortScience* **30**, 970–972.

Serek, M., Sisler, E.C. and Reid, M.S. (1994) Novel gaseous ethylene binding inhibitor prevents ethylene effects in potted flowering plants. *J Am Soc Horticultural Sci* **119**, 1230–1233.

Shibuya, K., Barry, K.G., Ciardi, J.A. *et al.* (2004) The central role of PhEIN2 in ethylene responses throughout plant development in petunia. *Plant Physiol* **136**, 2900–2912.

Shore, P. and Sharrocks, A.D. (1995). The MADS-box family of transcription factors. *Eur J Biochem* **229**, 1–13.

Smith, M.T., Saks, Y. and van Staden, J. (1992) Ultrastructural changes in the petals of senescing flowers of *Dianthus caryophyllus* L. *Ann Bot* **69**, 277–285.

Solomon, M., Belenghi, B., Delledonne, M., Menachem, E. and Levine, A. (1999) The involvement of cysteine proteases and protease inhibitor genes in the regulation of programmed cell death in plants. *Plant Cell* **11**, 431–443.

Stead, A.D. (1992) Pollination-induced flower senescence: a review. *Plant Growth Regul* **11**, 13–20.

Stead, A.D. and Reid, M.S. (1990) The effect of pollination and ethylene on the colour change of the banner spot of *Lupinus albifrons* (Bentham) flowers. *Ann Bot* **66**, 655–663.

Sugawara, H., Shibuya, K., Yoshioka, T., Hashiba, T. and Satoh, S. (2002) Is a cysteine proteinase inhibitor involved in the regulation of petal wilting in senescing carnation (*Dianthus caryophyllus* L.) flowers? *J Exp Bot* **53**, 407–413.

Sultan, S.M. and Farooq, S. (1999) Effect of sucrose and GA3 on the senescence of cut flowers of *Narcissus tazetta* cv. Kashmir local. *Adv Horticultural Sci* **13**, 105–107.

Takahashi, Y., Berberich, T., Yamashita, K., Uehara, Y., Miyazaki, A. and Kusano, T. (2004) Identification of tobacco HIN1 and two closely related genes as spermine-responsive genes and their differential expression during the Tobacco Mosaic Virus-induced hypersensitive response and during leaf- and flower-senescence. *Plant Mol Biol* **54**, 613–622.

Tang, X., Gomes, A., Bhatia, A. and Woodson, W.R. (1994) Pistil-specific and ethylene-regulated expression of 1-aminocyclopropane-1-carboxylate oxidase genes in petunia flowers. *Plant Cell* **6**, 1227–1239.

Taverner, E., Letham, D.S., Wang, J., Cornish, E. and Willcocks, D.A. (1999) Influence of ethylene on cytokinin metabolism in relation to *Petunia* corolla senescence. *Phytochemistry* **51**, 341–347.

ten Have, A. and Woltering, E.J. (1997) Ethylene biosynthetic genes are differentially expressed during carnation (*Dianthus caryophyllus* L.) flower senescence. *Plant Mol Biol* **34**, 89–97.

Thimann, K.V., Tetley, R.M. and Krivak, B.M. (1977) Metabolism of oat leaves during senescence: V. Senescence in light. *Plant Physiol* **59**, 448–454.

Thomas, H., Ougham, H.J., Wagstaff, C. and Stead, A.D. (2003) Defining senescence and death. *J Exp Bot* **54**, 1127–1132.

Thompson, J.E., Froese, C.D., Madey, E., Smith, M.D. and Hong, Y.W. (1998) Lipid metabolism during plant senescence. *Prog Lipid Res* **37**, 119–141.

Thompson, A.R. and Vierstra, R.D. (2005) Autophagic recycling: lessons from yeast help define the process in plants. *Curr Opin Plant Biol* **8**, 165–173.

Tiwari, B.S., Belenghi, B. and Levine, A. (2002) Oxidative stress increased respiration and generation of reactive oxygen species, resulting in ATP depletion, opening of mitochondrial permeability transition, and programmed cell death. *Plant Physiol* **128**, 1271–1281.

Valpuesta, V., Lange, N.E., Guerrero, C. and Reid, M.S. (1995) Up-regulation of a cysteine protease accompanies the ethylene-insensitive senescence of daylily (*Hemerocallis*) flowers. *Plant Mol Biol* **28**, 575–582.

van Doorn, W.G. (2001) Categories of petal senescence and abscission: A re-evaluation. *Ann Bot* **87**, 447–456.

van Doorn, W.G. (2004) Is petal senescence due to sugar starvation? *Plant Physiol* **134**, 35–42.

van Doorn, W.G., Balk, P.A., van Houwelingen, A.M. *et al.* (2003) Gene expression during anthesis and senescence in *Iris* flowers. *Plant Mol Biol* **53**, 845–863.

van Doorn, W.G. and Woltering, E.J. (2005) Many ways to exit? Cell death categories in plants. *Trends Plant Sci* **10**, 117–122.

van Staden, J. and Dimalla, G.G. (1980) The effect of silver thiosulfate preservative on the physiology of cut carnations: II. Influence of endogenous cytokinins. *Z Pflanzenphysiol* **99**, 19–26.

Veen, H. and van de Geijn, S.C. (1978) Mobility and ionic form of silver as related to longevity of cut carnations. *Planta* **140**, 93–96.

Vrebalov, J., Ruezinsky, D., Padmanabhan, V. *et al.* (2002). A MADS-box gene necessary for fruit ripening at the tomato *Ripening-inhibitor* (*Rin*) locus. *Science* **296**, 343–346.

Wagstaff, C., Leverentz, M.K., Griffiths, G. *et al.* (2002) Cysteine protease gene expression and proteolytic activity during senescence of *Alstroemeria* petals. *J Exp Bot* **53**, 233–40.

Waithaka, K., Dodge, L.L. and Reid, M.S. (2001) Carbohydrate traffic during opening of gladiolus florets. *J Hort Sci Biotechnol* **76**, 120–124.

Wang, H., Li, J., Bostock, R.M. and Gilchrist, D.G. (1996) Apoptosis: A functional paradigm for programmed plant cell death induced by a host-selective phytotoxin and invoked during development. *Plant Cell* **8**, 375–391.

Wang, K.L., Li, H. and Ecker, J.R. (2002) Ethylene biosynthesis and signaling networks. *Plant Cell* **14**, S131–151.

Watanabe, N. and Lam, E. (2005) Two Arabidopsis metacaspases AtMCP1b and AtMCP2b are arginine/lysine-specific cysteine proteases and activate apoptosis-like cell death in yeast. *J Biol Chem* **280**, 14691–14699.

Weiss, D., Van Der Luit, A., Knegt, E., Vermeer, E., Mol, J.N.M. and Kooter, J.M.(1995) Identification of endogenous gibberellins in petunia flowers: Induction of anthocyanin biosynthetic gene expression and the antagonistic effect of abscisic acid. *Plant Physiol* **107**, 695–702.

Whitehead, C.S., Halevy, A.H. and Reid, M.S. (1984) Role of ethylene and ACC in pollination and wound-induced senescence of *Petunia hybrida* flowers. *Plant Physiol Biochem* **61**, 643–648.

Wilkinson, J.Q., Lanahan, M.B., Clark D.G., *et al.* (1997) A dominant mutant receptor for *Arabidopsis* confers ethylene insensitivity in heterologous plants. *Nat Biotechnol* **15**, 444–447.

Woltering, E.J. (2004) Death proteases come alive. *Trends Plant Sci* **9**, 469–472.

Woltering, E.J., Smohorst, D. and Van der Veer, P. (1996) The role of ethylene in interorgan signaling during flower senescence. *Plant Physiol* **109**, 1219–1225.

Woltering, E.J., Van Der Bent, A. and Hoeberichts, F.A. (2002) Do plant caspases exist? *Plant Physiol* **130**, 1764–1769.

Woltering, E.J. and van Doorn, W.G. (1988) Regulation of ethylene in senescence of petals: morphological and taxonomic relationships. *J Exp Bot* **39**, 1605–1616.

Xie, Q., Sanz-Burgos, A.P., Guo, H., Garcia, J.A. and Gutierrez, C. (1999). GRAB proteins, novel members of the NAC domain family, isolated by their interaction with a geminivirus protein. *Plant Mol Biol* **39**, 647–656.

Xu, Y. and Hanson, M.R. (2000) Programmed cell death during pollination-induced petal senescence in petunia. *Plant Physiol* **122**, 1323–1333.

Xu, X., Hunter, D. and Reid, M.S. (2005) An efficient regeneration system for Four O'clocks (*Mirabilis jalapa*). *Acta Horticulturae* **669**, 153–156.

Yamakawa, H., Kamada, H., Satoh, M. and Ohashi, Y. (1998) Spermine is a salicylate-independent endogenous inducer for both tobacco acidic pathogenesis-related proteins and resistance against tobacco mosaic virus infection. *Plant Physiol* **118**, 1213–1222.

Yao, N., Eisfelder, B.J., Marvin, J. and Greenberg, J.T. (2004) The mitochondrion–an organelle commonly involved in programmed cell death in *Arabidopsis thaliana*. *Plant J* **40**, 596–610.

12 Fruit ripening and its manipulation

James J. Giovannoni

12.1 Introduction

Ripening represents the final stage of fruit development and is characterized by structural, biochemical and physiological events that make the fruit organ capable of seed dispersal and/or attractive to organisms that facilitate this process. From a purely developmental perspective, ripening has also been defined as a senescence process as it often results in tissue decay and death (Grierson, 1987). The degenerative nature of late- or over-ripening stages has lead some to propose that ripening represents, at least in consequence, a degradative process. An alternative view that is supported by accumulating molecular data is that while ripening is indeed ultimately directed toward tissue breakdown and senescence, it can more accurately be compared to other generative developmental processes involving *de novo* modification of gene expression and activation of specific biochemical processes and novel physiological attributes (Adams *et al.*, 2004a; Giovannoni, 2004).

Fruit ripening has become an increasingly important field of plant biology in recent decades due to the significance of fruit in the human diet. The model species of fruit development, tomato, has also become an increasingly important model system for plant biology providing numerous molecular and genomic tools for studying aspects of plant development and responsiveness including ripening (Alba *et al.*, 2004; Mueller *et al.*, 2005). These tools have also facilitated comparative analyses with additional species (Moore *et al.*, 2005; Wang *et al.*, 2006).

Plants direct considerable resources toward the development and final maturation of fruit organs (Gray *et al.*, 1992). The ripening of fruit can be generally defined as the summation of changes in texture, flavor, aroma and color which mark the complete maturation of the organ in terms of seed release or attraction for seed-dispersing organisms (Grierson *et al.*, 1992; Giovannoni, 2001, 2004; Adams *et al.*, 2004a). Lipid metabolism, fiber content and vitamin levels are also affected over the course of ripening, ultimately impacting the nutritional quality of mature fruit tissues. It has also been shown that levels of antioxidant compounds, capable of modifying enzyme activities and detoxifying potentially damaging free radical compounds, can be altered substantially during ripening, with effects on both fruit color and nutrient quality (Ronen *et al.*, 1999; Verhoeyen *et al.*, 2002). Indeed, research toward elucidation of biochemical pathways impacting nutritional quality have been areas of increasing activity in recent years (reviewed in Goff and Klee, 2006) with important advances also toward understanding and manipulation of nutrient content specifically in fruit species (Liu *et al.*, 2004; Davuluri *et al.*, 2005).

12.2 Physiologies of ripening fruit

12.2.1 Climacteric ripening

Ripening fruit has classically been divided into two physiological categories – climacteric and nonclimacteric – with the basis of these classifications originally linked to respiration patterns (Biale and Young, 1981). Climacteric fruits undergo a predictable elevation in respiration and typically ripen (on or off the plant (the latter if mature when picked). While not necessary by the classical definition, most climacteric fruits show increased ethylene production at or just prior to the onset of ripening and a necessity for ethylene to complete the process (Abeles *et al.* 1992). Careful time course studies suggest that the ethylene and respiration peaks can be detected prior to other measurable ripening traits and may represent the most accurate early marker of ripening in climacteric fruit (Lincoln *et al.*, 1987). Examples of important climacteric fruits include tomato, apple, pear, avocado, rough skinned melons, most stone fruits (cherry is an exclusion) and many tropical fruits including mango, papaya and banana.

The necessity of ethylene in climacteric ripening has been validated in studies using ethylene biosynthesis and action inhibitors (Yang, 1985). Silver thiosulfate prevented ripening in segments of tomato having vascular connections to the site of pedicel infiltration (Hobson *et al.* 1984). Aminoethoxyvinylglycine retards ripening through its action as an inhibitor of 1-aminocyclopropane-1-carboxylic acid (ACC) synthase (ACS), a rate-limiting step in fruit ethylene synthesis (Boller *et al.*, 1979; Yu and Yang, 1979). Recently, 1-methycyclopropene has been shown to block ethylene action by binding and thus blocking ethylene receptors (Sisler and Serek, 1997) and shows great potential to retard ripening in many agriculturally important fruit species including apple (Watkins *et al.*, 2000) and banana (Jiang *et al.*, 2002). Registration for commercial use is being sought or has been approved for many fruit species.

12.2.2 Nonclimacteric ripening

Nonclimacteric fruits are numerous and include many species of significance to agriculture and trade including strawberry, grape, watermelon, smooth skinned melons, cucumber, pineapple, cherry and citrus. Fruit that are nonclimacteric do not display the burst of respiration or the increase in autocatalytic ethylene characteristic of their climacteric counterparts (McChurchie *et al.*, 1972) and typically do not ripen as fully as climacteric fruit after harvest, suggesting a component of attachment that compensates for or provides regulatory influences or information not required in climacteric species. While these fruits generally do not require ethylene for ripening, note that many nonclimacteric fruit are responsive in terms of ripening characteristics to exogenous ethylene. Nonclimacteric fruit (cherries are a notable exception) display a rise in respiration upon addition of exogenous ethylene; however this increase reverts upon removal of the hormone (Lelievre *et al.*, 1997). In strawberry, addition of exogenous ethylene causes softening and enhances color development (Tian *et al.*, 2000). Pepper is interesting in that sweet bell types are typically climacteric while hot types such as habanero are nonclimacteric. In nonclimacteric pepper, ethylene has nevertheless been associated with induction of carotenoid synthesis (Ferrarese *et al.*, 1995; Harpster *et al.*, 1997) and is used in some commercial applications for color improvement.

Note that the ripening classification of a number of species is not completely clear. While some varieties of melon and pepper are clearly climacteric, others behave physiologically as nonclimacteric fruits. It remains unclear as to whether the nonclimacteric varieties display a truly different ripening program or simply produce and/or perceive less ethylene than their climacteric counterparts. In this regard, note that a genetic study of a cross between climacteric and nonclimacteric melons indicates that the climacteric/nonclimacteric trait is regulated by only one or two loci (Perin *et al.*, 2002). This result would suggest that nonclimacteic ripening in species with closely related climacteric counterparts may simply represent mutation in a locus or loci necessary for ethylene synthesis or perception in what is otherwise a climacteric species. The comparatively slow ripening of smooth versus netted skinned melons, along with the much more pronounced fruit abscission of the latter, would support this hypothesis, though conclusive molecular data is not yet avaible.

12.3 Model ripening systems

Morphologically, fruit are typically either fleshy (and nondehiscent) or dry (and usually dehiscent) with tomato emerging long ago as the primary model for fleshy fruit development, while *Arabidopsis* (siliques) are the best-studied dehiscent fruit at the molecular level (reviewed in White, 2002).

Arabidopsis has been used as a model genetic system for plants due in part to its small stature, compact genome, short life cycle, simple genetics, efficiency of transformation and availability of a complete genome sequence. Genetic studies have resulted in identification of a number of important fruit development genes, especially as related to carpel identity and development revealed through screens for floral mutations (Malcomber and Kellog, 2005). As this chapter deals with ripening, dry fruits will not be discussed further but additional information on *Arabidopsis* carpel and fruit development can be found in Giovannoni (2004 and references therein).

12.3.1 Tomato – the model for climacteric ripening

Tomato has long served as a system for studying plant development, physiology and responses in part due to its importance as a crop and short life cycle. Tomato has emerged as the primary model system for fleshy fruit development and ripening for many of the same reasons mentioned for *Arabidopsis*, especially as compared to other species yielding fleshy fruit. In addition, numerous tomato mutations have been identified in which normal ripening is impacted and often with few if any other plieotropic effects. These mutations include *Never-ripe* (*Nr*), *ripening inhibitor* (*rin*), *nonripening* (*nor*) (Tigchellar *et al.*, 1973), *Green ripe* (*Gr*) (Barry *et al.*, 2005) and *Colorless nonripening* (*Cnr*) (Thompson *et al.*, 1999), which have global effects on ripening. *Nr*, *rin*, *nor*, *Gr* and *Cnr* display the most complete inhibition of ripening, but numerous additional mutants altering subsets of ripening phenotypes, most notably pigment accumulation due to their ease of identification in screens, have been described (see Gray *et al.*, 1994; Giovannoni, 2001, 2004 and references therein). Tomato mutant collections can be accessed online at http://zamir.sgn.cornell.edu/mutants/ and http://tgrc.ucdavis.edu/.

Tomato belongs to the *Solanaceae*, or nightshade, family and is indigenous to central and south America (Kalloo and Bergh, 1993). The increasing importance of tomato as a crop is clearly shown by the amount of land under cultivation for tomato production (over 2 000 000 hectares not including intensive and expanding greenhouse production) with a steady trend upward world-wide (Kalloo and Bergh, 1993). Demand has been met via increased yield and shelf life in commercial lines though quality traits have not been targeted for breeding and have been lost in this process. Declining consumer satisfaction with the quality of many fruits and vegetables, including tomato, has driven recent research interests (Goff and Klee, 2006).

Tomato possesses a diploid genome ($n = 12$) and research on this organism is facilitated by a number of tools that contribute to its molecular and genetic characterization. Mutants involved in various aspects of tomato growth and development, including ripening, have been isolated and characterized (Giovannoni, 2004), along with numerous quantitative trait loci (QTL) including those responsible for variation in ripening time (Doganlar *et al.*, 2000). Nearly 2000 molecular markers have been mapped to the tomato genome at a spacing approaching 1 cM on average (Tansksley *et al.*, 1992; http://sgn.cornell.edu/). Together these tools have contributed to chromosome walks to a number of important loci including many that impact fruit traits from weight or shape to abscission to ripening and including fw2.2 (Frary *et al.*, 2000), *ovate* (Liu *et al.*, 2002), *jointless-1* (Mao *et al.*, 2000), *rin* (Vrebalov *et al.*, 2002) and *Gr* (Barry and Giovannoni, 2006).

A number of public-mapping populations have also been developed, and they serve as a common and integrated resource for mapping of molecular and morphological markers (http://sgn.cornell.edu/). Of particular utility are 50 ordered recombinant introgression lines (RILs) from a cross between *Solanum Lycopersicum* and the wild tomato species *S. pennellii*. This collection represents a library of the complete *S. pennellii* genome through single ordered and overlapping introgressions (Eshed and Zamir, 1995). Each introgression line contains a portion of a single chromosome of *S. pennellii* introgressed into the M82 (*S. lycopersicum)* parent, yielding abundant phenotypic variation including traits for fruit ripening and color resulting from the corresponding *S. pennellii* chromosomal segment. Additional RILs have been isolated via backcrossing of original RILs to the M82 parent to define smaller introgressions for a current total of 76 lines (Liu and Zamir, 1999). These RILs have been used to map QTLs for important fruit quality traits (Fridman *et al.*, 2002, 2004).

12.3.1.1 Tomato genomic resources facilitate ripening research

In recent years, tomato has been the subject of considerable expressed sequence tag (EST) sequencing resulting in a collection of sequences estimated to represent approximately half of all gene sequences in tomato (Van der Hoeven *et al.*, 2002; Fei *et al.*, 2004). Numerous cDNA libraries were constructed and sequenced from a comprehensive selection of tissues and physiological treatments to yield an extensive EST database (Moore *et al.*, 2002; Van der Hoeven *et al.*, 2002). The EST collection has been annotated into a unigene set based on combining multiple copy ESTs into contigs totaling approximately 30 000 unigenes which can be accessed through the World Wide Web (http://sgn.cornell.edu/ and www.tigr.org/). This EST collection has allowed the prediction of the number of genes within the tomato genome (Van der Hoeven *et al.*, 2002). Computational analysis of tomato EST prevalence has allowed for digital assessment of gene expression profiles across over 20 tissues and treatments and facilitated comparative expression analyses with *Arabidopsis* and grape (Fei *et al.*, 2004).

A cDNA microarray has been constructed using the above-mentioned EST libraries and has been shown to be useful both in tomato and related members of the family *Solanaceae* (Alba *et al.*, 2004; Moore *et al.*, 2005). cDNA microarrays are being employed to explore the expression profile of tomato developmental processes such as fruit ripening and have yielded novel insights into the regulation of ethylene synthesis, carotenoid metabolism and ascorbate accumulation during ripening (Alba *et al.*, 2005). Tomato EST and long oligonucleotide microarrays are available to the research public at http://bti.cornell.edu/CGEP/CGEP.html. Raw and analyzed expression data for numerous microarray experiments including many focused on development and ripening of normal and mutant tomato fruit varieties are available through the Tomato Expression Database at http://ted.bti.cornell.edu/.

12.3.2 Additional model systems for ripening research

Strawberry is probably the most widely studied model systems for nonclimacteric fruit ripening. Strawberry have a rapid growth cycle reaching full size approximately 30 days postanthesis depending on variety, while time to ripening can vary considerably due to genotype and growth conditions (Seymour *et al.*, 1993). The true fruit (ovaries) of the strawberry are the achenes lining the outside of the predominant fleshy receptacle tissue (Seymour *et al.*, 1993). Strawberry maturation is characterized by accumulation of anthocyanins, sugars (sucrose and various hexoses), aromatic volatiles and the simultaneous breakdown of cell-wall components resulting in softening and eventual overripening (Manning, 1998). Auxin released from the achenes has been shown to regulate maturation and development of the receptacle, and its loss in later fruit development is directly attributed to subsequent ripening (Civello *et al.*, 1999). Much like its climacteric counterparts, strawberry ripening involves a complex group of processes, which rely on numerous changes in the expression of ripening-related genes (Manning, 1998). The short life cycle of strawberry, diploid varieties (cultivated strawberry is octaploid), developing molecular and genomic resources and reasonably efficient transformation protocols have contributed to the utility of strawberry as a model system.

In contrast to tomato, strawberry germplasm is deficient in mutations with dramatic consequences on ripening for use in gene discovery or variety improvement. As a result, most gene identification has been performed through reverse rather than forward genetics approaches. Wilkinson *et al.* 1995 performed differential screens on ripe and unripe tissues to identify ripening-related genes and reported a cDNA (RJ4), recovered through differential display of unripe and ripe strawberry, with homology to the annexin super family. Annexins are calcium-dependent phospholipid-binding proteins associated with voltage-gated ion channels, signaling molecules and regulation of various enzyme activities (Andrawis *et al.*, 1993; Verma and Hong 2001). Annexins may contribute to changes in cell-wall structure and membrane properties, both of which are affected by calcium and may contribute to tissue integrity, pathogen tolerance and fruit texture (Redgwell and Fischer, 1996). (Wilkinson *et al.* 1995) also identified a cDNA (RJ5) encoding chalcone synthase (CHS). CHS catalyzes conversion of three molecules of 4-coumaroyl-CoA and 1 molecule of *p*-coumaroyl to naringenin chalcone, the first committed step in flavanoid biosynthesis and resulting in production of anthocyanins responsible for pigmentation of ripe strawberry (Clegg and Durbin, 2000; Hadacek, 2002).

Manning (1998) undertook the first extensive attempt at isolating strawberry-ripening genes via differential screening of two cDNA libraries constructed from white versus red strawberry. The gene families represented in the recovered sequences included homologs to enzymes present in phenylpropanoid synthesis and cell-wall metabolism. One member of the cell-wall class, EGases (endo-Beta-(1,4)-glucanase), represented enzymes which hydrolyze (1,4)-Beta linkages (Brummell and Harpster, 2001) and may contribute to softening of strawberry and other fleshy fruit. Color and flesh integrity are two of the major challenges in strawberry cultivar development, and these genes include some of the first molecular tools with potential to impact these traits.

Though genomic efforts in strawberry are minor as compared to tomato, a small strawberry microarray was constructed containing 1701 unsequenced cDNAs (Aharoni *et al.*, 2000). A number of these cDNAs were shown to reflect differential gene expression when comparing probes derived from ripe versus unripe fruit. Among the differentially expressed sequences, one putative acyltransferase was identified (*SAAT*, strawberry alcohol acyltransferase) which was shown via *in vitro* experiments to impact the synthesis of a key strawberry flavor compound (Aharoni *et al.*, 2000). The expression pattern of *SAAT* showed a dramatic increase in red versus green fruit and was limited to maturing receptacle tissue (Aharoni *et al.*, 2000). In subsequent use of the strawberry array to measure oxidative stress response, 20 ripening-related cDNAs were recovered including CHS (Aharoni *et al.*, 2002). Auxin treatment resulted in repression of flavonoid metabolism genes and the recovery of novel auxin-independent genes including annexin and thus produced molecular tools for further characterization of hormone-independent regulation of ripening processes.

Indeed, an emerging and important area of current molecular investigations into ripening regulation relates to how and at what level climacteric and nonclimacteric ripening regulatory mechanisms are related and conserved. The continued discovery of genes involved in ripening of both nonclimacteric and climacteric fruit will allow for cross-comparison among species of different ripening types. The recent development of large EST collections in many fruiting species has facilitated the computational comparison of gene expression through informatics-based approaches. Such strategies rely upon the fact that putative orthologs can be identified based on EST sequence and then compared for relative expression through analysis of EST abundance. This approach has been employed with tomato and grape (nonclimacteric). For each species approximately 20 ripening-related putative transcription factors were identified though only three are highly conserved between both species (Fei *et al.*, 2004). The ability to computationally compare phylogenetically distinct species should allow for the rapid identification of both species-specific and cross-species conserved regulatory gene candidates.

The recent cloning of a MADS-box transcription factor necessary for tomato ripening at the *rin* locus represents one such well-conserved regulator of ripening acting upstream of ethylene and with a homolog expressed in strawberry (Vrebalov *et al.*, 2002). Efforts are underway to repress the strawberry RIN homolog in transgenic strawberry to test for function in ripening (K. Manning, G. Seymour, J. Giovannoni, unpublished data). Conservation of RIN homologs across species representing diverse ripening types suggests a more global (ethylene-independent) mechanism of ripening control. The characterization of *RIN* genes from additional species will provide possible tools for controlling fruit ripening in a variety of agriculturally important plant species in a manner that does not focus on ethylene synthesis or responses (Vrebalov *et al.*, 2002). This is of critical importance in manipulation of ripening as ethylene repression alone has been shown to have adverse effects on final fruit quality (below). In contrast, better quality is achieved with more global ripening control as with the tomato *rin* mutation. *rin* is currently widely used in breeding long shelf-life tomato hybrids and in many western production programs in particular; such varieties represent the majority of fresh market tomato lines under cultivation.

Charentais melons (a netted skinned/climacteric variety) were transformed with antisense ACC oxidase (ACO), which resulted in melon fruit displaying reduction of ethylene production (Ayub *et al.*, 1996). These antisense fruits also showed inhibition of ripening regardless of whether they were on or off the vine. When crossed to lines of agronomic importance, the aroma-related volatile ester content of the resulting fruit was shown to be significantly reduced (Bauchot *et al.*, 1998). These results indicate that ethylene plays an important regulatory role in many aspects of climacteric ripening, and while one desirable trait may be achieved via manipulation of ethylene (shelf life), another equally important fruit quality trait (aroma/flavor) may be compromised. These results further suggest that dissociation of ethylene-mediated ripening programs from nonethylene-regulated ripening processes may have undesirable effects that are difficult to correct. Molecular tools, such as the *RIN* gene, for comprehensive regulation of ripening at a step upstream of ethylene will be more universal in application to both climacteric and nonclimacteric species and may result in climacteric fruit with better quality than those whose ripening is modified only at the level of ethylene.

12.4 Ripening processes and their manipulation

12.4.1 Cell-wall metabolism

Fruit ripening and cell-wall metabolism are interconnected processes as changes in cell-wall ultrastructure underscore ripening-related textural modifications. Fruit cell walls are typical in that they generally consist of cellulose microfibrils embedded in a hemicellulose matrix with a pectin network serving as the anchoring agent between adjacent cell walls (reviewed in Cosgrove, 2000b). Cell-wall modifications typical of maturing and ripening fruit tissues include decreased cell–cell adhesion, reduced apoplastic pH, depolymerization and solubilization of pectins and hemicelluloses, and a general increase in cell-wall flexibility and porosity (Brummell and Harpster, 2001). This decline of general cell-wall structural integrity, and especially the changes in degree of attachments to adjacent cells, governs the rigidity and related texture and oral sensation of typical maturing fruit tissues (Redgwell and Fischer, 1996).

During ripening, cell-wall compounds including polyuronides are solubilized resulting in cell-wall destabilization and loss of side-chain sugars (galactan and arabinan) (Gross, 1984; Seymour *et al.*, 1993). In tomato, this loss in cell-wall integrity specifically results in swelling of the wall volume and loss of cell-to-cell adhesion leading to softening of the pericarp and locule liquefication. A key enzyme long implicated in cell-wall degradation and associated textural changes during tomato ripening, and the ripening of many fruits, is polygalacturonase (PG). This enzyme remains one of the single-most studied activities associated with fleshy fruit maturation and ripening. There is a small family of tomato genes encoding PG enzymes involved in various developmental processes though a single member of this family is implicated in tomato ripening (Sitrit and Bennett, 1998). The PG involved in ripening is known to be expressed only in fruit with transcription limited to the ripening process and proposed to account for up to 1% of the total fruit mRNA in some cultivars (Bird *et al.*, 1988; Della-Penna *et al.*, 1989). It had been thought previously that PG alone was responsible for the majority of cell-wall degradation and associated softening in ripening fruit (Bennett and Della-Penna, 1987). In transgenic studies designed to yield firmer ripe fruit and the basis of the FlavrSavr transgeinc tomato product, repression of PG resulted in degradation of pectin but did not affect softening (Sheehy *et. al.*, 1988). Complementary to this result, it was also shown that transgenic PG expression in a cultivar homozygous for the *rin* allele (which normally blocks PG expression) yielded PG activity and *in vivo* pectin depolymerization (and solubilization), but did not result in significant fruit softening (Giovannoni *et al.*, 1989). This result, in concert with data from PG antisense experiments, provided clear molecular evidence that cell-wall modifications leading to fruit textural modifications are more complicated than activation of a single cell-wall modifying enzyme (Carrington *et al.*, 1993). Note that while commercial antisense PG lines failed to alter fruit texture and softening, they did realize extended shelf life through reduced susceptibility to postharvest pathogens and as such retained a marketable trait that allowed for some commercial production. For a limited time, processing varieties expressing a PG antisense gene were also produced and resulted in puree products that required less processing to reach desired viscosity levels. At present, neither of these products are on the market primarily due to marketing concerns related to consumer acceptance of genetically modified crops.

Additional cell-wall metabolizing enzymes have been cloned and characterized in ripening tomato, revealing a more complex view of cell-wall ultrastructural changes associated with ripening than pectin degradation alone. Pectin methylesterase (PME) has been found in a variety of plant species as well as bacteria and fungi that are pathogenic to plants (Rexova-Benkova and Markovic, 1976; Huber, 1983; Collmer and Keen, 1986). PME is involved in the removal of methyl groups from, and thus providing additional susceptibility to, ripening-related PG activities in maturing fruit (Brummell and Harpster, 2001). Antisense PME tomato lines with substantially reduced PME activity yielded cell-wall extracts with higher molecular weight (MW) pectins but neither had notable impact on texture nor other aspects of ripening, consistent with the results observed for PG repression (Tieman *et al.*, 1992). These results indicate that like PG, PME alone is not sufficient for textural changes associated with the maturation of tomato.

Endo-1,4-Beta-Glucanases (EGases) have been correlated with a number of cell-wall modification-associated development and response processes such as abscission and ripening (of both climacteric and nonclimacteric species). EGases have been isolated from strawberry (Harpster *et al.*, 1997; Manning, 1998; Llop-Tous *et al.*, 2000; Trainotti *et al.*, 1999; Spolaore *et al.*, 2003), pepper (Ferrarese *et al.*, 1997; Harpster *et al.*, 1997), and tomato (Lashbrook *et al.*, 1994; Rose and Bennett, 1999) among others. Two EGases, Cel1 (Lashbrook *et al.*, 1994) and Cel2 (Gonzalez-Bosch *et al.*, 1996), increase in accumulation upon ripening pointing to a possible role in cell-wall disassembly during ripening. Cel1 was placed under the direction of the CaMV 35s promoter in the antisense orientation to determine the effects of targeted gene suppression in tomato. Transgenic lines displayed a decrease in Cel1 expression with no measurable decrease in ripening-associated softening. However, a decrease of Cel1 mRNA in abscission zones caused a significant reduction in abscission in transgenic tomato flowers (Lashbrook *et al.*, 1998b). A second suppression study using Cel2 also under 35s regulation resulted in transgenic tomatoes whose Cel2 expression was reduced to less than 5% of control levels in fruit and to approximately 20% in abscission zones (Brummell *et al.*, 1999b). Transgenics exhibited no discernable changes in softening of ripening fruit, but did display in an increase in pedicel abscission zone break strength (Brummell *et al.*, 1999b). These transgenic studies indicate that during ripening, Cel1 and Cel2 may compensate for absence of the other and/or may be expressed at levels far beyond those needed for normal activity. Harpster *et al.* (2002a) showed that suppression of a ripening-related EGase (CaCel1) in pepper did not cause a decrease in the softening of pepper fruit upon ripening and did not affect depolymerization of cell-wall glycans. CaCel1 was also placed under CaMV 35s promoter control and transformed into tomato (Harpster *et al.*, 2002b). Transgenics tomatoes did not show any increases in depolymerization of xyloglucans nor any changes in softening of ripening fruit. These results once again indicate that suppression/overexpression of only one EGase is not sufficient for significant changes in cell-wall softening, suggesting either compensation by other family members and/or involvement of other activities in the dramatic textural changes associated with ripening.

An additional enzyme contributing to cell-wall metabolism and modifications in ripening fruit is xyloglucan endotransglycosylase (also termed XET or XTH) (Rose *et al.*, 2002). XTHs catalyze cleavage of xyloglucan backbone internal linkages in the (1 $\rightarrow$ 4) Beta-D orientation (Steele and Fry, 1999). Several genes encoding XTHs have been isolated from tomato (t*XET-B1*, t*XET-B2* and *LeExt2*; Arrowsmith and de Silva, 1995; Catala *et al.*, 2001), ripe kiwi fruit (Schroder *et al.*, 1998) and grape (Ishimaru and Kobayashi, 2002) suggesting functions in both climacteric and nonclimacteric species. *LeEXT*B1 (expressed in ripening fruit) and *LeEXGT1* (expressed in immature and expanding green fruit) have both been expressed in transgenic tomato lines (de Silva *et al.*, 1994). Through overexpression of *LeEXGT1*, the corresponding abundance of *LeEXGT1* mRNA was elevated in concert with the size of transgenic fruit confirming a role for the endogenous activity in cell and whole fruit expansion during development (Asada *et al.*, 1999). de Silva *et al.* (1994) demonstrated antisense suppression of *LeXET*B1 under a fruit specific (*PG*) promoter, but found no parallels between suppression and changes in mature fruit texture. As with many of the other cell-wall modifying activities, these results indicate possible redundancy by other family members and/or support for the strengthening belief that multiple enzymes must be involved in the metabolism and modification of cell-wall ultrastructure during fruit maturation and especially at ripening.

The cell wall is populated by numerous proteins and upwards of 10% of all plant peptides are thought to be targeted to the extracellular matrix with many of these peptides remaining poorly defined (Rose *et al.*, 2004a). As such, many cell-wall modifying proteins may have activities beyond the more heavily studied polysacharide hydrolases. Expansins are one such group of integral cell-wall proteins and were originally identified as contributing to cell-wall extension of cucumber hypocotyls (McQueen-Mason *et al.*, 1992). Expansins have been shown to be capable of loosening and stretching of isolated cell walls *in vitro* (Cosgrove, 1999) and two families of these peptides (alpha-expansins and beta-expansins) have been described (Cosgrove, 2000a). Alpha-expansins are well correlated with cell elongation and wall loosening (Cosgrove, 2000a), while beta-expansins are less well defined in terms of *in vivo* activity but are among the grass pollen allergens and are likely associated with cell-wall extension during pollen germination (Cosgrove, 1999;

Cosgrove, 2005). A key feature of expansin peptides is the presence of a putative cellulose-binding domain, though catalytic activity on cellulose microfibrils remains to be demonstrated (Rose *et al.*, 2004b).

The tomato alpha-expansin family has been extensively characterized (Brummell and Harpster, 2001). *LeEXP1* is the predominant family member expressed in ripening fruit, and homology to *LeEXP1* has been employed to isolate homologous genes from melon and strawberry (Rose *et al.*, 1997). *LeEXP1* expression is regulated by ethylene and is consistently reduced in expression in the firm fruit of the homozygous *rin* and *nor* nonripening mutants (Rose *et al.*, 1997). The tomato *Nr* mutation results from a lack of ethylene binding to a receptor encoded at the *Nr* locus rendering mutant tissues, including fruit, insensitive to ethylene (Wilkinson *et al.*, 1997). *LeEXP1* mRNA concentration was not increased upon addition of exogenous ethylene to *Nr* fruit (Rose *et al.*, 1997). These results correlate well with the decrease in fruit softening characteristic of all three mutations and indicate an influence on *LeEXP1* by the regulatory pathways defined by the *rin* and *nor* mutations in addition to ethylene (Rose *et al.*, 1997; Vrebalov *et al.*, 2002). Rose *et al.* (2000) defined a structurally distinct subclass of alpha-expansins including *LeEXP1*, *LeEXP4* (present in expanding flowers and fruit) (Brummell *et al.*, 1999a) and *LeEXP18* (expressed primarily in meristamatic tissues) with additional hydrolytic activities. The function of *LeEXP1* in maturing tomato has been studied via transgenic overexpression using the 35sCaMV promoter. Ectopic expression in fruit prior to ripening resulted in increased softening at all stages of ripening, including typically firm mature green (MG) fruit. Transgenic repression of *LeEXP1* resulted in inhibition of pectin depolymerization and an increase in fruit firmness at ripening, but did not impact the metabolism of other measured cell-wall components (Brummell *et al.*, 1999a).

Cell-wall modifying enzymes have been the most studied class of fruit-ripening activities as a result of their association with the biologically and practically important process of fruit cell wall ultrastructure modification and textural changes during ripening. Only PG has seen commercialization though many of these activities remain markers for texture- and fruit quality-associated breeding activities. Whether any of the corresponding genes will see implementation in transgenic plants for commercial applications will depend on the specific interplay of benefit, economy, consumer acceptance and regulatory constraints.

12.4.2 Ethylene biosynthesis and perception

Ethylene biosynthesis results from conversion of methionine to S-adenosine-L-methionine (SAM). The enzyme ACS catalyzes the conversion of SAM to ACC. ACS synthesis and its subsequent conversion to ethylene are singly, or together, rate limiting in ethylene production depending on species and tissue. ACS is encoded by multigene families varying in size among species and whose members demonstrate differential expression including members expressed solely or primarily in maturing fruit (Yip *et al.*, 1992). The final step in ethylene synthesis is catalyzed by ACO. A multigene family consisting of four tomato genes, *ACO1–ACO4*, which display variation in expression both temporally and spatially has been reported (Barry *et al.*, 1996; Nakatsuka *et al.*, 1998). Repression of ACS or ACO results in decreased ethylene synthesis and inhibition of ripening in tomato (Theologis *et al.*, 1993). Genes encoding both ACS and ACO have been cloned from a number of additional plant species including rice (ACS) (Zhou *et al.*, 2002), apple (ACS) (Lay-Yee and Knighton, 1995), apple (ACO) (Ross *et al.*, 1992; Atkinson *et al.*, 1999), banana (ACO) (Lopez-Gomez *et al.*, 1997) and melon (ACO) (Balague *et al.*, 1993) as ethylene and its effects are important well beyond ripening.

Genetic screens performed primarily in *Arabidopsis* were the main tools used in elucidating general mechanisms of plant ethylene signal transduction (Figure 12.1). This is due in large part to the ease of screening for seedling triple response phenotypes in mutagenized seed, which has revealed key steps within the signaling pathway (reviewed in Chang and Shockey, 1999). A number of ethylene-signaling genes identified in *Arabidopsis* have been shown to have homologs in crop species including tomato (reviewed in Watkins, 2002). These include homologs to the *CTR1* kinase, the *EIN3* family of transcription factors, as well as members of the ethylene receptor family (Tieman *et al.*, 2001; Adams-Phillips *et al.*, 2004b). While *Arabidopsis* has been used to study many aspects of general ethylene signaling, it clearly cannot be used to study the role of ethylene signaling in fleshy climacteric fruit development.

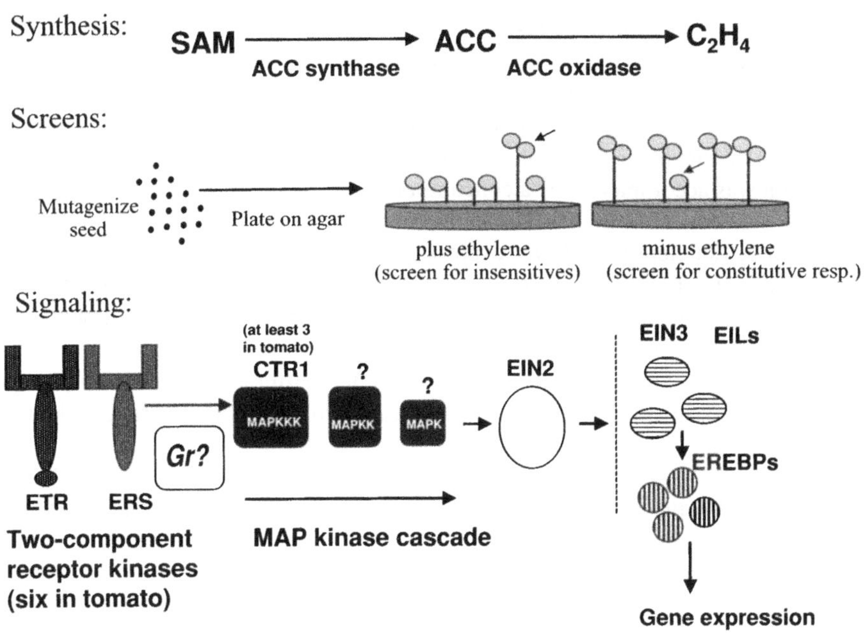

Figure 12.1 Summary of ethylene synthesis, mutant screens and perception and signal transduction pathway.

The dominant tomato *Nr* mutation was originally described for its delayed and incomplete ripening and firmer texture (Rick and Butler, 1956). *Nr* fruit are incapable of complete ripening, petioles do not display epinasty and *Nr* seedlings exposed to exogenous ethylene are inhibited in the seedling 'triple response' indicating whole plant impact on ethylene perception (Lanahan *et al.*, 1994; Yen *et al.*, 1995). In addition, the *Nr* mutant also demonstrated delayed floral abscission and impaired senescence of leaves and petals and a consistent reduction in expression of ethylene-regulated genes (Lanahan *et al.*, 1994; Yen *et al.*, 1995). The *NR* gene was isolated and is homologous to ethylene receptors found in *Arabidopsis* including *ETR1* though with the most similarity to *ERS* (Hua *et al.*, 1995; Wilkinson *et al.*, 1995). The *ETR1* gene encodes a protein with homology to bacterial two-component regulators (Chang *et al.*, 1993) and is sufficient to confer dominant ethylene insensitivity when mutated and transferred to *Arabidopsis* or petunia (Wilkinson *et al.*, 1997). Homologs for the ethylene receptors have also been cloned from nonclimacteric fruit including an *ERS* homolog from *Citrus sinensis* (Li *et al.*, 1998). The presence of ethylene receptor homologs in nonclimacteric fruit may indicate either use of ethylene for processes other than ripening or the presence of fruit ethylene influences that are exerted through lower ethylene concentrations.

In addition to *Nr*, additional genes belonging to the ethylene receptor family have been isolated from tomato (*LeETR1, 2, 4, 5*) (Klee and Tieman, 2002). *NR* and *LeETR4* are the primary fruit ripening receptors in tomato (Tieman *et al.*, 2000). Interestingly (as single receptor knock outs in *Arabidopsis* have only subtle phenotypes, Hua and Meyerowitz, 1998), transgenic lines showing repression of *LeETR4* exhibited a wide range of obvious phenotypes suggestive of increased hormone sensitivity and including extreme epinasty of leaves, increased senescence of flowers and accelerated ripening (Tieman *et al.*, 2000). Conversely, lines having reduced expression of *NR* showed normal sensitivity to ethylene. Decreased *Nr* mRNA concentration was accompanied by a concomitant increase in accumulation of *LeETR4* mRNA (Tieman *et al.*, 2000). The converse proved true in transgenic lines engineered to be deficient in *LeETR4* expression, suggesting that *NR* and *LeETR4* are redundant in function (Tieman *et al.*, 2000). Comparative functional and expression analyses of the tomato and *Arabidopsis* ethylene receptor families suggest similar genes and corresponding predicted peptide structures, yet these genes are clearly under regulatory constraints that presumably reflect differential selective pressures on the Solanaceae versus the Brassicaceae and corresponding differing biological manifestations.

In addition to *Nr*, a spontaneous tomato mutation called *epinastic (epi)* is believed to be involved in the ethylene-signaling pathway (Barry *et al.*, 2001). The mutation possesses profound epinasty of leaves, thickening of petioles and stems, a shorter yet highly branched root system, overproduction of ethylene in vegetative tissues and normal ripening. Constitutive seedling triple response in the absence of ethylene is also a characteristic of the *epi* mutant and is comparable to the *constitutive triple response (ctr1)* mutant of *Arabidopsis* (Kieber *et al.*, 1993). The CTR1 gene encodes a MAP KKK through which all measured ethylene responses reported to date have been shown to flow in *Arabidopsis* (Kieber *et al.*, 1993). Although homologs to

Arabidopsis CTR1 have been found in tomato (Adams-Phillips *et al.*, 2004b), *epi* does not appear to map to any TCTR loci and this may represent a distinct ethylene function.

The most recent tomato mutant to be shown to represent a lesion in ethylene response is the *Green-ripe* (*Gr*) mutant. This mutation was originally described as a ripening mutant but has more recently been shown to reflect the novel attribute of ethylene insensitivity restricted primarily to fruit tissue with a less severe impact on pedicel and petal abscission (Barry *et al.*, 2005). *Gr* fruit show climacteric ethylene induction and normal ethylene responsiveness in most nonfruit tissues. *Gr* has been recently isolated through positional cloning and may encode a novel component of ethylene signal transduction that has not been defined previously in any plant (C. Barry and J. Giovannoni, unpublished data).

The potential to manipulate ethylene responses such as ripening and associated postharvest shelf life is intriguing, and a number of genes to facilitate such manipulations are available and some have been tested in model species as described above. While modulation of ethylene synthesis or response has resulted in delayed ripening and extended shelf life in tomato and melon, quality attributes were also negatively impacted. In addition, it has been shown that horticultural performance can be negatively influenced making production untenable even if other desired attributes are attained (Clevenger *et al.*, 2004). While the number of examples of implementation of such technologies remains relatively few, results to date suggest that strategies that are directed specifically to the fruit may yield varieties with better field performance. This may be achieved through use of tissue-specific promoters or genes such as that encoded by the *Gr* locus, which appear to have tissue-specific activities. Quality may be minimally compromised through manipulations targeting more global ripening regulators that result in more uniform and complete ripening delay or through more specific targeting of pathways contributing to specialized ripening attributes.

12.4.3 Global ripening control

Several tomato genes defined originally by mutations appear to impact nonethylene signaling aspects of ripening control that occur prior to fruit maturation-related ethylene synthesis and response. For example, the phenotype of the recessive and nonripening *rin* mutation includes the inability of fruit to ripen when supplied with exogenous ethylene; yet ethylene responsive genes are induced in such ethylene-treated fruit (Giovannoni *et al.*, 1989). Furthermore, *rin* seedlings display the triple response to exogenous ethylene, indicating that nonfruit ethylene responses remain intact (Yen *et al.*, 1995). These phenotypes of *rin* have been interpreted to indicate a lesion in a regulatory component that impacts both ripening-related ethylene production and nonethylene-mediated components of climacteric fruit ripening (Vrebalov *et al.*, 2002). *RIN* encodes a MADS-box transcription factor, which, in plants, has been primarily associated with floral development (reviewed in Soltis *et al.*, 2002). A floral development function is consistent with the predicted developmental ripening regulation implied by the *rin* phenotype, and homologs of *RIN* which are expressed

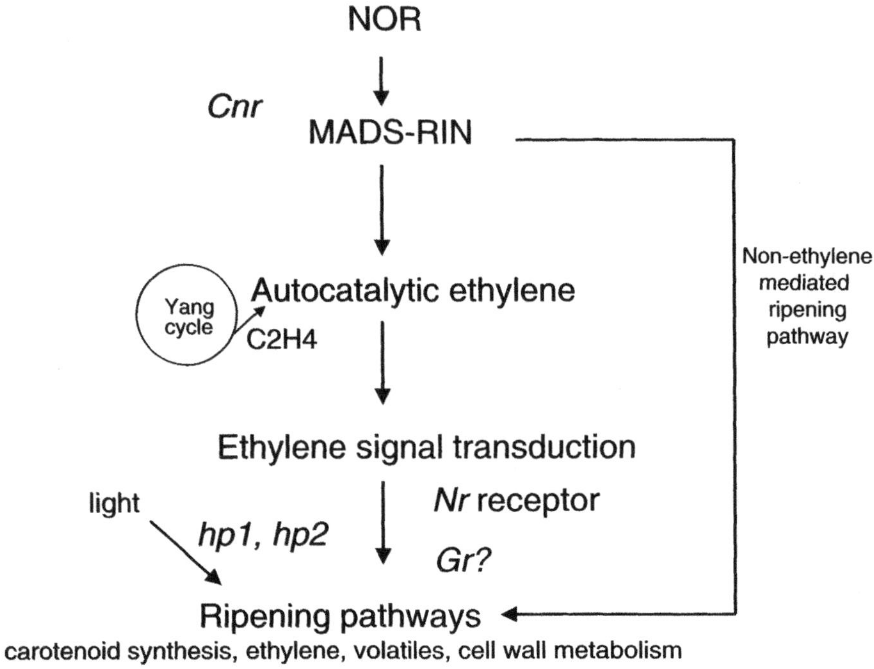

Figure 12.2 Model of ripening control as currently defined in tomato.

in ripening fruit have been isolated from both climacteric and nonclimacteric species suggesting the first common molecular regulatory link among fruits with the two classical ripening physiologies (Vrebalov *et al.*, 2002).

Another mutation closely resembling *rin* in phenotype is the recessive *nor* mutation (Tigchelaar *et al.*, 1973). *nor* displays ethylene synthesis/response phenotypes of the whole fruit and at the level of gene expression similar to those described above for *rin*, suggesting that both gene products may participate in the same or similar regulatory circuits (Figure 12.2). The corresponding *NOR* gene has been isolated and shown to reside upstream of *RIN* in the ripening regulatory cascade (J. Vrebalov and J. Giovannoni, unpublished data). As sated previously, the *rin* mutation is widely used in commercial tomato lines both demonstrating the utility of *RIN* manipulation and limiting the potential for transgenic tomato lines with modified *RIN* gene levels. However, *RIN* and additional global ripening regulators such as that encoded at the *nor* locus may prove useful as transgenic solutions for species where available germplasm resources do not include such mutations.

12.4.4 Modification of specific ripening pathways: pigmentation

Modification of individual pathways associated with ripening and influencing specific high-value targets for improvement may yield consequential quality enhancements with reduced secondary impacts on broader ripening attributes. For example, changes in fruit pigmentation are characteristic of fruit ripening in many species. Ripe fruit pigments have both quality and nutritional benefits and may be manipulated without broader ripening effects (Alba *et al.*, 2000). Genes representing steps involved in biosynthesis of certain pigments have been cloned from tomato and in some cases through screens for ripening-related genes. A number of fruit-color mutants in tomato have been shown to reflect lesions in various steps of the carotenoid biosynthesis pathway. Many of these mutants have been mapped and most of the corresponding genes have been cloned. Practical use of a number of carotenoid mutants in tomato breeding confirms the potential for manipulating fruit carotenoid levels for fruit quality and nutritional impact (reviewed in Hirschberg, 2001). In the latter regard, carotenoids are antioxidant compounds capable of quenching oxygen radicals produced as a consequence of photosynthesis. They include provitamin A compounds (such as beta-carotene) and thus have potential high-value nutritional benefits both as precursors to essential nutrients and as antioxidants. For example, high levels of dietary lycopene have been associated with lower prostate cancer cases in men (reviewed in Giovannucci, 2002).

Due to the endogenous role of carotenoids in light harvesting and photoprotection of plant tissues, carotenoid synthesis, including that associated with fruit ripening, has evolved to be strongly influenced by light quality and intensity (Alba *et al.*, 2000). Thus, in addition to targeting carotenoid synthesis genes as tools for fruit carotenoid improvement, manipulation of light perception and response may serve as an additional avenue through which carotenoid accumulation may be enhanced for fruit color and nutrient quality in crop species.

The tomato *high-pigment (hp1)* mutant represents a lesion in a repressor of light signal transduction resulting in light hypersensitivity and elevated carotenoid accumulation in *hp1/hp1* fruit (Peters *et al.*, 1989). The seedling phenotypes of *hp1* have been shown to be red-light enhanced and far-red reversible, serving as the basis of the hypothesis of a lesion resulting in hyperactivity of phytochrome responses (Kerckhoffs and Kendrick, 1997). In further support of a role in phytochrome hypersignaling, overexpression of oat phytochrome (phyA) in tomato resulted in phenotypes similar to those observed in *hp1* (Boylan and Quail, 1989), indicating that responses mediated by phytochromes are impacted by *hp* mutations (Mustilli *et al.*, 1999). Although to date there have been no reported mutants in *Arabidopsis* corresponding directly to *hp1* at the phenotypic level, *Arabidopsis cop, det* and *fus* light-signal transduction mutants share some similar seeding characteristics (reviewed in Schwechheimer and Deng, 2000). The gene for *HP1* has recently been cloned and shown to be a tomato homolog of the *Arabidopsis damaged DNA-binding protein 1* (*DDB1*) gene and a plausible single gene target for manipulation of fruit carotenoid levels (Liu *et al.*, 2004). A second nonallelic mutation, phenotypically similar to *hp1* and termed *hp2*, has also been described (Soressi, 1975). *HP2* was cloned and determined to be the tomato homolog of the *Arabidopsis DET1* gene (Mustilli *et al.*, 1999). DET1 is a nuclear-encoded protein implicated in regulation of gene expression and development in response to light (Pepper *et al.*, 1994), and it also forms a complex with DDB1 (Schroeder *et al.*, 2002). While both *hp1* and *hp2* have proven difficult to use in tomato breeding because of associated detrimental phenotypes resulting from whole plant alterations in light response (e.g. stunting and low fruit yield), fruit-specific manipulation of *HP2/DET1* was recently shown to yield high-carotenoid fruit without negative pleiotropic phenotypes (Davuluri *et al.*, 2005).

Finally, it has recently been shown that changes in fruit color and carotenoid levels may also be achieved by influencing levels of polyamines (Mehta *et al.*, 2002). Polyamines are organic cations of low molecular weight derived from amino acids via decarboxylation and associated with numerous physiological processes across diverse taxa (Kakkar and Sawhney, 2002). In Mehta *et al.* (2002), a yeast *S*-adenosylmethionine decarboxylase (ySAMdc) gene was introduced into tomato under the fruit-specific E8 promoter (Deikman *et al.*, 1992). The resulting transgenic plants displayed increased levels of spermine and spermidine throughout fruit ripening, with corresponding reduction in levels of putrescine. Lycopene was also increased by 200–300% yielding deeper red transgenic fruit that displayed additional fruit quality attributes including increased shelf life and enhanced texture (Mehta *et al.*, 2002).

12.5 Summary

Fruit ripening is a complex process that includes changes in physiology, biochemistry, color, flavor and nutritional content. Many fruit species are of great economic importance and thereby are targets for extensive study and manipulation at both the molecular and biochemical levels. While the division between fruits that ripen climacterically and those that are nonclimacteric is clear at the physiological level, current work on genes present in both types of fruit is beginning to piece together possible universal regulation mechanisms for nonethylene-mediated ripening. Recent technical advances such as microarrays and EST databases are allowing for further comparison between the ripening profiles of model ripening systems for both climacteric tomato and nonclimacteric strawberry. Numerous homologs for genes involved in ethylene biosynthesis have been cloned in several species allowing for a broader understanding of ethylene biosynthesis, while comprehension of ethylene signal transduction has benefited greatly from both *Arabidopsis* and tomato. Cloning of genes encoding enzymes thought to be involved in ripening has lead to better understanding of cell-wall breakdown during this process. These enzymes have become useful in the quest to manipulate softening of many fruits. Understanding of the function of light during fruit ripening has been advanced through considerable innovations in the field of light signaling and color development. Several genes involved in light signaling and fruit pigment accumulation have provided targets for manipulation of ripening and color as well as nutritional content in fruit. Further elucidation of the biosynthesis of ripening pigments and light signaling, as well as those pathways discussed above, will increase the capacity to increase nutritional value and attractiveness to the consumer.

References

Abeles, F.B., Morgan, P.W. and Saltveit, M.E. (1992) *Ethylene in Plant Biology,* 2nd edn, Academic Press Inc., CA.

Adams, D.O. and Yang, S.F. (1979) Ethylene biosythesis: identification of 1-aminocyclopropane-1-carboxylic acid is an intermediate in the conversion of methionine to ethylene. *Proc Natl Acad Sci USA* **76**, 170–174.

Adams-Phillips, L., Barry, C. and Giovannoni, J. (2004a) Signal transduction systems regulating fruit ripening. *Trends Plant Sci* **9**, 331–338.

Adams-Phillips, L., Barry, C., Kannan, P., Leclercq, J., Bouzayen, M. and Giovannoni, J. (2004b) Evidence that CTR1-mediated ethylene signal transduction in tomato is encoded by a multigene family whose members display distinct regulatory features. *Plant Mol Biol* **54**, 387–404.

Aharoni, A., Leopold, C.P., Keizer, H.C., *et al.* (2000) Identification of the *SAAT* gene involved in strawberry biogenesis by use of DNA microarrays. *Plant Cell* **12**, 647–661.

Aharoni, A., Leopold, C.P., Keizer, H.C., *et al.* (2002) Novel insight into vascular, stress, and auxin-dependent and –independent gene expression programs in strawberry, a non-climacteric fruit. *Plant Physiol* **129**, 1019–1031.

Alba, R., Cordonnier-Pratt, M.M. and Pratt, L.H. (2000) Fruit-localized phytochromes regulate lycopene accumulation independently of ethylene production in tomato. *Plant Physiol* **123**, 363–370.

Alba, R., Fei, Z., Payton, P., *et al.* (2004) ESTs, cDNA microarrays, and gene expression profiling: tools for dissecting plant physiology and development. *Plant J* **39**, 697–714.

Alba, R., Payton, P., Fei, Z., *et al.* (2005) Transcriptome and selected fruit metabolite analysis reveal multiple points of ethylene regulatory control during tomato fruit development. *Plant Cell* **17**, 2954–2965.

Alexander, L. and Grierson, D. (2002) Ethylene biosynthesis and action in tomato: a model for climacteric ripening. *J Exp Bot* **53**, 2039–2055.

Andrawis, A., Solomon, M. and Delmer, D.P. (1993) Cotton fiber annexins: a potential role in regulation of callose synthase. *Plant J* **3**, 763–772.

Arabidopsis Initiative (2000) Analysis of the genome sequence of the flowering plant *Arabidopsis thaliana*. *Nature* **406**, 796–815.

Arrowsmith, D.A. and de Silva, J. (1995) Characterization of two tomato fruit-expressed cDNAs encoding xyloglucan endotransglycosylates. *Plant Mol Biol* **28**, 391–403.

Atkinson, R.G., Bolitho, K.M., Wright, M.A., Iturriagagoitia, B.T., Reid, S.J. and Ross, G.S. (1999) Apple ACC-oxidase and polygalacturonase, Ripening-specific gene expression and promoter analysis in transgenic tomato. *Plant Mol Biol* **38**, 449–460.

Ayub, R., Guis, M., Ben Amor, M. *et al.* (1996) Expression of ACC oxidase antisense gene inhibits ripening of cantaloupe melon fruit. *Nat Biotechnol* **14**, 862–866.

Balague, C., Watson, C.F., Turner, A.J. *et al.* (1993) Isolation of a ripening and wound-induced cDNA from *Cucumis melo* L. encoding a protein with homology to the ethylene-forming enzyme. *Eur J Biochem* **212**, 27–34.

Barry, C.S., Blume, B., Bouzayen, M., Cooper, W., Hamilton, A.J. and Grierson, D. (1996) Differential expression of the 1-aminocyclopropane-1-carboxylate oxidase gene family of tomato. *Plant J* **9**, 525–535.

Barry, C.B., Fox, E.A., Yen, H.C. *et al.* (2001) Analysis of the ethylene response in the *epinastic (epi)* mutant of tomato. *Plant Physiol* **127**, 58–66.

Barry, C. and Giovannoni, J. (2006) Ripening inhibition in the tomato *Green-ripe* mutant results from ectopic expression of a novel protein which disrupts ethylene signal transduction. *PNAS USA* **103**, 7923–7928.

Barry, C., McQuinn, R., Thompson, A., Seymour, G., Grierson, D. and Giovannoni, J. (2005) Ethylene insensitivity conferred by the *Green-ripe* (*Gr*) and *Never-ripe 2* (*Nr-2*) ripening mutants of tomato. *Plant Physiol* **138**, 267–275.

Bauchot, A.D., Mottram, D.S., Dodson, A.T. and John, P. (1998) Effect of aminocyclopropane-1-carboxylic acid oxidase antisense gene on the formation of volatile esters in cantaloupe Charentais melon (Cv. Vedranais). *J Agric Food Chem* **46**, 4787–4792.

Beecher, G.R. (1998) Nutrient content of tomatoes and tomato products. *Proc Soc Exp Biol Med* **218**, 98–100.

Bennett, A.B. and Della-Penna, D. (1987) Polygalacturonase gene expression in ripening tomato fruit. In: *Tomato Biotechnology* (eds Nevins, D. and Jones, R.). Alan R. Liss, NY, pp. 299–308.

Biale, J.B. and Young, R.E. (1981) Respiration and ripening in fruits – retrospect and prospect. In: *Recent Advances in the Biochemistry of Fruits and Vegetables* (eds Friend, J. and Rhodes, M.J.C.). Academic Press, London, UK, pp. 1–39.

Bird, C.R., Smith, C.J.S., Ray, J.A., *et al.* (1988) The tomato polygalacturonase gene and ripening-specific expression in transgenic plants. *Plant Mol Biol* **11**, 651–662.

Boller, T., Herner, R.C. and Kende, H. (1979) Assay for an enzymatic formation of ethylene precursor 1-aminocyclopropane-1-carboxylic acid. *Planta* **145**, 293–304.

Boylan, M.T. and Quail, P.H. (1989) Oat phytochrome is biologically active in transgenic tomatoes. *Plant Cell* **1**, 765–773.

Brummell, D.A., Harpster, M.H., Civello, P.M., Palys, J.M. and Bennett, A.B. (1999a) Modification of expansin protein abundance in tomato fruit alters softening and cell wall polymer metabolism during ripening. *Plant Cell* **11**, 2203–2216.

Brummell, D.A., Hall, B.D. and Bennett, A.B. (1999b) Antisense suppression of tomato endo-1,4-β-glucanase Cel2 mRNA accumulation increases the force required to break fruit abscission zones but does not affect fruit ripening. *Plant Mol Biol* **40**, 615–622.

Brummell, D.A. and Harpster, M.H. (2001) Cell wall metabolism in fruit softening and quality and its manipulation in transgenic plants. *Plant Mol Biol* **47**, 311–340.

Buchanan, B., Gruissem, W. and Jones, R.L. (2000) *Biochemistry and Molecular Biology of Plants.* American Society of Plant Physiologists, MD.

Carrington, C.M.S., Greve, L.C. and Labavitch, J.M. (1993) Cell wall metabolism in ripening fruit. VI. Effects of the antisense polygalacturonase gene on cell wall changes accompanying ripening in transgenic tomatoes. *Plant Physiol* **103**, 429–434.

Catala, C., Rose, J.K.C. and Bennett, A.B. (2001) Auxin-regulated genes encoding cell-wall modifying proteins are expressed during early tomato fruit growth. *Plant Physiol* **122**, 527–534.

Chang, C., Kwok, S.F., Bleecker, A.B. and Meyerowitz, E.M. (1993) *Arabidopsis* ethylene-response gene *ETR1*, Similarity of product to two-component regulators. *Science* **262**, 539–545.

Chang, C. and Shockey, J.A. (1999) The ethylene-response pathway, signal perception to gene regulation. *Curr Opin Plant Biol* **2**, 352–358.

Chory, J. (1997) Light modulation of vegetative development. *Plant Cell* **9**, 1225–1234.

Civello, P.M., Powell, A.L.T., Sabehat, A. and Bennett, A.B. (1999) An expansin gene expressed in ripening strawberry fruit. *Plant Physiol* **121**, 1273–1279.

Clayberg, C.D., Butler, L., Kerr, E.A., Rick, C.M. and Robinson, R.W. (1966) Third list of known genes in tomato. *J Hered* **57**, 189–196.

Clayberg, C.D., Butler, L., Kerr, E.A., Rick, C.M. and Robinson, R.W. (1967) Supplementary list of tomato genes as of January, 1967. *Rep Tomato Genet Coop* **17**, 2–11.

Clegg, M.T. and Durbin, M.L. (2000) Flower color variation, a model for the experimental study of evolution. *Proc Natl Acad Sci USA* **97**, 7016–7023.

Clevenger, D., Barrett, J., Klee, H. and Clark, D. (2004) Factors affecting seed production in transgenic ethylene-insensitive petunias. *J American Soc Hort Sci* **129**, 401–406.

Collmer, A. and Keen, N.T. (1986) The role of pectic enzymes in plant pathogenesis. *Annu Rev Phytopath* **24**, 383–409.

Cosgrove, D.J. (1999) Enzymes and other agents that enhance cell wall extensibility. *Annu Rev Plant Physiol Plant Mol Biol* **50**, 391–417. Annual Reviews Inc. (ed. Jones, R.L.).

Cosgrove, D.J. (2000a) New genes and new biological roles for expansins. *Curr Opin Plant Biol* **3**, 73–78.

Cosgrove, D.J. (2000b) Expansive growth of plant cell walls. *Plant Physiol Biochem* **38**, 109–124.

Cosgrove, D. (2005) Growth of the plant cell wall. *Nat Rev Mol Cell Biol* **6**, 850–861.

Davuluri, R., van-Tuinen, A., Fraser, P., *et al.* (2005) Fruit-specific RNAi-mediated suppression of DET1 enhances carotenoid and flavonoid content in tomatoes. *Nat-Biotechnol* **23**, 890–895.

Deikman, J., Kline, R. and Fischer, R.L. (1992) Organization of ripening and ethylene regulatory regions in a fruit-specific promoter from tomato. *Plant Physiol* **100**, 2013–2017.

Della-Penna, D., Lincoln, J.E., Fischer, R.L. and Bennett, A.B. (1989) Transcriptional analysis of polygalacturonase and other ripening associated genes in rutgers, *rin, nor*, and *Nr* tomato fruit. *Plant Physiol* **90**, 1372–1277.

de Silva, J., Arrowsmith, D., Hellyer, A., Whiteman, S. and Robinson, S. (1994) Xyloglucan endo-transglycosylase and plant growth. *J Exp Bot* **45**, 1693–1701.

Doganlar, S., Tanksley, S.D. and Mutschler, M.A. (2000) Identification and molecular mapping of loci controlling fruit ripening time in tomato. *Theor Appl Genet* **100**, 249–255.

Dogbo, O., Laferriere, A., D'harlingue, A. and Camara, B. (1988) Carotenoid biosynthesis, Isolation and characterization of a bifunctional enzyme catalyzing the synthesis of phytoene. *Proc Natl Acad Sci USA* **85**, 7054–7058.

Eshed, Y. and Zamir, D. (1995) An introgression line population of Lycopersicon pennellii in the cultivated tomato enables the identification and fine mapping of yield-associated QTL. *Genetics* **141**, 1147–1162.

Fei, Z., Tang, X., Alba, R., *et al.* (2004) Comprehensive EST analysis of tomato and comparative genomics of fruit ripening. *Plant J* **40**, 47–59.

Fei, Z., Tang, X., Alba, R. and Giovannoni, J. (2006) Tomato Expression Database (TED), A suite of data presentation and analysis tools. *Nucl Acids Res* **34** (Database issue), D766–D770.

Ferrarese, L., Trainotti, L., Moretto, P., Polverino de Laureto, P., Rascio, N. and Casadoro, G. (1995) Differential ethylene-inducible expression of cellulase in pepper plants. *Plant Mol Biol* **29**, 735–747.

Frary, A., Nesbitt, C., Frary, A., *et al.* (2000) *fw2.2*, A quantative trait locus key to the evolution of tomato fruit size. *Science* **289**, 85–88.

Fridman, E., Carrari, F., Liu, Y., Fernie, A. and Zamir, D. (2004) Zooming in on a quantitative trait for tomato yield using interspecific introgressions. *Science* **305**, 1786–1789.

Fridman, E., Liu, Y.-S., Carmel-Goren, L. *et al.* (2002) Two tightly linked QTLs modify tomato sugar content via different physiological pathways. *Mol Genet Genom* **266**, 821–826.

Fujino, D.W., Nissen, S.J., Jones, D.A., Burger, D.W. and Bradford, K.J. (1988) Quantification of indole-3-acetic acid in dark-grown seedlings of the *Diageotropica* and *Epinastic* mutants of tomato (*Lycopersicon esculentum* MILL.). *Plant Physiol* **88**, 780–784.

Giovannoni, J.J., DellaPenna, D., Bennett, A.B. and Fischer, R.L. (1989) Expression of a chimeric polygalacturonase in transgenic *rin* (*ripening inhibitor*) tomato fruit results in polyuronide degradation but not fruit softening. *Plant Cell* **1**, 53–64.

Giovannoni, J.J. (1993) Molecular and genetic analysis of tomato fruit development and ripening. In: *Methods in Plant Biochemstry* (ed. Byrant, J.). Academic Press, London, UK, pp. 253–287.

Giovannoni, J.J. (2001) Molecular biology of fruit maturation and ripening. *Annu Rev Plant Physiol Plant Mol Biol* **52**, 725–749.

Giovannoni, J. (2004) Genetic regulation of fruit development and ripening. *Plant Cell* **16**, S170–S180.

Giovannucci, E. (2002) A review of epidemiologic studies of tomatoes, lycopene, and prostate cancer. *Exp Biol Med* **227**, 852–859.

Goff, S. and Klee, H. (2006) Plant volatile compounds, sensory cues for health and nutritional value. *Science* **311**, 815–819.

Gonzalez-Bosch, C., Brummell, D.A. and Bennett, A.B. (1996) Differential expression of two endo-1,4-β-glucanase genes in pericarp and locules of wild-type and mutant tomato fruit. *Plant Physiol* **111**, 1313–1319.

Gray, J.E., Picton, S., Giovannoni, J.J. and Grierson, D. (1994) The use of transgenic and naturally occurring mutants to understand and manipulate tomato fruit ripening. *Plant Cell Env* **17**, 557–571.

Gray, J.E., Picton, S., Shabbeer, J., Schuch, W. and Grierson, D. (1992) Molecular biology of fruit ripening and its manipulation with antisense genes. *Plant Mol Biol* **19**, 69–87.

Grierson, D. (1984) The appearance of polygalacturonase messenger RNA in tomatoes Lycopersicon 1 of a series of changes in gene expression during development and ripening. *Planta* **163**, 263–271.

Grierson, D., Tucker, G.A., Keen, J., *et al.* (1986) Sequencing and identification of a cDNA clone for tomato polygalacturonase. *Nucl Acids Res* **14**, 8595–8603.

Grierson, D. (1987) Senescence in fruits. *Hort Sci* **22**, 859–862.

Grierson, D., Hamilton, A.J., Bouzayen, M., Kock, M., Lycett, G.W. and Barton, S. (1992) Regulation of gene expression, ethylene synthesis and ripening in transgenic tomatoes. In: *Inducible Plant Proteins* (ed. Wray, J.L.). Cambridge University Press, UK, pp. 155–174.

Gross, K.C. (1984) Fractionation and partial characterization of cell walls from normal and non-ripening mutant tomato fruit. *Physiol Plant* **62**, 25–32.

Hadacek, F. (2002) Secondary metabolites as plant traits, current assessment and future perspectives. *Crit Rev Plant Sci* **21**, 273–322.

Hall, L.N., Tucker, G.A., Smith, C.J.S., *et al.* (1993) Antisense inhibition of pectin esterase gene expression in transgenic tomatoes. *Plant J* **3**, 121–129.

Harpster, M.H., Brummell, D.A. and Dunsmuir, P. (2002a) Suppression of a ripening-related endo-1,4-β-glucanase in transgenic pepper fruit does not prevent depolymeriztion of cell wall polysaccharides during ripening. *Plant Mol Biol* **50**, 345–355.

Harpster, M.H., Dawson, D.M., Nevins, D.J., Dunsmuir, P. and Brummell, D.A. (2002b) Constitutive overexpression of a ripening-related pepper endo-1,4-β-glucanase in transgenic tomato fruit does not increase xyloglucan depolymerization or fruit softening. *Plant Mol Biol* **50**, 357–369.

Harpster, M.H., Lee, K.L. and Dunsmuir, P. (1997) Isolation and characterization of a gene encoding endo-β-1,4-glucanase from pepper (*Capiscum annuum* L.). *Plant Mol Biol* **33**, 47–59.

Harpster, M., Brummell, D. and Dunsmuir, P. (1997) Expression analysis of a ripening-specific, auxin-repressed endo-1,4-beta-glucanase gene in strawberry. *Plant Phys* **118**, 1307–1316.

Hauser, B.A., Cordonnier-Pratt, M-M., Daniel-Vedele, F. and Pratt, L.H. (1995) The phytochrome gene family in tomato includes a novel subfamily. *Plant Mol Biol* **29**, 1143–1155.

Hirschberg, J. (2001) Carotenoid biosynthesis in flowering plants. *Curr Opin Plant Biol* **4**, 210–218.

Hobson, G.E., Nichols, R., Davies, J.N. and Atkey, P. (1984) The inhibition of tomato fruit ripening by silver. *J Plant Physiol* **116**, 21–29.

Hua, J., Chang, C., Sun, Q. and Meyerowitz, E.M. (1995) Ethylene insensitivity conferred by *Arabdiopsis* ERS gene. *Science* **269**, 1712–1714.

Hua, J. and Meyerowitz, E.M. (1998) Ethylene responses are negatively regulated by a receptor gene family in *Arabidopsis*. *Cell* **94**, 261–271.

Huber, D.J. (1983) The role of cell wall hydrolases in fruit softening. *Hortic Rev* **5**, 169–219.

Isaacson, T., Ronen, G., Zamir, D. and Hirschberg, J. (2002) Cloning of *tangerine* from tomato reveals a carotenoid isomerase essential for the production of β-carotene and xanthophylls in plants. *Plant Cell* **14**, 333–342.

Ishimaru, M. and Kobayashi, S. (2002) Expression of a xyloglucan endo-transglycosylase gene is closely related to grape berry softening. *Plant Sci* **161**, 621–628.

Jeffrey, D., Smith, C., Goodenough, P., Prosser, I. and Grierson, D. (1984) Ethylene-independent and ethylene-dependent biochemical changes in ripening tomatoes. *Plant Physiol* **74**, 32–38.

Jenkins, J.A. and MacKinney, G. (1995) Carotenoids of the apricot tomato and its hybrids with yellow and tangerine. *Genetics* **40**, 715–720.

Jiang, Y., Joyce, D.C. and Macnish, A.J. (2002) Softening response of banana fruit treated with 1-methylcyclopropene to high temperature exposure. *Plant Growth Reg* **36**, 7–11.

Kakkar, R. and Sawhney, V. (2002) Polyamine research in plants: A changing perspective. *Physiologia Plantarum* **116**, 281–292.

Kalloo, G. and Bergh, B.O. (1993) *Genetic Improvement of Vegetable Crops*. Pergamon Press, NY.

Kerckhoffs, L.H.J. and Kendrick, R.E. (1997) Photocontrol of anthocyanin biosynthesis in tomato. *J Plant Res* **110**, 141–149.

Kieber, J.J., Rothenberg, M., Roman, G., Feldman, K.A. and Ecker, J.R. (1993) *CTR1*, a negative regulator of the ethylene response pathway in *Arabidopsis*, encodes a member of the raf family of protein kinases. *Cell* **72**, 411–427.

Kinet, J.M. and Peet, M.M. (1997) Tomato. In: *The Physiology of Vegetable Crops* (ed. Wien, H.C.). CAB International, England, UK, pp. 207–258.

Klee, H. and Tieman, D. (2002) The tomato ethylene receptor gene family, form and function. *Physiol Plant* **115**, 336–341.

Klee, H. (2002) Control of ethylene-mediated processes in tomato at the level of receptors. *J Exp Bot* **53**, 2057–2063.

Knapp, J., Moureau, P., Schuch, W. and Grierson, D. (1989) Organisation and expression of polygalacturonase and other ripening genes in Alisa Craig '*Never-ripe*' and 'ripening-inhibitor' tomato mutants. *Plant Mol Biol* **12**, 105–116.

Kumar, A., Taylor, M.A., Arif, S.A.M. and Davies, H.V. (1996) Potato plants expressing antisense and sense *S*-adenosylmethionine decarboxylase (SAMDC) transgenes show altered levels of polyamines and ethylene, antisense plants display abnormal phenotypes. *Plant J* **9**, 147–158.

Lanahan, M.B., Yen, H.C., Giovannoni, J.J. and Klee, H.J. (1994) The *Never-ripe* mutation blocks ethylene preception in tomato. *Plant Cell* **6**, 521–530.

Lashbrook, C., Gonzalez-Bosch, C. and Bennett, A. (1994) Two divergent endo-beta-1,4-glucanase genes exhibit overlapping expression in ripening fruit and abscising flowers. *Plant-Cell* **6**(10), 1485–1493.

Lashbrook, C.C., Tieman, D.M. and Klee, H.J. (1998a) Differential regulation of the tomato *ETR* gene family throughout plant development. *Plant J* **15**, 243–252.

Lashbrook, C., Giovannoni, J., Hall, B., Fischer, R. and Bennett, A. (1998b) Transgenic analysis of tomato endo-beta-1,4-glucanase gene function. Role of cell in floral abscission. *Plant Journal* **13**, 303–310.

Lay-Yee, M. and Knighton, M.L. (1995) A full-length cDNA encoding 1-aminocyclopropane-1-carboxylate synthase from apple. *Plant Physiol* **107**, 1017–1918.

Lelievre, J.M., Latche, A., Jones, B., Bouzayen, M. and Pech, J.C. (1997) Ethylene and fruit ripening. *Physiol Plantarum* **101**, 727–739.

Leclercq, J., Adams-Phillips, L.C., Zegzouti, H., *et al.* (2002) *LeCTR1*, a tomato *CTR1*-like gene, demonstrates ethylene signaling ability in *Arabidopsis* and novel expression patterns in tomato. *Plant Physiol* **130**, 1132–1142.

Li, C.Y., Jacob-Wilk, D., Zhong, G.Y., Goron, R. and Holland, D. (1998) A full-length cDNA encoding an ethylene receptor ERS homologue from Citrus (Ascession No . AF092088). *Plant Physiol* **118**, 1534.

Lincoln, J.E., Cordes, S., Read, E. and Fischer, R.L. (1987) Regulation of gene expression ethylene during *Lycoperisicon esculentum* (tomato) fruit development. *Proc Natl Acad Sci USA* **84**, 2793–2797.

Liu, Y., Roof, S., Ye, Z., *et al.* (2004) Manipulation of light signal transduction as a means of modifying fruit nutritional quality in tomato. *Proc Natl Acad Sci USA* **26**, 9897–9902.

Liu, J., Van Eck, J., Cong, B. and Tanksley, S.D. (2002) A new class of regulatory genes underlying the cause of pear-shaped tomato fruit. *Proc Natl Acad Sci* **99**, 13302–13306.

Liu, Y.-S. and Zamir, D. (1999) Second generation *L. pennellii* introgression lines and the concept of bin mapping. *Tomato Genet Coop* **49**, 26–30.

Llop-Tous, I., Barry, C.S. and Grierson, D. (2000) Regulation of ethylene biosynthesis in response to pollination in tomato flowers. *Plant Physiol* **123**, 971–978.

Lopez-Gomez, R., Cambell, A., Dong, J.G., Yang, S.F. and Gomez-Lim, M.A. (1997) Ethylene biosynthesis in banana fruit, isolation of a genomic clone to ACC oxidase and expression studies. *Plant Sci* **123**, 123–131.

Lyons, J.M. and Pratt, H.K. (1964) Effect of stage of maturity and ethylene treatment on respiration and ripening of tomato fruits. *Proc Am Soc Hort Sci* **84**, 491–500.

Malcomber, S. and Kellogg, E. (2005) SEPALLATA gene diversification, brave new whorls. *Trends Plant Sci* **10**, 427–435.

Mann, V., Pecker, I. and Hirschberg, J. (1994) Cloning and characterization of the gene for phytoene desaturase (Pds) from tomato (Lycopersicon esculentum). *Plant Mol Biol* **24**, 429–434.

Manning, K. (1998) Identification of a set of ripening-related genes from strawberry, their identification and possible relationship to fruit quality traits. *Planta* **205**, 622–631.

Mansson, P.E., Hsu, D. and Stalker, D. (1985) Characterization of fruit specific cDNAs from tomato. *Mol Gen Genet* **200**, 356–361.

Mao, L., Begum, D., Chuang, H.W., *et al.* (2000) *JOINTLESS* is a MADS-box gene controlling tomato flower abscission zone development. *Nature* **406**, 910–913.

Marin-Rodriguez, M.C., Orchard, J. and Seymour, G.B. (2002) Pectate lyases, cell wall degredation and fruit softening. *J Experiment Bot* **53**, 2115–2119.

Martin, G.B., Brommonschenkel, S.H., Chunwongse, J., *et al.* (1993) Map-based cloning of a protein kinase gene conferring disease resistance in tomato. *Science* **262**, 1432–1436.

McChurchie, E.J., McGlasson, W.B. and Eaks, I.L. (1972) Treatment of fruit with propylene gives information about the biogenesis of ethylene. *Nature* **237**, 235–236.

McGlasson, W.B. (1985) Ethylene and fruit ripening. *Hort Sci* **20**, 51–54.

McQueen-Mason, S.J., Durachko, D.M. and Cosgrove, D.J. (1992) Two endogenous proteins that induce cell wall extension in plants. *Plant Physiol* **107**, 87–100.

Medina-Escobar, N., Cardenas, J., Moyano, E., Caballero, J.L. and Munoz-Blanco, J. (1997) Cloning, molecular characterization and expression pattern of a strawberry ripening-specific cDNA with sequence homology to pectate lyase from higher plants. *Plant Mol Biol* **34**, 867–877.

Mehta, R.A., Cassol, T., Li, N., Ali, N., Handa, A.K. and Mattoo, A.K. (2002) Engineered polyamine accumulation in tomato enhances phytonutrient content, juice quality, and vine life. *Nat Biotech* **20**, 613–618.

Moore, S., Vrebalov, J., Payton, P. and Giovannoni, J. (2002) Use of genomic tools to isolate key ripening genes and analyse fruit maturation in tomato. *J Exp Botany* **53**, 2023–2030.

Moore, S., Payton, P., Wright, M., Tanksley, S. and Giovannoni, J. (2005) Utilization of tomato microarrays for comparative gene expression analysis in the Solanaceae. *J Exp Bot* **56**, 2885–2895.

Mueller, L.A., Tanksley, S.D., Giovannoni, J.J., *et al.* (2005) The Tomato Sequencing Project, the first cornerstone of the International Solanaceae Project (SOL). *Comp Funct Genom* **6**, 153–158.

Mustilli, A.C., Fenzi, F., Ciliento, R., Alfano, F. and Bowler, C. (1999) Phenotype of the tomato *high-pigment-2* mutant is caused by a mutation in the tomato homolog of *DETIOLATE1*. *Plant Cell* **11**, 145–157.

Nakatsuka, A., Murachi, S., Okunishi, H., *et al.* (1998) Differential expression and internal feedback regulation of 1-aminocyclopropane-1-carboxylate synthase, 1-aminocyclopropane-1-carboxylate oxidase, and ethylene receptor genes in tomato fruit ripening during development and ripening. *Plant Physiol* **118**, 1295–1305.

Oeller, P.W., Wong, L.M., Taylor, L.P., Pike, D.A. and Theologis, A. (1991) Reversible inhibition of tomato fruit ripening. *Science* **254**, 437–439.

Pepper, A., Delaney, T., Washburn, T., Poole, D. and Chory, J. (1994) DET1, a negative regulator of light-mediated development and gene expression in *Arabidopsis*, encodes a novel nuclear-localized protein. *Cell* **78**, 109–116.

Perin, C., Gomez-Jimenez, M., Hagen, L., *et al.* (2002) Molecular and genetic characterization of a non-climacteric phenotype in melon reveals two loci conferring altered ethylene response in fruit. *Plant Physiol* **129**, 300–309.

Peters, J.L., Schreuder, M.E.L., Verduin, S.J.W. and Kendrick, R.E. (1992) Physiological characterization of a high-pigment mutant of tomato. *Photochem Photobiol* **56**, 75–82.

Peters, J.L., van Tuinen, A., Adamse, P., Kendrick R.E. and Koorneef, M. (1989) High-pigment mutants of tomato exhibit high sensitivity for phytochrome action. *J Plant Physiol* **134**, 661–666.

Picton, S., Barton, S.L., Bouzayen, M., Hamilton, A.J. and Grierson, D. (1993) Altered fruit ripening and leaf senescence in tomatoes expressing an antisense ethylene-forming enzyme transgene. *Plant J* **3**, 469–481.

Redgwell, R.J. and Fischer, M. (1996) Fruit texture, cell wall metabolism and consumer perceptions. In: *Physiology and Biochemistry of Plant Cell Walls* (eds Brett, C.T. and Waldron, K.W.). Chapman Hall, London, UK, pp. 47–88.

Rexova-Benkova, L. and Markovic, O. (1976) Pectic enzymes. In: *Advances in Carbohydrate Chemistry and Biochemistry*, Vol. 33 (eds Tipton, R.S. and Horton, D.). New York Academic Press, NY, pp. 323–385.

Rick, C.M. and Butler, L. (1956) Phytogenetics of the tomato. *Adv Genet* **8**, 267–382.

Rick, C.M. (1978) The tomato. *Sci Am* **239**, 67–76.

Ronen, G., Cohen, M., Zamir, D. and Hirschberg, J. (1999) Regulation of carotenoid biosynthesis during tomato fruit development, expression of the gene for lycopene epsilon-cyclase is down-regulated during ripening and is elevated in the mutant delta. *Plant J* **17**, 341–351.

Rose, J.K.C., Lee, H.H. and Bennett, A.B. (1997) Expression of a divergent expansin gene is fruit-specific and ripening-regulated. *Proc Natl Acad Sci USA* **94**, 5955–5960.

Rose, J.K.C and Bennett, A.B. (1999) Cooperative disassembly of the cellulose-xyloglucan network of plant cell walls, parallels between cell expansion and fruit ripening. *Trends Plant Sci* **4**, 176–183.

Rose, J., Bashir, S., Giovannoni, J., Jahn, M. and Saravanan, S. (2004a) Tackling the plant proteome, practical approaches, hurdles and experimental tools. *Plant J* **39**, 715–733.

Rose, J.K.C, Braam, J., Fry, S.C. and Nishitani, K. (2002) The XTH family of enzymes involved in xyloglucan endtransglycosylation and endohydrolysis, current perspectives and a new unifying nomenclature. *Plant Cell Physiol* **43**, 1421–1435.

Rose, J., Cosgrove, D., Albersheim, P., Darvill, A. and Bennett, A. (2000) Detection of expansin proteins and activity during tomato fruit ontogeny. *Plant Phys* **123**, 1583–1592.

Rose, J.K.C., Saladié, M. and Catalá, C. (2004b) The plot thickens, new perspectives of primary cell wall modification. *Curr Opin Plant Biol* **7**, 296–301.

Ross, G.S., Knighton, M.L. and Lay-Yee, M. (1992) An ethylene regulated cDNA from ripening apples. *Plant Mol Biol* **19**, 231–238.

Schaller, G.E. and Bleecker, A.B. (1995) Ethylene-binding sites generated in yeast expressing the *Arabidopsis ETR1* gene. *Science* **270**, 1086–1089.

Schwechheimer, C. and Deng, X.-W. (2000) The COP/DET/FUS proteins-regulators of eukaryotic growth and development. *Sem Cell Dev Biol* **11**, 495–503.

Schroder, R., Atkinson, R.G., Langenkamper, G. and Redgwell, R.J. (1998) Biochemical and molecular characterisation of xyloglucan endotransglycosylase from ripe kiwifruit. *Planta* **204**, 242–251.

Schroeder, D., Gahrtz, M., Maxwell, B., *et al.* (2002) De-etiolated 1 and Damaged DNA Binding Protein 1 interact to regulate *Arabidopsis* photomorphogenesis. *Curr Biol* **12**, 1462–1472.

Seymour, G.B., Taylor, J.E. and Tucker, G.A. (1993) *Biochemistry of Fruit Ripening*. Chapman & Hall, London, UK.

Sheehy, R.E., Kramer, M. and Hiatt, W.R. (1988) Reduction of polygalacturonase activity in tomato fruit by antisense RNA. *Proc Natl Acad Sci USA* **85**, 8805–8809.

Sisler, E.C., Serek, M. and Dupille, E. (1995) Comparison of cyclopropene, 1-methylcyclopropene, and 3,3-dimethylcyclopropene as ethylene antagonists in plants. *Plant Growth Reg* **17**, 1–6.

Sisler, E.C. and Serek, M. (1997) Inhibitors of ethylene responses in plants at the receptor level, recent developments. *Physiol Plant* **100**, 577–582.

Sitrit, Y. and Bennett, A.B. (1998) Regulation of tomato fruit polygalacturonase mRNA accumulation by ethylene, a re-examination. *Plant Physiol* **116**, 1145–1150.

Smith, C.J.S., Watson, C.F., Morris, P.C., *et al.* (1990) Inheritance and effect on ripening of antisense polygalacturonase genes in transgenic tomatoes. *Plant Mol Biol* **14**, 369–379.

Smith, C.J.S., Watson, C.F., Ray, J., *et al.* (1988) Antisense RNA inhibition of polygalacturonase gene expression in transgenic tomatoes. *Nature* **334**, 724–726.

Soltis, D.E., Soltis, P.S., Albert, V.A., *et al.* (2002) Missing links, the genetic architecture of flower and floral diversification. *Trends Plant Sci* **7**, 22–31.

Soressi, G.P. (1975) New spontaneous or chemically-induced fruit ripening tomato mutants. *Tomato Genet Coop Rep* **25**, 21–22.

Spolaore, S., Trainotti, L., Pavanello, A. and Casadoro, G. (2003) Isolation and promoter analysis of two genes encoding different endo-β-1,4-glucanases in the non-climacteric strawberry. *J Exp Bot* **54**, 271–277.

Steele, N.M. and Fry, S.C. (1999) Purification of xyloglucan endotransglycosylases (XETs), a generally applicable and simple method based on reversible formation of an enzyme-substrate complex. *Biochem J* **340**, 207–211.

Takada, N. and Nelson, P. (1983). New consistency method for tomato products-the precipitate weight ratio. *J Food Sci* **48**, 1460–1462.

Tanksley, S.D., Ganal, M.W., Prince, J.P., de Vincente, M.C. and Bonierbale, M.W. (1992) High density molecular linkage maps of the tomato and potato genomes. *Genetics* **132**, 1141–1160.

Theologis, A., Oeller, P.W., Wong, L.M., Rottmann, W.H. and Gantz, D.M. (1993) Use of a tomato mutant constructed with reverse genetics to study fruit ripening, a complex developmental process. *Dev Gen* **14**, 282–295.

Thompson, A.J., Tor, M., Barry, C.S., *et al.* (1999) Molecular and genetic characterization of a novel pleiotropic tomato-ripening mutant. *Plant Physiol* **120**, 383–389.

Tian, M.S., Prakash, S., Elgar, H.J., Young, H., Burmeister, D.M. and Ross, G.S. (2000) Responses of strawberry fruit to 1-Methylcyclopropene (1-MCP) and ethylene. *Plant Growth Reg* **32**, 83–90.

Tieman, D.M., Ciardi, J.A., Taylor, M.G. and Klee, H.J. (2001) Members of the tomato *LeEIL (EIN3-like)* gene family are functionally redundant and regulate ethylene responses throughout plant development. *Plant J* **26**, 47–58.

Tieman, D.M., Harriman, R.W., Ramamohan, G. and Handa, A.K. (1992) An antisense pectin methylesterase gene alters pectin chemistry and soluble solids in tomato fruit. *Plant Cell* **4**, 667–679.

Tieman, D.M., Taylor, M.G., Ciardi, J.A. and Klee, H.J. (2000) The tomato ethylene receptors NR and LeETR4 are negative regulators of ethylene response and exhibit functional compensation within a multigene family. *Proc Natl Acad Sci USA* **97**, 5663–5668.

Tigchelaar, E.C., McGlasson, W.B. and Buescher, R.W. (1978) Genetic regulation of tomato fruit ripening. *Hort Sci* **13**, 508–513.

Tigchelaar, E.C., Tomes, M.L., Kerr, E.A. and Barman, R.J. (1973) A new fruit ripening mutant, non-ripening *(nor)*. *Rep Tomato Genet Coop* **23**, 33.

Trainotti, L., Spolaore, S., Pavanello, A., Baldan, B. and Casadoro, G. (1999) A novel E-type endo β-1,4-glucanase with putative cellulose binding domain is highly expressed in ripening strawberry fruits. *Plant Mol Biol* **40**, 323–332.

Tucker, G.A. and Grierson, D. (1982) Synthesis of polygalacturonase during tomato fruit ripening. *Planta* **155**, 64–67.

Van der Hoeven, R., Ronning, C., Giovannoni, J., Martin, G. and Tanksley, S. (2002) Deductions about the number, organization, and evolution of genes in the tomato genome based on analysis of a large expressed sequence tag collection and selective genomic sequencing. *Plant Cell* **14**, 1441–1456.

Van der Knaap, E. and Tanksley, S.D. (2001) Identification and characterization of a novel locus controlling early fruit development in tomato. *Theor Appl Genet* **103**, 353–358.

van Tuinen, A., Cordonnier-Pratt, M-M., Pratt, L.H., Verkerk, R., Zabel, P. and Koorneef, M. (1997) The mapping of phytochrome genes and photomorphogenic mutants in tomato. *Theor Appl Genet* **94**, 115–122.

Verhoeyen, M.E., Bovy, A., Collins, G., *et al.* (2002). Increasing antioxidant levels in tomatoes through modification of the flavanoid biosynthesis pathway. *J Exp Bot* **53**, 2099–2106.

Verma, D.P.S. and Hong, Z. (2001) Plant callose synthase complexes. *Plant Mol Biol* **47**, 693–701.

Vrebalov, J., Ruezinsky, D., Padmanabhan, V., *et al.* (2002) A MADS-box gene necessary for fruit ripening at the tomato *Ripening-inhibitor* (*Rin*) *Locus. Science* **296**, 343–346.

Watkins, C.B., Nock, J.F. and Whitaker, B.D. (2000) Responses of early, mid and late season apple cultivars to postharvest application of 1-methylcyclopropene(1-MCP) under air and controlled atmosphere storage conditions. *Postharvest Biol Technol* **19**, 17–32.

Watkins, C.B. (2002) Ethylene synthesis, mode of action, consequences and control. In: *Fruit Quality and its Biological Basis* (ed. Knee, M.). CRC Press. Boca Raton, FL, pp. 180–224.

Wang, Y., Tang, X., Cheng, Z., Mueller, L., Giovannoni, J. and Tanksley, S. (2006) Euchromatin and pericentric heterochromatin, comparative composition in the tomato genome. *Genetics* **172**, 2529–2540.

Wann, E.V., Jourdain, E.L., Pressey, R. and Lyon, B.G. (1985) Effect of mutant genotypes *hp* og^c and *dg* og^c on tomato fruit quality. *J Am Soc Hort Sci* **110**, 212–215.

White, P.J. (2002) Recent advances in fruit development and ripening, an overview. *J Exp Bot* **53**, 1995–2000.

Wilkinson, J.Q., Lanahan, M.B., Conner, T.W. and Klee, H.J. (1995) Identification of mRNAs with enhanced expression in ripening strawberry fruit using polymerase chain reaction differential display. *Plant Mol Biol* **27**, 1097–1108.

Wilkinson, J.Q., Lanahan, M.B., Clark, D.G., Bleecker, A.B. and Chang, C. (1997) A dominant mutant receptor from *Arabidopsis* confers ethylene insensitivity in heterologous plants. *Nat Biotechnol* **15**, 444–447.

Yang, S.F. and Baur, A.H. (1969) Pathways of ethylene biosynthesis. *Qualitas Plantarum et Materiae Vegetabiles* **19**, 201–220.

Yang, S.F. (1985) Biosynthesis and action of ethylene. *Hort Sci* **20**, 41–45.

Yang, S.F. (1987) The role of ethylene and ethylene synthesis in fruit ripening. In: *Plant Senescence, its Biochemistry and Physiology* (eds Thompson, W.W., Nothnagel, E.A. and Huffaker, R.C.). American Society of Plant Physiology, Rockville, Maryland, pp. 156–166.

Yen, H.C., Lee, S., Tanksley, S., Lanahan, M.B., Klee, H.J. and Giovannoni, J.J. (1995) The tomato *Never-ripe* locus regulates ethylene-inducible gene expression and is linked to a homolog of the *Arabidopsis ETR1* gene. *Plant Physiol* **107**, 1343–1353.

Yen, H.C., Shelton, A., Howard, L., Vrebalov, J. and Giovannoni, J.J. (1997) The tomato *high-pigment (hp)* locus maps to chromosome 2 and influences plastome copy number and fruit quality. *Theor Appl Genet* **95**, 1069–1079.

Yip, W.-K., Moore, T. and Yang, S.F. (1992) Differential accumulation of transcripts for four tomato 1-aminocyclopropane-1-carboxylate synthase homologs under various conditions. *Proc Natl Acad Sci USA* **89**, 2475–2479.

Yu, Y.B. and Yang, S.F. (1979) Auxin-induced ethylene production and its inhibition by aminoethoxyvinylglycine and cobalt ion. *Plant Physiol* **64**, 1074–1077.

Zhang, H.B., Budiman, M.A. and Wing, R.A. (2000) Genetic mapping of *jointless-2* to tomato chromosome 12 using RFLP and RAPD markers. *Theor Appl Genet* **100**, 1183–1189.

Zhou, Z., de Almeida Engler, J., Rouan, D., Michiels, F., Van Montagu, M. and Van Der Straeten, D. (2002) Tissue localization of submergence-induced 1-aminocyclopropane-1-carboxylic acid synthase in rice. *Plant Physiol* **129**, 72–84.

13 Genetic manipulation of leaf senescence

Yongfeng Guo and Susheng Gan

13.1 Introduction

Leaves are the primary site where solar energy is absorbed and used to fix CO_2 and produce carbohydrates via a process called photosynthesis. Leaves are also the factory where amino acids, antioxidants and some other nutrients are synthesized. Once a leaf undergoes senescence, its photosynthetic capability decreases sharply; therefore, leaf senescence limits crop yields and biomass accumulation. During leaf senescence, macromolecules (such as proteins), antioxidants and other nutrient components are degraded. Leaf senescence also limits storage, transportation and shelf-life of many vegetable crops after harvest, and devalues ornamental plants. In addition, a senescing leaf becomes more susceptible to pathogens and some of the pathogens may produce toxins, rendering our vegetable foods unsafe. On the other hand, leaf senescence is sometimes wanted. For example, cotton bolls are generally harvested mechanically; green leaves are easily damaged and the leaf trash and green staining (by chlorophylls) will reduce lint/fiber quality. Therefore, manipulation (either inhibition or promotion) of leaf senescence is highly desirable in practice.

13.2 Strategies of manipulating leaf senescence

The initiation and progress of senescence are controlled by many external and internal factors (He *et al.*, 2001). Environmental stresses such as extreme temperature, drought, nutrient deficiency, insufficient light/shadow/darkness and pathogen infection can induce senescence (He and Gan, 2003). Internal factors influencing senescence include age, developmental processes, such as reproductive growth, and levels of plant hormones/growth regulators (Gan, 2003). Manipulation of any of these factors can affect senescence. Surgical removal of inflorescence, e.g., can delay leaf senescence (Noodén, 1988). Low temperature has been widely used to prolong storage and shelf-life of many foliar vegetables and fruits since it reduces metabolic rate, and thus delays senescence (Noodén, 1988). Foliar application of nutrients such as nitrogen and phosphorus slows down the senescence process of wheat plants (Benbella and Paulsen, 1998). Plant growth regulators ethylene, jasmonic acid, salicylic acid, brassinosteroids and abscisic acid are believed to be inducers/promoters, while cytokinins and polyamines are antagonists of senescence (Gan, 2003). Exogenous applications of cytokinins such as dihydrozeatin and benzyladenine have been used to effectively delay senescence of freshly harvested vegetables (Clarke *et al.*, 1994) and flowers (Taverner *et al.*, 1999). Antagonists of ethylene action such as Ag^+ and 1-MCP are commonly used in postharvest storage to prevent vegetables, flowers and fruits from senescing (Blankenship and Dole, 2003).

Development of molecular biology and plant transformation technology in 1980s makes it possible to manipulate senescence via changing endogenous hormone levels in transgenic plants. Delayed plant senescence can be achieved by either suppressing senescence-promoting hormones such as ethylene or overproducing senescence-inhibiting cytokinins (Gan, 2003). Transgenic tomato plants with suppressed expression of two genes encoding for ethylene biosynthetic enzymes, ACC (1-aminocyclopropane-1-carboxylic acid) synthase (Oeller *et al.*, 1991) and ACC oxidase (John *et al.*, 1995; Aida *et al.*, 1998), showed significantly reduced ethylene production and retarded fruit senescence. The overproduction of cytokinins by expressing a foreign gene encoding cytokinin-synthesizing enzyme, isopentenyl transferase (IPT), has been a predominant approach to inhibiting senescence in transgenic plants (Gan and Amasino, 1996; Li *et al.*, 2000).

13.3 IPT-based transgenic techniques for manipulation of cytokinin production

Cytokinins are a class of plant hormones including such adenine derivatives as kinetin, isopentenyl adenine and zeatins. This class of hormones have been implicated in almost all stages of plant growth and development, ranging from seed germination to floral induction to senescence (Brault and Maldiney, 1999). The role of cytokinins in preventing plant senescence was first observed through external application of kinetin on detached cocklebur leaves (Richmond and Lang, 1957). There is an inverse correlation between cytokinin levels and the progression of senescence (Van Styaden *et al.*, 1988). Regulatory role of cytokinins in plant senescence is further supported by delayed senescence of transgenic plants expressing the cytokinin-biosynthesizing enzyme IPT (Gan and Amasino, 1996).

IPT is an enzyme that catalyzes the first committed and rate-limiting step of cytokinin biosynthesis, condensation of dimethylallyl pyrophosphate (DMAPP) and 5′-AMP to isopentenyladenosine (IPA) 5′-phosphate (Barry *et al.*, 1984; Gan and Amasino, 1996). Because plant genes involved in cytokinin biosynthesis were not identified until recently (Takei *et al.*, 2001; Sun *et al.*, 2003), a bacterial version of *IPT* gene was used to increase cytokinin levels in transgenic plants. The enzyme encoded by the *IPT* gene from plant pathogen *Agrobacterium tumefaciens* catalyzes the rate-limiting step of cytokinin biosynthesis in tumorous plant tissues (Akiyoshi *et al.*, 1984; Barry *et al.*, 1984). When *A. tumefaciens* infects a plant, the T-DNA region of the Ti-plasmid is introduced into the plant chromosome. The T-DNA carrying *IPT* gene is expressed in a plant cell and is responsible for deregulated production of cytokinins that function in tumor formation together with auxin. The IPT protein produced in *Escherichia coli* exhibits IPT enzyme activity (Akiyoshi *et al.*, 1984; Barry *et al.*, 1984). Recent studies showed that unlike the bacterial IPT; which catalyzes the transfer of the isopentenyl moiety from DMAPP to AMP, several plant IPT catalyze the transfer of the isopentenyl moiety from DMAPP preferentially to ATP and to ADP (Kakimoto, 2003). Nevertheless, the bacterial *IPT* gene appears to work well in producing cytokinins in a wide range of plant species including tobacco, tomato, potato, *Arabidopsis*, cucumber, soybean, (Gan and Amasino, 1996), lettuce (Kunkel *et al.*, 1999), cauliflower (Nguyen *et al.*, 1998) and peach (Hammerschlag and Smigocki, 1998).

Various promoters have been used to drive expression of *IPT* in transgenic plants. Constitutive promoters, including native IPT promoter (Beinsberger *et al.*, 1991; Ooms *et al.*, 1991; Yusibov *et al.*, 1991; Zhang *et al.*, 1995; Alekseeva *et al.*, 2000) and the strong cauliflower mosaic virus 35S promoter (Smigocki and Owens, 1988; Faiss *et al.*, 1997; Lee *et al.*, 1999), generally encountered problems of regeneration of transgenic plants because of the inhibition of root formation by overproduced cytokinins. In some cases transgenic plants were regenerated, most likely from low expression cells due to position effects, but these plants had dramatic morphological changes. Inducible and tissue-specific promoters have been used to generate relatively normal-looking plants with higher potential for agricultural application. For example, heat-inducible promoters of heatshock genes have been used to direct *IPT*; upon heatshock treatment, cytokinins were overproduced in transgenic plants (Medford *et al.*, 1989; Schmulling *et al.*, 1989; Smart *et al.*, 1991; Smigocki, 1991; Ainley *et al.*, 1993; Synkova *et al.*, 1997). Light-inducible (Beinsberger *et al.*, 1991; Hamdi *et al.*, 1995; Thomas *et al.*, 1995; Garnier *et al.*, 1996; Carpin *et al.*, 1997; Synkova *et al.*, 1997), wound-inducible (Thomas *et al.*, 1995), copper-inducible (McKenzie *et al.*, 1998), dexamethasone-inducible (Faiss *et al.*, 1997; Kunkel *et al.*, 1999) and tissue-specific (Li *et al.*, 1992; Martineau *et al.*, 1994; Faiss *et al.*, 1997; Ma *et al.*, 1998; Guivarc'h *et al.*, 2002; Sa *et al.*, 2002) promoters have also been used to direct *IPT* expression in transgenic plants.

Compared with wild type, the levels of cytokinins in transgenic plants expressing *IPT* increased 3- to 500-fold (Gan and Amasino, 1996). These transgenic plants displayed various phenotypes including retarded degradation of photosynthetic apparatus (Synkova *et al.*, 1997), delayed senescence of leaves (Ooms *et al.*, 1991; Smart *et al.*, 1991; Hewelt *et al.*, 1994; Banowetz, 1997; Bultynck *et al.*, 1997; McKenzie *et al.*, 1998; Lee *et al.*, 1999), fruits (Martineau *et al.*, 1994), inflorescence (Nguyen *et al.*, 1998) and whole plants (Hammerschlag and Smigocki, 1998). One of the other phenotypes associated with these transgenics is the abnormal development, such as a stunted stature, shortened internodes, small leaves, less-developed root systems, reduced apical dominance and increased stem diameter (Schmulling *et al.*, 1999), which might have resulted from 'leaky' expression from the promoters used (Medford *et al.*, 1989). To avoid these abnormalities, systems with precisely controlled expression of *IPT* are needed.

13.4 Development of the SAG12-*IPT* autoregulatory cytokinin production system

Leaf senescence is a genetically controlled self-destruction process that is associated with changes in gene expression. At the onset of senescence, the majority of the genes that are expressed in non-senescent tissues are down-regulated. These genes include photosynthesis-related genes. In contrast, a subset of genes, generally referred to as senescence-associated genes (*SAGs*), are up-regulated (Gan, 2003). *SAG* expression is required for senescence because both transcription and translation inhibitors may inhibit the leaf sagging process (Noodén, 1988). For the past decade much effort has been made to isolate *SAGs*, and many *SAGs* have been cloned from various plant species including *Arabidopsis*, barley, *Brassica*, maize and soybean. The isolation of highly senescence-specific genes including *SAG12* not only provided a platform for dissecting regulatory mechanisms underlying leaf senescence but also made it possible to genetically manipulate the senescence process.

SAG12 was first isolated by differential screen of *Arabidopsis* leaf senescence cDNA libraries by Susheng Gan in 1991 (Gan, 1995). It encodes an apparent cysteine proteinase and its expression is highly senescence specific (Lohman *et al.*, 1994; Gan, 1995). The SAG12 promoter was fused to *IPT* to form an autoregulatory cytokinin production system (Gan and Amasino, 1995). As shown in Figure 13.1A, at the onset of leaf senescence, the senescence-specific promoter activates the expression of *IPT*, resulting in an increase in the cytokinin levels, which in turn prevents the leaf from senescing. The inhibition of leaf senescence will render the senescence-specific promoter inactive to prevent cytokinins from accumulating to very high levels; overproduction of cytokinins may interfere with other aspects of plant development. Because cytokinin production is targeted to senescing leaves, overproduction of cytokinins before senescence will be avoided. Leaf senescence in transgenic tobacco plants containing this autoregulatory cytokinin production system was efficiently retarded without any other developmental abnormalities (Figures 13.1B and C) (Gan and Amasino, 1995). The delayed senescence phenotype in tobacco plants has also been observed by other scientists (Gan and Amasino, 1995; Wingler *et al.*, 1998; Jordi *et al.*, 1999, 2000; Ludewig and Sonnewald, 2000; Soejima, 2000; Dertinger *et al.*, 2003; Cowan *et al.*, 2005). In addition to the significantly delayed leaf senescence phenotype, the transgenic tobacco plants lived longer: the whole plants of wild type were senescent 20 weeks after transplanting, while the transgenic plants were still producing flowers. The transgenic plants produced 80% more flowers than did wild-type plants (Gan and Amasino, 1995). The photosynthetic longevity in leaves of transgenics was significantly prolonged such that both the seed yield and biomass increased more than 40% (Gan and Amasino, 1995).

13.5 Use of the SAG12-*IPT* to manipulate senescence in crops

Perhaps because the autoregulatory senescence inhibition system described above does not cause any noticeable effect on other aspects of growth and development in transgenic tobacco plants, it has soon been applied to other plant species, including some important agronomic and horticultural crops. Among them are rice (Hildebrand *et al.*, 1998; Rubia *et al.*, 1999; Cao, 2001; Lin *et al.*, 2002), rape (Zhang *et al.*, 2005), sugarcane-grass (Weng *et al.*, 2002), tomato (Zhang *et al.*, 1999; Swartzberg *et al.*, 2006), cassava (Zhang and Gruissem, 2005), broccoli (Chen *et al.*,

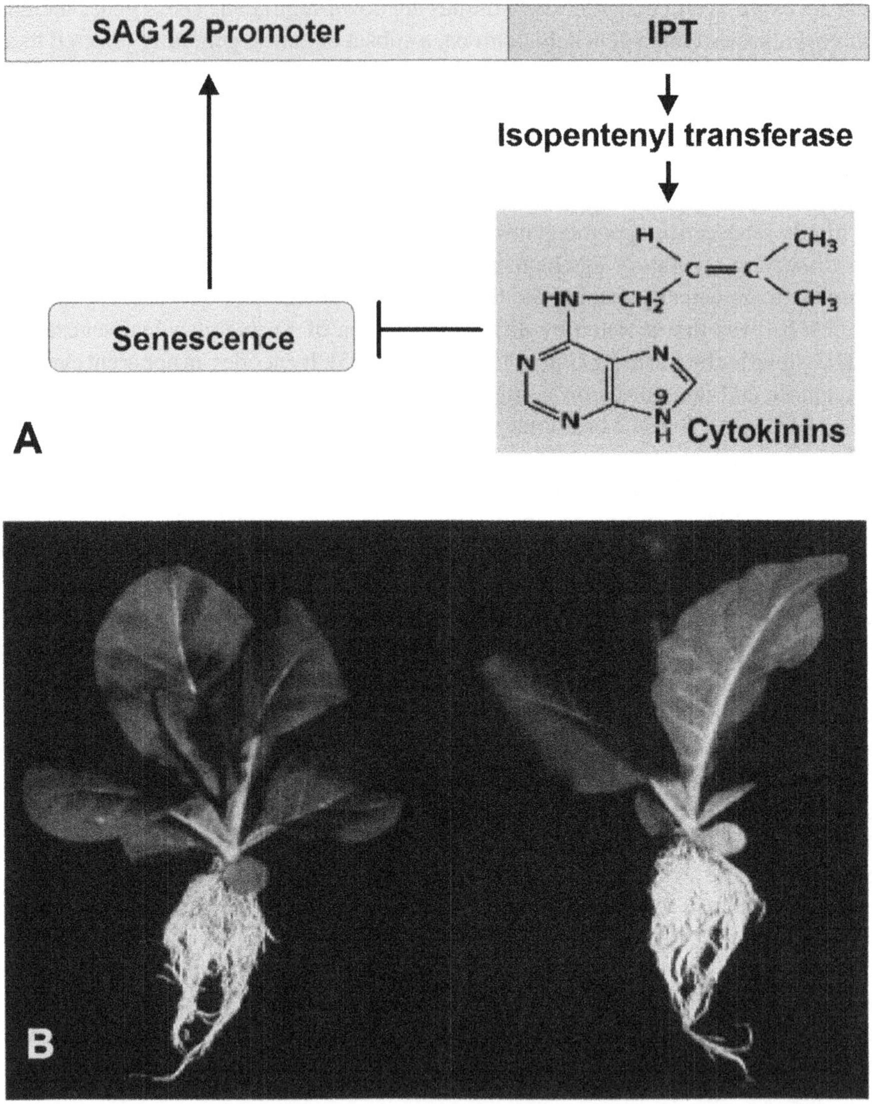

Figure 13.1 Inhibition of leaf senescence by autoregulated production of cytokinins. (A) The senescence-specific SAG12 promoter was fused to *IPT*, which encodes isopentenyl transferase that catalyzes the rate-limiting step in cytokinin biosynthesis. The onset of senescence activates this promoter to direct *IPT* expression, resulting in the biosynthesis of cytokinins. Cytokinins in turn inhibit senescence, thus attenuating the SAG12 promoter activity to prevent overproduction of cytokinins. (B) Normal development of the transgenic tobacco plant harboring the SAG12-*IPT* chimeric gene (left) compared with wild type at early stages. (C) Comparison of transgenic SAG12-*IPT* plants (left) with wild type. Note the significantly delayed leaf senescence in the transgenic plant (Gan and Amasino, 1995). (*Continued*)

Figure 13.1 *(Continued)*

2001; Gapper *et al.*, 2002), lettuce (McCabe *et al.*, 1998, 2001), cauliflower (Nguyen *et al.*, 1998), bok choy (Yuan *et al.*, 2002), petunia (Jandrew and Clark, 2001; Chang *et al.*, 2003; Clark *et al.*, 2004), ornamental tobacco *Nicotiana alata* (Schroeder and Stimart, 1998; Schroeder *et al.*, 2001), alfalfa (D. Johnson, personal communication) and *Arabidopsis* (Gan, 1995; Zhang *et al.*, 2000; Huynh *et al.*, 2005) (Figure 13.2). Transgenic maize (Robson *et al.*, 2004; Young *et al.*, 2004) and ryegrass (Li *et al.*, 2004) have also been generated; in these experiments, the promoter of a maize senescence-enhanced cysteine-protease gene, *SEE1*, in stead of the *Arabidopsis* SAG12 promoter, was used.

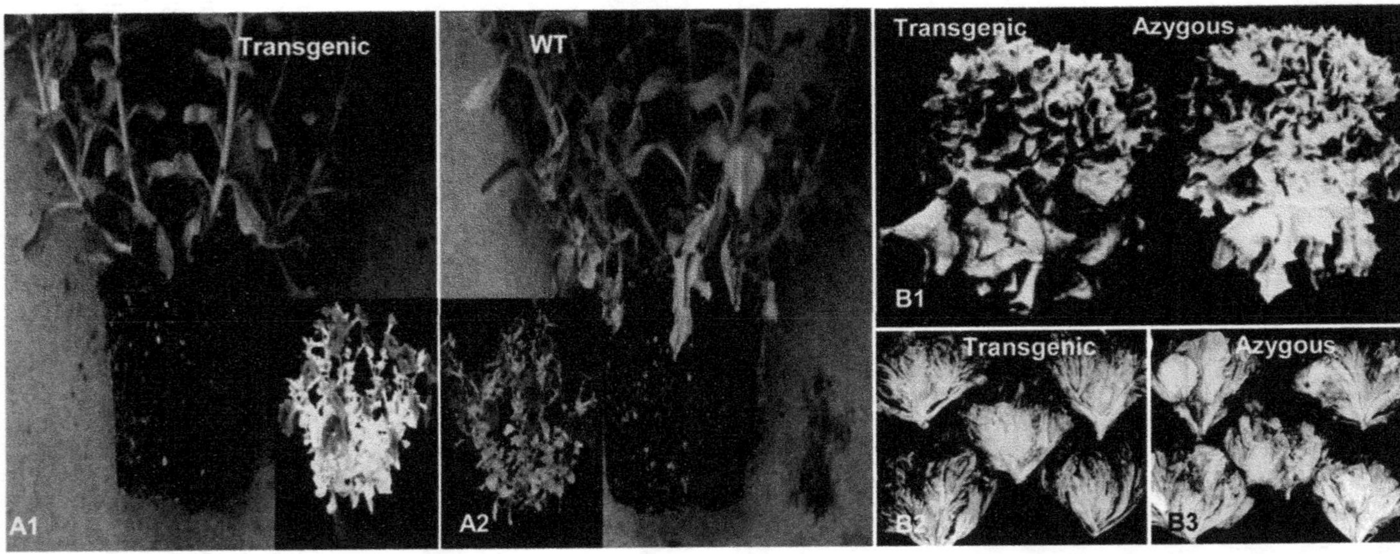

Figure 13.2 Examples of SAG12-*IPT* transgenic plants with delayed leaf senescence phenotypes. (A1, A2) Petunia (Clark *et al.*, 2004). (B) Lettuce plants (B1) and harvested lettuce (B2, B3) (McCabe *et al.*, 2001). (C) Cassava plants (C1) and detached leaves (C2) (Zhang and Gruissem, 2005). (D) Broccoli plants (D1), detached leaves (D2) and harvested heads (D3) (Chen *et al.*, 2001). With kind permission of Springer Science and Business Media. (E) Bok choy (Yuan *et al.*, 2002). (F) *Arabidopsis* (Gan, 2005). WT = wild type. *(Continued)*

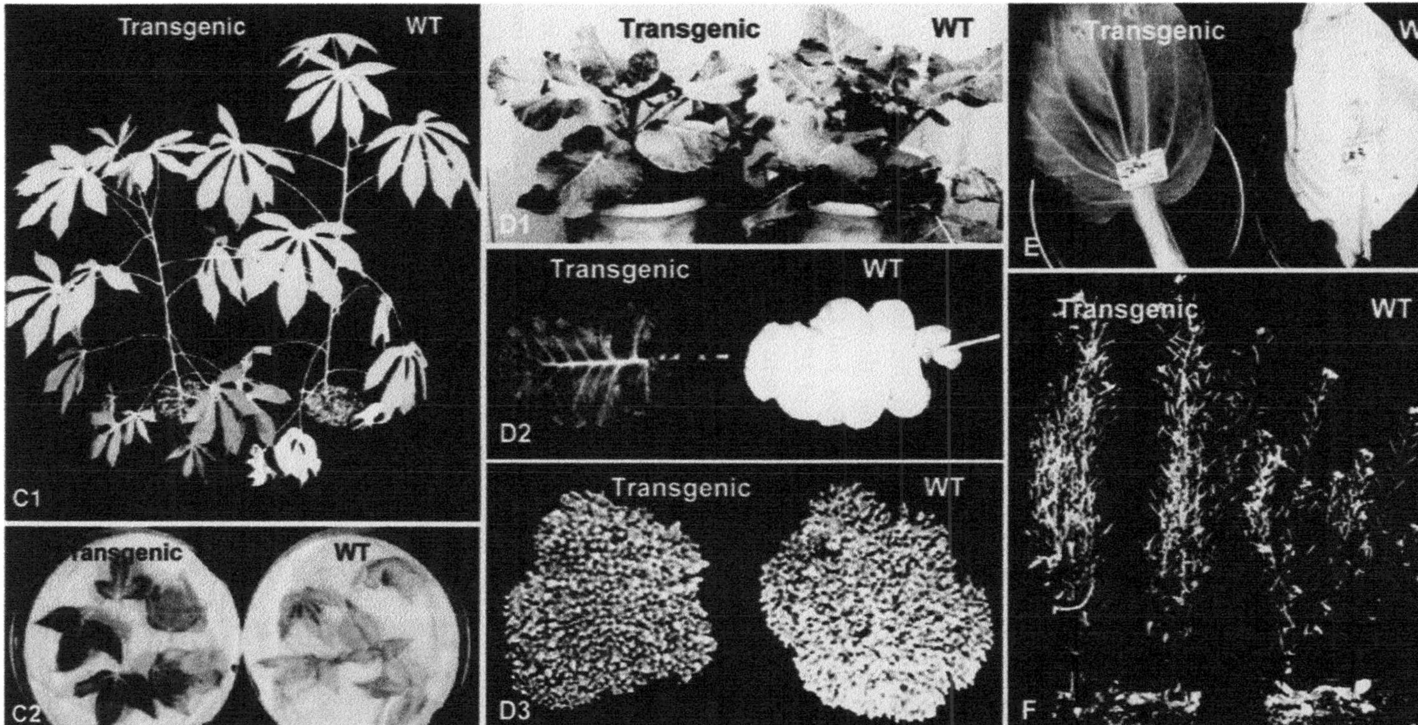

Figure 13.2 *(Continued)*

13.5.1 IPT *expression and cytokinin production in transgenic plants*

Because of the autoregulatory nature of the chimeric gene, and because very low amount of cytokinins can readily inhibit senescence, expression of the foreign *IPT* in plants is of very low abundance and has been difficult to be detected (Jandrew and Clark, 2001; Chang *et al.*, 2003). Medford *et al.* (1989) used 20 μg of poly A^+ RNA per lane and extensive exposure of autoradiograms, but still could not detect any *IPT* transcript although the level of ribosyl zeatin 5′-monophosphate (a species of cytokinins) increases eightfold due to leaky *IPT* expression from a heatshock promoter. The very low level of *IPT* expression in transgenic plants harboring this autoregulatory senescence inhibition system was demonstrated in experiments involving transgenic tobacco plants double hemizygous for SAG12-*IPT* and SAG12-*GUS*. In plants hemizygous for SAG12-*GUS*, the GUS activity increased exponentially with the progression of leaf senescence. In contrast, the GUS activity in the double hemizygotes was extremely low. Because *IPT* was subject to the same SAG12 promoter as *GUS* was, the expression of *IPT* should be at an extremely low level (Gan and Amasino, 1995). Although the *IPT* transcript was beyond the detection by Northern analysis, it was detectable in 'would-be senescing' tissues by the sensitive reverse transcription (RT) PCR technology (Jandrew and Clark, 2001; Chang *et al.*, 2003).

Consistent with the very low levels of *IPT* expression, endogenous cytokinin levels in SAG12-*IPT* transgenic plants were elevated moderately; in some cases the increases were not detectable (Table 13.1), which is much less dramatic than the changes in cytokinin levels of transgenic plants with *IPT* driven by some other promoters (Gan and Amasino, 1996). Unlike wild-type plants, where cytokinin levels dropped with the progression of leaf senescence, the SAG12-*IPT* plants showed an increase in cytokinin levels in old leaves with retarded senescence (Jordi *et al.*, 2000; Yuan *et al.*, 2002; Chang *et al.*, 2003; Zhang *et al.*, 2005). The increase of total cytokinin levels in the SAG12-*IPT* plants, as in other transgenic plants involving the agrobacterial *IPT*, were mainly due to the increase of Z-type cytokinins including zeatin riboside and zeatin (Z) (Jordi *et al.*, 2000; Chang *et al.*, 2003; Huynh *et al.*, 2005). The IPA-type cytokinins, on the other hand, did not change significantly in the SAG12-*IPT* transgenic plants (Jordi *et al.*,

Table 13.1 Levels of cytokinins in aged tissues of SAG12-*IPT* transgenic plants

Plant species	Fold of total cytokinins increase in transgenic plants	Tissue type	Reference
Tobacco	6–15 (free ZR equivalents)	Oldest Leaves	Jordi et al. (2000)
Petunia	27.6	Corollas	Chang et al. (2003)
Rice	1	Leaves	Lin et al. (2002)
Rape	3–5	Old leaves	Zhang et al. (2005)
Bok choy	9	Old Leaves	Yuan et al. (2002)
Lettuce	1	Bottom Leaves	McCabe et al. (2001)
Arabidopsis	2.5–3	Leaves	Huynh et al. (2005)

2000; Chang *et al.*, 2003; Huynh *et al.*, 2005). The IPA-type cytokinins are believed to be rapidly converted into the Z-type cytokinins (Smart *et al.*, 1991; Zhang *et al.*, 1995). It should be noted that several studies observed an increase in cytokinin levels in young tissues of the SAG12-*IPT* lettuce and bok choy plants (McCabe *et al.*, 2001; Yuan *et al.*, 2002). Whether this resulted from leaky expression of the *Arabidopsis* SAG12 promoter or transgene position effects is unknown.

*13.5.2 Delayed leaf senescence in the SAG-*IPT *plants*

Significant delay of leaf senescence without any developmental or morphological abnormality is the most striking phenotype of transgenic plants harboring the SAG12-*IPT* chimeric gene. Transgenic plants had two- to eightfold fewer senescent leaves in comparison with the age-matched wild type (McCabe *et al.*, 2001; Schroeder *et al.*, 2001; Yuan *et al.*, 2002; Zhang and Gruissem, 2005). Chlorophyll content of old leaves from transgenic plants was much higher than that of the leaves from the control plants (Table 13.2, Figure 13.2). The ratio of chlorophyll *a* to chlorophyll *b* was higher in old leaves of transgenic ornamental tobacco (*N. alata*) plants than in the control leaves of wild-type plants (Schroeder *et al.*, 2001).

In addition to the staygreen phenotype, old leaves of the SAG12-*IPT* plants contained high concentration of total soluble proteins (Wingler *et al.*, 1998; Jordi *et al.*, 2000; McCabe *et al.*, 2001; Dertinger *et al.*, 2003; Garratt *et al.*, 2005; Zhang and Gruissem, 2005), high levels of photosynthetic enzymes such as ribulose-1, 5-bisphosphate carboxylase/oxygenase (Rubisco) (Jordi *et al.*, 2000; McCabe *et al.*, 2001; Zhang and Gruissem, 2005) and hydroxypyruvate reductase (Wingler *et al.*, 1998). Most importantly, leaves with retarded senescence of the SAG12-*IPT* transgenics remained photosynthetically active (Gan and Amasino, 1995; Wingler

Table 13.2 Chlorophyll retention in some SAG12-*IPT* transgenic plants

	Chlorophyll content of the oldest leaf			
Plant species	Wild type	SAG12-*IPT*	Unit	Reference
Tobacco	70	177	mg m^{-2}	Wingler et al. (1998)
Tobacco	0.2	4.0	SPAD-502 reading	Jordi et al. (2000)
N. alata	11	22	μg cm^{-2}	Schroeder et al. (2001)
Broccoli	0.91	3.39	mg/g FW[a]	Chen et al. (2001)
Rice	37.659	43.622	SPAD-502 reading[b]	Lin et al. (2002)
Rice	0.456	0.644	%[b]	Cao (2001)
Rape	0.8	1.6	mg/g FW	Zhang et al. (2005)
Tobacco	120	260	mg m^{-2}	Ludewig and Sonnewald (2000)
Lettuce	0	20	mg m^{-2}	McCabe et al. (2001)
Bok choy	30	150	μg/g	Yuan et al. (2002)
Arabidopsis	678±38	745±24	μg g^{-1} FW[c]	Huynh et al. (2005)
Cassava	0.7	0	mg/g	Zhang and Gruissem (2005)

a: leaves after 4 days of detachment; b: flag leaf; c: leaves under waterlogging stress for 3 days.

et al., 1998; Jordi *et al.*, 2000; Cao, 2001). These leaves produced more carbohydrates (Gan and Amasino, 1995; Wingler *et al.*, 1998; Jordi *et al.*, 2000; Cao, 2001) and accumulated more nitrogen than did the controls (Jordi *et al.*, 2000).

In contrast to the above-mentioned reports, it has been observed that chlorophyll contents of young leaves in the SAG12-*IPT* transgenic plants were slightly lower than those in the counterpart leaves of the wild-type plants (Jordi *et al.*, 2000; McCabe *et al.*, 2001). This likely resulted from limited nutrient supplies. Under nutrient stress, old leaves of wild type will undergo senescence and the nutrients released from the old leaves will be remobilized to actively growing young leaves. In contrast, senescence in old leaves of transgenic plants is retarded and less nutrients (especially N) will be recycled to the young leaves.

*13.5.3 Delayed floral senescence in the SAG12-*IPT *plants*

SAG12 is a highly senescence-specific gene. It is expressed not only in senescing leaves but also in senescing stems, floral sepals and petals (Gan, 1995). Therefore, the SAG12-*IPT* autoregulatory senescence inhibition system operates in flowers. Not only has the flowering period been prolonged in the SAG12-*IPT* transgenic tobacco plants so that more flowers were produced as discussed above, but also the longevity (from petal opening to senescence) of individual flowers was extended by more than 50% from the 3 days in wild type to at least 4.5 days in the transgenic plants (Gan and Amasino, 1996). The delay in floral senescence was also apparent by the fact that there were more nonsenescent flowers on the floral stalks in transgenic plants than on the corresponding stalks of wild type; in these plants, no difference was observed in the rate of flower production (Gan, 1995). Transgenic *Arabidopsis* plants containing the SAG12-*IPT* transgene started to flower at the same time as wild-type plants did, but the period from flowering to senescence was increased by 7–12 days (Zhang *et al.*, 2000). In another study, the floral longevity of the ornamental tobacco plants (*N. alata*) containing SAG12-*IPT* increased 29% compared with wild type (Schroeder *et al.*, 2001). In SAG12-*IPT* transgenic petunia, the *IPT* transcript abundance and levels of cytokinins concurrently increased in corolla after pollination, and floral senescence was delayed for 6–10 days compared with wild type (Chang *et al.*, 2003).

*13.5.4 Delayed postharvest senescence in the SAG12-*IPT *plants*

Postharvest senescence contributes much to the loss of vegetable crops (especially those perishable produce) during transportation, storage and on shelf. The SAG12-*IPT* system has been used to manipulate senescence in such vegetable crops as lettuce (McCabe *et al.*, 1998, 2001; Garratt *et al.*, 2005), broccoli (Chen *et al.*, 2001; Gapper *et al.*, 2002), cauliflower (Nguyen *et al.*, 1998), bok choy (Yuan *et al.*, 2002) and tomato (Zhang *et al.*, 1999). After 4 days of postharvest storage at 25°C, 31% of the SAG12-*IPT* broccoli plants showed more than 50% chlorophyll retention in detached leaves, 16% of them had the same effect on florets and 7.2% of the transformants showed significantly delayed senescence on both leaves and florets (Figures 13.2D2 and D3) (Chen *et al.*, 2001). The retardation of senescence was correlated with the abundance of *IPT* transcript (Chen *et al.*, 2001). Harvested SAG12-*IPT* lettuce heads stored for 7 days retained their chlorophyll, while the basal leaves of control plants became yellow (Figures 13.2B2 and B3) (McCabe *et al.*, 2001; Garratt *et al.*, 2005). Leaves of wild-type bok choy became yellow 3 days after detachment. In contrast, there were no visible sign of yellowing in leaves of the SAG12-*IPT* bok choy plants until 10 days after detachment (Figure 13.2E) (Yuan *et al.*, 2002).

*13.5.5 Increased yield and biomass production in the SAG12-*IPT *plants*

Because leaves with delayed senescence were photosynthetically active (Gan and Amasino, 1995; Wingler *et al.*, 1998; Jordi *et al.*, 2000; Lin *et al.*, 2002) and because the flowering period was extended (Gan and Amasino, 1996; Zhang *et al.*, 2000), transgenic tobacco plants containing SAG12-*IPT* were shown to set more seeds and accumulate more biomass (dry weight) than did wild-type plants (Gan and Amasino, 1995). In the SAG12-*IPT* transgenic rice, leaf senescence was also significantly delayed (Hildebrand *et al.*, 1998). The transgenic flag leaf, the uppermost leaves on the stem that contributes most to rice seed filling, retained chlorophyll and photosynthetic capacity longer than did nontransgenic ones (Cao, 2001; Lin *et al.*, 2002). The SAG12-*IPT* rice plants showed improved agronomic traits including increased total grains per hill (Cao, 2001; Lin *et al.*, 2002), seed setting rate (Cao, 2001; Lin *et al.*, 2002) and 1000-grain weight (Cao, 2001). Seed yield of transgenic rice increased up to 70% in some rice cultivars transformed with SAG12-*IPT* (Lin *et al.*, 2002).

*13.5.6 Increased stress tolerance in the SAG12-*IPT *plants*

Senescence can be induced by many types of environmental stresses and biological insults. Many stress-induced genes are expressed even during natural senescence. The overlapping between senescence and stress responses led to the hypothesis that the SAG12-*IPT* system may delay stress-induced senescence and may increase stress tolerance in the transgenic plants. One general consequence of many stresses is the increased accumulation of reactive oxygen species (ROS); ROS can induce or accelerate plant senescence (Polle and Rennenberg, 1993) (see also Chapters 4 and 6). The extent of the degenerative effects of ROS depends on the effectiveness of the antioxidative system of the plant cells, including low-molecular-weight antioxidants and enzymatic defense (Mittler, 2002). The SAG12-*IPT* transgenic tobacco plants were resistant to drought stress (S. Gan and R.M. Amasino, unpublished data). Dertinger *et al.* (2003) studied the antioxidative system in SAG12-*IPT* tobacco plants. Old leaves with retarded senescence had a higher concentration of antioxidants ascorbate and glutathione, as well as higher activities of antioxidative enzymes including ascorbate peroxidase, glutathione reductase and superoxide dismutase (SOD) than did wild-type counterpart (Dertinger *et al.*, 2003). Senescing leaves of SAG12-*IPT* rape plants also showed higher SOD activity and lower lipid oxidation products malondialdehyde compared with wild type (Zhang *et al.*, 2005). Transgenic *Arabidopsis* plants containing SAG12-*IPT* were tested for flooding tolerance; these plants were consistently more tolerant to water logging and complete submergence than were wild type (Zhang *et al.*, 2000; Huynh *et al.*, 2005). Transgenic cassava plants harboring the SAG12-*IPT* transgene showed higher tolerance to drought-induced leaf senescence (Figure 13.2C) (Zhang and Gruissem, 2005).

13.6 Other strategies for manipulation of leaf senescence

As discussed in previous chapters, significant progress has been made in deciphering the physiological, cellular, biochemical and molecular mechanisms underlying leaf senescence, which makes it possible to design strategies for manipulating leaf senescence. Some genes that play important roles in controlling leaf senescence have been identified, and altering expression of these genes, individually or in combination, shall lead to new strategies to inhibit or promote leaf senescence.

For examples, *Arabidopsis* null mutant of *ORE9*, an F-box gene, displayed prolonged leaf longevity (Woo *et al.*, 2001); *SAG101* is an *Arabidopsis* gene encoding an acyl hydrolase and leaf senescence was delayed for 4–5 days in the *SAG101* antisense plants (He and Gan, 2002). More recently, an NAC family transcription factor gene called *NAP* was shown to play a critical role in leaf senescence. Leaf senescence in *Arabidopsis nap* null mutants was delayed for up to 10 days (Guo and Gan, 2006). The *NAP* homologs have been found in variety of plant species including rice, maize, wheat, soybean, kidney bean, peach, tomato, petunia, potato and poplar, and, like *Arabidopsis NAP*, the homologs of rice (a monocot) and kidney bean (a dicot) have been shown to be expressed in senescing leaves but not in nonsenescing ones, and more importantly, the homologs function as *Arabidopsis NAP* because they complemented the *Arabidopsis nap* null mutants (Guo and Gan, 2006). It is likely that knocking *NAP* out in various crops will cause a significant delay of leaf senescence, and thus has a great potential for agricultural application.

Another example of altering expression of a single gene to delay leaf senescence was demonstrated in transgenic *Arabidopsis* plants in which expression of *deoxyhypusine synthase* (*DHS*) was suppressed using an antisense technique. Leaf senescence in *DHS* antisense plants was delayed significantly. Drought-induced senescence was also delayed in the antisense *Arabidopsis* plants (Wang *et al.*, 2003). DHS catalyzes the first step of hypusinating eukaryotic initiation factor 5A (eIF-5A) protein, which activates the translation initiation factor (Thompson *et al.*, 2004). Hypusine-modified eIF-5A is thought to act as a nucleocytoplasmic transporter, translocating newly synthesized mRNAs from the nucleus to the cytoplasm for subsequent translation (Thompson *et al.*, 2004). Knocking-down of *DHS* therefore indirectly reduced translation of *SAG*s, and thus slowed down the senescence process.

On the other hand, inducible overexpression of either *SAG101* (He and Gan, 2002) or *NAP* (Guo and Gan, 2006) could cause precocious leaf senescence. This may be a new strategy for promoting leaf senescence. For example, chemical-inducible expression of *NAP* in cotton may promote defoliation to facilitate mechanical harvest of cotton bolls.

13.7 Closing remarks

Leaf senescence is an important trait in agriculture. The ultimate goal of studying leaf senescence is, based on the regulatory mechanisms of senescence, to design ways to manipulate this process for agricultural improvement. The current molecular genetic approaches that have been used in delaying senescence are based on plant hormone biology: either blocking ethylene production or enhancing cytokinin production. The former strategy is primarily used for prevention of fruit senescence. The latter strategy encountered some difficulty associated with developmental abnormality. But the cloning and use of the highly senescence-specific SAG12 promoter from *Arabidopsis* to direct *IPT* in transgenic plants has made it possible to use the cytokinin-based strategy to manipulate senescence. The SAG12-*IPT* is an autoregulatory system. The *IPT* expression is targeted spatially (in senescing cells), temporally (at senescing stage) and quantitatively (no overproduction of cytokinins). The system has been demonstrated to work effectively in delaying leaf senescence in various plant species including some important agronomic and horticultural crops. Recently several senescence-specific genes have been identified from other plant species, and it is likely that the promoters of the new genes will be used to fuse with *IPT* to form the autoregulatory senescence inhibition systems similar to SAG12-*IPT*. This system is approaching commercialization.

As discussed in previous chapters, significant progress has been made in unraveling molecular regulatory mechanism underlying leaf senescence, and some important regulators of leaf senescence have been identified. We expect that new strategies involving altering these key factors, individually or in combination, will soon be developed and used to manipulate leaf senescence.

Acknowledgment

The SAG12-*IPT* system was developed in Dr. Richard Amasino's Lab. Our research on leaf senescence has been supported by grants to Susheng Gan from the National Science Foundation, US Department of Energy's Basic Energy Bioscience Program, US Department of Agriculture National Research Initiative, US–Israel Binational Agriculture Research and Development and by a new faculty startup fund from Cornell University.

References

Aida, R., Yoshida, T., Ichimura, K., Goto, R. and Shibata, M. (1998) Extension of flower longevity in transgenic torenia plants incorporating ACC oxidase transgene. *Plant Sci* **138**, 91–101.

Ainley, W.M., McNeil, K.J., Hill, J.W., *et al.* (1993) Regulatable endogenous production of cytokinins up to toxic levels in transgenic plants and plant tissues. *Plant Mol Biol* **22**, 13–23.

Akiyoshi, D.E., Klee, H., Amasino, R.M., Nester, E.W. and Gordon, M.P. (1984) T-DNA of *Agrobacterium tumefaciens* encodes an enzyme of cytokinin biosynthesis. *Proc Natl Acad Sci USA* **81**, 5994–5998.

Alekseeva, V.V., Rukavtsova, E.B., Shutova, T.V., Khorobrykh, A.A. and Bur'yanov, Y.I. (2000) Physiological and biochemical traits of tobacco plants carrying an agrobacterial isopentenyltransferase gene. *Russ J Plant Physiol* **47**, 360–366.

Banowetz, G.M. (1997) Cultivars of hexaploid wheat of contrasting stature and chlorophyll retention differ in cytokinin content and responsiveness. *Ann Bot* **79**, 185–190.

Barry, G.F., Rogers, S.G., Fraley, R.T. and Brand, L. (1984) Identification of a cloned cytokinin biosynthetic gene. *Proc Natl Acad Sci USA* **81**, 4776–4780.

Beinsberger, S.E.I., Valcke, R.L.M., Deblaere, R.Y., Clijsters, H.M.M., Degreef, J.A. and Vanonckelen, H.A. (1991) Effects of the introduction of *Agrobacterium Tumefaciens* t-DNA ipt gene in *Nicotiana tabacum*-L. cv Petit Havana Sr1 plant cells. *Plant Cell Physiol* **32**, 489–496.

Benbella, M. and Paulsen, G. (1998) Efficacy of treatments for delaying senescence of wheat leaves. I: senescence under controlled conditions. *Agron J* **90**, 329–332.

Blankenship, S.M. and Dole, J.M. (2003) 1-methylcyclopropene: a review. *Postharvest Biol Technol* **28**, 1–25.

Brault, M. and Maldiney, R. (1999) Mechanisms of cytokinin action. *Plant Physiol Biochem* **37**, 403–412.

Bultynck, L., Geuns, J.M.C., VanGinkel, G. and Caubergs, R.J. (1997) Properties of plasma membranes of Phsp 70-ipt transformed to bacco. *(Nicotiana tabacum) Phytochemistry* **45**, 1337–1341.

Cao, M.L. (2001) Performance of autoregulatory senescence-inhibition gene in rice. *Human Agric Sci Technol Newsletter* **2**, 17–24.

Carpin, S., Garnier, F., Andreu, F., Chenieux, J.C., Rideau, M. and Hamdi, S. (1997) Changes of both polypeptide pattern and sensitivity to cytokinin following transformation of periwinkle tissues with the isopentenyl transferase gene. *Plant Physiol Biochem* **35**, 603–609.

Chang, H., Jones, M.L., Banowetz, G.M. and Clark, D.G. (2003) Overproduction of cytokinins in petunia flowers transformed with PSAG12-IPT delays corolla senescence and decreases sensitivity to ethylene. *Plant Physiol* **132**, 2174–2183.

Chen, L.F.O., Hwang, J.Y., Charng, Y.Y., Sun, C.W. and Yang, S.F. (2001) Transformation of broccoli (*Brassica oleracea* var. *italica*) with isopentenyltransferase gene via *Agrobacterium tumefaciens* for post-harvest yellowing retardation. *Mol Breed* **7**, 243–257.

Clark, D.G., Dervinis, C., Barret, J.E., Klee, H. and Jones, M. (2004) Drought-induced leaf senescence and horticultural performance of transgenic P-SAG12-IPT petunias. *J Am Soc Hortic Sci* **129**, 93–99.

Clarke, S.F., Jameson, P.E. and Downs, C. (1994) The influence of 6-benzylaminopurine on postharvest senescence of floral tissues of broccoli (*Brassica oleracea* var. *italica*). *Plant Growth Regul* **14**, 21–27.

Cowan, A.K., Freeman, M., Björkman, P.-O., Nicander, B., Sitbon, F. and Tillberg, E. (2005) Effects of senescence-induced alteration in cytokinin metabolism on source-sink relationships and ontogenic and stress-induced transitions in tobacco. *Planta* **221**, 801–814.

Dertinger, U., Schaz, U. and Schulze, E.D. (2003) Age-dependence of the antioxidative system in tobacco with enhanced glutathione reductase activity or senescence-induced production of cytokinins. *Physiol Plant* **119**, 19–29.

Faiss, M., Zalubilova, J., Strnad, M. and Schmulling, T. (1997) Conditional transgenic expression of the ipt gene indicates a function for cytokinins in paracrine signaling in whole tobacco plants. *Plant J* **12**, 401–415.

Gan, S. (1995) *Molecular characterization and genetic manipulation of plant senescence*. PhD Dissertation . University of Wisconsin-Madison, Madison, WI.

Gan, S. (2003) Mitotic and postmitotic senescence in plants. *Sci SAGE KE* **38**, RE7.

Gan, S. (2005) Hormonal regulation of senescence. In: *Plant Hormones: Biosynthesis, Signal Transduction, Action* (ed. Davies, P.J.). Kluwer Academic, The Netherlands, pp. 561–581.

Gan, S. and Amasino, R.M. (1995) Inhibition of leaf senescence by autoregulated production of cytokinin. *Science* **270**, 1986–1988.

Gan, S. and Amasino, R.M. (1996) Cytokinins in plant senescence: from spray and pray to clone and play. *Bioessays* **18**, 557–565.

Gapper, N.E., McKenzie, M.J., Christey, M.C., *et al.* (2002) *Agrobacterium tumefaciens*-mediated transformation to alter ethylene and cytokinin biosynthesis in broccoli. *Plant Cell Tissue Organ Cult* **70**, 41–50.

Garnier, F., Carpin, S., Label, P., Creche, J., Rideau, M. and Hamdi, S. (1996) Effect of cytokinin on alkaloid accumulation in periwinkle callus cultures transformed with a light-inducible ipt gene. *Plant Sci* **120**, 47–55.

Garratt, L.C., Linforth, R., Taylor, A.J., Lowe, K.C., Power, J.B. and Davey, M.R. (2005) Metabolite fingerprinting in transgenic lettuce. *Plant Biotechnol J* **3**, 165–174.

Guivarc'h, A., Rembur, J., Goetz, M., *et al.* (2002) Local expression of the *ipt* gene in transgenic tobacco (*Nicotiana tabacum* L. cv. SR1) axillary buds establishes a role for cytokinins in tuberization and sink formation. *J Exp Bot* **53**, 621–629.

Guo, Y. and Gan, S. (2006) AtNAP, a NAC family transcription factor, has an important role in leaf senescence. *Plant J* **46**, 601–612.

Hamdi, S., Creche, J., Garnier, F., *et al.* (1995) Cytokinin involvement in the control of coumarin accumulation in *Nicotiana tabacum* – Investigations with normal and transformed tissues carrying the isopentyl transferase gene. *Plant Physiol Biochem* **33**, 283–288.

Hammerschlag, F.A. and Smigocki, A.C. (1998) Growth and in vitro propagation of peach plants transformed with the shooty mutant strain of *Agrobacterium tumefaciens*. *HortScience* **33**, 897–899.

He, Y. and Gan, S. (2002) A gene encoding an acyl hydrolase is involved in leaf senescence in *Arabidopsis*. *Plant Cell* **14**, 805–815.

He, Y. and Gan, S. (2003) Molecular characteristics of leaf senescence. In: *Recent Research Developments in Plant Molecular Biology*, Vol. 1. Research Signpost, Kerala, India, pp. 1–17.

He, Y., Tang, W., Swain, J.D., Green, A.L., Jack, T.P. and Gan, S. (2001) Networking senescence-regulating pathways by using *Arabidopsis* enhancer trap lines. *Plant Physiol* **126**, 707–716.

Hewelt, A., Prinsen, E., Schell, J., Vanonckelen, H. and Schmulling, T. (1994) Promoter tagging with a promoterless *ipt* gene leads to cytokinin-induced phenotypic variability in transgenic tobacco plants: implications of gene dosage effects. *Plant J* **6**, 879–891.

Hildebrand, D.F., Fukushige, H., Afitlhile, M. and Wang, C. (1998) Lipoxygenases in plant development and senescence. In: *Eicosanoids and Related Compounds in Plants and Animals* (eds Rowley, A.F., Kuhn, H. and Schewe, T.). Portland Press, London, Chap. 8, pp. 151–181.

Huynh, L.N., VanToai, T., Streeter, J. and Banowetz, G. (2005) Regulation of flooding tolerance of SAG12:IPT *Arabidopsis* plants by cytokinin. *J Exp Bot* **56**, 1397–1407.

Jandrew, J. and Clark, D.G. (2001) Selectively induced nutrient deficiency in transgenic PSAG12-IPT, PSAG13-IPT and PSAG12-Kn1 petunias. *HortScience* **36**, 518–519.

John, I., Drake, R., Farrell, A., *et al.* (1995) Delayed leaf senescence in ethylene-deficient Acc-oxidase antisense tomato plants – molecular and physiological analysis. *Plant J* **7**, 483–490.

Jordi, W., Schapendonk, A., Davelaar, E., *et al.* (2000) Increased cytokinin levels in transgenic P_{SAG12}-*IPT* tobacco plants have large direct and indirect effects on leaf senescence, photosynthesis and N partitioning. *Plant Cell Environ* **23**, 279–289.

Jordi, W., Schapendonk, A.H.C.M., Davelaar, E., *et al.* (1999) Leaf senescence of transgenic PSAG12-IPT tobacco, consequences for N remobilisation. *Biol Plantarum* **42**, S84.

Kakimoto, T. (2003) Biosynthesis of cytokinins. *J Plant Res* **116**, 233–239.

Kunkel, T., Niu, Q.W., Chan, Y.S. and Chua, N.H. (1999) Inducible isopentenyl transferase as a high-efficiency marker for plant transformation. *Nat Biotechnol* **17**, 916–919.

Lee, C.-K., Klopp, R.G., Weindruch, R. and Prolla, T.A. (1999) Gene expression profile of aging and its retardation by caloric restriction. *Science* **285**, 1390–1393.

Li, Q., Bettany, A.J.E., Donnison, I., Griffiths, C.M., Thomas, H. and Scott, I.M. (2000) Characterisation of a cysteine protease cDNA from Lolium multiflorum leaves and its expression during senescence and cytokinin treatment. *Biochim Biophys Acta-Gene Struct Expression* **1492**, 233–236.

Li, Q., Robson, P.R.H., Bettany, A.J.E., Donnison, I.S., Thomas, H. and Scott, I.M. (2004) Modification of senescence in ryegrass transformed with IPT under the control of a monocot senescence-enhanced promoter. *Plant Cell Reports* **22**, 816–821.

Li, Y., Hagen, G. and Guilfoyle, T.J. (1992) Altered morphology in transgenic tobacco plants that overproduce cytokinins in specific tissues and organs. *Dev Biol* **153**, 386–395.

Lin, Y.J., Cao, M.L., Xu, C.G., Chen, H., Wei, J. and Zhang, Q.F. (2002) Cultivating rice with delayed leaf senescence by P-SAG12-IPT gene transformation. *Acta Botanica Sinica* **44**, 1333–1338.

Lohman, K.N., Gan, S.S., John, M.C. and Amasino, R.M. (1994) Molecular analysis of natural leaf senescence in *Arabidopsis thaliana*. *Physiol Plant* **92**, 322–328.

Ludewig, F. and Sonnewald, U. (2000) High CO_2-mediated down-regulation of photosynthetic gene transcripts is caused by accelerated leaf senescence rather than sugar accumulation. *FEBS Lett* **479**, 19–24.

Ma, Q.H., Zhang, R., Hocart, C.H., Letham, D.S. and Higgins, T.J.V. (1998) Seed-specific expression of the isopentenyl transferase gene (*ipt*) in transgenic tobacco. *Aust J Plant Physiol* **25**, 53–59.

Martineau, B., Houck, C.M., Sheehy, R.E. and Hiatt, W.R. (1994) Fruit-specific expression of the *A. tumefaciens* isopentenyl transferase gene in tomato – effects on fruit ripening and defense-related gene expression in leaves. *Plant J* **5**, 11–19.

McCabe, M., Mohapatra, U., Power, J.B., Davey, M.R., Schepers, F. and Van-Dun, K. (1998) Delayed senescence in transgenic lettuce using an autoregulated ipt gene. *J Exp Bot* **49**, 49.

McCabe, M.S., Garratt, L.C., Schepers, F., *et al.* (2001) Effects of P-SAG12-IPT gene expression on development and senescence in transgenic lettuce. *Plant Physiol* **127**, 505–516.

McKenzie, M.J., Mett, V., Reynolds, P.H.S. and Jameson, P.E. (1998) Controlled cytokinin production in transgenic tobacco using a copper-inducible promoter. *Plant Physiol* **116**, 969–977.

Medford, J.I., Horgan, R., El-Sawi, Z. and Klee, H.J. (1989) Alterations of endogenous cytokinins in transgenic plants using a chimeric isopentenyl transferase gene. *Plant Cell* **1**, 403–413.

Mittler, R. (2002) Oxidative stress, antioxidants and stress tolerance. *Trends Plant Sci* **7**, 405–410.

Nguyen, K., Kane, E. and Dix, P. (1998) Hormonal regulation of senescence in cauliflower (*Brassica oleracea* var. *botrytis*) (abstract no. 96). In: *Plant Biotechnology and In Vitro Biology in the 21st Century* (eds Altman, A., Ziv, M. and Izhar, S.). IX International Congress Plant Tissue Culture. Kluwer Academic Publishers, Dordrecht, The Netherlands, p 164.

Noodén, L.D. (1988) The phenomenon of senescence and aging. In: *Senescence and Aging in Plants* (eds Noodén, L.N. and Leopold, A.C.). Academic Press, San Diego, CA, pp 1–50.

Oeller, P.W., Wong, L.M., Taylor, L.P., Pike, D.A. and Theologis, A. (1991) Reversible inhibition of tomato fruit senescence by antisense RNA. *Science* **254**, 437–439.

Ooms, G., Risiott, R., Kendall, A., *et al.* (1991) Phenotypic changes in t-cyt-transformed potato plants are consistent with enhanced sensitivity of specific cell-types to normal regulation by root-derived cytokinin. *Plant Mol Biol* **17**, 727–743.

Polle, A. and Rennenberg, H. (1993) Significance of antioxidants in plant adaptation to environmental stress. In: *Plant Adaptation to Environmental Stress* (eds Fowden, L., Mansfield, T. and Stoddart, J.). Chapman Hall, London, pp. 263–273.

Richmond, A.E. and Lang, A. (1957) Effect of Kinetin on protein content and survival of detached Xanthium leaves. *Science* **125**, 650–651.

Robson, P.R.H., Donnison, I.S., Wang, K., *et al.* (2004) Leaf senescence is delayed in maize expressing the *Agrobacterium* IPT gene under the control of a novel maize senescence-enhanced promoter. *Plant Biotechnol J* **2**, 101–112.

Rubia, L., De-Palma, J., Elliott, M.C., Khush, G.S. and Bennett, J. (1999) Transgenic rice plants containing an isopentenyl transferase gene driven by a senescence-specific promoter. *Biol Plantarum* **42**, S24.

Sa, G., Mi, M., He-Chun, Y. and Guo-Feng, L. (2002) Anther-specific expression of *ipt* gene in transgenic tobacco and its effect on plant development. *Transgenic Res* **11**, 269–278.

Schmulling, T., Beinsberger, S., Degreef, J., Schell, J., Vanonckelen, H. and Spena, A. (1989) Construction of a heat-inducible chimaeric gene to increase the cytokinin content in transgenic plant-tissue. *FEBS Lett* **249**, 401–406.

Schmulling, T., Rupp, H.-M., Frank, M. and Schafer, S. (1999) Recent advances in cytokinin research: receptor candidates, primary response genes, mutants and transgenic plants. In: *Advances in Regulation of Plant Growth and Development* (eds Strnad, M., Pec, P. and Beck, E.). Peres, Prague, pp. 85–96.

Schroeder, K.R. and Stimart, D.P. (1998) Effects of an autoregulatory senescence-inhibitor gene construct on *Nicotiana alata* Link and Otto. *Hortscience* **33**, 519.

Schroeder, K.R., Stimart, D.P. and Nordheim, E.V. (2001) Response of *Nicotiana alata* to insertion of an autoregulated senescence-inhibition gene. *J Am Soc Hortic Sci* **126**, 523–530.

Smart, C.M., Scofield, S.R., Bevan, M.W. and Dyer, T.A. (1991) Delayed leaf senescence in tobacco plants transformed with tmr, a gene for cytokinin production in *Agrobacterium*. *Plant Cell* **3**, 647–656.

Smigocki, A.C. (1991) Cytokinin content and tissue distribution in plants transformed by a reconstructed isopentenyl transferase gene. *Plant Mol Biol* **16**, 105–115.

Smigocki, A.C. and Owens, L.D. (1988) Cytokinin gene fused with a strong promoter enhances shoot organogenesis and zeatin levels in transformed plant cells. *Proc Natl Acad Sci USA* **85**, 5131–5135.

Soejima, H. (2000) Importance of retarded leaf-senescence on crop productivity, in which cytokinins are possibly involved. *Chem Regul Plants* **35**, 180–193.

Sun, J., Niu, Q.W., Tarkowski, P., *et al.* (2003) The *Arabidopsis* AtIPT8/PGA22 gene encodes an isopentenyl transferase that is involved in de novo cytokinin biosynthesis. *Plant Physiol* **131**, 167–176.

Swartzberg, D., Dai, N., Gan, S., Amasino, R., and Granot, D. (2006) Effects of cytokinin production under two SAG promoters on senescence and development of tomato plants. *Plant Biol* **8**, 579–586.

Synkova, H., VanLoven, K. and Valcke, R. (1997) Increased content of endogenous cytokinins does riot delay degradation of photosynthetic apparatus in tobacco. *Photosynthetica* **33**, 595–608.

Takei, K., Sakakibara, H. and Sugiyama, T. (2001) Identification of genes encoding adenylate isopentenyltransferase, a cytokinin biosynthesis enzyme, in *Arabidopsis thaliana*. *J Biol Chem* **276**, 26405–26410.

Taverner, E., Letham, D.S., Wang, J., Cornish, E. and Willcocks, D.A. (1999) Influence of ethylene on cytokinin metabolism in relation to Petunia corolla senescence. *Phytochemistry* **51**, 341–347.

Thomas, J.C., Smigocki, A.C. and Bohnert, H.J. (1995) Light-induced expression of ipt from *Agrobacterium tumefaciens* results in cytokinin accumulation and osmotic stress symptoms in transgenic tobacco. *Plant Mol Biol* **27**, 225–235.

Thompson, J.E., Hopkins, M.T., Taylor, C. and Wang, T.W. (2004) Regulation of senescence by eukaryotic translation initiation factor 5A: implications for plant growth and development. *Trends Plant Sci* **9**, 174–179.

Van Styaden, J., Cook, E. and Nooden, L.D. (1988) Cytokinins and senescence. In: *Senescence and Aging in Plants* (eds Noodén, L.N. and Leopold, A.C.). Academic Press, San Diego, CA, pp. 281–328.

Wang, T.-W., Lu, L., Zhang, C.-G., Taylor, C. and Thompson, J.E. (2003) Pleiotropic effects of suppressing deoxyhypusine synthase expression in *Arabidopsis thaliana*. *Plant Mol Biol* **52**, 1223–1235.

Weng, P., Chen, L., Chen Mu Z., Lu Chuan, B. and Hong Yue, Y. (2002) Study on transformation of sugarcane-grass 42 mediated by *Agrobacterium*. *J Xiamen Univ* **41**, 536–540.

Wingler, A., von Schaewen, A., Leegood, R.C., Lea, P.J. and Quick, W.P. (1998) Regulation of leaf senescence by cytokinin, sugars, and light – effects on NADH-dependent hydroxypyruvate reductase. *Plant Physiol* **116**, 329–335.

Woo, H.R., Chung, K.M., Park, J.H., *et al.* (2001) ORE9, an F-box protein that regulates leaf senescence in *Arabidopsis*. *Plant Cell* **13**, 1779–1790.

Young, T.E., Giesler-Lee, J. and Gallie, D.R. (2004) Senescence-induced expression of cytokinin reverses pistil abortion during maize flower development. *Plant J* **38**, 910–922.

Yuan, Z., Pan, A., Jian, Z., *et al.* (2002) Senescence delay characterization of transgenic *Brassica chinensis* L. containing an anti-senescence chimeric gene SAG12-IPT. *J Plant Physiol Mol Biol* **28**, 379–384.

Yusibov, V.M., Il, P.C., Andrianov, V.M. and Piruzian, E.S. (1991) Phenotypically normal transgenic T-cyt tobacco plants as a model for the investigation of plant gene-expression in response to phytohormonal stress. *Plant Mol Biol* **17**, 825–836.

Zhang, J., Van Toai, T., Huynh, L. and Preiszner, J. (2000) Development of flooding-tolerant *Arabidopsis thaliana* by autoregulated cytokinin production. *Mol Breed* **6**, 135–144.

Zhang, P. and Gruissem, W. (2005) Leaf senescence-inducible expression of isopentenyl transferase in cassava rendering it resistant to drought stress. In: *The Global Food & Product Chain – Dynamics, Innovations, Conflicts, Strategies* (eds Tielkes, E., Hülsebusch, C., Häuser, I., Deininger, A. and Becker, K.). MDD Media Digitaldruck Copy Shop Büromaschinen GmbH, Stuttgart, Germany, p. 277.

Zhang, R., Zhang, X., Wang, J., Letham, D.S., McKinney, S.A. and Higgins, T.J.V. (1995) The effect of auxin on cytokinin levels and metabolism in transgenic tobacco tissue expressing an *ipt* gene. *Planta* **196**, 84–94.

Zhang, S.Q., Ye, Z.B., Wu, C.Y. and Zhang, Q.F. (1999) Genetic transformation of isopentenyl transferase gene to tomato and regeneration of transgenic plants. *Acta Hortic Sinica* **26**, 376–379.

Zhang, Z.L., Zheng, X.Q. and Lu, Y.T. (2005) Delaying leaf senescence by regulating cytokinin biosynthesis in *Brassica napus*. *Acta Agron Sinica* **31**, 1–6.

Index

Printed and bound by CPI Group (UK) Ltd, Croydon, CR0 4YY

15/09/2026

14962390-0003